Neurodegenerative Diseases

ADVANCES IN EXPERIMENTAL MEDICINE AND BIOLOGY

Neurodegenerative Diseases

Edited by

Shamim I. Ahmad, BSc, MSc, PhD
*School of Science and Technology, Nottingham Trent University,
Nottingham, United Kingdom*

Springer Science+Business Media, LLC

Landes Bioscience

Springer Science+Business Media, LLC
Landes Bioscience

Copyright ©2012 Landes Bioscience and Springer Science+Business Media, LLC

Printed in the USA.

Springer Science+Business Media, LLC, 233 Spring Street, New York, New York 10013, USA
http://www.springer.com

Please address all inquiries to the publishers:
Landes Bioscience, 1806 Rio Grande, Austin, Texas 78701, USA
Phone: 512/ 637 6050; FAX: 512/ 637 6079
http://www.landesbioscience.com

The chapters in this book are available in the Madame Curie Bioscience Database.
http://www.landesbioscience.com/curie

Neurodegenerative Diseases, edited by Shamim I. Ahmad. Landes Bioscience / Springer Science+Business Media, LLC dual imprint / Springer series: Advances in Experimental Medicine and Biology.

ISBN: 978-1-4614-0652-5

While the authors, editors and publisher believe that drug selection and dosage and the specifications and usage of equipment and devices, as set forth in this book, are in accord with current recommendations and practice at the time of publication, they make no warranty, expressed or implied, with respect to material described in this book. In view of the ongoing research, equipment development, changes in governmental regulations and the rapid accumulation of information relating to the biomedical sciences, the reader is urged to carefully review and evaluate the information provided herein.

Library of Congress Cataloging-in-Publication Data

Neurodegenerative diseases / edited by Shamim I. Ahmad.
    p. ; cm. -- (Advances in experimental medicine and biology ; 724)
  Includes bibliographical references and index.
  ISBN 978-1-4614-0652-5
  1. Nervous system--Degeneration. I. Ahmad, Shamim I. II. Series: Advances in experimental medicine and biology ; v. 724. 0065-2598
  [DNLM: 1. Neurodegenerative Diseases. W1 AD559 v.724 2011 / WL 358.5]
  RC365.N473 2011
  616.8'0442--dc23
                    2011021159

DEDICATION

The editor dedicates this book to three Professors who have played very important roles in his academic life including their love and care while working under their supervision: Professor Robert H. Pritchard, retired Professor from University of Leicester, Leicester, England; Professor Abe Eisenstark of Cancer Research Centre, University of Missouri, Columbia, Missouri, USA; and Professor Fumio Hanaoka of University of Osaka, Osaka, Japan (now working in Tokyo).

This book is also dedicated to the late Professor Mark Smith, a dedicated and famous scientist and a senior author of Chapter 16 in this book.

Finally, dedication goes to the sufferers of neurodegenerative disease and their parents, relations and nurses who painstakingly look after them throughout their suffering.

PREFACE

Neurodegeneration can be defined as progressive loss of structure and/or function of neurons, including their death. The loss can be inherited, determined at birth or soon after birth or can be due to an injury.[1] Many inherited neurodenegerative diseases (NDs) stay with the patients for life and many lead to premature death. For some NDs environmental factors may also play important roles such as in autism described in Chapter 7. In addition, impairments in protein degradation, folding and unfolding, and in axonal transport, and mitochondrial function have been associated with NDs.

It is very difficult to quantify the number of NDs, as many have not yet been fully classified, and several overlap in clinical features and therefore may have been classified into classes and subclasses. According to Medline dated 18th April 2011 the number of research papers published on NDs is 183,165, since 1945; and since 1948, 36,426 review papers have been published on this subject. Based on that it is clear that NDs cover a very wide group of diseases and research on most diseases is on-going. Hence for this volume, it has only been possible to cover a selection of diseases.

Since most NDs stand on their own in terms of molecular and genetic basis, and impairment of neuronal functions, the diseases in the volume have been presented in alphabetical order. However readers are advised to read certain chapters together such as Creutzfeldt-Jakob disease (Chapter 6), Gerstmann-Straussler-Scheinker disease (Chapter 10) and kuru disease (Chapter 12). Likewise spinocerebellar ataxia (Chapter 27) may be combined with Machado-Joseph disease (Chapter 14). Autism in Chapter 4 and epigenetic disorders in Chapter 7 may also be combined. For most common NDs such as Alzheimer's disease and Huntington's disease, readers are advised to refer to the book, *Diseases of DNA Repair*,[2] and in this volume other important NDs such as autism, Creutzfeldt-Jakob disease, epilepsy, frontotemporal lobar degeneration, multiple sclerosis, Down syndrome, premature ageing, and spinocerebellar ataxia and many more have been covered.

Chapters 16 in this book may draw special attraction: Mitochondrial importance in Alzheimer's, Huntington's and Parkinson's diseases, and amyotrophic lateral sclerosis (published in *Diseases of DNA Repair*).[2] These have been shown to be tightly linked to mitochondrial dysfunction affected by reactive oxygen species (ROS). ROS have also been implicated in xeroderma pigmentosum,[3] Cockayne syndrome,[4] ataxia telangiectasia,[5] Fanconi anemia,[6] neuronal ceroid-lipofuscinosis mucopolysaccharidoses, Hunter syndrome, mitochondrial myopathy, encephalopathy, lactic acidosis and stroke-like episodes (MELAS), spinal muscular atrophy, dentatorubral pallidoluysian atrophy, Lafora disease, Rett syndrome and Down syndrome covered in this volume. The genetic causes of Huntington's disease[2] and spinal and bulbar muscular atrophy[2] have been shown to be uncontrolled repeats of the CAG nucleotide triplet. CAG encodes glutamine and a repeat of CAG results in polyglutamine tracts (polyQ). Extra glutamine in the protein causes irregular protein folding, its degradation, alteration in subcellular localization and abnormal interactions with other proteins leading to the diseases.[7] A total of nine NDs have been shown to be linked with polyQ repeats.[8]

ROS have been identified to play major roles in our life; ROS, including superoxide ($O_2^{-\cdot}$), hydroxyl radicals ($\cdot OH$) and hydrogen peroxide (H_2O_2) are produced endogenously and play certain positive roles in cellular functions such as in eliminating invading harmful pathogens and foreign elements by phagocytosis. However, ROS play a role of double edged sword in that they participate in harmful events such as in neurodegeneration, induction of mutation through DNA damage, and in ageing processes via damage to various cellular components, DNA, proteins, and lipids. Although nature has provided defensive mechanisms to eliminate $O_2^{-\cdot}$ and H_2O_2, as far as the editor's knowledge goes there are no known naturally produced scavengers that would exclusively remove $\cdot OH$. Melatonin has been shown to have a scavenging effect on various free radicals including $\cdot OH$.[9]

The editor of this volume, having research interests in the field of ROS production and the damage to cellular systems, has identified a number of enzymes showing $\cdot OH$ scavenging activities details of which are anticipated to be published in the near future as confirmatory experiments are awaited.

Savant syndrome is a rare disease as only 50-100 cases are possibly present worldwide. Described in Chapter 25 Savant cases display certain outstanding brain functions and each person can perform one or more of these functions at a level that exceeds that of normal persons; these include calendar calculation, musical savants with perfect pitch and the hyperlexics, 3-D drawings, map memory, poetry, painting and sculpturing. One savant could recite without error the value of Pi to 22,154 places. Savant syndrome has been associated with epilepsy and because exact reason(s) for these phenomena has yet to be worked out, a number of plausible theories have been presented in the chapter.

It is ironic that most, if not all, NDs are incurable. For many NDs drugs are being used to control the pathophysiology such as gait, epileptic fits and trembling. Research to find curative treatment is going on, but since most NDs are genetically linked the most promising approaches may be successful gene replacement therapy and therapies arising from stem cell research. Which of these will win the race, only time will tell.

Acute disseminated encephalomyelitis (ADEM) is a disorder of the central nervous system which usually occurs in young children. Alarming information associated with ADEM (Chapter 1) is that there may be a link between vaccination for smallpox (which no longer exists), measles, mumps, rubella, Japanese B encephalitis, pertussis, influenza, diphtheria/pertussis/tetanus, and hepatitis B and Hog vaccine. Since most children born in the first world are immunized using most of these vaccines, more studies are needed for this finding to avoid vaccination related medical complications.

Autism, one of the most common neurodegenerative diseases described in Chapter 4, has also been associated with MMR vaccines but later studies could not show any such link. Since the exact molecular mechanism of autism has not been established yet, the authors in this chapter have presented excellent guidelines for pediatricians, nurses and caregivers supporting families of young children suffering from this disease. In addition, the addresses of a large number of societies and organizations have been provided from which parents and caregivers can benefit directly.

Important information associated with autosomal recessive Charcot-Marie-Tooth neuropathy (Chapter 5) is that it comprises a complex group of more than 50 diseases and is the most commonly inherited neuropathy. Twenty-nine genes and more than 30 loci have been identified.

Epilepsy and epileptic syndrome, also very common neurological disorders affecting around 50 million people worldwide, have been excellently presented in Chapter 8. At the molecular level they have been shown to be complex diseases and for temporal lobe epilepsy alone, more than 2000 genes have been linked with it. Like many other NDs there is no cure for epilepsy, but many epileptic drugs are available on the market that show neuroprotective properties, antiepileptic effects, targeting molecular pathway (such as rapamycin showing antiepileptogenic effects), anti-inflammatory and antioxidants to reduce the inflammation of neurons and development of epilepsy respectively. Associated with epilepsy, Premi et al have addressed fronto-temporal lobar degeneration (FTLD) in Chapter 8. As mentioned above for epilepsy, FTLD is a challenging NDs and research in this field has been ongoing to find reliable biomarkers and new therapeutic drugs.

Leukodystrophy in Chapter 13 covers a large number of different types of diseases including X-linked adrenoleukodystrophy, Krabbe disease, metachromatic leukodystrophy, Pelizaeus-Merzbacher disease, Alexander disease, Cavanan disease, megalencephalic leukoencephalopathy with subcortical cysts and vanishing white matter disease. Furthermore a number of these leukodystrophies have been subclassified resulting in a large number of diseases in this area of neurodegeneration. Hence understanding the diseases at the molecular level and finding treatments and/or cures have been challenging and to some extent frustrating.

Spinocerebellar ataxia and Machado-Joseph diseases are another set of complex diseases, and these have been exhaustively covered in Chapters 27 and 14 respectively. Lack of motor coordination is the major symptom identified and the genetic basis of many of these diseases have been worked out but no cure is available, although the control of certain pathophysiological features such as ataxia is possible by drugs and training programmes.

Environmental contamination and certain man-made chemicals have long been associated with neuronal damage. In Chapter 15, Hargreaves has addressed this issue by citing examples of organophosphates and has described the molecular mechanisms of some of the compounds which induce neurodegeneration. The presentation of this alarming information should alert us to be careful in the use of such compounds and also make sure of their safe release into the environment.

In the editor's view, Chapter 16 on mitochondrial importance in certain NDs is one of the key chapter presented by the late professor Mark Smith (see the In Memoriam on page xxxi) and his colleagues. Although the work covers three key NDs, viz Alzheimer's, Huntington's and Parkinson's diseases, there are many more diseases (addressed in Chapter 22) that have been linked with mitochondrial dysfunction and ROS. Oxidative stress inducing brain disorder has also been covered in a separate chapter (Chapter 21) by Hayashi et al who shows that lipid peroxidation is a major reason for induction of apoptosis and cell death in neuron systems.

Multiple sclerosis, which is an autoimmune disease, has also been associated with oxidative stress and this has been addressed by Elizabeth Miller (Chapter 17) who shows that the major alteration in MS is the loss of axons, its demyelination and disruption in blood brain barriers. Mechanisms of oxidative stress and targets for these ROS are excellently presented in the chapter.

Neuropathy is a well known phenomenon in diabetes, and this has been fully explained in Chapter 19 by Umegaki who also describes a number of measures that can be taken for future direction of the control of neuropathy in Type 2 diabetic patients. In the editor's view, the best control for neuropathy in diabetic patients should be to try to reduce the glycaemic index (HbA1c) to its basal level.

Chapter 22 covers the important and commonly prevailing disease, Down syndrome, caused partly due to mitochondrial dysfunction induced by oxidative stress. An explanation presented is that complete or partial triosomy of chromosome 21 displays various pleiotropic phenotypes in the patients. The triosomy leads to over-expression of Cu-Zn superoxide dismutase (SOD-1). This imbalances the ratio of SOD-1 to catalase and glutathione peroxidase, thus more H_2O_2 is generated giving rise to oxidative stress.

Premature ageing syndrome includes two NDs: Werner's syndrome and progeria. These have been exquisitely described in Chapter 24 by Fabio Copppede.

Sjogren–Larsson syndrome (SLS) belongs to a set of diseases coming under the ichthyosis diseases; these include Tay syndrome, Refsum, Dorfman-Chanarin and Rud Passwell. SLS is an inborn error of lipid metabolism and these have been excellently described in Chapter 26.

It is hoped that the information presented in *Neurodegenerative Diseases* will stimulate both expert and novice researchers in the field with excellent overviews of the current status of research and pointers to future research goals. Clinicians, nurses as well as families and caregivers should also benefit from the material presented in handling and treating their specialised cases. Also the insights gained should be valuable for further understanding of the diseases at molecular levels and should lead to development of new biomarkers, novel diagnostic tools and more effective therapeutic drugs to treat the clinical problems raised by these devastating diseases.

Shamim I. Ahmad, BSc, MSc, PhD
School of Science and Technology, Nottingham Trent University,
Nottingham, United Kingdom

REFERENCES

1. Bredesen DE, Rao RV, Mehlen P. Cell death in nervous system. Nature 2006; 443:796-802
2. Ahmad SI, ed. Diseases of DNA Repair. Landes Bioscience/Springer Science+Business Media, 2010.
3. Ahmad SI, Hanaoka F, eds. Molecular Mechanisms of Xeroderma Pigmentosum. Austin/New York: Landes Bioscience/Springer Science+Business Media, 2008.
4. Ahmad SI, ed. Molecular Mechanisms of Cockayne Syndrome. Austin/New York: Landes Bioscience/Springer Science+Business Media, 2009.
5. Ahmad SI, ed. Molecular Mechanisms of Ataxia Telangiectasia, Austin/New York: Landes Bioscience/Springer Science+Business Media, 2009.
6. Ahmad SI, Kirk SH, eds. Molecular Mechanisms of Fanconi anemia. Austin/New York: Landes Bioscience/Springer Science+Business Media, 2007.
7. Thompson LM. Neurodegeneration: a question of balance. Nature 2008; 443:707-708.
8. Zoghbi HY, Orr HT. Pathogenic mechanisms of a polyglutamine-mediated neurodegenerative disease, spinocerebellar ataxia type 1. J Biol Chem 2009; 284:7425-7429.
9. Galano A. On the direct scavenging activity of melatonin hydroxyl and a series of peroxyl radicals, Phys Chem Chem Phts 2011; 13:7178-7188.

ABOUT THE EDITOR...

SHAMIM I. AHMAD after obtaining his Master's degree in Botany from Patna University, Bihar, India and his PhD in Molecular Genetics from Leicester University, England, joined Nottingham Polytechnic as Grade 1 lecturer and was subsequently promoted to SL post. Nottingham Polytechnic subsequently became Nottingham Trent University where, after serving for about 37 years, he took early retirement to spend the remaining time writing books and conducting full-time research. For more than three decades he worked on different areas of biology including thymineless death in bacteria, genetic control of nucleotide catabolism, development of anti-AIDS drugs, control of microbial infection of burns, phages of thermophilic bacteria and microbial flora of Chernobyl after the nuclear accident. But his primary interest, which started 27 years ago, is DNA damage and repair, particularly near UV photolysis of biological compounds, production of reactive oxygen species and their implications on human health including skin cancer and xeroderma pigmentosum. He is also investigating photolysis of non-biological compounds such as 8-methoxypsoralen+UVA, mitomycin C, and nitrogen mustard and their importance in psoriasis treatment and in Fanconi anemia. In 2003 he received a prestigious "Asian Jewel Award" in Britain for "Excellence in Education". He is also the Editor of *Molecular Mechanisms of Fanconi Anaemia, Molecular Mechanisms of Xeroderma Pigmentosum, Molecular Mechanisms of Ataxia Telangiectasia, Molecular Mechanisms of Cockayne Syndrome* and *Diseases of DNA Repair* published by Landes Bioscience.

PARTICIPANTS

Shamim I. Ahmad
School of Science and Technology
Nottingham Trent University
Nottingham
UK

Orlando Graziani Povoas Barsottini
Department of Neurology
  and Neurosurgery
Federal University of São Paulo
São Paulo
Brazil

Barbara Borroni
Centre for Ageing Brain
  and Neurodegenerative Disorders
Neurology Unit
University of Brescia
Brescia
Italy

Pedro Braga-Neto
Department of Neurology
  and Neurosurgery
Federal University of São Paulo
São Paulo
Brazil

Paul Brown
CEA/DSV/iMETI/SEPIA
Fontenay-aux-Roses
France

Eduardo Calpena
Unit 732
Centro de Investigación Biomédica en Red
  de Enfermedades Raras (CIBERER)
and
Unit of Genetics and Molecular Medicine
Instituto de Biomedicina de Valencia (IBV)
Consejo Superior de Investigaciones
  Científicas (CSIC)
Valencia
Spain

Giuseppe Castello
CROM
Cancer Research Center
Mercoglianao
Italy

Andrea E. Cavanna
Department of Neuropsychiatry
University of Birmingham and BSMHFT
Birmingham
and
Sobell Department of Movement Disorders
Institute of Neurology
University College London
London
UK

Jocelyn Cherry
Department of Clinical Neurosciences
Southampton General Hospital
Southampton
UK

Fabio Coppedè
Department of Human and Environmental
   Sciences
Section of Medical Genetics
University of Pisa
Rome
Italy

Marc Corral-Juan
Basic, Translational and Neurogenetics
   Research Unit
Department of Neurosciences
Health Sciences Research Institute
   Germans Trias i Pujol (IGTP)
Universitat Autònoma de Barcelona
Barcelona
Spain

Sónia C. Correia
Center for Neuroscience and Cell Biology
and
Faculty of Sciences and Technology
Department of Life Sciences
University of Coimbra
Coimbra
Portugal

Lívia Almeida Dutra
Department of Neurology
   and Neurosurgery
Federal University of São Paulo
São Paulo
Brazil

Ralf-Ingo Ernestus
Department of Neurosurgery
University of Würzburg
Würzburg
Germany

Carmen Espinós
Unit 732
Centro de Investigación Biomédica en Red
   de Enfermedades Raras (CIBERER)
Valencia
Spain

Aristea S. Galanopoulou
Saul R. Korey Department of Neurology
and
Dominick P. Purpura Department
   of Neuroscience
Albert Einstein College of Medicine
Bronx, New York
USA

Alan J. Hargreaves
School of Science and Technology
Nottingham Trent University
Nottingham
UK

Masaharu Hayashi
Department of Clinical Neuropathology
Tokyo Metropolitan Institute
   for Neuroscience
Tokyo
Japan

Takae Hirasawa
Department of Epigenetics Medicine
Faculty of Medicine
Interdisciplinary Graduate School
   of Medicine and Engineering
University of Yamanashi
Yamanashi
Japan

John R. Hughes
Department of Neurology
University of Illinois Medical Center
Chicago, Illinois
USA

James W. Ironside
National CJD Surveillance Unit
Western General Hospital
Edinburgh
UK

Michaela Jelen
University of Alberta
Edmonton, Alberta
Canada

Sam Khandhadia
Department of Clinical Neurosciences
Southampton General Hospital
Southampton
UK

Yuki Kishimoto
Department of Neuropsychiatry
Okayama University Graduate School
    of Medicine
Dentistry and Pharmaceutical Sciences
Shikata-cho, Okayama
Japan

Richard Knight
National CJD Surveillance Unit
Western General Hospital
Edinburgh
UK

Tsuyoshi Koide
Department of Mouse Genomics Resource
    Laboratory
National Institute of Genetics
Shizuoka
Japan

Takeo Kubota
Department of Epigenetics Medicine
Faculty of Medicine
Interdisciplinary Graduate School
    of Medicine and Engineering
University of Yamanashi
Yamanashi
Japan

Paweł P. Liberski
Laboratory of Electron Microscopy
    and Neuropathology
Department of Molecular Pathology
    and Neuropathology
Medical University Lodz
Lodz
Poland

Andrew John Lotery
Department of Clinical Neurosciences
Southampton General Hospital
Southampton
UK

Vincenzo Lupo
Unit 732
Centro de Investigación Biomédica en Red
    de Enfermedades Raras (CIBERER)
and
Unit of Genetics and Molecular Medicine
Instituto de Biomedicina de Valencia (IBV)
Consejo Superior de Investigaciones
    Científicas (CSIC)
Valencia
Spain

Soe Mar
Division of Pediatric and Developmental
    Neurology
Department of Neurology
Washington University School of Medicine
Saint Louis, Missouri
USA

Dolores Martínez-Rubio
Unit 732
Centro de Investigación Biomédica
    en Red de Enfermedades Raras
(CIBERER)
and
Unit of Genetics and Molecular Medicine
Instituto de Biomedicina de Valencia (IBV)
Consejo Superior de Investigaciones
    Científicas (CSIC)
Valencia
Spain

Antoni Matilla-Dueñas
Basic, Translational and Molecular
    Neurogenetics Research Unit
Department of Neurosciences
Health Sciences Research Institute
    Germans Trias i Pujol (IGTP)
Universitat Autònoma de Barcelona
Barcelona
Spain

Elżbieta Miller
Neurorehabilitation Ward
III General Hospital of Lodz
Lodz
and
Department of Chemistry and Clinical
    Biochemistry
University of Bydgoszcz
Bydgoszcz
Poland

Kunio Miyake
Department of Epigenetics Medicine
Faculty of Medicine
Interdisciplinary Graduate School
    of Medicine and Engineering
University of Yamanashi
Yamanashi
Japan

Rie Miyata
Department of Clinical Neuropathology
Tokyo Metropolitan Institute
    for Neuroscience
Tokyo
Japan

Paula I. Moreira
Center for Neuroscience and Cell Biology
and
Faculty of Medicine
Institute of Physiology
University of Coimbra
Coimbra
Portugal

Tomonori Ono
Saul R. Korey Department of Neurology
Albert Einstein College of Medicine
Bronx, New York
USA

Alessandro Padovani
Centre for Ageing Brain
    and Neurodegenerative Disorders
Neurology Unit
University of Brescia
Brescia
Italy

Giovanni Pagano
CROM
Cancer Research Center
Mercogliano
Italy

Joy B. Parrish
Jacobs Neurological Institute
Department of Neurology
School of Medicine and Biomedical
    Sciences
University at Buffalo
Buffalo, New York
USA

Stephanie Patterson
University of California Los Angeles
Los Angeles, California
USA

José Luiz Pedroso
Department of Neurology
    and Neurosurgery
Federal University of São Paulo
São Paulo
Brazil

Seth J. Perlman
Division of Pediatric and Developmental
    Neurology
Department of Neurology
Washington University School of Medicine
Saint Louis, Missouri
USA

George Perry
Department of Pathology
Case Western Reserve University
Cleveland, Ohio
and
UTSA Neurosciences Institute
    and Department of Biology
University of Texas at San Antonio
San Antonio, Texas
USA

Scott R. Plotkin
Massachusetts General Hospital
Stephen E. and Catherine Pappas Center
  for Neuro-Oncology
Boston, Massachusetts
USA

Enrico Premi
Centre for Ageing Brain
  and Neurodegenerative Disorders
Neurology Unit
University of Brescia
Brescia
Italy

Vincenzo Romeo
Department of Neurosciences
University of Padova
Padova
Italy

Ivelisse Sanchez
Basic, Translational and Neurogenetics
  Research Unit
Department of Neurosciences
Health Sciences Research Institute
  Germans Trias i Pujol (IGTP)
Universitat Autònoma de Barcelona
Barcelona
Spain

Renato X. Santos
Center for Neuroscience and Cell Biology
and
Faculty of Sciences and Technology
Department of Life Sciences
University of Coimbra
Coimbra
Portugal

Tilmann Schweitzer
Department of Neurosurgery
University of Würzburg
Würzburg
Germany

Beata Sikorska
Laboratory of Elecron Microscopy
  and Neuropathology
Department of Moleculary Pathology
  and Neuropathology
Medical University of Lodz
Lodz
Poland

Mark A. Smith
Department of Pathology
Case Western Reserve University
Cleveland, Ohio
USA

Veronica Smith
University of Alberta
Edmonton, Alberta
Canada

Naoya Takeda
Department of Neuropsychiatry
Okayama University Graduate School
  of Medicine
Dentistry and Pharmaceutical Sciences
Shikata-cho, Okayama
Japan

Naoyuki Tanuma
Department of Clinical Neuropathology
Tokyo Metropolitan Institute
  for Neuroscience
Tokyo
Japan

Cristiano Termine
Child Neuropsychiatry Unit
Department of Experimental Medicine
University of Insubria
Varese
Italy

Erik J. Uhlmann
Massachusetts General Hospital
Stephen E. and Catherine Pappas Center
  for Neuro-Oncology
Boston, Massachusetts
USA

Hiroyuki Umegaki
Department of Geriatrics
Nagoya University Graduate School
    of Medicine
Nagoya, Aichi
Japan

Victor Volpini
Molecular Diagnosis Center
    of Inherited Diseases
Institut d'Investigacions Biomèdiques
    de Bellvitge (IDIBELL)
L'Hospitalet de Llobregat
Barcelona
Spain

Shiyao Wang
Peking University
People's Hospital
Beijing
China

Thomas Westermaier
Department of Neurosurgery
University of Würzburg
Würzburg
Germany

E. Ann Yeh
Jacobs Neurological Institute
Department of Neurology
School of Medicine and Biomedical
    Sciences
University at Buffalo
Buffalo, New York
USA

Osamu Yokota
Department of Neuropsychiatry
Okayama University Graduate School
    of Medicine
Dentistry and Pharmaceutical Sciences
Shikata-cho, Okayama
Japan

Xiongwei Zhu
Department of Pathology
Case Western Reserve University
Cleveland, Ohio
USA

CONTENTS

7. EPIGENETICS IN AUTISM AND OTHER NEURODEVELOPMENTAL DISEASES 91

Kunio Miyake, Takae Hirasawa, Tsuyoshi Koide and Takeo Kubota

8. EPILEPSY AND EPILEPTIC SYNDROME 99

Tomonori Ono and Aristea S. Galanopoulou

9. FRONTOTEMPORAL LOBAR DEGENERATION 114

Enrico Premi, Alessandro Padovani and Barbara Borroni

10. GERSTMANN-STRÄUSSLER-SCHEINKER DISEASE............................128

Paweł P. Liberski

Shiyao Wang

Paweł P. Liberski, Beata Sikorska and Paul Brown

Seth J. Perlman and Soe Mar

Antoni Matilla-Dueñas

15. NEURODEGENERATIONS INDUCED BY ORGANOPHOSPHOROUS COMPOUNDS 189

Alan J. Hargreaves

16. MITOCHONDRIAL IMPORTANCE IN ALZHEIMER'S, HUNTINGTON'S AND PARKINSON'S DISEASES 205

Sónia C. Correia, Renato X. Santos, George Perry, Xiongwei Zhu,
Paula I. Moreira and Mark A. Smith

17. MULTIPLE SCLEROSIS 222

Elżbieta Miller

26. SJOGREN-LARSSON SYNDROME 344

Lívia Almeida Dutra, Pedro Braga-Neto, José Luiz Pedroso
    and Orlando Graziani Povoas Barsottini

27. THE SPINOCEREBELLAR ATAXIAS: CLINICAL ASPECTS
AND MOLECULAR GENETICS...351

Antoni Matilla-Dueñas, Marc Corral-Juan, Victor Volpini and Ivelisse Sanchez

28. TOURETTE SYNDROME 375

Andrea E. Cavanna and Cristiano Termine

ACKNOWLEDGEMENTS

The editor is cordially thankful and acknowledges all the authors for their contributions of chapters by using their dedication, in-depth knowledge and professional skill used for the production of this book. Without their cooperation it would not have been possible to produce this unique piece of work. I would also like to acknowledge Cynthia Conomos, Celeste Carlton, Erin O'Brien and Daniel Olasky of Landes Bioscience for their hard work and patience in the preparation of this book. Also I am thankful to Ron Landes of Landes Bioscience who commissioned this work. Finally, I wish to thank my family members, Riasat Jan, Farhin Ahmad, Mahrin Ahmad and Tamsin Ahmad for their love, support and encouragement during the production of this book.

IN MEMORIAM

MARK ANTHONY SMITH
(August 15, 1965 – December 19, 2010)

An Inspiration for Alzheimer's Disease

Mark Smith was the first proponent of Alzheimer disease (AD) being governed by the laws of physiology. While he is best known for questioning the amyloid cascade hypothesis,[1] fundamentally he had distrust of any mechanism that did not rely on the biology of normal aging. Amyloid-β (Aβ), tau, oxidative stress and other mechanisms recapitulate the response of the brain to injury.

Owing to his originality, his efforts were concentrated on connecting AD to biology, leading to the two-hit hypothesis[2] and other theories, which all focused on the pleiotropic threshold between pathology and physiology. The miracle of AD is that neurons filled with tau and in an environment inundated by Aβ can live for decades. He proposed that Aβ forms which are in dynamic equilibrium, along with tau and numerous other changes, are in fact critical homeostatic responses to AD rather than the cause. This was the foundation of Mark's dissatisfaction with the Aβ, tau, and other pathology-based theories; his focus on biology as the route to understand and reverse disease. Indeed, it is not surprising that Mark suggested over a decade ago the failure of the Aβ vaccine,[3] which he elegantly described as disturbing the balance of a critical response to the underlying pathogenesis of AD. His dogma persists today, as no disease mechanism clearly indicts monomers, dimers, oligomers, or other Aβ forms. He envisioned that only working with the biology to modulate the response of aging and to restore physiological balance would offer the hope of more years of health.

Besides questioning the amyloid cascade hypothesis, Mark's contributions span from defining oxidative stress, cell cycle re-entry of neurons, gonadotropins, mitochondria, and many more. Over the past 10 years he was the most prolific author in AD research, and the fifth most cited.[4]

Mark won numerous awards including the Jordi Folch-Pi Award from the American Society for Neurochemistry, Outstanding Investigator Award from the American Society for Investigative Pathology, Denham Harman Research Award from the American Aging Association, Zenith Award from the Alzheimer's Association, Hermann-Esterbauer Award from the HNE Society, and the Goudie Lecture and Medal from the Pathological Society of Great Britain and Ireland. Additionally, Mark was a Fellow of the American Association for the Advancement of Science, American Aging Association, Royal Society of Medicine, and the Royal College of Pathologists as well as Daland Fellow of the American Philosophical Society.

Mark's greatest pride was his service, ranging from Editor-in-Chief of the *Journal of Alzheimer's Disease*, Executive Director of the American Aging Association, organizer of numerous meetings, participant in National Institutes of Health, Veteran's Administration, other grant reviews, and member of over 200 editorial boards. The outpouring of remembrance from AD researchers throughout the world has been astounding, so many saying that Mark's courage to deliver a message as he saw it was an inspiration in their own lives (http://www.j-alz.com/marksmith.html).

Mark's meteoric rise in the mid-1990s was as much marked by his innovative thoughts, exceptional enthusiasm, and communication skills, as by his unlimited giving to collaborators and coworkers. When he found and defined the first oxidative modifications of AD pathology, pyrraline and pentosidine,[5] and the first oxidative response protein induced, heme oxygenase,[6] he revolutionized the field. Mark understood oxidative stress was not the end but an opening into the beginning of AD.

His boyish charm and brotherly demeanor made everyone feel as though they were his true friend. He put everyone else's needs above his own and made a great effort to help wherever he saw the need. Mark's lively spirit is still felt by each of us, as we find ourselves smiling when remembering his antics. We, along with all of those he touched, will continue to be inspired by Mark's courage to trust in his own science, by his creativity in developing novel hypotheses, and by his charisma, which allowed his ideas to be heard and understood.

Xiongwei Zhu, Hyoung-gon Lee and George Perry

REFERENCES

1. Joseph J, Shukitt-Hale B, Denisova NA et al. Copernicus revisited: amyloid beta in Alzheimer's disease. Neurobiol Aging 2001; 22:131-146.
2. Zhu X, Raina AK, Perry G, Smith MA. Alzheimer's disease: the two-hit hypothesis. Lancet Neurol 2004; 3:219-226.
3. Perry G, Nunomura A, Raina AK, Smith MA. Amyloid-beta junkies. Lancet 2000; 355:757.
4. Sciencewatch.com, Special Topic: Alzheimer's Disease. Top 20 Authors, http://sciencewatch.com/ana/st/alz2/authors/, Accessed March 1, 2011.
5. Smith MA, Taneda S, Richey PL et al. Advanced Maillard reaction end products are associated with Alzheimer disease pathology. Proc Natl Acad Sci USA 1994; 91:5710-5714.
6. Smith MA, Kutty RK, Richey PL et al. Heme oxygenase-1 is associated with the neurofibrillary pathology of Alzheimer's disease. Am J Pathol 1994; 145:42-47.

CHAPTER 1

ACUTE DISSEMINATED ENCEPHALOMYELITIS

Joy B. Parrish* and E. Ann Yeh

Jacobs Neurological Institute, Department of Neurology, School of Medicine and Biomedical Sciences, University at Buffalo, Buffalo, New York, USA
Corresponding Author: Joy B. Parrish—Email: jparrish@thejni.org

Abstract: Acute disseminated encephalomyelitis (ADEM) is a disorder of the central nervous system (CNS) characterized by an acute event, typically with encephalopathy, in which diffuse CNS involvement occurs. It may follow an infectious event and occurs more commonly in young children. Pulse steroid treatment is frequently used to treat ADEM. Although ADEM is typically described as a benign condition, with children generally recovering motor function and resolution of lesions on magnetic resonance imaging (MRI), residual cognitive deficits may occur. This chapter aims to review the clinical features, typical presentation, differential diagnosis, treatment and prognosis of ADEM.

INTRODUCTION

Acute disseminated encephalomyelitis (ADEM) is a demyelinating disorder affecting the central nervous system (CNS). It typically presents in early childhood, but has been documented throughout the lifespan.[1,2] According to the consensus definition of the International Pediatric MS Study Group (IPMSSG), ADEM is a "first clinical event with a presumed inflammatory or demyelinating cause, with acute or subacute onset that affects multifocal areas of the CNS".[3] This definition also states that encephalopathy (behavioral change and/or alteration in consciousness) is required to make the diagnosis of ADEM. However, other literature has demonstrated that encephalopathy may be associated with a first episode of what is later diagnosed as multiple sclerosis (MS).[4] Further, not all patients with ADEM present with encephalopathy.[5]

Neurodegenerative Diseases, edited by Shamim I. Ahmad.
©2012 Landes Bioscience and Springer Science+Business Media.

Other features of ADEM included in the IPMSSG definition are: (1) neuroimaging of the brain showing focal or multifocal lesions, particularly involving white matter; (2) no history of prior demyelinating episode; (3) absence of other underlying pathology to explain the event (e.g., infection, tumor); and (4) improvement either clinically and/or on magnetic resonance imaging (MRI). The definition goes on to specify that the brain MRI may reveal large ($\geq$1-2 cm), multifocal, hyperintense lesions in the supratentorial or infratentorial white matter regions. Gray matter may also be involved, especially the thalamus and basal ganglia and, in rare cases, there is one large lesion ($\geq$1-2 cm) predominately involving white matter. In addition to abnormalities on brain MRI, MRI of the spinal cord may show confluent intramedullary lesions with variable enhancement.[3]

Although viral infection may precede symptoms of ADEM, the presence of prior infection is not required for diagnosis. Common laboratory findings include elevated white blood cell count and CSF protein. Oligoclonal bands in the CSF may be present but are less frequently found in ADEM compared to MS (see below under laboratory testing).[6]

Relapses have been described in children with an initial diagnosis of acute disseminated encephalomyelitis, with reported rates ranging from 5-21%.[2,7-10] According to consensus definitions proposed by the IPMSSG, symptoms and/or MRI findings associated with an episode of ADEM occurring within 3 months of the initial ADEM event are considered part of the initial event.[3]

Recurrent ADEM (R-ADEM) is defined as the recurrence of initial symptoms and signs of the first ADEM event that occur three or more months after the initial event, without involvement of new clinical areas. The symptoms must occur at least one month after completion of steroid treatment and with no new lesions on MRI.[3]

Multiphasic ADEM (MDEM) is an event that meets criteria for diagnosis of ADEM but involves new areas of the CNS. Again, this event must occur three or more months after the initial event and one month following completion of steroid treatment.[3] It is noted that more than two such events should raise suspicion of MS.[3]

These diagnostic categories were formulated in order to highlight a distinct group of patients that may experience relapses, but whose relapses will be limited to a small number and who will not progress to chronic relapses and neurodegeneration, as in MS. Although controversy regarding recurrent ADEM and MDEM exists, long term follow-up of a small group of children (n = 13) with relapsing forms of ADEM suggests low rates of conversion to MS (mean follow-up 9 years). In this study, only two patients who were classified as MDEM following international definitions experienced further recurrences and received alternate diagnoses (MS and CNS vasculitis).[11] Further systematic evaluation of these definitions using large cohorts of patients is necessary to determine their validity.

See Table 1 for IPMSSG definitions.

INCIDENCE AND PREVALENCE

The annual incidence of ADEM is estimated to be between 0.2 and 0.8 per 100,000.[9,12-15] Incidence varies with geographical location, with estimates of 0.4 per 100,000 (under the age of 20 years) in California,[9] 0.64 per 100,000 (under the age of 15 years) in Japan,[14] 0.07 per 100,000 in Germany,[13] and 0.2 per 100,000 in Canada.[12] Several studies have suggested a slight predominance in males, with M to F ratios ranging from 1:0.8 to 2.3:1.[2,12,14-17] Onset usually occurs early in childhood.[2,5,7-9,16,18-21] Eighty percent of patients with ADEM have onset before the age of 10 years.[15] Adult onset ADEM occurs but is rare.[22-26]

Table 1. International Pediatric MS Study Group (IPMSSG) Definitions[3]

ADEM (monophasic)	• First inflammatory or demyelinating clinical event
	• Acute onset
	• Multiple areas of the CNS affected
	• Polysymptomatic presentation
	• Presence of encephalopathy (e.g., behavioral change, alteration in consciousness)
	• Neuroimaging shows focal or multifocal lesions primarily involving white matter
	• No radiological evidence of previous destructive white matter changes
	• Improvement either clinically, on MRI, or both, with possible residual deficits
	• New or fluctuating symptoms within 3 months of initial event considered part of the initial event
	• No other explanation for presenting symptoms
Recurrent ADEM (R-ADEM)	• New event of ADEM with recurrence of initial symptoms occurring 3 or more months after the initial event
	• No involvement of new areas
	• Event occurs at least 1 month after completing steroid therapy
	• No new lesions on MRI (there can be enlargement of original lesions)
Multiphasic ADEM (MDEM)	• New clinical event of ADEM involving new anatomical areas
	• Occurring at least 3 months after the initial ADEM event and at least 1 month after completing steroid treatment
	• Event must have polysymptomatic presentation including encephalopathy
	• Neurologic symptoms (other than mental status changes) must differ from initial event
	• Brain MRI must show new areas of involvement with complete or partial resolution of lesions associated with the initial event
	• **More than 2 events should raise suspicion of MS**

ADEM has been described in children who have recently experienced infections or who have received vaccinations. It is estimated that 70-77% of patients with ADEM are reported to have had clinically evident infection or vaccination during the few weeks prior to onset.[15] It is important to note that ADEM rarely occurs postvaccination (0.1 to 0.2 cases per 100,000).[27] It may occur more frequently after primary vaccination rather than revaccination.[28] Case reports of ADEM after the following vaccines have been published: smallpox,[28,29] measles,[2,30,31] mumps,[32] rubella,[31] Japanese B encephalitis,[33] diphtheria/ pertussis/tetanus,[31] pertussis,[15,31] influenza,[34,35] hepatitis B[19] and the Hog vaccine.[36]

Post-infectious ADEM has been described in association with viral infections including measles,[31] mumps,[2,37] rubella virus,[2,31] varicella-zoster,[38] Epstein-Barr virus,[18,39] cytomegalovirus,[40] herpes simplex virus,[2,41] hepatitis A or C,[42,43] Coxsackie B virus,[44] influenza A[45] or B,[7,18] H1N1 influenza,[46] HIV,[47] human T-cell lymphotropic virus-1,[48] human herpes virus 6,[49] Rocky Mountain spotted fever virus,[50] and human coronavirus.[51] Other pathogens include *Mycoplasma pneumonia*,[52] *Borrelia burgdorferi*,[53] *Campylobacter*,[54] *Leptospira*,[55,56] *Chlamydia*,[57] *Legionella*,[58] *Rickettsia rickettsii*,[50] *Mycoplasma pneumonia*,[52] *Streptococcus*,[59] and group A beta-haemolytic streptococci.[60] ADEM has also been reported after organ transplantation, leukemia and non-Hodgkin's lymphoma.[61-65]

Systematic studies evaluating the relationship between specific infectious agents and ADEM have not been published. One study evaluating the relationship between vaccine and inflammatory demyelination has been published. Mikaeloff and colleagues reported no association between the Hepatitis B vaccination and inflammatory demyelination, although when subgroup analysis was performed on specific brands of vaccine, the Engerix B vaccine was found to increase risk, especially in children with multiple sclerosis (OR 2.77, 1.23-6.24).[66] Further studies of the relationship between vaccination and inflammatory demyelination are needed to confirm the presence or absence of an association.

DIFFERENTIAL DIAGNOSIS OF MULTIPLE SCLEROSIS AND ACQUIRED CENTRAL NERVOUS SYSTEM DEMYELINATING DISORDERS IN CHILDREN AND ADOLESCENTS

The differential diagnosis of white matter abnormalities on MRI in childhood is broad. Inflammatory, infectious, metabolic and neurodegenerative disorders may present with white matter abnormalities on MRI.[67,68] Careful history and physical examination are necessary, but not sufficient to distinguish between children with demyelinating disorders, acute infectious or vascular processes, or chronic degenerative/metabolic processes.[69-71] In many cases, MRI evaluation together with a detailed clinical history may help to distinguish metabolic white matter disorders such as Alexander's disease, Canavan's disease, Vanishing White Matter disease, Adrenoleukodystrophy and Metochromatic leukodystrophy from ADEM, particularly if attention is paid to the distribution of white matter lesions.[72] Specifically, ADEM is typically acute to sub- acute in onset and causes diffuse involvement of the deep gray matter and white matter, with prominent involvement of the cerebellum in many cases. Lesions are usually large with bilateral involvement (see Fig. 1).

Importantly, the diagnosis of neuromyelitis optica (NMO) should be considered in all cases of acute demyelination in childhood, particularly in children who present with longitudinally extensive transverse myelitis and/or severe optic neuritis. Cerebral involvement with ADEM-like lesions is frequently seen in these cases; one case describes brain lesions in children who were later diagnosed with NMO.[73]

Even after the diagnosis of an acute demyelinating process has been established, in many cases, it is difficult to distinguish monophasic demyelinating conditions from a first attack of pediatric MS, especially in younger patients. As noted above, proposed definitions suggest that encephalopathy is required for one to diagnose ADEM, but recent reports suggest that encephalopathy may also occur at onset of neuromyelitis optica and MS, especially in younger children.[4] Patients with an ADEM-like presentation whose MRI is suggestive of MS (e.g., periventricular lesions) at onset have an increased risk of developing a second episode.[74]

The sections below review diagnostic testing that may be of utility in distinguishing ADEM from other diseases and predicting whether children with a first presentation of ADEM will later receive the diagnosis of multiple sclerosis.

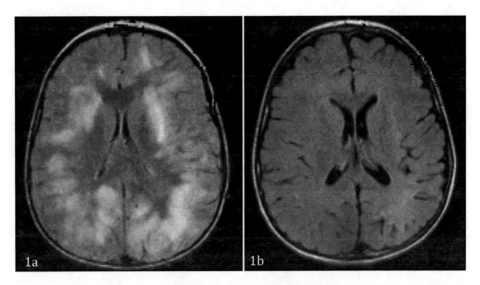

Figures 1. MRI scan of a 5 year-old child with acute disseminated encephalomyelitis at onset (1a) and 1 year later (1b). Note the diffuse involvement of both cerebral hemispheres (1a) and almost complete resolution of the previously apparent lesions (1b).

DIAGNOSTIC TESTING

Diagnostic testing for acute disseminated encephalomyelitis includes serum and CSF evaluation, visual testing and magnetic resonance imaging (MRI).

Magnetic Resonance Imaging

MRI is of central importance in making the diagnosis of ADEM. In general, as noted above, MRI lesions, associated with ADEM, include those in the deep grey nuclei, with widespread, bilateral involvement of subcortical white matter together with marked cerebellar and brainstem involvement.[7,75] The spinal cord is also frequently involved. These lesions often disappear on repeated imaging.

Much work has been devoted to distinguishing ADEM, from MS. One study has suggested that the presence of well-defined periventricular and corpus callosum lesions may be specific but very insensitive predictors of MS after a first attack of CNS demyelination in childhood (KIDMUS criteria).[74] More recent studies have confirmed these general findings and have found that the presence of any of the adult Barkhof criteria for MS, as well as small, well-defined lesions that are perpendicular to the corpus callosum may be effective in differentiating children with MS from those with a monophasic disease at the time of disease presentation.[76] Another study reported that a combination of two of the following criteria may distinguish a first attack of pediatric MS from ADEM[77]: (1) presence of T1-black holes; (2) presence of two or more periventricular lesions; and (3) absence of diffuse bilateral lesions. Importantly, this study failed to analyze the presence of gadolinium-enhancing foci that may also appear dark on T1-weighted imaging. Recently, these diagnostic criteria have been independently tested and found to be highly specific (95% specificity) and moderately sensitive (75%) in distinguishing ADEM from MS.[78]

CSF Analysis

CSF analysis is central to the diagnosis and treatment of childhood demyelinating disorders. As children with ADEM often present with fever and encephalopathy, standard CSF testing is necessary, including evaluation of cell counts, protein, glucose and testing for infectious agents. Importantly, 64% of children with ADEM are reported to have lymphocytosis on CSF analysis and 60% have CSF protein elevation.[7]

IgG index and oligoclonal bands (OCB) are also useful in distinguishing ADEM from MS in many cases. Up to 29% children with ADEM are reported to have oligoclonal bands in the CSF,[2,6,7,19] while the majority of children with MS are positive for oligoclonal bands in the CSF.[6] Within the French KIDMUS cohort, 94% of children with positive OCB (69/72) went on to develop MS. However, only 40% of established MS patients in this study had OCB, suggesting that this test has a low sensitivity but high specificity for the development of MS when present at disease onset. Similarly, another study reported OCB to be present in the CSF of 92% of children with MS.[6]

It is important to note that the CSF profile in childhood-onset demyelinating disorders may vary by age. OCB may be less frequent in younger children with MS (43% versus 63% in adolescents).[79] These results must be interpreted with caution as information regarding timing of lumbar puncture and of OCB detection was not provided in this paper, nor is it clear how many children included in this study had actually confirmed testing of OCB and/or CSF.[21] The IgG index had been found to be elevated in 68% of adolescents with MS (>11 years), but in 35% of younger children (<11 years),[79] numbers similar to the rate of OCB positivity in younger children with MS and ADEM. The absence of neutrophils in the CSF at onset is predictive of an earlier second neurological episode.[79]

Serum Testing

No serum biomarkers have been found to be sensitive or specific for the diagnosis of ADEM. However, in a small study, children with ADEM and clinically isolated syndrome (CIS) were found to have high serum titres to nMOG (native myelin oligodendrocyte glycoprotein) in comparison to children with other neurological diseases and healthy controls.[80]

Serum testing may reveal leukocytosis in pediatric ADEM: almost 2/3 of patients with ADEM will present with elevated WBC as compared to 22% of children with MS.[7]

For children with severe optic neuritis and/or longitudinally extensive transverse myelitis, in whom the diagnosis of neuromyelitis optica is suspected, NMO IgG antibody testing should be performed. NMO-IgG testing has been found to be positive in 78%[73,81] of children who were diagnosed clinically with relapsing NMO and only 12.5% of those with monophasic NMO.[81]

Visual Testing

Optic neuritis may be seen in up to 60% of children presenting with ADEM,[82] and almost a quarter (23%) of children with a first time demyelinating event.[12] Importantly, optic neuritis in younger children with ADEM is frequently bilateral. One study reported that 23% of children with ADEM present with bilateral optic neuritis.[7] Approximately one-third of children who are later diagnosed with MS experience optic neuritis as an initial presenting symptom.[74,83] Even a higher proportion of children with demyelinating disorders may experience subclinical abnormalities of the visual pathway.[84] The limited ability of standard Snellen charts to distinguish subtle visual dysfunction is well documented in the adult MS population.[85] Low contrast letter acuity charts (LCLA, Sloan charts) have been shown to provide a sensitive and reliable assessment of visual acuity in cases of pediatric demyelination.[86]

Other tests, such as visual evoked potential (VEP), or pattern reversal visual evoked potentials (PRVEP), have been shown to be of diagnostic utility in childhood demyelinating disorders, with almost half of such patients showing prolonged visual latency.[84]

More recently, optical coherence tomography (OCT), previously used for patients with glaucoma, has been applied to pediatric patients with demyelinating disorders. This procedure uses near infrared light to quantify the thickness of the retinal nerve fiber layer (RNFL) (which contains only nonmyelinated axons). It has been shown to provide a sensitive evaluation of the RNFL thickness in this population, a correlation of optic atrophy.[86] Taken together, VEP, OCT and LCLA testing can provide objective evidence of previous inflammatory insult to the optic nerve in the pediatric population with demyelinating disorders. They may help to establish a diagnosis of MS and may also be used for disease monitoring on follow-up.

CLINICAL AND DEMOGRAPHIC PREDICTORS OF THE RISK TO DEVELOP MS AFTER AN INITIAL DEMYELINATING EVENT

At present, outside of MRI features, no clinical features at the time of presentation can accurately predict whether a child with an acute demyelinating event will develop MS. Clinical studies have been hampered in part by the lack of consistent definitions used across publications and the small numbers of subjects at any one site. In available studies, the risk of developing MS after ADEM has been reported to be 0% in a study from Argentina,[2] 9.5% in a study from San Diego,[9] and 18% to 29% in studies from France.[21,74] Variability in the criteria used to define ADEM and pediatric MS may have contributed to the wide range of published incidence figures.

The KIDMUS study group from France examined pediatric patients with an initial demyelinating event, including CIS-like and ADEM-like events and showed that overall, 57% developed a second attack during a mean follow-up period of 5.4 years.[21] Of patients presenting initially with optic neuritis, 86% developed MS, while 50% of those initially with a brainstem syndrome developed MS. Overall, positive predictive factors for the development of MS were: age at onset 10 years or older and optic nerve involvement. A lower risk of developing MS was found in patients with mental status change at presentation, suggesting that the presence of encephalopathy may be a negative predictive factor. Of patients with an initial diagnosis of ADEM, 29% developed MS. In a subsequent publication by this group, when the diagnosis of ADEM was redefined by the KIDSEP study to include "change in mental status" as a qualifying criterion, 18% of children were found to develop MS, as defined by the development of a second event during follow-up.[10]

Another recent study described presenting characteristics of 89 patients who experienced an acute clinical demyelinating event, and compared those who had a monophasic course and those who ultimately experienced relapses.[5] Age of onset was higher in patients with MS. Family history of MS was present in approximately 23% of MS patients and none of the ADEM cohort. Encephalopathy was present in approximately 41% of ADEM patients and seizure in 25% of these patients, while neither (seizure or encephalopathy) were present in any of the MS patients. Children without encephalopathy had a significantly higher likelihood of converting to MS. There was no difference between race, sex, history of preceding infection or immunization, or other features of clinical presentation. The most frequent symptom reported in MS patients was paresthesia, while weakness was most frequent in ADEM patients. CSF's white cell count and protein contents did not differ between the groups. Oligoclonal bands were positive in 8/18 MS patients and none of the ADEM (n = 13) patients. IgG index was more frequently elevated in MS patients. MRI abnormalities commonly seen in MS patients included periventricular white matter lesions (generally characterized by periventricular perpendicular ovoid lesions (PVPOLs)). No ADEM patients had PVPOLs.[5] Predominance of periventricular lesions in pediatric MS and relative sparing of this region in ADEM have been previously reported.[7]

Similar findings were reported in another study comparing pediatric MS and ADEM patients after a mean follow-up period of 5.6 years. ADEM patients more frequently experienced the following symptoms: infection prior to disease onset, polysymptomatic presentation, pyramidal signs, encephalopathy and bilateral optic neuritis (ON).[7] In this study, seizures only occurred in ADEM patients and unilateral ON occurred only in MS patients. ADEM patients were also more likely to have blood leucocytosis and nearly half had onset between the ages of 3-5 years, whereas only 23% of MS patients presented under the age of 5 years. ADEM was also more prevalent during the winter months.[7]

Adult onset ADEM may be characterized by a higher likelihood of relapse. One small study found that 35% of adults initially diagnosed with ADEM went on to develop MS,[1] with all relapses occurring within one year of initial presentation.[1] Patients eventually diagnosed with MS were more likely to have periventricular lesions (n = 14, 54%).[1] Another studies identified several factors associated with increased risk of relapse including age over 55 years, elevated CSF albumin (>100 mg/dl), female sex, spinal cord and/or PNS involvement.[23]

Distinguishing characteristics of monophasic ADEM in adults include preceding infection, acute onset, brainstem symptoms (e.g., ocular motor deficits, dysarthria), alteration of consciousness, aphasia, hemiplegia, paraplegia, tetraplegia, seizure, vomiting, bilateral ON, confusion, no oligoclonal bands in the CSF and gray matter involvement.[1,25,87] Overall, adult patients with monophasic ADEM appear to be younger, their onset of symptoms more acute, they experience more severe initial symptoms and have more widespread CNS disturbance, but they respond more favorably to treatment.[1] Information regarding ADEM in adults is limited and is based on small cohorts; therefore, definitive conclusions regarding presentation, prognosis and likelihood for relapse are limited.

See Table 2 for a description of initial presentation in ADEM versus MS in children and adolescents.

TREATMENT

Steroids

There have been no randomized, double blind studies of the treatment of ADEM. However, several retrospective analyses have suggested that high dose steroids may be used with reasonable success,[88] either in the form of oral dexamethasone[2] or high

Table 2. Features of ADEM and MS in children and adolescents

	ADEM	MS
Age of onset	Predominantly younger (<10 yrs)	Adolescent to early adulthood
Sex	Likely greater prevalence in males	Greater prevalence in females
MRI	Larger lesions; resolution of lesions over time; involvement of grey and white matter	Periventricular lesions/ periventricular perpendicular ovoid lesions; new lesions on follow-up
CSF	Increased WBC (neutrophils, lymphocytes)	Oligoclonal bands, elevated IgG index
Visual Testing	Bilateral ON	Unilateral ON
Serum Testing	Leukocytosis, elevated WBC	Normal WBC
Presentation	Polysymptomatic, encephalopathy, acute, at times severe presentation, associated with infection	Monosymptomatic, acute to sub-acute onset

dose IV solumedrol (10-30 mg/kd/d).[88,89] In many cases, the decision to intervene with steroid therapy is a clinical one. Treatment is sometimes reserved for patients with severe neurologic deficits, including visual loss, severe weakness with bowel/bladder involvement, severe encephalopathy/coma or other focal neurologic deficits, including cerebellar and brainstem deficits. One series of cases evaluating 16 patients with ADEM described response to high dose solumedrol within 10 days in 10/16 patients (63%).[88]

IVIg

IVIg therapy has been described in the treatment of ADEM, usually in the setting of steroid resistant cases.[88,90,91] One retrospective study examined the use of IVIg in severe steroid-resistant cases of acute disseminated encephalomyelitis. Cases received steroids as first-line therapy. A small group (n = 6) received IVIg as first line therapy because of contraindications for steroid use. Cases which were steroid resistant were more likely to involve peripheral nervous system damage (89%) and myelitis (95%); 53% of patients (10/19) experienced clinical improvement over the course of the treatment period.[90] Dosing of IVIg for acute treatment of demyelination follows other pediatric IVIg treatment protocols of 2 g/kg total; this is frequently divided over 5 days.[90]

Plasmapheresis

The use of plasmapheresis for ADEM has been described in small retrospective case series of patients who are resistant to steroid therapy and/or IVIg; its benefit in ADEM has not been clearly established. One study has suggested limited recovery, which only occurred months after treatment.[89] In a group of adult patients diagnosed with ADEM (n = 3), neurologic improvement was not seen after the administration of plasmapheresis.[92] Importantly, plasmapheresis is usually reserved for the most treatment-resistant patients and is usually performed days after the onset of symptoms. It is unclear if the effect of this therapy is beneficial if given earlier in the course of the disease.

PROGNOSIS

As mentioned earlier, the prognosis for ADEM is thought to be favorable, with more than half of patients showing complete resolution of clinical symptoms and lesions on MRI.[2,18,60,93] Average time to recovery is generally over several weeks to months.[7,18,94] Prognosis in adults with ADEM has been shown to be somewhat poorer than pediatric patients.[94] Mortality rates are higher,[1,24,25,87] up to 25%, compared to less than 5% in pediatric patients. Persistent deficits (e.g., functional state, sensory or motor disability, cranial nerve abnormality or cognitive impairment)[1,25,87] may also be more common in adult onset ADEM patients (in 35-90%). However, differences noted between adult and pediatric ADEM populations may be due to differences in specific areas tested.[1,24,87] One study describes nearly 90% of adult patients having persistent deficits, mainly mild cognitive impairment;[87] studies of pediatric patients have rarely included neuropsychological data.

Over half of children diagnosed with ADEM are reported to have complete recovery.[2,7-9,18-21] However, most of these studies are limited by a lack of detailed follow-up data. One study described that fewer than 17% (3/18) of ADEM patients had continuing deficits on follow-up (2-60 months post episode).[18] In this study, deficits included urinary symptoms and gait problems. A larger prospective study of 84 patients with ADEM suggested that the use of high-dose corticosteroid treatment was associated with good recovery and resolution of lesions on MRI. The majority of these patients (89%) had either complete recovery with normal neurologic examinations or abnormal signs without disability at follow-up (mean follow-up of 6.6 years). Deficits in the remaining children included hemiparesis, partial epilepsy, decreased visual acuity, paraparesis and mental handicap.[2] Within this cohort, no association could be made between initial MRI findings and outcomes.[2]

Another longitudinal study of 24 patients with ADEM (mean follow-up of 52.8 months) showed that only 3 had mild persistent neurologic signs. Normalization on MRI was seen in 59% of the patients and 36% showed improvement on MRI. Only one patient showed no improvement on MRI.[93] A similar study including 28 pediatric ADEM patients found that 57% made a complete recovery after a mean follow-up duration of 5.8 years. Ninety percent had partial or complete resolution on MRI and no new lesions. Residual symptoms included motor impairment (17%), parasthesia (6%), visual impairment (11%), cognitive impairment (11%) and behavior problems (11%).[7]

Neurocognitive Functioning

Many outcome studies in patients with ADEM fail to include assessment of cognitive and psychosocial functioning, or include only broad outcomes (e.g., IQ) and qualitative descriptions. Studies that have examined neurocognitive functioning in patients with ADEM are limited by their small sample size.

A variety of residual mild cognitive deficits in areas of visuospatial and visuomotor functioning, attention, executive function, mood, behavior and social skills have been found in children with prior diagnosis of ADEM.[75,95-97] Specifically, one case-control study (ADEM n = 19, controls n = 19) showed that MRI lesion load did not correlate with cognitive performance; however, disease severity and earlier onset were associated with poorer cognitive performance.[95] This study showed that early onset ADEM patients (<5 years) had lower IQ and lower academic skills compared to controls, whereas older onset ADEM patients (>5 years) had poorer verbal processing compared to controls.[95] In a somewhat larger follow-up study of a group of MS, ADEM and MDEM patients (ADEM, n = 28, MDEM n = 7, MS n = 13, with a mean follow-up 5.8 years) Dale et al (2000) found that within the ADEM/MDEM group, 11% had cognitive deficits and 11% had residual behavioral problems. Cognitive and behavioral outcomes were described as diminished IQ (IQ of 70), aggression and obsessive compulsive disorder.[7]

A recent study examining neurocognitive features in children with ADEM suggests mild impairment in divided attention and cognitive flexibility.[96] Another small study showed ongoing cognitive deficits in all patients (n = 6) 2-5 years post ADEM episode, particularly in attention and executive function. This was despite complete (n = 4) or partial resolution on MRI.[97] Cognitive and behavioral sequelae including low IQ, impaired speech and language, low academic achievement and aggressive behavior have also been described in a case series of pediatric ADEM patients with cerebellar involvement.[75]

CONCLUSION

ADEM is a rare condition with a generally favorable outcome which occurs in children, adolescents and adults. However, there is some evidence for poorer outcome in adults, with a higher risk for relapse and mortality. Approximately one-fifth of children go on to have relapses. Some of these children are eventually diagnosed with multiple sclerosis.

No single diagnostic test can reliably predict which patients with ADEM will go on to have relapses, although recently proposed MRI criteria may be helpful in this regard. Earlier detection of patients who will likely go on to have recurrent demyelination may lead to earlier treatment intervention, which may, in turn, have the potential to improve outcomes.

Finally, until recently, ADEM was assumed to be a condition with relatively few lingering deficits. However, there is growing evidence to suggest that neurocognitive impairment may persist even in the face of resolution of lesions on MRI and minimal to no lasting physical deficits. Ongoing research with larger patient samples and greater breadth in evaluation of functioning, including more sensitive measures of neurocognitive functioning, are needed.

REFERENCES

1. Schwarz S, Mohr A, Knauth M et al. Acute disseminated encephalomyelitis: A follow-up study of 40 adult patients. Neurology 2001; 56:1313-1318.
2. Tenembaum S, Cahamoles N, Fejerman N. Acute disseminated encephalomyelitis: a long-term follow-up study of 84 pediatric patients. Neurology 2002; 59(8):1224-1231.
3. Krupp LB, Banwell B, Tenembaum S. Consensus definitions proposed for pediatric multiple sclerosis and related disorders. Neurology 2007; 68(16 Suppl 2):S7-S12.
4. McKeon A, Lennon VA, Lotze T et al. CNS aquaporin-4 autoimmunity in children. Neurology 2008; 71(2):93-100.
5. Alper G, Heyman R, Wang L. Multiple sclerosis and acute disseminated encephalomyelitis diagnosed in children after long-term follow-up: Comparison of presenting features. Developmental Medicine and Child Neurology 2009; 51:480-486.
6. Pohl D, Rostasy K, Reiber H, Hanefeld F. CSF characteristics in early-onset multiple sclerosis. Neurology 2004; 63(10):1966-1967.
7. Dale RC, de Sousa C, Chong WK et al. Acute disseminated encephalomylitis, multiphasic disseminated encephalomyelitis and multiple sclerosis in children. Brain 2000; 12:2407-2422.
8. Anlar B, Basaran C, Kose G et al. Acute disseminated encephalomyelitis in children: outcome and prognosis. Neuropediatrics 2003; 34(4):194-199.

9. Leake JA, Albani S, Kao AS et al. Acute disseminated encephalomyelitis in childhood: epidemiologic, clinical and laboratory features. Pediatr Infect Dis J 2004; 23(8):756-764.
10. Mikaeloff Y, Caridade G, Husson B et al. Acute disseminated encephalomyelitis cohort study: prognostic factors for relapse. Eur J Paediatr Neurol 2007; 11(2):90-95.
11. Olgac Dundar N, Anlar B, Guven A et al. Relapsing Acute Disseminated Encephalomyelitis in Children: Further Evaluation of the Diagnosis. J Child Neurol 2010; [Epub ahead of print]
12. Banwell B, Kennedy J, Sadovnick D et al. Incidence of acquired demyelination of the CNS in Canadian children. Neurology 2009; 72(3):232-239.
13. Tenembaum SN. Disseminated encephalomyelitis in children. Clin Neurol Neurosurg 2008; 110(9):928-938.
14. Torisu H, Kira R, Ishizaki Y et al. Clinical study of childhood acute disseminated encephalomyelitis, multiple sclerosis and acute transverse myelitis in Fukuoka Prefecture, Japan. Brain and Development 2010; 32:454-462.
15. Tenembaum S, Chitnis T, Ness J et al. Acute disseminated encephalomyelitis. Neurology 2007; 68(16 Suppl 2):S23-S36.
16. Tenembaum S. Disseminated encephalomyelitis in children. Clinical Neurology and Neurosurgery 2008; 110:928-938.
17. Atzori M, Battistella PA, Perini P et al. Clinical and diagnostic aspects of multiple sclerosis and acute monophasic encephalomyelitis in pediatric patients: A single centre prospective study. Mult Sclerosis 2009; 15:363-370.
18. Murthy KSN, Faden HS, Cohen ME et al. Acute disseminated encephalomyelitis in children. Pediatrics 2002; 110(2).
19. Hynson JL, Kornberg AJ, Coleman LT et al. Clinical and neuroradiologic features of acute disseminated encephalomyelitis in children. Neurology 2001; 56(10):1308-1312.
20. Gupte G, Stonehouse M, Wassmer E et al. Acute disseminated encephalomyelitis: a review of 18 cases in childhood. J Paediatr Child Health 2003; 39(5):336-342.
21. Mikaeloff Y, Suissa S, Vallee L et al. First episode of acute CNS inflammatory demyelination in childhood: prognostic factors for multiple sclerosis and disability. J Pediatr 2004; 144(2):246-252.
22. Schwarz S, Mohr A, Knauth M et al. Acute disseminated encephalomyelitis: a follow-up study of 40 adult patients. Neurology 2001; 56(10):1313-1318.
23. Marchioni E, Tavazzi E, Minoli L et al. Acute disseminated encephalomyelitis. Neurol Sci 2008; 29:S286-S288.
24. Sonneville R, Demeret S, Klein I et al. Acute disseminated encephalomyelitis in the intensive care unit: clinical features and outcome of 20 adults. Intensive Care Med 2008; 34(3):528-532.
25. Sonneville R, Klein I, de Broucker T et al. Post-infectious encephalitis in adults: diagnosis and management. J Infect 2009; 58(5):321-328.
26. Axer H, Ragoschke-Schumm A, Bottcher J et al. Initial DWI and ADC imaging may predict outcome in acute disseminated encephalomyelitis: report of two cases of brain stem encephalitis. J Neurol Neurosurg Psychiatry 2005; 76(7):996-998.
27. Menge T, Kiesseier BC, Nessler S et al. Acute disseminated encephalomyelitis: an acute hit against the brain. Curr Opin Neurol 2007; 20:247-254.
28. Booss J, Davis LE. Smallpox and smallpox vaccination: neurological implications. Neurology 2003; 60(8):1241-1245.
29. Sejvar JJ, Labutta RJ, Chapman LE et al. Neurologic adverse events associated with smallpox vaccination in the United States, 2002-2004. JAMA 2005; 294(21):2744-2750.
30. Bennetto L, Scolding N. Inflammatory/post-infectious encephalomyelitis. J Neurol Neurosurg Psychiatry 2004; (75 Suppl 1):i22-i28.
31. Fenichel GM. Neurological complications of immunization. Ann Neurol 1982; 12(2):119-128.
32. Nalin DR. Mumps, measles and rubella vaccination and encephalitis. BMJ 1989; 299(6709):1219.
33. Plesner AM, Arlien-Soborg P, Herning M. Neurological complications to vaccination against Japanese encephalitis. Eur J Neurol 1998; 5(5):479-485.
34. Cheong JH, Bak KH, Kim CH et al. Acute disseminated encephalomyelitis associated with influenza vaccination. J Korean Neurosurg Soc 2004; 35:223-225.
35. Izurieta HS, Haber P, Wise RP et al. Adverse events reported following live, cold-adapted, intranasal influenza vaccine. JAMA 2005; 294(21):2720-2725.
36. Dodick DW, Silber MH, Noseworthy JH et al. Acute disseminated encephalomyelitis after accidental injection of a hog vaccine: successful treatment with plasmapheresis. Mayo Clin Proc 1998; 73(12):1193-1195.
37. Sonmez FM, Odemis E, Ahmetoglu A et al. Brainstem encephalitis and acute disseminated encephalomyelitis following mumps. Pediatr Neurol 2004; 30(2):132-134.
38. Miller DH, Kay R, Schon F et al. Optic neuritis following chickenpox in adults. J Neurol 1986; 233(3):182-184.
39. Fujimoto H, Asaoka K, Imaizumi T et al. Epstein-Barr virus infections of the central nervous system. Intern Med 2003; 42(1):33-40.
40. Revel-Vilk S, Hurvitz H, Klar A et al. Recurrent acute disseminated encephalomyelitis associated with acute cytomegalovirus and Epstein-Barr virus infection. J Child Neurol 2000; 15(6):421-424.

41. Kaji M, Kusuhara T, Ayabe M et al. Survey of herpes simplex virus infections of the central nervous system, including acute disseminated encephalomyelitis, in the Kyushu and Okinawa regions of Japan. Mult Scler 1996; 2(2):83-87.
42. Sacconi S, Salviati L, Merelli E. Acute disseminated encephalomyelitis associated with hepatitis C virus infection. Arch Neurol 2001; 58(10):1679-1681.
43. Tan H, Kilicaslan B, Onbas O et al. Acute disseminated encephalomyelitis following hepatitis A virus infection. Pediatr Neurol 2004; 30(3):207-209.
44. David P, Baleriaux D, Bank WO et al. MRI of acute disseminated encephalomyelitis after coxsackie B infection. J Neuroradiol 1993; 20(4):258-265.
45. Wang YH, Huang YC, Chang LY et al. Clinical characteristics of children with influenza A virus infection requiring hospitalization. J Microbiol Immunol Infect 2003; 36(2):111-116.
46. Rellosa N, Bloch KC, Shane AL et al. Neurologic Manifestations of Pediatric Novel H1n1 Influenza Infection. Pediatr Infect Dis J.
47. Silver B, McAvoy K, Mikesell S et al. Fulminating encephalopathy with perivenular demyelination and vacuolar myelopathy as the initial presentation of human immunodeficiency virus infection. Arch Neurol 1997; 54(5):647-650.
48. Tachi N, Watanabe T, Wakai S et al. Acute disseminated encephalomyelitis following HTLV-I associated myelopathy. J Neurol Sci 1992; 110(1-2):234-235.
49. Kamei A, Ichinohe S, Onuma R et al. Acute disseminated demyelination due to primary human herpesvirus-6 infection. Eur J Pediatr 1997; 156(9):709-712.
50. Wei TY, Baumann RJ. Acute disseminated encephalomyelitis after Rocky Mountain spotted fever. Pediatr Neurol 1999; 21(1):503-505.
51. Yeh EA, Collins A, Cohen ME et al. Detection of coronavirus in the central nervous system of a child with acute disseminated encephalomyelitis. Pediatrics 2004; 113(1 Pt 1):e73-e76.
52. Riedel K, Kempf VA, Bechtold A et al. Acute disseminated encephalomyelitis (ADEM) due to Mycoplasma pneumoniae infection in an adolescent. Infection 2001; 29(4):240-242.
53. van Assen S, Bosma F, Staals LM et al. Acute disseminated encephalomyelitis associated with Borrelia burgdorferi. J Neurol 2004; 251(5):626-629.
54. Nasralla CA, Pay N, Goodpasture HC et al. Postinfectious encephalopathy in a child following Campylobacter jejuni enteritis. AJNR Am J Neuroradiol 1993; 14(2):444-448.
55. Lelis SS, Fonseca LF, Xavier CC et al. Acute disseminated encephalomyelitis after leptospirosis. Pediatric Neurology 2009; 40(6):471-473.
56. Chandra SR, Kalpana D, Anilkumar TV et al. Acute disseminated encephalomyelitis following leptospirosis. J Assoc Physicians India 2004; 52:327-329.
57. Heick A, Skriver E. Chlamydia pneumoniae-associated ADEM. Eur J Neurol 2000; 7(4):435-438.
58. Spieker S, Petersen D, Rolfs A et al. Acute disseminated encephalomyelitis following Pontiac fever. Eur Neurol 1998; 40(3):169-172.
59. Dale RC, Church AJ, Cardoso F et al. Poststreptococcal acute disseminated encephalomyelitis with basal ganglia involvement and auto-reactive antibasal ganglia antibodies. Ann Neurol 2001; 50(5):588-595.
60. Huynh W, Cordato DJ, Kehdi E et al. Post-vaccination encephalomyelitis: Literature review and illustrative case. Journal of Clinical Neuroscience 2008; 15:1315-1322.
61. Au WY, Lie AK, Cheung RT et al. Acute disseminated encephalomyelitis after para-influenza infection post bone marrow transplantation. Leuk Lymphoma 2002; 43(2):455-457.
62. Horowitz MB, Comey C, Hirsch W et al. Acute disseminated encephalomyelitis (ADEM) or ADEM-like inflammatory changes in a heart-lung transplant recipient: a case report. Neuroradiology 1995; 37(6):434-437.
63. Re A, Giachetti R. Acute disseminated encephalomyelitis (ADEM) after autologous peripheral blood stem cell transplant for non-Hodgkin's lymphoma. Bone Marrow Transplant 1999; 24(12):1351-1354.
64. Tomonari A, Tojo A, Adachi D et al. Acute disseminated encephalomyelitis (ADEM) after allogeneic bone marrow transplantation for acute myeloid leukemia. Ann Hematol 2003; 82(1):37-40.
65. Jaster JH, Niell HB, Dohan FC, Jr et al. Demyelination in the brain as a paraneoplastic disorder: candidates include some cases of leukemia and non-Hodgkin's lymphoma. Ann Hematol 2003; 82(11):714-715.
66. Mikaeloff Y, Caridade G, Suissa S et al. Hepatitis B vaccine and the risk of CNS inflammatory demyelination in childhood. Neurology 2008; 72(10):873-80
67. Belman A, Chabas D, Chitnis T et al. Clinical Spectrum of Disorders Masquerading as Pediatric Multiple Sclerosis. Annals of Neurology 2007; Supplement: Abstracts CNS.
68. van der Knaap MS, Salomons GS, Li R et al. Unusual variants of Alexander's disease. Ann Neurol 2005; 57(3):327-338.
69. Harris MO, Walsh LE, Hattab EM et al. Is it ADEM, POLG, or both? Arch Neurol 67(4):493-496.
70. Gorman MP, Golomb MR, Walsh LE et al. Steroid-responsive neurologic relapses in a child with a proteolipid protein-1 mutation. Neurology 2007; 68(16):1305-1307.

71. Weinstock A, Giglio P, Cohen ME et al. Diffuse magnetic resonance imaging white-matter changes in a 15-year-old boy with mitochondrial encephalomyopathy. J Child Neurol 2002; 17(1):47-49.
72. van der Knaap MS, Valk J, de Neeling N et al. Pattern recognition in magnetic resonance imaging of white matter disorders in children and young adults. Neuroradiology 1991; 33(6):478-493.
73. Lotze TE, Northrop JL, Hutton GJ et al. Spectrum of pediatric neuromyelitis optica. Pediatrics 2008; 122(5):e1039-e1047.
74. Mikaeloff Y, Adamsbaum C, Husson B et al. MRI prognostic factors for relapse after acute CNS inflammatory demyelination in childhood. Brain 2004; 127(Pt 9):1942-1947.
75. Parrish JB, Weinstock-Guttman B, Yeh EA. Cerebellar mutism in pediatric acute disseminated encephalomyelitis. Pediatr Neurol 2010; 42(4):259-266.
76. Neuteboom RF, Boon M, Catsman Berrevoets CE et al. Prognostic factors after a first attack of inflammatory CNS demyelination in children. Neurology 2008; 71(13):967-973.
77. Callen DJ, Shroff MM, Branson HM et al. Role of MRI in the differentiation of ADEM from MS in children. Neurology 2008.
78. Ketelslegers IA, Neuteboom RF, Boon M et al. A comparison of MRI criteria for diagnosing pediatric ADEM and MS. Neurology 2010; 74(18):1412-5
79. Chabas D, Ness J, Belman A et al. Younger children with pediatric MS have a distinct CSF inflammatory profile at disease onset. Neurology 2010; 74(5):399-405.
80. Selter RC, Brilot F, Grummel V et al. Antibody responses to EBV and native MOG in pediatric inflammatory demyelinating CNS diseases. Neurology 2010; 74(21):1711-1715.
81. Banwell B, Tenembaum S, Lennon VA et al. Neuromyelitis optica-IgG in childhood inflammatory demyelinating CNS disorders. Neurology 2008; 70(5):344-352.
82. Kotlus BS, Slavin ML, Guthrie DS et al. Ophthalmologic manifestations in pediatric patients with acute disseminated encephalomyelitis. J AAPOS 2005; 9(2):179-183.
83. Simone IL, Carrara D, Tortorella C et al. Course and prognosis in early-onset MS: comparison with adult-onset forms. Neurology 2002; 59(12):1922-1928.
84. Pohl D, Rostasy K, Treiber-Held S et al. Pediatric multiple sclerosis: detection of clinically silent lesions by multimodal evoked potentials. J Pediatr 2006; 149(1):125-127.
85. Frohman E, Costello F, Zivadinov R et al. Optical coherence tomography in multiple sclerosis. Lancet Neurol 2006; 5(10):853-863.
86. Yeh E, Weinstock-Guttman B, Lincoff N et al. Retinal nerve fiber thickness in inflammatory demyelinating diseases of childhood onset. Mult Scler 2009; 15(7):802-810.
87. Hollinger P, Sturzenegger M, Mathis J et al. Acute disseminated encephalomyelitis in adults: a reappraisal of clinical, CSF, EEG and MRI findings. J Neurol 2001; 249:320-329.
88. Shahar E, Andraus J, Savitzki D et al. Outcome of severe encephalomyelitis in children: effect of high-dose methylprednisolone and immunoglobulins. J Child Neurol 2002; 17(11):810-814.
89. Khurana DS, Melvin JJ, Kothare SV et al. Acute disseminated encephalomyelitis in children: discordant neurologic and neuroimaging abnormalities and response to plasmapheresis. Pediatrics 2005; 116(2):431-436.
90. Ravaglia S, Piccolo G, Ceroni M et al. Severe steroid-resistant post-infectious encephalomyelitis: general features and effects of IVIg. J Neurol 2007; 254(11):1518-1523.
91. Straussberg R, Schonfeld T, Weitz R et al. Improvement of atypical acute disseminated encephalomyelitis with steroids and intravenous immunoglobulins. Pediatr Neurol 2001; 24(2):139-143.
92. Kaynar L, Altuntas F, Aydogdu I et al. Therapeutic plasma exchange in patients with neurologic diseases: retrospective multicenter study. Transfus Apher Sci 2008; 38(2):109-115.
93. Suppiej A, Vittorini R, Fontanin M et al. Acute disseminated encephalomyelitis in chidlren: Focus on Relapsing Patients. Pediatr Neurol 2008; 39:12-17.
94. Tenembaum S, Chitnis T, Ness J et al. Acute disseminated encephalomyelitis. Neurology 2007; 68(Suppl 2):S23-S36.
95. Jacobs RK, Anderson VA, Neale JL et al. Neuropsychological outcome after acute disseminated encephalomyelitis: impact of age at illness onset. Pediatr Neurol 2004; 31(3):191-197.
96. Deery B, Anderson V, Jacobs R et al. Childhood MS and ADEM: Investigation and comparison of neurocognitive features in children. Developmental Neuropsychology 2010; 35(5):506-521.
97. Hahn CD, Miles BS, MacGregor DL et al. Neurocognitive outcome after acute disseminated encephalomyelitis. Pediatr Neurol 2003; 29:117-123.

CHAPTER 2

AGE-RELATED MACULAR DEGENERATION

Sam Khandhadia, Jocelyn Cherry and Andrew John Lotery*

Department of Clinical Neurosciences, Southampton General Hospital, Southampton, UK
*Corresponding Author: Andrew John Lotery—Email: a.j.lotery@soton.ac.uk

Abstract: Age-related macular degeneration (AMD) is the leading cause of irreversible blindness in the developed world. Despite recent advances in treatment, AMD causes considerable morbidity. For the non-ophthalmologist, a brief background on retinal structure is provided, followed by a description of the characteristic changes seen in AMD. Subsequently the typical clinical features of AMD are discussed with an outline of present management, followed by the current theories of AMD pathogenesis. The similarities between AMD and another neurodegenerative disease are then highlighted. Finally, we review the on-going clinical trials of potential treatments for the future. Since it is clear that multiple risk factors are involved in the pathogenesis of AMD, a multi-faceted approach will most likely be required in order to prevent further patients progressing to blindness as a result of this devastating condition.

INTRODUCTION

The retina is of neural origin and is considered an end-organ of the central nervous system. The centre of the retina, the macula, is particularly prone to age related degenerative changes, a process known as age-related macular degeneration (AMD). Unfortunately its effects can lead to impairment of central vision, which may be severe enough in some individuals to lead to legal blindness. Although some forms of AMD can be treated, the majority of patients sustain some loss of vision and there are as yet no definitive preventative measures known.

In this chapter a brief description of AMD including a summary of retinal anatomy is presented, followed by clinical features and current management. We then compare AMD with other neurodegenerative diseases and conclude with ongoing developments in research.

Neurodegenerative Diseases, edited by Shamim I. Ahmad.
©2012 Landes Bioscience and Springer Science+Business Media.

EPIDEMIOLOGY

AMD is the leading cause of blindness in Europe, USA and Australia accounting for up to 50% of all cases.[1] It is a common disease—the prevalence in adults is around 3%.[2] The prevalence inevitably increases with age, as is the case with several other neurodegenerative diseases. Almost two-thirds of the population over 80 years old will have some signs of AMD.[3,4] The prevalence of visual impairment (defined by the World Health Organisation, as a visual acuity of less than 6/18 in the better eye) in the over 65s due to AMD is up to 3%.[5] AMD is more common in Caucasians than in other ethnic groups.[6] There is no obvious sex preponderance, although some studies suggest women may be more susceptible.[7]

NORMAL ANATOMY

The macula is defined as the centre of the retina. The retina forms the innermost of the three layers constituting the wall of the eyeball. The tough fibrous white sclera forms the outermost wall and the vascular choroid sits in between the two. The retina consists of two delicate, thin layers of tissue. The innermost layer is the neurosensory retina, which consists of sensory cells, neurons and their supporting cells. The outer layer is the retinal pigment epithelium (RPE).

The outer layer of the neurosensory retina consists of photoreceptor cells, which detect light from the external sources and convert this to neuronal signals. These are then relayed for processing via the optic nerve in the visual cortex. This is located in the posteriorly situated occipital cortex. There are two types of photoreceptor cells named after their histological shape. Cones are utilised for colour vision and fine detail in bright light conditions. Rods are utilised for night vision. Cones are distributed mainly within the macular region, whereas rods are distributed more peripherally. However, overall rods are more numerous (comprising 95% of all photoreceptors) and indeed the macula consists mostly of rods.[8]

Each photoreceptor cell consists of an outer and inner segment, nucleus (the "cell body") and axon. The outer segment is in close contact with the RPE and consists of discs containing photosensitive pigments. These discs are constantly renewed. The used discs are shed and phagocytosed by the RPE (see below). As a photoreceptor cell is exposed to light, a renewable chain of chemical reactions is triggered in the outer segment discs, (the "visual cycle"), leading to the creation of electrical neuronal impulses ("phototransduction"). The inner segment, attached to the outer segment by a connecting cilium, contains multiple mitochondria and other supporting organelles. The axon synapses with bipolar cells deeper in the neurosensory retinal structure. The bipolar cells in turn synapse with ganglion cells. The axons of the ganglion cells make up the innermost nerve fibre layer, which carry signals generated by the photoreceptor cells to the optic nerve. Other cells provide a supporting role (e.g., Horizontal, Amacrine and Muller cells).

The RPE layer, despite being only one cell thick, performs important supporting functions, including phagocytosis of discarded photoreceptor cell outer segments, absorption of excess heat from incoming light, regulation of transport to and from the retina, maintenance of the extracellular matrix and metabolism of retinol (a vitamin A derivative involved in the visual cycle). The inner surface of an RPE cell contains microvilli, which are in contact with the outer segments of up to 45 photoreceptor cells. Approximately 1 million discs per year are phagocytosed by each RPE cell. These discs are engulfed in phagosomes within the RPE cells, which on fusion with lysosomes are exposed to lysosomal enzymes.[9] Each RPE cell contains numerous granules of the pigment melanin, called melanosomes.[10]

The optic disc is a pink-yellow circular structure, which acts as a landmark during retinal examination. It represents the beginning of the optic nerve and is the point at which neurons from the neurosensory retina converge and then leave the eye. Sandwiched between the RPE and choroid, lies Bruch's membrane (BM). This consists of collagen, with a central layer of elastin fibres. The choroid itself sits just external to BM. It is a highly vascular tissue and consists of an inner capillary layer, the choriocapillaris and an outer large vessel layer. The capillaries within the choriocapillaris are fenestrated, allowing free passage of substances up to a certain size between the intra-capillary and the extravascular space.

The retina has a dual blood supply. The inner retina is supplied by its own vascular system, situated within the neurosensory retina. The retinal arteries emanate from the centre of the optic disc, then branch out in four diagonal directions to end in capillaries. These then drain into retinal veins that also form four branches running alongside the retinal arteries. These leave the eye through the centre of the optic disc. The endothelial cells of this vascular system together with tight junctions compose the inner blood–retinal barrier, corresponding to the blood brain barrier.

The choroid supplies the outer retina, via passive diffusion and active transport through BM and the RPE layer. Tight junctions between the RPE cells restrict movement and constitute the outer blood–retinal barrier. These barriers are important to ensure regulation of transport of fluids, nutrients and waste products. BM also acts as a physical barrier to restrict abnormal neovascularisation from the choroid (see below).

The macula is identifiable as an area about 3 mm in diameter, temporal to the optic disc. The temporal retinal artery branches arc around the macula (the 'arcades'). Histologically, the macula is defined as the area of the retina where the ganglion cell layer is more than one cell thick. The centre of the macula comprises the fovea. This is the point on which light rays from the external sources are focussed by the eye's optical system. The fovea appears darker clinically than the surrounding retina, since pigment levels are higher here. These include lutein and melanin. Lutein is a diet-derived yellow pigment found principally at the fovea. Melanin, found throughout the retina is maximal at the fovea in conjunction with larger RPE cells.

The fovea has evolved to allow maximal resolution of incoming images. Cone photoreceptor cells are at their highest concentration in the retina at the fovea.[8] Most other neurons within the neurosensory retina, including the ganglion cells, bipolar cells and nerve fibre layer, have been displaced away from the foveal region to reduce interference with incoming light. Furthermore, there are no retinal blood vessels at the fovea, the blood supply being derived from the outer-lying adjacent choroid.

AMD AND THE RETINA

Drusen are the earliest manifestation of AMD. These result from accumulation of extracellular material normally phagocytosed by the RPE cells. Drusen are of several types: (i) Hard drusen (HD) which appear as discrete yellow spots, (ii) Soft drusen (SD) which are larger, more confluent and (iii) Reticular drusen (RD) which appear as a yellow interlacing network. HD and SD are located between the RPE and BM whereas RD are located above the RPE. Drusen are composed of a selective multitude of proteins and lipids many of which are associated with inflammation or the complement pathway.[11;12] SD and RD are associated with AMD progression.[13,14]

Both RPE atrophy and hypertrophy/hyperplasia often co-exist with drusen. RPE atrophy occurs due to apoptosis of RPE cells presumably as a result of local insult. RPE atrophy appears clinically as pale areas within the retina, with discrete edges. The underlying choroidal vessels may be visible. Large areas of RPE atrophy are termed "geographic atrophy" (GA). RPE hypertrophy/hyperplasia may be a reactive response to local insult and appears as irregular hyperpigmentation often located around areas of RPE atrophy or choroidal neovascular membrane (CNV—see below). These changes are associated with "dry" AMD. Most patients with this form of AMD will retain useful vision but may experience gradual visual deterioration. The advanced form of dry AMD (GA) can lead to blindness.

Unfortunately a small proportion of patients with AMD (the exact proportion varies according to how AMD is defined and the population studied) will develop sudden reduction of vision. This can be due to growth of abnormal blood vessels (CNV) under the retina, constituting "wet" AMD. CNV originates from the highly vascular choroidal layer. The proliferating vessels breach the normal barrier function of BM to proliferate under the RPE and neurosensory retina. The endothelia of these neovascular membranes are highly leaky and as a result these vessels are prone to extravasation of fluid and blood. Alternatively, neovascularisation can originate from within the retina ("retinal angiomatous proliferation," or RAP). Potential spaces for fluid and blood to collect include under the RPE (sub-RPE), between the RPE and neurosensory retina (sub-retinal) and within the neurosensory retina (intra-retinal). Haemorrhage is easy to identify clinically, whereas fluid leaking into the retinal layers needs careful stereoscopic examination to detect increased retinal thickening. If left untreated, CNV will eventually undergo fibrosis. The resulting macular scar is often referred to as "disciform" due to its characteristic disc-like shape. GA and macular scarring are the end-points of dry and wet AMD respectively; this occurs in about 3% of patients with AMD.[15] (Refer to Fig. 1 for a simplified diagrammatic cross-section of the macula and changes seen in AMD. Also refer to Fig 2 for the clinical findings seen in AMD.)

SYMPTOMS OF AMD

Early AMD may be asymptomatic and may only be picked up as an incidental finding on routine optometrist review. Dry AMD is associated with increased difficulty with central vision, especially with reading, and this tends to progress gradually. Wet AMD usually presents suddenly, with impaired central vision, distortion, metamorphopsia (change in size of objects) and a central scotoma (blind spot). Late AMD is associated with permanent impairment of central vision and may be sufficiently reduced to fall within the criteria of severe sight impairment (the legal term for "blindness"). However, it is important to note that patients with late AMD are not completely blind, since peripheral vision is usually preserved. This can provide useful navigational vision.

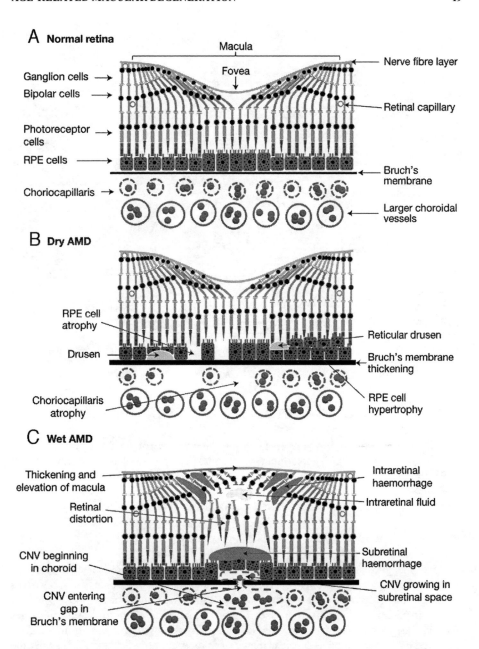

Figure 1. Changes seen in the retina in dry and wet AMD. A) Normal retina. B) Dry AMD. C) Wet AMD Reprinted with permission from: Khandhadia S, Lotery AJ. Expert Rev Mol Med 2010; 12, e34 (October 2010) ©2010 Cambridge University Press.

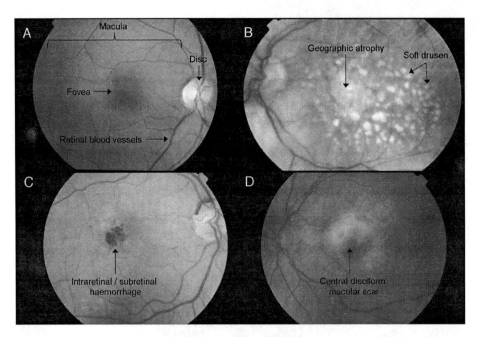

Figure 2. Retinal photographs depicting age-related macular degeneration (AMD). A) Normal. B) Dry AMD. C) Wet AMD. D) End-stage disciform macular scar. Reprinted with permission from: Khandhadia S, Lotery AJ. Expert Rev Mol Med 2010; 12, e34 (October 2010) ©2010 Cambridge University Press.

CLINICAL MANAGEMENT OF AMD

There is no definitive preventative measure for either dry or wet AMD. Furthermore, until recently no satisfactory treatment existed and patients inevitably irretrievably lost sight. In the last few years, the development of intravitreal injections of growth factor inhibitors has radically changed our management of wet AMD. This has meant the deteriorating sight for many patients can be stabilised, if not partially restored.

Management of Wet AMD

Early diagnosis and treatment of wet AMD is associated with a much better outcome. Fundus fluorescein angiogram (FFA) is performed to confirm the presence of CNV. This investigation involves injection of fluorescein (a fluorescent dye) intravenously, usually via the antecubital fossa. This dye passes through the systemic circulation and as it passes through the eye, photos are taken with a fundus camera equipped with barrier filters. Thus reflected light is excluded and a clear image of fluorescent light is obtained. Usually the dye passes through the retinal vessels without leaking. However, the presence of CNV is demonstrated by the presence of leakage from choroidal vessels at the macula. Currently the FFA is usually augmented by a non-invasive test called the Optical Coherence Tomogram (OCT). This uses low intensity laser to produce high resolution cross-sectional images of the macula. Evidence of CNV is inferred by observing the presence of intraretinal or subretinal fluid (Fig. 3).

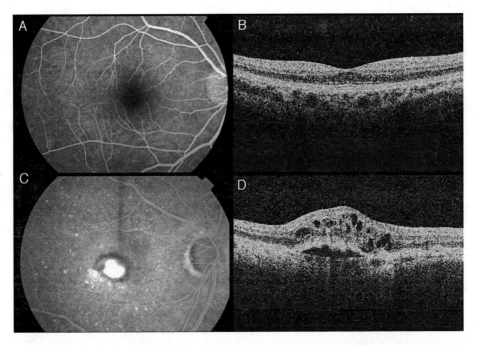

Figure 3. Fundus fluorescein angiograms (FFA) and Optical coherence tomograms (OCT) of wet AMD compared to normal. A) Normal FFA of macula. B) Normal OCT of macula. (see Fig. 1 for anatomy). C) FFA of eye with wet AMD. Note hyperfluorescent area at fovea, resulting from fluorescein dye leaking from choroidal neovascular membrane. D) OCT of eye with wet AMD. Note thickened retina, presence of intraretinal cysts full of fluid and sub-retinal pigment epithelial collection of fluid.

Once the presence of CNV is confirmed, current treatment consists of regular intravitreal injections of monoclonal antibodies which inhibit vascular endothelial growth factor (VEGF). Three injections are given as a preliminary course at four weekly intervals. Following this, patients are given further injections depending on the presence of active CNV, as demonstrated on ocular examination and OCT. Either ranibizumab (Lucentis®) or bevacizumab (Avastin®) are being used. Both work in similar ways, although comparative clinical trials are still ongoing. 0.05ml of the drug is injected 3.5-4 mm from the limbus (the junction between the cornea and sclera), perpendicular to the plane of the sclera. Serious adverse effects are rare (incidence <1%), but endophthalmitis (severe internal infection of the eye), internal haemorrhage, traumatic cataract and retinal detachment may occur. The number of injections required are variable, with some patients only requiring the minimum initial three injections, whereas others may require regular treatment for 2 years or more. All patients are usually monitored for a prolonged period of time even after stabilisation to ensure there is no recurrence of CNV.[16-18]

Patients may initially present with a large sub-macular haemorrhage, especially if on anticoagulant medication. In these cases, anti-VEGF intravitreal injections may not be effective. In addition, the presence of blood is toxic to the retinal cells and may lead to permanent extensive scarring. Removal of this blood is therefore beneficial. Intravitreal tissue plasminogen activator (TPA) and gas injections followed by head posturing, can liquefy and displace any clotted blood away from the macula, although clinical improvement in this situation is rare.[19] A few patients develop vitreous haemorrhage due to bleeding breaking through the retina. If dense, this can be cleared with a vitrectomy, a surgical procedure involving removal of the vitreous.

Management of Dry AMD

As mentioned above, dry AMD is more insidious and is often picked up as a coincidental finding in the asymptomatic patient. At present there is no established way of preventing this condition (although there are several investigational products undergoing clinical trials, see below). Recently, clinical trials demonstrated oral anti-oxidant supplements may reduce the risk of progression. The Age-Related Eye Disease Study (AREDS) found patients with high risk dry AMD (presence of multiple SD and RPE changes) were less likely to exhibit AMD progression if taking high doses of oral vitamins A, C, E and zinc.[20] Other beneficial dietary modification includes increasing consumption of fresh fruit, vegetables and oily fish.[21-23]

Control of risk factors associated with cardiovascular disease may also be beneficial in reducing risk of AMD progression. These include reduction of smoking, serum cholesterol levels, systemic blood pressure and obesity.[24-28] In addition sunlight exposure may be implicated in AMD, especially since ultraviolet (UV) light is associated with RPE and retinal cell damage in vitro. However, this is contentious and there is no evidence that UV-blocking sunglasses are beneficial.[29-31] Lastly, when exposed to low-intensity laser, drusen can fade; however a clinical trial of more than a 1000 patients did not reveal any significant visual benefit.[32]

Unfortunately some patients with dry AMD can develop CNV, especially if high risk features are present.[14] Therefore patients are advised to report any new central visual symptoms, especially distortion. The Amsler grid, consisting of a grid of small squares, can be used regularly by the patient to detect new distortion and metamorphosia.

Management of End-Stage AMD

About 3% of all patients with AMD will reach "end stage" AMD and may fall within the legal criteria for partial or severe sight impairment. Impaired vision in AMD is associated with a reduced quality of life and these patients require socio-psychological support.[5] Magnifying aids, telescopes, closed circuit television (CCTV), talking books and domestic modifications may be beneficial. Some patients report benefit from intra-ocular placement of a magnifying intraocular lens.[5] This sits just behind the iris and can have the effect of increasing magnification as well as displacing images away from a damaged macula onto healthier peri-macular retinal tissue. Extensive retinal rotation surgery has also been attempted, but is associated with a high rate of complications.[33]

PROGNOSIS OF AMD

Wet AMD

More than 90% of patients with wet AMD treated with anti-VEGF intravitreal injections will retain stable vision, and out of these one-third will improve.[16,17] However, wet AMD is often bilateral. The risk of a fellow eye being affected with a similar lesion is approximately 30% over a 6 year period.[34]

Dry AMD

Hard drusen is a common finding and is not associated with AMD progression. However soft drusen and RPE changes are associated with a 6.5% and 7.1% risk respectively of progression to late AMD over a 5 year period.[35] From the onset of GA, the mean duration to legal blindness is 5-9 years.[36,37] Furthermore, greater than 50% of fellow eyes are also afflicted with GA.[37]

PATHOLOGY OF AMD

AMD is a complex multi-factorial disease. It is likely that AMD develops clinically once an accumulation of risk factors exceeds disease threshold, based on an individual's repair and regenerative capabilities. Risk factors include advancing age, the effect of ongoing biological processes, genetic susceptibility and environmental/modifiable factors mentioned above (Fig. 4). Ageing is the most significant risk factor and is accompanied by alteration in gene expression and cellular functions. These include deteriorating mitochondrial function, increased protein degradation and apoptosis.[38,39] The main biological processes likely to contribute to the pathogenesis of AMD are described below.

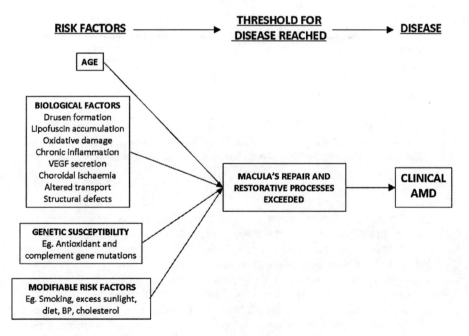

Figure 4. A model of AMD pathogenesis Modified from Swaroop et al 2009.[38]

Drusen Formation

At present it is unclear whether drusen contribute to the pathogenesis of AMD or if they are simply a manifestation of the underlying pathological processes. Drusen contain a range of lipids and proteins, the origins of which are as yet unknown.[11,12] Many of these are components of the inflammatory and complement system, suggesting drusen may play a key role as a focus of inflammation within the retina. Alternatively, drusen may accumulate as a result of ongoing inflammation elsewhere.

Lipofuscin Accumulation

A key function of RPE cells is phagocytosis of shed photoreceptor outer segments. This function diminishes with age.[40] As a result, undigested material accumulates within RPE and photoreceptor cells.[41,42] This material, called "lipofuscin", is composed of several toxic substances including retinoids, produced as by-products of the visual cycle, along with modified proteins and lipids.[43] N-retinylidene-N-retinylethanolamine (A2E), a fluorescent retinoid is a major component of lipofuscin and is associated with oxidative damage and complement activation.[44,45]

Inflammation

CNV development is associated with increased vascularisation, exudation and leukocyte chemotaxis; all these are features of the inflammatory response.[46] In particular, the complement system appears to play a central role. Mutations in genes coding for a range of complement proteins, factors and regulators have been associated with AMD. These include *Complement Factor H* (coding for a regulator of the alternate complement pathway), *CFHR1, CFHR3* (*CFH*-related genes), *Complement Factor B*, *Complement Factor I*, *Complement C3 and* the *SERPING1* gene (coding for C1 inhibitor, a regulator of the classic complement pathway).[47-56] Complement products have also been found in drusen and CNV membranes.[12,57]

Other inflammatory pathways may be implicated. Certain human leucocyte antigen (HLA) haplotypes have been associated with AMD.[58] Toll-like receptors (TLR) are involved in the innate immune system and *TLR* gene single nucleotide polymorphisms (SNPs) may be associated with AMD, although this is contentious.[59,60] Furthermore, chronic infections may provide a rationale for ongoing activation of inflammatory processes leading to macular damage. *Chlamydia pneumoniae*, for example, has been associated with increased AMD progression, although the link with AMD incidence is tentative.[61-63]

Oxidative Stress

The macula is particularly prone to oxidative stress especially generated by the UV wavelengths of incoming light. Furthermore, high metabolic demand at the macula produces further reactive oxygen species (ROS). This occurs from aerobic respiration within the electron cycle in mitochondria and from phagocytosis of outer segments by RPE cells. Antioxidant systems include the xanthophylls lutein and zeaxanthin, as well as antioxidant enzymes such as superoxide dismutase (SOD), glutathione peroxidase (GPX) and catalase. Oxidative damage increases with age due to the body's declining repair and regenerative capabilities and build-up of toxic substances such as lipofuscin. This may be compounded by environmental factors such as smoking. SNPs in the *ARMS2* gene are associated with AMD, and this may be via an increase oxidative stress, although this is speculative.[64-66] Finally, as mentioned above, antioxidant supplements in the form of vitamins A, C, E and zinc, decrease risk of AMD progression, as shown in the Age-Related Eye Disease Study (AREDS).[20] The dietary effects of lutein, zeaxanthin and omega-3 fatty acid are being investigated in an ongoing clinical trial (the "AREDS2" study).[67]

VEGF

Wet AMD is characterised by the development of CNV. Various growth factors may be involved, but the predominant stimulus appears to be VEGF-A, a cytokine produced by RPE cells which increases angiogenesis and vascular permeability.[68] Indeed as discussed above, monoclonal antibody fragments, blocking the action of VEGF-A at their receptors, are the present mainstay of treatment for wet AMD.[16,17] As yet it is unclear how VEGF production is stimulated, but it is likely to be a result of inflammatory activity.

Choroidal Ischaemia

The main watershed area of the posterior choroid lies between the fovea and optic disc and therefore the macula is vulnerable to ischaemia.[69] Choroidal blood flow reduces with age, presumably as a result of atherosclerosis or atrophy.[70] Choroidal ischaemia may be one factor contributing to the risk of AMD. Patients with lower choroidal blood flow are more likely to develop CNV.[71]

Alteration in Transport

With age, BM increases in thickness due to collagen remodelling and lipid deposition. This leads to decreased hydraulic conductivity impairing transport of solutes between the choroid and RPE.[72] Reduced RPE function can result, especially since RPE cells may already be compromised due to lipofuscin accumulation.

Structural Defects

Breaks in BM are a key factor in CNV development. The evidence for this is compelling. CNV growth is a risk of therapeutic thermal laser applied on the retina, due to BM disruption. Indeed, animal models of CNV are often created experimentally using retinal laser.[73] Breaks in BM are also features of non-AMD conditions associated with CNV growth, including high myopia, angioid streaks (seen, for example, in pseudoxanthoma elasticum) and eye trauma.[74] As mentioned above, the central component of BM is elastin. Interestingly, genetic variations in the fibulin-5 gene, which controls elastogenesis, have been associated with AMD.[75,76] This may lead to weakening of BM, thus increasing susceptibility to CNV.

The exact mechanism for the breakdown of BM in CNV formation is yet unknown. Present studies suggest that with age, BM may become increasingly fragile due to lipid-build-up, calcification and fragmentation.[77,78] Increased degradation of BM may occur from increased matrix metalloproteinase (MMP) activity, reduced activity of tissue inhibitors of MMP (TIMP) and from increased cytokine release from macrophage build-up.[79-81] Of note, SNPs in the *TIMP-3* gene have been associated with AMD.[82] In contradiction, decreased MMP activity is observed in dry AMD and may explain the characteristic accumulation in extracellular matrix and BM thickening.[83] However CNV formation may be associated with a more localised increase in MMP activity to initiate BM rupture.

AMD AND ALZHEIMER'S DISEASE

Alzheimer's disease (AD) is a common heterogeneous neurodegenerative disease strongly associated with age. Several population based studies suggest intriguing associations between AMD and AD.

The Blue Mountains Eye Study by Pham et al (n = 3509, age 49-97) used a Mini- Mental State Examination, modified to exclude the 5 items needing vision. They found a multivariate adjusted association between late AMD and cognitive impairment with an odds ratio (OR) of 2.2 (95% Confidence Interval (C.I.) 1.0-5.0). They concluded that there was a significant cross-sectional association between late AMD and cognitive impairment, independent of visual impairment in this population.[84] The Atherosclerosis Risk in Communities Study by Wong et al (n = 9286, age 51-70) showed an OR of 1.6 (95%C.I. 1.1-2.2) for patients with severe cognitive impairment having early AMD. The grading of cognition was based on Word Fluency Test scores. However, two other cognitive test scores used by the study did not show an association between severe cognitive impairment and early AMD. The authors suggested a weak association existed between AMD and AD.[85] The population based Rotterdam Study by Klaver et al (n = 1438, age 75+) found patients with advanced AMD at baseline had a controlled relative risk (RR) of 1.5 (95%C.I.0.6-3.5) of incident AD. They concluded some overlap of pathogenesis between AD and AMD.[86]

Amyloid and Activation of the Immune System

A hallmark of AD is the deposition and aggregation of misfolded protein as amyloid fibrils or plaques, including amyloid-associated protein tau, beta-amyloid (beta-A) and amyloid precursor protein (APP).[87] Interestingly, the drusen in AMD eyes also contain beta-A fibrils and prefibrillar amyloid oligomers, with none found in age-matched control eyes without drusen.[88] Where GA is present, the distribution of beta-A is at the periphery of the lesions.[88-90] Deposition of beta-A is thought to be toxic to the local RPE and photoreceptors, causing degeneration and may induce damaging levels of inflammation in them. Beta-A increases RPE monocyte chemoattractant protein-1 (MCP-1) and thereby recruits macrophages and microglia.[91] Beta-A also increases both macrophage and microglia IL-1 beta and TNF-alpha production. These cytokines result in significant RPE cell-mediated up regulation of factor B, the main alternative complement pathway activator.[91]

Proteins found in drusen also suggest an association with AD. These include vitronectin (known to produce amyloid fibrils when misfolded), the complement membrane attack complex (C5b-9), serum APP and apolipoprotein E (ApoE)..[92,93] Primary complement cascade activation at sites of beta-A deposition and the presence of C5b-9 suggests that chronic inflammation is a mechanism of neuro-degeneration in both AMD and AD.[94]

Apolipoprotein

Late onset AD is a multi-factorial disease. Its strongest genetic linkage is with the *ApoE* gene. Apolipoprotein is a major protein of the central nervous system and regulates cholesterol and lipid transport. It is ubiquitously found in AMD drusen.

The *ApoE* gene has 3 major isoforms, namely ApoE 2, 3 and 4, coded by the alleles epsilon 2, 3 and 4 respectively. Most common is the ApoE epsilon 3 allele, which is not linked to AD nor to AMD. ApoE epsilon 4 is associated with an increased risk of AD but appears protective for AMD.[95,96] In addition, ApoE epsilon 2 is associated with an increased risk of AMD.[96] Further research is needed to investigate the mechanism of the allelic associations of ApoE with both AMD and AD.

ONGOING RESEARCH AND ANTICIPATED DEVELOPMENTS

The breakthrough of anti-VEGF therapy has improved the prognosis of many patients with AMD. Furthermore this appears to have stimulated research interest in AMD as is evident in the recent proliferation of clinical trials. Tables 1 and 2 present a range of the ongoing/recently completed trials. Of note is the wide spectrum of mechanisms of action, including targeting the immune system, reducing growth factor activity, reducing oxidative stress, reducing lipofuscin deposition, neuroprotection, plasmapheresis, direct radiation therapy and replacing damaged RPE tissue. Methods of administration are also diverse, and include topical eye drops, subcutaneous/intravitreal/intravenous injections, oral medication and laser treatment. A likely development is the use of combination and adjuvant therapy with anti-VEGF agents, with the intention of reducing the frequency of injections required.

Genetics is likely to play an increasing role in the management of human disease. Genetic analysis may predict response to specific treatments. This implies future treatment could be tailor-made to an individual's genotype ("personalised medicine"). Gene therapy is another intriguing prospect. Patients could be screened for AMD-associated genes and faulty genes replaced with normal copies to reduce overall AMD risk. Alternatively gene therapy could enable sustained drug/monoclonal antibody delivery. Presently clinical trials are underway with the aim of delivering gene therapy to patients with Leber's Congenital Amaurosis, a rare retinal dystrophy.[97] As experience with ocular gene therapy increases, the hope is that more common and complex diseases such as AMD may benefit.

One of the exciting results of the AMD/AD copathology theory is the investigation of the use of putative neurodegenerative disease therapies in AMD. For example, systemic anti-beta-A immunotherapy of an AMD mouse model demonstrated attenuation of pathology and improvement in electroretinogram readings.[98] Reduced drusen area has also been noted in a murine AMD model treated with Copaxone, a T-cell-based vaccination undergoing clinical trials in AMD (Table 1).[99] Another area under investigation is the use of direct optical imaging of the retinal beta-A plaques to evaluate AD non-invasively. Potentially AD could then be diagnosed at an early sub-clinical phase, creating an opportunity for pre symptomatic treatment. Such imaging could also be used as a biomarker to evaluate response in clinical trials.[100] This could significantly reduce the cost of such clinical trials. A recent animal study by Koronyo-Hamaoui et al administered a safe beta-A-labeling systemic fluorochrome to an AD mouse model. It showed that retinal disease was detectable earlier than brain disease and was undetectable in controls. Beta-A-labeling progressed with the disease and was significantly reduced on administration of an immune-based therapy effective in reducing brain disease.[101]

As greater understanding develops of both the pathogenesis of AMD and systemic neurodegenerative diseases, further common aetiological factors may be discovered. This may lead to novel treatments for either of these conditions. The need for interdisciplinary research is therefore compelling.

Table 1. Clinical trial of investigational products for dry AMD

Therapeutic Agent	Mechanism of Action	Details	Route	Phase	Sponsor	CT Identifier
Sirolimus	Anti-inflammatory	mTOR (mammalian target of rapamycin) inhibitor	Subconjunctival injection	1/2	NEI	NCT00766649
Glatiramer (Copaxone)	Anti-inflammatory	T-cell immono-modulation	Subcutaneous injection	1	The New York Eye and Ear Infirmary	NCT00541333
Fluocinolone Acetonide Intravitreal Insert	Anti-inflammatory	Steroid	Intravitreal insertion of implant	2	Alimera Sciences	NCT00695318
Eculizumab	Complement inhibition	Binds and inhibits cleavage of C5	Intravenous	2	University of Miami/Alexion Pharmaceuticals	NCT00935883
ARC1905	Complement inhibition	Anti-C5 Aptamer—C5 inhibitor	Intravitreal	1	Ophthotech Corporation	NCT00950638 NCT00709527
Rheopheresis	Reduction of immune-related proteins	Double filtration plasmapheresis of high molecular weight proteins	External removal/ filtration with subsequent return of plasma	4	Apheresis Research Institute	NCT00751361
Fenretinide	Reduction of lipofuscin accumulation	Retinoic acid analogue—reduces accumulation of A2E by reducing systemic levels of retinol (vitamin A), a precursor of A2E	Oral	2	Sirion Therapeutics	NCT00429936
ACU-4429	Reduction of lipofuscin accumulation	"Visual cycle modulator" Inhibition of RPE65, a key enzyme involved in visual cycle in rods, reducing A2E build-up	Oral	2	Acucela/Otsuka Pharmaceutical	NCT01002950

continued on next page

Table 1. Continued

Therapeutic Agent	Mechanism of Action	Details	Route	Phase	Sponsor	CT Identifier
Retinal Transplantation	Replacement of lost tissue	Transplantation of human foetal neural retinal tissue and RPE	Subretinal surgical procedure	2	National Neurovision Research Institute	NCT00346060
Encapsulated Human NTC-201 Cell Implant	Neuro-protection	A small capsule containing RPE cells genetically engineered to release Ciliary Neurotrophic Factor (CNTF) through the capsule membrane	Intravitreal insertion	2	Neurotech Pharmaceuticals	NCT00447954
Brimonidine	Neuro-protection	Slow-releasing brimonidine implant	Intravitreal injection of implant	2	Allergan	NCT00658619
OT551	Anti-oxidant	Anti-oxidant eye drops	Topical eye drops	2	NEI	NCT00306488 NCT00485394
Lutein/ zeaxanthin/ Omega-3 long-chain polyunsaturated fatty acids	Anti-oxidant	Increases antioxidant capacity	Oral	3	NEI	NCT00345176
RNG65	Reduction of amyloid accumulation	Anti-amyloid beta antibody	Intravenous	1	Pfizer	NCT00877032
MC-1101	Increase of choroidal blood flow	Vasodilator—increases choroidal blood flow	Topical eye drops		MacuCLEAR	NCT01013376

CT = ClinicalTrials.gov
NEI = National Eye Institute, USA
siRNA = Short interfering ribo-nuclease acid
Information obtained from www.clinicaltrials.org on 25th September 2010

Table 2. Clinical trial of investigational products for wet AMD

Therapeutic Agent	Mechanism of Action	Details	Route	Phase	Sponsor	CT Identifier
Everolimus	Anti-inflammatory	mTOR inhibitor	Oral	2	Novartis	NCT00852259
Infliximab	Anti-inflammatory	Monoclonal antibody against TNF-alpha	Intravenous infusion	2	NEI	NCT00304954
Sirolimus	Anti-inflammatory	mTOR inhibitor	Oral	2	NEI	NCT00304954
Daclizumab	Anti-inflammatory	Monoclonal antibody vs interleukin-2, therefore limiting T-cell signalling. Binds to CD25, the alpha chain of the IL2 receptor	Intravenous infusion	2	NEI	NCT00304954
Volociximab	Anti-inflammatory	Alpha 5 Beta 1 Integrin Antagonist	Intravitreal injection	1	Ophthotech Corporation	NCT00782093
POT-4	Complement inhibitor	C3 inhibitor A derivative of the cyclic peptide Compstatin, which binds and inhibits C3.	Intravitreal injection	1	Potentia Pharmaceuticals	NCT00473928
AAV2-sFLT01	Anti-VEGF	Gene delivery of sFLT01, a novel chimeric VEGF-binding molecule, using adeno-associated virus (AAV) serotype 2	Intravitreal injection	1	Genzyme	NCT01024998
VEGF Trap	Anti-VEGF	A soluble VEGF receptor fusion protein which binds all forms of VEGF-A and placental growth factor (PGF).	Intravitreal injection	3	Bayer	NCT00637377
Bevasiranib	Anti-VEGF	siRNA of VEGF	Intravitreal injection	2	Opko Health	NCT00259753
PTK787	Anti-VEGF	VEGF receptor tyrosine kinase inhibitor	Oral	1,2	Pfizer	NCT00138632
AL-39324	Anti-VEGF	VEGF receptor tyrosine kinase inhibitor	Intravitreal injection	2	Alcon Research	NCT00992563
MP0112	Anti-VEGF	DARPin (Designed ankyrin repeat proteins)—based VEGF-A antagonist	Intravitreal injection	1	Molecular Partners AG	NCT01086761
Sirna-027	Anti-VEGF	siRNA vs VEGF receptor 1	Intravitreal injection	1,2	Allergan/Sirna Therapeutics Inc.	NCT00363714

continued on next page

Table 2. Continued

Therapeutic Agent	Mechanism of Action	Details	Route	Phase	Sponsor	CT Identifier
Vaccination by VEGFR1 and VEGFR2 – derived peptide	Anti-VEGF	Vaccination induces cytotoxic T-lymphocytes (CTLs) with potent cytotoxicity against cells expressing VEGFR1 and 2	Subcutaneous injection	1	Osaka University/ Human Genome Centre, Tokyo	NCT00791570
AdGV-PEDF.11D	Increases level of Pigment Epithelial Derived Growth Factor (PEDF)	E1, E3 and E4 deleted adenovirus vector containing PEDF gene	Intravitreal injection	1	GenVec	NCT00109499
E10030	Anti-platelet-derived growth factor (PDGF)	Anti-PDGF Pegylated Aptamer	Intravitreal injection	1,2	Ophthotech Corporation	NCT01089517 NCT00569140
iSONEP (Sonepcizumab LT1009)	Multiple growth factor inhibitor	A monoclonal antibody, which blocks action of sphingosine 1-phosphate (S1P), reducing levels of a range of growth and survival factors for multiple cells	Intravitreal	1	Lpath	NCT00767949
Pazopanib	Multiple growth factor inhibitor	Tyrosine kinase inhibitor of multiple receptors (including VEGF and platelet-derived growth factor receptors.)	Topical/oral	1,2	GlaxoSmith-Kline	NCT01134055 NCT01154062
ATG003 (mecamylamine)	Reduction of angiogenesis	Nicotinic acetylcholine receptor inhibitor	Topical eye drops	1	CoMentis	NCT00607750
CGC-11047	Reduction of cell proliferation	A polyamine analogue. Targets hyper-proliferating cells and halts cell growth/induces apoptosis	Subconjunctival injection	1	Cellgate	NCT00446654
Palomid 529	Reduction of cell proliferation	Dual TORC1/2 inhibitor of the PI3K/Akt/mTOR pathway. Suppresses angiogenesis and cell proliferation.	Intravitreal/subconjunctival injection	1	Paloma Pharmaceuticals	NCT01033721

continued on next page

Table 2. Continued

Therapeutic Agent	Mechanism of Action	Details	Route	Phase	Sponsor	CT Identifier
JSM6427	Reduction of cell adhesion/migration	Integrin α5β1-antagonist (this integrin normally mediates cell adhesion and migration)	Intravitreal injection	1	Jerini Ophthalmic	NCT00536016
PF-04523655	Reduces photoreceptor apoptosis	siRNA against RTP801	Intravitreal injection	2	Quark Pharmaceuticals/Pfizer	NCT00713518
Strontium90 Beta radiation	Radiation	Intravitreal radiation delivery by Epi-Rad system	Intravitreal delivery	3	NeoVista	NCT00454389
X-ray radiation	Radiation	Noninvasive robotically-guided delivery of radiation via 3 beams	External, via contact lens	1	Oraya Therapeutics	NCT01016873
WST11 (Stakel®)	Activation of photosensitiser	Becomes active on exposure to infrared laser light, causing occlusion of CNV	Intravenous, with external laser activation (photodynamic therapy)	2	Steba Biotech S.A.	NCT01021956

CT = ClinicalTrials.gov
NEI = National Eye Institute, USA
siRNA = Short interfering ribo-nuclease acid
Information obtained from www.clinicaltrials.org on 25th September 2010

CONCLUSION

AMD is still a distressing disease for many of the elderly population. A multi-faceted approach is required in the future to cumulatively reduce both intrinsic and extrinsic risk factors, to prevent onset of this blinding condition. With the current interest in research and a wealth of clinical trials, the future is promising. The intriguing overlap between AMD and AD suggests that therapies targeted at treating AD may also result in treatment for AMD and vice versa. Such developments are urgently needed for this devastating and prevalent disease.

ACKNOWLEDGEMENT

"The authors are thankful to TFC Frost Charitable Trust for providing funding supporting this publication".

REFERENCES

1. Resnikoff S, Pascolini D, Etya'ale D et al. Global data on visual impairment in the year 2002. Bull World Health Organ 2004; 82:844-851.
2. Klein R, Cruickshanks KJ, Nash SD et al. The prevalence of age-related macular degeneration and associated risk factors. Arch Ophthalmol 2010; 128:750-758.
3. Friedman DS, O'Colmain BJ, Munoz B et al. Prevalence of age-related macular degeneration in the United States. Arch Ophthalmol 2004; 122:564-572.
4. de Jong PT. Age-related macular degeneration. N Engl J Med 2006; 355:1474-1485.
5. Seland JH, Vingerling JR, Augood CA et al. Visual Impairment and quality of life in the Older European Population, the EUREYE study. Acta Ophthalmol 2009.
6. Klein R, Klein BE, Knudtson MD et al. Prevalence of age-related macular degeneration in 4 racial/ethnic groups in the multi-ethnic study of atherosclerosis. Ophthalmology 2006; 113:373-380.
7. Klein R, Klein BE, Jensen SC et al. The five-year incidence and progression of age-related maculopathy: the Beaver Dam Eye Study. Ophthalmology 1997; 104:7-21.
8. Jonas JB, Schneider U, Naumann GO. Count and density of human retinal photoreceptors. Graefes Arch Clin Exp Ophthalmol 1992; 230:505-510.
9. Ryan SJ, Ryan SJ. Retina 4th Ed, Elsevier/Mosby, Philadelphia, Pa, USA. Retina . 2006. Philadelphia, Pa.: Elsevier/Mosby.
10. Boulton M, Dayhaw-Barker P. The role of the retinal pigment epithelium: topographical variation and ageing changes. Eye 2001; 15:3-9.
11. Johnson LV, Leitner WP, Staples MK et al. Complement activation and inflammatory processes in Drusen formation and age related macular degeneration. Exp Eye Res 2001; 73:887-896.
12. Mullins RF, Russell SR, Anderson DH et al. Drusen associated with aging and age-related macular degeneration contain proteins common to extracellular deposits associated with atherosclerosis, elastosis, amyloidosis and dense deposit disease. FASEB J 2000; 14:835-846.
13. Klein R, Meuer SM, Knudtson MD et al. The epidemiology of retinal reticular drusen. Am J Ophthalmol 2008; 145:317-326.
14. Knudtson MD, Klein R, Klein BE et al. Location of lesions associated with age-related maculopathy over a 10-year period: the Beaver Dam Eye Study. Invest Ophthalmol Vis Sci 2004; 45:2135-2142.
15. Augood CA, Vingerling JR, de Jong PT et al. Prevalence of age-related maculopathy in older Europeans: the European Eye Study (EUREYE). Arch Ophthalmol 2006; 124:529-535.
16. Brown DM, Kaiser PK, Michels M et al. Ranibizumab versus verteporfin for neovascular age-related macular degeneration. N Engl J Med 2006; 355:1432-1444.
17. Rosenfeld PJ, Brown DM, Heier JS et al. Ranibizumab for neovascular age-related macular degeneration. N Engl J Med 2006; 355:1419-1431.
18. Lalwani GA, Rosenfeld PJ, Fung AE et al. A variable-dosing regimen with intravitreal ranibizumab for neovascular age-related macular degeneration: year 2 of the PrONTO Study. Am J Ophthalmol 2009; 148:43-58.

19. Hassan AS, Johnson MW, Schneiderman TE et al. Management of submacular hemorrhage with intravitreous tissue plasminogen activator injection and pneumatic displacement. Ophthalmology 1999; 106:1900-1906.
20. Age-Related Eye Disease Study Research Group. A randomized, placebo-controlled, clinical trial of high-dose supplementation with vitamins C and E, beta carotene and zinc for age-related macular degeneration and vision loss: AREDS report no. 8.[Erratum appears in Arch Ophthalmol. 2008;126(9):1251]. Arch Ophthalmol 2001; 119:1417-1436.
21. Montgomery MP, Kamel F, Pericak-Vance MA et al. Overall diet quality and age-related macular degeneration. Ophthalmic Epidemiol 2010; 17:58-65.
22. Augood C, Chakravarthy U, Young I et al. Oily fish consumption, dietary docosahexaenoic acid and eicosapentaenoic acid intakes and associations with neovascular age-related macular degeneration. Am J Clin Nutr 2008; 88:398-406.
23. Cho E, Seddon JM, Rosner B et al. Prospective study of intake of fruits, vegetables, vitamins and carotenoids and risk of age-related maculopathy. Arch Ophthalmol 2004; 122:883-892.
24. Khan JC, Thurlby DA, Shahid H et al. Smoking and age related macular degeneration: the number of pack years of cigarette smoking is a major determinant of risk for both geographic atrophy and choroidal neovascularisation. Br J Ophthalmol 2006; 90:75-80.
25. Klein R, Klein BE, Linton KL et al. The Beaver Dam Eye Study: the relation of age-related maculopathy to smoking. Am J Epidemiol 1993; 137:190-200.
26. Hogg RE, Woodside JV, Gilchrist SE et al. Cardiovascular disease and hypertension are strong risk factors for choroidal neovascularization. Ophthalmology 2008; 115:1046-1052.
27. Klein R, Klein BE, Tomany SC et al. The association of cardiovascular disease with the long-term incidence of age-related maculopathy: the Beaver Dam Eye Study. Ophthalmology 2003; 110:1273-1280.
28. Peeters A, Magliano DJ, Stevens J et al. Changes in abdominal obesity and age-related macular degeneration: the Atherosclerosis Risk in Communities Study. Arch Ophthalmol 2008; 126:1554-1560.
29. Tomany SC, Cruickshanks KJ, Klein R et al. Sunlight and the 10-year incidence of age-related maculopathy: the Beaver Dam Eye Study. Arch Ophthalmol 2004; 122:750-757.
30. Cruickshanks KJ, Klein R, Klein BE. Sunlight and age-related macular degeneration. The Beaver Dam Eye Study. Arch Ophthalmol 1993; 111:514-518.
31. Fletcher AE, Bentham GC, Agnew M et al. Sunlight exposure, antioxidants and age-related macular degeneration. Arch Ophthalmol 2008; 126:1396-1403.
32. Laser treatment in patients with bilateral large drusen: the complications of age-related macular degeneration prevention trial. Ophthalmology 2006; 113:1974-1986.
33. Giansanti F, Eandi CM, Virgili G. Submacular surgery for choroidal neovascularisation secondary to age-related macular degeneration. Cochrane Database Syst Rev 2009; CD006931.
34. Pauleikhoff D, Radermacher M, Spital G et al. Visual prognosis of second eyes in patients with unilateral late exudative age-related macular degeneration. Graefes Arch Clin Exp Ophthalmol 2002; 240:539-542.
35. Klein R, Klein BE, Linton KL. Prevalence of age-related maculopathy. The Beaver Dam Eye Study. Ophthalmology 1992; 99:933-943.
36. Maguire P, Vine AK. Geographic atrophy of the retinal pigment epithelium. Am J Ophthalmol 1986; 102:621-625.
37. Sarks JP, Sarks SH, Killingsworth MC. Evolution of geographic atrophy of the retinal pigment epithelium. Eye (Lond) 1988; 2 (Pt 5):552-577.
38. Swaroop A, Chew EY, Rickman CB et al. Unraveling a multifactorial late-onset disease: from genetic susceptibility to disease mechanisms for age-related macular degeneration. Annu Rev Genomics Hum Genet 2009; 10:19-43.
39. Yoshida S, Yashar BM, Hiriyanna S et al. Microarray analysis of gene expression in the aging human retina. Invest Ophthalmol Vis Sci 2002; 43:2554-2560.
40. Katz ML, Robison WG Jr. Age-related changes in the retinal pigment epithelium of pigmented rats. Exp Eye Res 1984; 38:137-151.
41. Wing GL, Blanchard GC, Weiter JJ. The topography and age relationship of lipofuscin concentration in the retinal pigment epithelium. Invest Ophthalmol Vis Sci 1978; 17:601-607.
42. Iwasaki M, Inomata H. Lipofuscin granules in human photoreceptor cells. Invest Ophthalmol Vis Sci 1988; 29:671-679.
43. Ng KP, Gugiu B, Renganathan K et al. Retinal pigment epithelium lipofuscin proteomics. Molecular and Cellular Proteomics 2008; 7:1397-1405.
44. Sparrow JR, Vollmer-Snarr HR, Zhou J et al. A2E-epoxides damage DNA in retinal pigment epithelial cells. Vitamin E and other antioxidants inhibit A2E-epoxide formation. Journal of Biological Chemistry 2003; 278:18207-18213.
45. Zhou J, Kim SR, Westlund BS et al. Complement activation by bisretinoid constituents of RPE lipofuscin. Investigative Ophthalmology and Visual Science 2009; 50:1392-1399.

46. Tsutsumi-Miyahara C, Sonoda KH, Egashira K et al. The relative contributions of each subset of ocular infiltrated cells in experimental choroidal neovascularisation. Br J Ophthalmol 2004; 88:1217-1222.
47. Klein RJ, Zeiss C, Chew EY et al. Complement factor H polymorphism in age-related macular degeneration. Science 2005; 308:385-389.
48. Haines JL, Hauser MA, Schmidt S et al. Complement factor H variant increases the risk of age-related macular degeneration. Science 2005; 308:419-421.
49. Edwards AO, Ritter R III, Abel KJ et al. Complement factor H polymorphism and age-related macular degeneration. Science 2005; 308:421-424.
50. Yates JR, Sepp T, Matharu BK et al. Complement C3 variant and the risk of age-related macular degeneration. N Engl J Med 2007; 357:553-561.
51. Gold B, Merriam JE, Zernant J et al. Variation in factor B (BF) and complement component 2 (C2) genes is associated with age-related macular degeneration. Nat Genet 2006; 38:458-462.
52. Ennis S, Gibson J, Cree AJ et al. Support for the involvement of complement factor I in age-related macular degeneration. Eur J Hum Genet 2009.
53. Fagerness JA, Maller JB, Neale BM et al. Variation near complement factor I is associated with risk of advanced AMD. Eur J Hum Genet 2009; 17:100-104.
54. Hageman GS, Hancox LS, Taiber AJ et al. Extended haplotypes in the complement factor H (CFH) and CFH-related (CFHR) family of genes protect against age-related macular degeneration: characterization, ethnic distribution and evolutionary implications. Ann Med 2006; 38:592-604.
55. Hughes AE, Orr N, Esfandiary H et al. A common CFH haplotype, with deletion of CFHR1 and CFHR3, is associated with lower risk of age-related macular degeneration. Nat Genet 2006; 38:1173-1177.
56. Ennis S, Jomary C, Mullins R et al. Association between the SERPING1 gene and age-related macular degeneration: a two-stage case-control study. Lancet 2008.
57. Baudouin C, Peyman GA, Fredj-Reygrobellet D et al. Immunohistological study of subretinal membranes in age-related macular degeneration. Jpn J Ophthalmol 1992; 36:443-451.
58. Goverdhan SV, Howell MW, Mullins RF et al. Association of HLA class I and class II polymorphisms with age-related macular degeneration. Investigative Ophthalmology and Visual Science 46(5):1726-34, 2005.
59. Cho Y, Wang JJ, Chew EY et al. Toll-like receptor polymorphisms and age-related macular degeneration: replication in three case-control samples. Invest Ophthalmol Vis Sci 2009; 50:5614-5618.
60. Yang Z, Stratton C, Francis PJ et al. Toll-like receptor 3 and geographic atrophy in age-related macular degeneration. N Engl J Med 2008; 359:1456-1463.
61. Robman L, Mahdi O, McCarty C et al. Exposure to Chlamydia pneumoniae infection and progression of age-related macular degeneration. Am J Epidemiol 2005; 161:1013-1019.
62. Kalayoglu MV, Galvan C, Mahdi OS et al. Serological association between Chlamydia pneumoniae infection and age-related macular degeneration. Arch Ophthalmol 2003; 121:478-482.
63. Robman L, Mahdi OS, Wang JJ et al. Exposure to Chlamydia pneumoniae infection and age-related macular degeneration: the Blue Mountains Eye Study. Invest Ophthalmol Vis Sci 2007; 48:4007-4011.
64. Kanda A, Chen W, Othman M et al. A variant of mitochondrial protein LOC387715/ARMS2, not HTRA1, is strongly associated with age-related macular degeneration. Proc Natl Acad Sci USA 2007; 104:16227-16232.
65. Fritsche LG, Loenhardt T, Janssen A et al. Age-related macular degeneration is associated with an unstable ARMS2 (LOC387715) mRNA. Nat Genet 2008; 40:892-896.
66. Wang G, Spencer KL, Court BL et al. Localization of age-related macular degeneration-associated ARMS2 in cytosol, not mitochondria. Invest Ophthalmol Vis Sci 2009; 50:3084-3090.
67. National Eye Institute (NEI). Age-Related Eye Disease Study 2 (AREDS2). In: ClinicalTrials.gov [Internet]. Bethesda (MD): National Library of Medicine (US). 2000- [cited 2010]. Available from: http://clinicaltrials.gov/ct2/show/NCT00345176. NLM Identifier: NCT00345176. 2010.
68. Frank RN. Growth factors in age-related macular degeneration: pathogenic and therapeutic implications. Ophthalmic Res 1997; 29:341-353.
69. Giuffre G. Main posterior watershed zone of the choroid. Variations of its position in normal subjects. Doc Ophthalmol 1989; 72:175-180.
70. Straubhaar M, Orgul S, Gugleta K et al. Choroidal laser Doppler flowmetry in healthy subjects. Arch Ophthalmol 2000; 118:211-215.
71. Metelitsina TI, Grunwald JE, Dupont JC et al. Foveolar choroidal circulation and choroidal neovascularization in age-related macular degeneration. Invest Ophthalmol Vis Sci 2008; 49:358-363.
72. Starita C, Hussain AA, Pagliarini S et al. Hydrodynamics of ageing Bruch's membrane: implications for macular disease. Exp Eye Res 1996; 62:565-572.
73. Francois J, De Laey JJ, Cambie E et al. Neovascularization after argon laser photocoagulation of macular lesions. Am J Ophthalmol 1975; 79:206-210.
74. Pruett RC, Weiter JJ, Goldstein RB. Myopic cracks, angioid streaks and traumatic tears in Bruch's membrane. Am J Ophthalmol 1987; 103:537-543.

75. Mullins RF, Olvera MA, Clark AF et al. Fibulin-5 distribution in human eyes: relevance to age-related macular degeneration. Exp Eye Res 2007; 84:378-380.
76. Stone EM, Braun TA, Russell SR et al. Missense variations in the fibulin 5 gene and age-related macular degeneration. N Engl J Med 2004; 351:346-353.
77. Spraul CW, Lang GE, Grossniklaus HE et al. Histologic and morphometric analysis of the choroid, Bruch's membrane and retinal pigment epithelium in postmortem eyes with age-related macular degeneration and histologic examination of surgically excised choroidal neovascular membranes. Surv Ophthalmol 1999; 44 Suppl 1:S10-S32.
78. Sheraidah G, Steinmetz R, Maguire J et al. Correlation between lipids extracted from Bruch's membrane and age. Ophthalmology 1993; 100:47-51.
79. Cherepanoff S, McMenamin P, Gillies MC et al. Bruch's membrane and choroidal macrophages in early and advanced age-related macular degeneration. Br J Ophthalmol 2010; 94:918-925.
80. Steen B, Sejersen S, Berglin L et al. Matrix metalloproteinases and metalloproteinase inhibitors in choroidal neovascular membranes. Invest Ophthalmol Vis Sci 1998; 39:2194-2200.
81. Janssen A, Hoellenriegel J, Fogarasi M et al. Abnormal vessel formation in the choroid of mice lacking tissue inhibitor of metalloprotease-3. Invest Ophthalmol Vis Sci 2008; 49:2812-2822.
82. Chen W, Stambolian D, Edwards AO et al. Genetic variants near TIMP3 and high-density lipoprotein-associated loci influence susceptibility to age-related macular degeneration. Proc Natl Acad Sci USA 2010; 107:7401-7406.
83. Guo L, Hussain AA, Limb GA et al. Age-dependent variation in metalloproteinase activity of isolated human Bruch's membrane and choroid. Invest Ophthalmol Vis Sci 1999; 40:2676-2682.
84. Pham TQ, Kifley A, Mitchell P et al. Relation of age-related macular degeneration and cognitive impairment in an older population. Gerontology 2006; 52:353-358.
85. Wong TY, Klein R, Nieto FJ et al. Is early age-related maculopathy related to cognitive function? The Atherosclerosis Risk in Communities Study. Am J Ophthalmol 2002; 134:828-835.
86. Klaver CC, Ott A, Hofman A et al. Is age-related maculopathy associated with Alzheimer's Disease? The Rotterdam Study. Am J Epidemiol 1999; 150:963-968.
87. Loffler KU, Edward DP, Tso MO. Immunoreactivity against tau, amyloid precursor protein and beta-amyloid in the human retina. Invest Ophthalmol Vis Sci 1995; 36:24-31.
88. Luibl V, Isas JM, Kayed R et al. Drusen deposits associated with aging and age-related macular degeneration contain nonfibrillar amyloid oligomers. J Clin Invest 2006; 116:378-385.
89. Isas JM, Luibl V, Johnson LV et al. Soluble and mature amyloid fibrils in drusen deposits. Invest Ophthalmol Vis Sci 2010; 51:1304-1310.
90. Dentchev T, Milam AH, Lee VM et al. Amyloid-beta is found in drusen from some age-related macular degeneration retinas, but not in drusen from normal retinas. Mol Vis 2003; 9:184-190.
91. Wang J, Ohno-Matsui K, Yoshida T et al. Amyloid-beta up-regulates complement factor B in retinal pigment epithelial cells through cytokines released from recruited macrophages/microglia: Another mechanism of complement activation in age-related macular degeneration. J Cell Physiol 2009; 220:119-128.
92. Anderson DH, Talaga KC, Rivest AJ et al. Characterization of beta amyloid assemblies in drusen: the deposits associated with aging and age-related macular degeneration. Exp Eye Res 2004; 78:243-256.
93. Shin TM, Isas JM, Hsieh CL et al. Formation of soluble amyloid oligomers and amyloid fibrils by the multifunctional protein vitronectin. Mol Neurodegener 2008; 3:16.
94. Johnson LV, Leitner WP, Rivest AJ et al. The Alzheimer's A beta -peptide is deposited at sites of complement activation in pathologic deposits associated with aging and age-related macular degeneration. Proc Natl Acad Sci USA 2002; 99:11830-11835.
95. Friedman DA, Lukiw WJ, Hill JM. Apolipoprotein E epsilon4 offers protection against age-related macular degeneration. Med Hypotheses 2007; 68:1047-1055.
96. Klaver CC, Kliffen M, van Duijn CM et al. Genetic association of apolipoprotein E with age-related macular degeneration. Am J Hum Genet 1998; 63:200-206.
97. Bainbridge JW, Smith AJ, Barker SS et al. Effect of gene therapy on visual function in Leber's congenital amaurosis. N Engl J Med 2008; 358:2231-2239.
98. Ding JD, Lin J, Mace BE et al. Targeting age-related macular degeneration with Alzheimer's disease based immunotherapies: anti-amyloid-beta antibody attenuates pathologies in an age-related macular degeneration mouse model. Vision Res 2008; 48:339-345.
99. Landa G, Butovsky O, Shoshani J et al. Weekly vaccination with Copaxone (glatiramer acetate) as a potential therapy for dry age-related macular degeneration. Curr Eye Res 2008; 33:1011-1013.
100. Guo L, Duggan J, Cordeiro MF. Alzheimer's disease and retinal neurodegeneration. Curr Alzheimer Res 2010; 7:3-14.
101. Koronyo-Hamaoui M, Koronyo Y, Ljubimov AV et al. Identification of amyloid plaques in retinas from Alzheimer's patients and noninvasive in vivo optical imaging of retinal plaques in a mouse model. Neuroimage 2010.

CHAPTER 3

ARACHNOID CYSTS

Thomas Westermaier,* Tilmann Schweitzer and Ralf-Ingo Ernestus
Department of Neurosurgery, University of Wuerzburg, Wuerzburg, Germany
*Corresponding Author: Thomas Westermaier—Email: westermaier.t@nch.uni-wuerzburg.de

Abstract: Arachnoid cysts are fluid-filled duplications or splittings of the arachnoid layer with a content which is similar but not equal to the cerebrospinal fluid. Arachnoid cysts are not actual neurodegenerative disorders, rather the underlying defect of the texture of the arachnoid layer is probably congenital in nature. They can occur sporadically or can be associated with other malformations or diseases. Arachnoid cysts may be discovered in early childhood. However, they can develop de novo, grow or decrease in size. They may be diagnosed by ultrasound screening in the fetal period or be discovered during childhood or adulthood. Many arachnoid cysts are asymptomatic.
    Treatment strategies are discussed controversially. If they are diagnosed incidentally or are correlated with only very mild symptoms, a conservative management with follow-up imaging may be favored. If they grow, they can cause headaches, seizures or other neurological symptoms and require neurosurgical treatment. This chapter addresses aspects of pathogenesis, clinical symptoms, indication for neurosurgical treatment and treatment options.

INTRODUCTION

Cystic malformations of the arachnoid layer had first been reported about 180 years ago.[1] In 1979, the English anatomist Cunningham published an autopsy report on a young acromegalic patient who had died from diabetes insipidus. He described the coincidence of a right hemispheric arachnoid cyst (AC) with a pituitary adenoma.[2] During the opening of the cyst, a "large quantity of sero-sanguilent fluid escaped". This autoptic finding and a previous episode of strong headache, suggested intracystic hemorrhage. In the first half of the 20th century, after X-ray and cerebral angiography had been introduced, further reports on this pathological entity were published.[3] Since this time, the nature and most favorable treatment of these space-occupying lesions has been discussed controversially.

Pathogenesis

Cystic fluid retentions within the arachnoid layer, which may develop secondary to intracranial hemorrhage or meningitis, have to be differentiated from true, primary AC. In addition, they have to be distinguished from other cystic lesions which involve the outer part of the cerebral cortex and the meninges, like glioependymal cysts or cystic lesions of infectious or neoplastic origin.[4,5] True AC are benign lesions and considered to be developmental or at least based on a congenital mistake of the arachnoid architecture.

It has to be noticed that, particularly in the area of the skull-base and the Sylvian fissure, arachnoid septations of varying strength are woven through the subarachnoid space. A variety of inner arachnoid membranes can be defined which physiologically form the subarachnoid cisterns.[6]

Ultrastructurally, AC walls are similar to the normal arachnoid membrane. Analyzing operatively resected AC walls, by light and electron microscopy, Miyagami and Tsubokawa found that the inner surface of the AC wall consisted of one or several layers of arachnoid cells with slender processes and large extracellular spaces.[7] Rengachary and Watanabe observed structural differences from the normal arachnoid: (1) the splitting of the arachnoid membrane at the margin of the cyst, (2) a thicker layer of collagen in the cyst wall, (3) the absence of traversing trabecular processes within the cyst and (4) the presence of hyperplastic arachnoid cells in the cyst wall, which presumably participate in collagen synthesis.[8]

The pathogenesis of AC is still not entirely known. Meningeal tissue is of mesectodermal origin and derives from the neural crest. AC, as well as meningeomas, may be located intraventricularly, which indicates that mesectodermal tissue may be dislocated into the neural tube during the folding of the latter, which begins around embryonic day 20. A network of mesenchymal cells develops between the brain and the epidermis and eventually differentiates into a dense outer layer and a reticular inner layer to form the meninges. The inner layer further differentiates to form the dura mater and the leptomeninges.[9]

The improvement of ultrasound technology and the practice of routine prenatal ultrasound diagnostic provided new insight regarding the time of appearance of AC, their in utero development and the co-incidence with other brain malformations. Intracranial cysts now can be detected as early as in the first trimester of pregnancy.[10] AC may occur together with anomalies of the corpus callosum,[11] Chiari malformations,[12] skeletal malformations,[13] and complex vascular malformations.[14] Furthermore, they may appear in the frame of syndromes and genetic disorders such as neurofibromatosis Type 1,[15] Pallister-Hall-Syndrome or Aicardi's syndrome.[4] Their frequent occurrence in early childhood, histological features with absent postinflammatory or posthemorrhagic changes and the occurrence together with genetic disorders strongly suggest a developmental nature. Therefore, it is the current opinion that AC are congenital malformations.

However, this opinion has recently been challenged. In children and adults, temporal and Sylvian location is highly predominant[16,17] while in a prenatal analysis most AC are located interhemispherically and temporal AC are extremely rare[11] (Table 1). This observation raises some doubt on the hypothesis that the formation of AC is strictly congenital. This might be confirmed by the observations, that they may develop de novo in adults,[18] may grow over time and, the cyst size on first diagnosis is larger with increasing age.[19] In fetuses, the subarachnoid spaces covering the temporal lobes are very deep. Under normal circumstances, these spaces regress in size from the end of gestation until the first few months of life. It may be hypothesized that any peri- or postnatal event, such

Table 1. Distribution of arachnoid cysts

Localization	Prenatal (n = 54)[11]	Pediatric (n = 67)[16]	Elderly (n = 23)[17]
Supratentorial	34	31	20
Middle fossa	1	13	11
Interhemispheric	16	9	
Hemispheric	1	5	9
Suprasellar	5	4	
Skull base	7		
Intraventricular	5		
Infratentorial	12	31	3
CPA	1	1	
Hemispheric	1	22	
Other	10	8	
Tentorial incisure	8	5	

Distribution of arachnoid cysts discovered by (a) routine ultrasound screening in pregnancy, (b) in children and (c) in patients older than 65 years. Arachnoid cysts rarely become symptomatic in old patients. Published cases were collected and reviewed by Caruso et al.[17] One of the infratentorial cysts was located within the fourth ventricle causing symptoms of normal pressure hydrocephalus. The differences in the distribution between pre- and postnatally diagnosed arachnoid cysts is apparent. Arachnoid cysts in the middle cranial fossa do practically not appear in the prenatal period but make up the majority of lesions in the pediatric and adult age. On the contrary, the predominant localization in prenatal diagostics is interhemisheric. These lesions are less frequently found in infants and rare in adults.

as even minor hemorrhage or inflammatory processes, could lead to the formation of a cyst by inhibiting this process of regression.[11]

In summary, most reports suggest that there is at least a congenital mistake in the texture of the arachnoid layer. Minor irregularities of the developing CSF flow through the mesenchymal network can result in a splitting of the developing arachnoid in the fetus. The development or filling of these arachnoid pockets, however, seems to be dynamic and further changes can occur at any stage of life.

NATURAL BEHAVIOUR OF ARACHNOID CYSTS

The knowledge about the spontaneous behaviour of AC is not only essential for assessing the indication for surgical treatment but also, in case of prenatal diagnosis, in order to counsel parents on the further course of pregnancy. Pierre-Kahn et al thoroughly analyzed the outcome of 56 fetuses in whom AC were diagnosed by routine ultrasound screening. In that study, 55% of the cysts were diagnosed between gestational week 20 and 30 and 45% after gestational week 30. Interestingly, the latter had normal ultrasound examinations between week 20 and 30, supporting that AC are developing dynamically. After their detection, serial follow-up ultrasound and MRI examinations were performed. In utero, the cysts grew in 20% of the cases, remained stable in 76% and decreased in 4%. After delivery, the cysts grew in 24%, remained stable in 52% and decreased in 24%.[11]

Focussing on AC of the middle cranial fossa and the Sylvian fissure, Galassi et al classified the lesions into three categories, class 1 and 2 representing small and medium-sized cyst with none or only little mass effect and a more or less open communication to the subarachnoid space. Class 3 represented voluminous cysts with a distinct mass effect and dislocation of midline structures and no communication to the subarachnoid space as assessed by intracisternal contrast application.[20] (Fig. 1) Becker et al found that only in class 3 cysts there is a positive correlation between the age at which a patient becomes symptomatic and the size of the cyst. This observation suggests that there is an obvious growth of these large lesions. Smaller lesions, however, do not show a correlation of size and age at first admission.[19] In the same study, a subgroup of

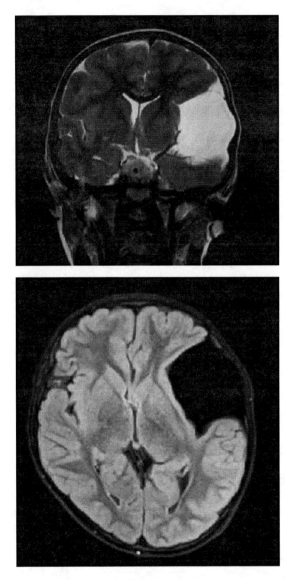

Figure 1. Coronar (T2 TSE) and transverse (T2 TIRM) MRI of a large left temporal/Sylvian anrachnoid cyst of an 8-year-old boy. An asymmetric bulging of the skull was the only symptom of the otherwise normally developed child who was not surgically treated.

patients (11 patients) was not operated on but followed by CT-scans. In 3 of these, the cyst grew in follow-up CT-scans, 6 remained unchanged while 2 decreased in size. There are several reports on a decreasing AC's volume without surgical intervention. This may occur during traumatic brain injury, allegedly by a rupture of the cyst wall with consecutive drainage into the subarachnoid space. In some cases, however, spontaneous reduction or even disappearance of AC has been observed without a history of head injury.[21,22] In summary, the spontaneous course of an AC is uncertain. This has a direct implication for treatment decisions. If the finding is truly incidental or the symptoms are only mild, a conservative management might be preferred at first to observe the further development.

MECHANISMS OF CYST-GROWTH

In a number of patients, the therapeutical strategy changes with the growth of the cyst. If this occurs, it may produce neurological symptoms. In 1964, Robinson reported a different consistence of the content of the cyst and the normal cerebrospinal fluid.[3] Berle and coworkers systematically analyzed the intracystic fluid and found different contents of phosphate, protein, ferritin and lactate dehydrogenase as compared to normal CSF. Possible mechanisms of cyst-growth, therefore, have to be consistent with these findings.[23] A variety of mechanisms has been proposed. Active secretion by the cells of the cyst-wall, expansion following an osmotic gradient or a unidirectional valve-mechanism may be possible mechanisms.

Active Fluid Secretion

In an ultrastructural and cytochemical study, Go et al analyzed the resected walls of AC. They observed similarities between the neurothelial lining of arachnoid granulations such as intercellular clefts and sinusoid dilatations and desmosomal intercellular junctions, pinocytic vesicles, lysosomal structures and a basal lamina. On the luminal surface, the cyst wall is densely covered with microvilli, thus, suggesting active secretion.[24] Enzyme cytochemistry demonstrated Na-K-ATPase at the luminal plasma membrane and alkaline phosphatase at the opposite membrane. This architecture might indicate fluid transport towards the lumen. Different consistence compared to CSF may be explained by this theory.

Osmotic Gradient

The hypothesis that AC grow by osmotic gradients has been proposed due to the minor differences of substance concentrations compared to CSF. A thin arachnoid wall with loose connective tissue could allow a penetration of water. However, the formation of an osmotic gradient has to be triggered. In secondary cysts, hemorrhage with products of blood degradation or inflammatory reactions could act as osmotic triggers. In primary AC, there is no evidence that these mechanisms exist.[25]

Valve Mechanism

Small and medium-sized cysts have been found to remain constant in size and to communicate readily with the normal subarachnoid space.[19,20] In larger cysts or in growing cysts, a valve-mechanism may develop with unidirectional fluid transport. Under circumstances with a positive extra-/intracystic pressure, the cyst may increase in volume. In fact, valve-like foldings of the cyst wall or intracystic membranes have actually been observed during endoscopic or open operative procedures.[26,27]

Other Mechanisms Leading to Clinical Symptoms

Apart from symptoms due to an enlargement of the cyst, several other mechanisms like spontaneous intracystic hemorrhage due to aneurysm rupture or traumatic brain injury have been reported to cause clinical symptoms.[28-30] Patients with AC appear to be prone to develop chronic subdural hematomas even after minor head trauma which can make the previously asymptomatic AC clinically apparent.[31,32] Clinical symptoms due to a spontaneous or traumatic rupture of the cyst and consecutive formation of a subdural hygroma have also been described.[33]

CLINICAL SYMPTOMS

The clinical symptoms are determined by the location of the cyst. AC can occur at any location in the cranium and spine where an arachnoid layer is present. In addition, they can be found in the ventricular system. In this case, it has to be noted, that tight adherance to the choroid plexus has been observed intraoperatively. It has been proposed that intraventricular AC arise from arachnoid tissue which has been displaced into the ventricles by in-growing vascular mesenchyma at the time of formation of the choroid plexus.[34-37] The prevalence of AC in a group of healthy adults has been found to be 1.1-1.5%.[38,39] Familial cases of AC have been described with[40,41] or without association to other hereditary diseases.[42-44] A predominance of the male gender for temporal AC is highly significant while cerebellopontine angle AC more often occur in females.[45] To date, there is no firm explanation for this sex distribution. Typical symptoms due to AC are described below.

Temporal and Sylvian Arachnoid Cysts

The middle fossa is the most common location of AC. In larger series of pediatric and adult patients, approximately two third of them sustain AC in the middle fossa or in the Sylvian fissure. Rarely, temporal AC are found bilaterally; these patients may suffer from glutaric aciduria.[46] In general, however, there is a predilection for the left hemisphere.[45] Symptoms either result from a mass effect causing increased intracranial pressure or from an irritation of the temporal or frontal cortex. Most often, these patients present with headache and seizures as well as contralateral motor deficits.[47,48] Developmental delay and retardation are more common in patients with AC.[48] An asymmetry of the skull may develop as well as a facial asymmetry or exophthalmus due to a misdevelopment of the sphenoid ridge.[3,49,50]

There has been an ongoing debate about the association of temporal AC with psychiatric disorders and neuropsychiological deficits. Various cases of psychiatric disorders have been reported to be associated with AC. These include attention deficit hyperactivity syndrome, depressions, hallucinations, dementia, schizophrenia, anorexia nervosa, insomnia and other psychotic disorders. Some of those patients were surgically treated for their AC and an improvement after surgery was reported in some cases.[51]

Systematic pre- and postoperative evaluation of cognitive function has been performed by Wester et al. They found that many patients with left temporal AC show significant deficits in various cognitive functions such as verbal perception, memory function, complex verbal tasks, visuospatial functions and visual attention. These deficits were, in fact, ameliorated in follow-up neuropsychological testing after surgical treatment. In many of these patients, however, these cysts did not cause complaints and neuropsychololgical deficits had not been clinically apparent until specific testing had been performed. These cysts can be interpreted as true incidental findings.

Intraventricular Arachnoid Cysts

Intraventricularly located AC are rare. They have been found in the lateral, third and fourth ventricles (Fig. 2). Classically, these cysts become clinically apparent by development of an acute or chronic hydrocephalus in case of increasing cyst size. Symptoms due to a compression of the surrounding structures have also been reported. AC in the fourth ventricle, for example, may cause cerebellar dysfunction[27] or by depression-like symptoms.[52]

Cerebellopontine Angle and Retrocerebellar Arachnoid Cysts

Infratentorial location accounts for approximately 10% of all AC. The majority of these are located in the cerebellopontine angle or in retrocerebellar location.[53,54] Symptoms of retrocerebellar cysts include hydrocephalus and compression of cerebellar or brainstem structures. Patients predominantly present with headaches or gait disturbances.

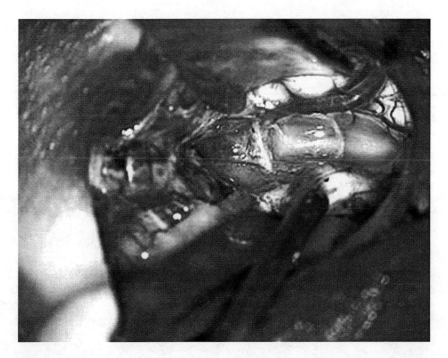

Figure 2. Intraoperative view after median suboccipital craniotomy and retraction of the cerebellar tonsils. The inferior vermis was slightly incised to allow sufficient overview of the fourth ventricle. A complex network of intracystic septations was found intraoperatively allowing free CSF flow after complete resection of the cyst.

Cerebellopontine angle AC may occlude the CSF circulation in the posterior fossa and cause obstructive hydrocephalus. Cerebrllopontine angle structures may be directly compressed or affected in terms of a neurovascular compression. Cranial nerve symptoms may evolve, including hearing loss, trigeminal neuralgia, hemifacial spasm, diplopia, hoarseness, dysphagia or diplopia.[55-61]

Supra- and Intrasellar Arachnoid Cysts

Intra- and suprasellar AC arise in the basal subarachnoid space in close vicinity to the optic nerve, pituitary gland and stalk, hypothalamus and mesencephalon. In case of massive suprasellar extension, they may obstruct the physiological drainage of cerebrospinal fluid at the level of the third ventricle or Foramen of Monroi, thus causing obstructive hydrocephalus. Visual impairment may be caused by compression of the optic nerve and endocrine disturbances by a compression of the pituitary gland or stalk. A compression of hypothalamic structures may cause developmental abnormalities like precocious puberty.[62] A bobble-head doll syndrome with an involuntary back and forward movement of the head at a frequency of 2-3 per second may be due to a compression of third ventricle structures or thalamic nuclei. It is a clinical feature highly suggestive for suprasellar AC. However, this typical feature can only be found in a minority of these patients.

Spinal Arachnoid Cysts

Similar to cranial AC, spinal AC typically cause symptoms due to a compression of the spinal cord and nerve roots. Myelopathic symptoms like ataxia, re-appearance of primitive reflexes or progressive paralysis may occur due to cord compression. Radicular symptoms due to involvement of exiting nerve roots may include pain, sensory deficits or weakness of the effector muscles (Fig. 3).

DIAGNOSTIC PROCEDURES

In many patients with intracranial and spinal AC, it is a challenging to evaluate the indication for surgical treatment. The success of the treatment strategy is based on the correlation of the lesion with the clinical symptoms. A detailed interview of the patient or the patient's parents and a thorough neurological examination are indispensable in order to correlate signs and symptoms with the radiological finding. This particularly counts for presumably incidental cysts.

If headache is the only symptom, its course should be documented and be brought in line with the results of serial imaging studies, if a conservative and observing strategy with follow-up imaging studies is chosen. The studies of Wester et al have demonstrated that patients with AC might have distinct cognitive deficits. In adult patients or older children, a neuropsychological testing should be recommended in order to detect more obscure clinical symptoms. Laboratory investigations for screening of glutaric aciduria must be recommended in patients with bilateral AC.[46] In fetuses with AC detected by routine ultrasound screening, karyotype evaluation and fetal MRI should be performed to rule out a possible association with other malformations.[11]

Electroencephalography is essential in patients with supratentorial AC in order to determine the necessity for antiepileptic medication. In spinal AC a cord compression may

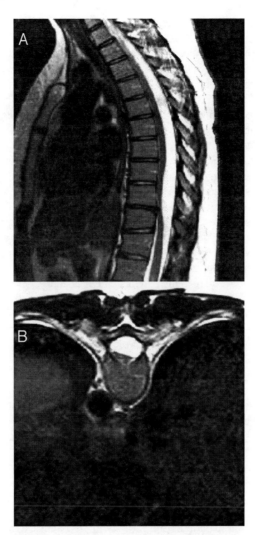

Figure 3. T2-weighted sagittal (A) and axial (Th 8) (B) magnetic resonance images of a large spinal arachnoid cyst in a 20-year-old female who presented with a progressive paraparesis. The spinal cord is thinned and shifted anteriorly by the space occupying cyst. After hemilaminectomy, fenestration and excision of intracystic membranes, the patient recovered completely.

be further evaluated by somatosensory and motor evoked potentials. Magnetic resonance imaging (MRI) is the mainstay of radiological diagnostics. Due to their worse imaging quality and because of X-ray exposure, computed tomography scan (CTS) is, especially in children, not acceptable as a diagnostic tool. Modern phase contrast MRI and functional MRI are not only capable of demonstrating structural abnormalities but also able to depict CSF-flow patterns and functional organization of the brain respectively. Cisternography with contrast filling of the subarachnoid space followed by CTS, sometimes at more than one time-point to evaluate the time-course of contrast-filling of CSF-compartments, is reserved for very few cases of diagnostic uncertainty.

INDICATION FOR TREATMENT

As discussed above, the successful treatment of AC depends on a thorough patient evaluation and determination of the appropriate treatment strategy. Rarely, patients present with signs of an acute obstructive hydrocephalus or compression of vital structures caused by an expansion of an AC requiring emergent decompression.[52,63] Hemorrhage into an AC, e.g., by rupture of an intracranial aneurysm, might be another emergent situation, requiring rapid surgical intervention.[28] The obvious association of AC with chronic subdural hematomas necessitates evacuation of the hematoma but not primarily a treatment of the asymptomatic AC.[31] In contrast, subdural hygroma associated with an AC has been reported to disappear by a therapy with acetazolamind, avoiding surgery.[64,65] In cases of progressive chronic hydrocephalus and focal neurological deficits like aphasia, motor deficits, cranial nerve deficits, spinal cord compression, surgery is to be recommended if the AC is the obvious cause of the respective deficit. Concerning epilepsy, the benefit of surgery is not clear. Most patients seem to improve as the frequency of seizures decreases. However, medical therapy might be as effective hence is the first choice of treatment.[47]

Headaches are the most frequent complaint of patients with an AC. However, the origin of headaches is manifold and in many patients with headaches, AC must be regarded as an incidental finding. The results of surgical treatment are discussed controversially. Some authors are restrictive to operate on patients with headaches as the sole symptom,[47] others report very good results of surgery for headaches as the leading symptom.[66,67] Headaches might be a sign of an intermittent elevation of intracranial pressure. The diagnostic challenge is the diagnosis of elevated ICP. In some patients, clear signs are visible on imaging studies. In cases of doubt, invasive ICP monitoring might be performed as reported by di Rocco et al.[68] For patients with only moderate and infrequent headaches, responding to medication, a non-operative strategy is adequate since the fate of AC is uncertain and a decrease in size is possible.[19] Continuous follow-up examinations with assessment of the course of complaints, neurological symptoms and intake of analgetics are necessary in these cases.

In systematic assessment of cognitive functions, patients with AC have been found to have reproducible deficits. Moreover, these deficits have been reported to improve after surgical therapy.[69] However, many of the patients assessed did not report specific subjective complaints. A too aggressive surgical strategy bears the danger to impose operative risks on the patient without improving the quality of life. In addition, the cysts may spontaneously decrease in size. Similar to patients presenting with headaches, a follow-up strategy may be chosen including imaging studies, neurological and neuropsychological examinations.

TREATMENT

There are several options for treatment of AC. Resection of the cyst-walls requires open surgery. Fenestration of the cyst with creation of an opening to the subarachnoid cisterns can be performed endoscopically or in an open microsurgical technique. Various shunting techniques have been reported including cystoperitoneal and cystosubdural shunts. There is an ongoing debate on the optimal surgical strategy. The form of treatment which is individually chosen, depends on various factors: The patient's symptoms, the location and shape of the cyst, the extent of intracystic septations, the vicinity to relevant neural and vascular structures and the surgeon's experience with different techniques. For patients without relevant symptoms and without related neurological deficits, observation may be recommended. If there is an indication for treatment, endoscopic fenestration can be recommended as the treatment of first choice for many AC, in particular if there are no relevant intracystic septations. For suprasellar arachnoid cysts, very good results have been reported. Ventriculocystostomy and ventriculocystocisternostomy seem to be the most successful long-term treatment. Temporal and temporobasal AC appear to be more difficult to be treated by endoscopic techniques.[70] If there are intracystic septations or multiple fine neural structures like in the cerebellopontine angle, the recurrence rate might be higher if treatment only consists of cyst-fenestration. Complete or near-complete excision of the cyst wall or adherences supplies the most successful treatment at this locations (Fig. 4).[53]

For small or medium-sized temporal cysts, excision or fenestration supplies a good chance of complete disappearance of the cyst and complete re-expansion of the adjacent temporal lobe. For large cysts, an only incomplete expansion of the adjacent brain has been

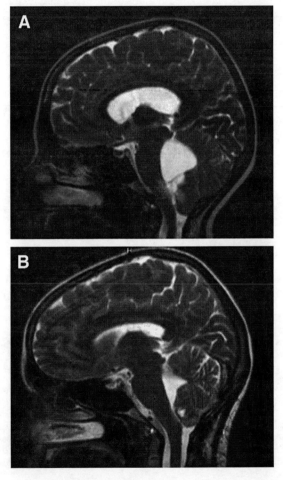

Figure 4. Preoperative (A) T2-weighted sagittal MRI depicting a large septated arachnoid cyst in the fourth ventricle containing multiple septations. Those were confirmed intraoperatively (see Fig. 4). Brainstem structures are displaced anteriorly. The lateral ventricles appear slightly enlarged as compared to postoperative imaging (B). Postoperative MRI shows a regular flow-void in the aquaeductus mesencephali.

demonstrated. Additional shunt placement may be required.[71] Shunting procedures are only little invasive, supply good control of the cyst volumes by pressure-controlled drainage, but may have a relatively high recurrence and revision-rate.[72] Fenestration or excision of the cyst may bear a higher operative risk, but leaves the patient shunt-independent. A long-term success-rate of 76-90% has been reported for these ways of treatment.[70,72] The necessity to treat and the choice of the optimal way of treatment have to be chosen deliberately. All forms of treatment have more or less operative risks and recurrence rates. Retreatment, again, imposes a recurrent treatment-risk on the patient and patients may eventually become discontented.

CONCLUSION

Arachnoid cysts are benign malformations of the arachnoid texture. They can occur in any location in the cranium or spinal canal with an arachnoid layer. They can become symptomatic by growth causing a mass effect or by intracystic hemorrhage. The clinical symptoms mainly depend on the location of the cyst. With increasing availability of diagnostic facilities, in particular MRI, many arachnoid cysts are discovered incidentally. Therapeutically, endoscopic fenestration of the cyst, open resection of the cyst-walls or permanent drainage of the intracystic fluid via a cystosubarachnoid or cystoperitoneal shunt are possible. The most challenging task is to evaluate the indication and the appropriate way of surgical treatment.

REFERENCES

1. Bright R. Serous cysts in the arachnoid diseases of the brain and nervous system, part I: reports of medical cases selected with a view of illustrating the symptoms and cure of diseases by a reference to morbid anatomy. 1831; 2:437-439.
2. Cunningham DJ. A Large sub-arachnoid cyst involving the greater part of the parietal lobe of the brain. J Anat Physiol 1879; 13:508-517.
3. Robinson RG. The temporal lobe agenesis syndrome. Brain 1964; 87:87-106.
4. Bannister CM, Russell SA, Rimmer S et al. Fetal arachnoid cysts: their site, progress, prognosis and differential diagnosis. Eur J Pediatr Surg 1999; 9 Suppl 1:27-28.
5. Shukla-Dave A, Gupta RK, Roy R et al. Prospective evaluation of in vivo proton MR spectroscopy in differentiation of similar appearing intracranial cystic lesions. Magn Reson Imaging 2001; 19:103-110.
6. Inoue K, Seker A, Osawa S et al. Microsurgical and endoscopic anatomy of the supratentorial arachnoidal membranes and cisterns. Neurosurgery 2009; 65:644-664.
7. Miyagami M, Tsubokawa T. Histological and ultrastructural findings of benign intracranial cysts. Noshuyo Byori 1993; 10:151-160.
8. Rengachary SS, Watanabe I. Ultrastructure and pathogenesis of intracranial arachnoid cysts. J Neuropathol Exp Neurol 1981; 40:61-83.
9. Angelov DN, Vasilev VA. Morphogenesis of rat cranial meninges. A light- and electron-microscopic study. Cell Tissue Res 1989; 257:207-216.
10. Bretelle F, Senat MV, Bernard JP et al. First-trimester diagnosis of fetal arachnoid cyst: prenatal implication. Ultrasound Obstet Gynecol 2002; 20:400-402.
11. Pierre-Kahn A, Hanlo P, Sonigo P et al. The contribution of prenatal diagnosis to the understanding of malformative intracranial cysts: state of the art. Childs Nerv Syst 2000; 16:619-626.
12. Piatt JHJ, D'Agostino A. The Chiari II malformation: lesions discovered within the fourth ventricle. Pediatr Neurosurg 1999; 30:79-85.
13. Holmes LB, Redline RW, Brown DL et al. Absence/hypoplasia of tibia, polydactyly, retrocerebellar arachnoid cyst, and other anomalies: an autosomal recessive disorder. J Med Genet 1995; 32:896-900.
14. Ito J, Yamazaki Y, Honda H et al. Angiographic appearance of a huge retrocerebellar arachnoid cyst in an infant. Neuroradiology 1977; 13:115-119.

15. Martinez-Lage JF, Poza M, Rodriguez CT. Bilateral temporal arachnoid cysts in neurofibromatosis. J Child Neurol 1993; 8:383-385.
16. Pascual-Castroviejo I, Roche MC, Martinez BA et al. Primary intracranial arachnoidal cysts. A study of 67 childhood cases. Childs Nerv Syst 1991; 7:257-263.
17. Caruso R, Salvati M, Cervoni L. Primary intracranial arachnoid cyst in the elderly. Neurosurg Rev 1994; 17:195-198.
18. Struck AF, Murphy MJ, Iskandar BJ. Spontaneous development of a de novo suprasellar arachnoid cyst. Case report. J Neurosurg 2006; 104:426-428.
19. Becker T, Wagner M, Hofmann E et al. Do arachnoid cysts grow? A retrospective CT volumetric study. Neuroradiology 1991; 33:341-345.
20. Galassi E, Tognetti F, Gaist G et al. CT scan and metrizamide CT cisternography in arachnoid cysts of the middle cranial fossa: classification and pathophysiological aspects. Surg Neurol 1982; 17:363-369.
21. Russo N, Domenicucci M, Beccaglia MR et al. Spontaneous reduction of intracranial arachnoid cysts: a complete review. Br J Neurosurg 2008; 22:626-629.
22. Yamauchi T, Saeki N, Yamaura A. Spontaneous disappearance of temporo-frontal arachnoid cyst in a child. Acta Neurochir (Wien) 1999; 141:537-540.
23. Berle M, Wester KG, Ulvik RJ et al. Arachnoid cysts do not contain cerebrospinal fluid: a comparative chemical analysis of arachnoid cyst fluid and cerebrospinal fluid in adults. Cerebrospinal Fluid Res 2010; 7:8.
24. Go KG, Houthoff HJ, Blaauw EH et al. Arachnoid cysts of the sylvian fissure. Evidence of fluid secretion. J Neurosurg 1984; 60:803-813.
25. Cagnoni G, Fonda C, Pancani S et al. Intracranial arachnoid cyst in pediatric age. Pediatr Med Chir 1996; 18:85-90.
26. Schroeder HW, Gaab MR. Endoscopic observation of a slit-valve mechanism in a suprasellar prepontine arachnoid cyst: case report. Neurosurgery 1997; 40:198-200.
27. Westermaier T, Vince GH, Meinhardt M et al. Arachnoid cysts of the fourth ventricle—short illustrated review. Acta Neurochir (Wien) 2010; 152:119-124.
28. Zanini MA, Gabarra RC, Faleiros AT et al. Cerebral aneurysm and arachnoid cyst: about a case with intracystic hemorrhage. Arq Neuropsiquiatr 2000; 58:330-335.
29. Gunduz B, Yassa MI, Ofluoglu E et al. Two cases of arachnoid cyst complicated by spontaneous intracystic hemorrhage. Neurol India 2010; 58:312-315.
30. Bilginer B, Onal MB, Oguz KK et al. Arachnoid cyst associated with subdural hematoma: report of three cases and review of the literature. Childs Nerv Syst 2009; 25:119-124.
31. Parsch CS, Krauss J, Hofmann E et al. Arachnoid cysts associated with subdural hematomas and hygromas: analysis of 16 cases, long-term follow-up, and review of the literature. Neurosurgery 1997; 40:483-490.
32. Gelabert-Gonzalez M, Castro-Bouzas D, Arcos-Algaba A et al. Chronic subdural hematoma associated with arachnoid cyst. Report of 12 cases. Neurocirugia 2010; 21:222-227.
33. Gupta R, Vaishya S, Mehta VS. Arachnoid cyst presenting as subdural hygroma. J Clin Neurosci 2004; 11:317-318.
34. Martinez-Lage JF, Poza M, Sola J et al. Congenital arachnoid cyst of the lateral ventricles in children. Childs Nerv Syst 1992; 8:203-206.
35. Goda K, Tsunoda S, Sakaki T et al. Intraventricular arachnoid cyst appearing with attacks of orbital pain: case report and review of the literature. No Shinkei Geka 1990; 18:757-760.
36. Okamura K, Watanabe M, Inoue N et al. Intraventricular arachnoid cyst—on the origin of intraventricular arachnoid cysts. No To Shinkei 1996; 48:1015-1021.
37. Fandino J, Garcia-Abeledo M. Giant intraventricular arachnoid cyst: report of 2 cases. Rev Neurol 1998; 26:763-765.
38. Weber F, Knopf H. Cranial MRI as a screening tool: findings in 1,772 military pilot applicants. Aviat Space Environ Med 2004; 75:158-161.
39. Vernooij MW, Ikram MA, Tanghe HL et al. Incidental findings on brain MRI in the general population. N Engl J Med 2007; 357:1821-1828.
40. Alehan FK, Gurakan B, Agildere M. Familial arachnoid cysts in association with autosomal dominant polycystic kidney disease. Pediatrics 2002; 110:13.
41. Jadeja KJ, Grewal RP. Familial arachnoid cysts associated with oculopharyngeal muscular dystrophy. J Clin Neurosci 2003; 10:125-127.
42. Pomeranz S, Constantini S, Lubetzki-Korn I et al. Familial intracranial arachnoid cysts. Childs Nerv Syst 1991; 7:100-102.
43. Sinha S, Brown JI. Familial posterior fossa arachnoid cyst. Childs Nerv Syst 2004; 20:100-103.
44. Arriola G, de Castro P, Verdu A. Familial arachnoid cysts. Pediatr Neurol 2005; 33:146-148.
45. Helland CA, Lund-Johansen M, Wester K. Location, sidedness, and sex distribution of intracranial arachnoid cysts in a population-based sample. J Neurosurg 2010; 113(5):934-9.
46. Hald JK, Nakstad PH, Skjeldal OH et al. Bilateral arachnoid cysts of the temporal fossa in four children with glutaric aciduria type I. AJNR Am J Neuroradiol 1991; 12:407-409.

47. Koch CA, Voth D, Kraemer G et al. Arachnoid cysts: does surgery improve epileptic seizures and headaches? Neurosurg Rev 1995; 18:173-181.

48. Arai H, Sato K, Wachi A et al. Arachnoid cysts of the middle cranial fossa: experience with 77 patients who were treated with cystoperitoneal shunting. Neurosurgery 1996; 39:1108-1112.

49. Smith RA, Smith WA. Arachnoid cysts of the middle cranial fossa. Surg Neurol 1976; 5:246-252.

50. Brugger G. Cerebral arachnoid cysts in children with local bulging of the skull. Wien Klin Wochenschr 1976; 88:764-769.

51. Wester K. Intracranial arachnoid cysts—do they impair mental functions? J Neurol 2008; 255:1113-1120.

52. Szucs A, Varady P, Pestality P et al. Occlusive hydrocephalus caused by a fourth ventricle arachnoid cyst. Ideggyogy Sz 2008; 61:54-58.

53. Samii M, Carvalho GA, Schuhmann MU et al. Arachnoid cysts of the posterior fossa. Surg Neurol 1999; 51:376-382.

54. Di Rocco C. Comment on: Samii M, Carvalho CA, Schuhmann MU, et al. Arachnoid cysts of the posterior fossa. Surg Neurol 1999; 51:376-382.

55. Cartwright MJ, Eisenberg MB, Page LK. Posterior fossa arachnoid cyst presenting with an isolated twelfth nerve paresis. Case report and review of the literature. Clin Neurol Neurosurg 1991; 93:69-72.

56. Mastronardi L, Taniguchi R, Caroli M et al. Cerebellopontine angle arachnoid cyst: a case of hemifacial spasm caused by an organic lesion other than neurovascular compression: case report. Neurosurgery 2009; 65:1205.

57. Ashker L, Weinstein JM, Dias M et al. Arachnoid cyst causing third cranial nerve palsy manifesting as isolated internal ophthalmoplegia and iris cholinergic supersensitivity. J Neuroophthalmol 2008; 28:192-197.

58. Hayden MG, Tornabene SV, Nguyen A et al. Cerebellopontine angle cyst compressing the vagus nerve: case report. Neurosurgery 2007; 60:1150.

59. Chao TK. Middle cranial fossa arachnoid cysts causing sensorineural hearing loss. Eur Arch Otorhinolaryngol 2005; 262:925-927.

60. Pirotte B, Morelli D, Alessi G et al. Facial nerve palsy in posterior fossa arachnoid cysts: report of two cases. Childs Nerv Syst 2005; 21:587-590.

61. Achilli V, Danesi G, Caverni L et al. Petrous apex arachnoid cyst: a case report and review of the literature. Acta Otorhinolaryngol Ital 2005; 25:296-300.

62. Starzyk J, Kwiatkowski S, Urbanowicz W et al. Suprasellar arachnoidal cyst as a cause of precocious puberty—report of three patients and literature overview. J Pediatr Endocrinol Metab 2003; 16:447-455.

63. Pillai LV, Achari G, Desai S et al. Acute respiratory failure as a manifestation of an arachnoid cyst. Indian J Crit Care Med 2008; 12:42-45.

64. Choong CT, Lee SH. Subdural hygroma in association with middle fossa arachnoid cyst: acetazolamide therapy. Brain Dev 1998; 20:319-322.

65. Longatti P, Marton E, Billeci D. Acetazolamide and corticosteroid therapy in complicated arachnoid cyst. Childs Nerv Syst 2005; 21:1061-1064.

66. Kandenwein JA, Richter HP, Borm W. Surgical therapy of symptomatic arachnoid cysts—an outcome analysis. Acta Neurochir (Wien) 2004; 146:1317-1322.

67. Helland CA, Wester K. A population based study of intracranial arachnoid cysts: clinical and neuroimaging outcomes following surgical cyst decompression in adults. J Neurol Neurosurg Psychiatry 2007; 78:1129-1135.

68. Di Rocco C, Tamburrini G, Caldarelli M et al. Prolonged ICP monitoring in Sylvian arachnoid cysts. Surg Neurol 2003; 60:211-218.

69. Raeder MB, Helland CA, Hugdahl K et al. Arachnoid cysts cause cognitive deficits that improve after surgery. Neurology 2005; 64:160-162.

70. Oertel JM, Wagner W, Mondorf Y et al. Endoscopic treatment of arachnoid cysts: a detailed account of surgical techniques and results. Neurosurgery 2010; 67:824-836.

71. Galassi E, Piazza G, Gaist G et al. Arachnoid cysts of the middle cranial fossa: a clinical and radiological study of 25 cases treated surgically. Surg Neurol 1980; 14:211-219.

72. Raffel C, McComb JG. To shunt or to fenestrate: which is the best surgical treatment for arachnoid cysts in pediatric patients? Neurosurgery 1988; 23:338-342.

CHAPTER 4

AUTISM SPECTRUM DISORDERS:
Information for Pediatricians Supporting Families of Young Children on the Spectrum

Stephanie Patterson,*,[1] Veronica Smith[2] and Michaela Jelen[2]

[1]University of California Los Angeles, Los Angeles, California, USA; [2]University of Alberta, Edmonton, Alberta, Canada
*Corresponding Author: Stephanie Patterson—sypatterson@ucla.edu

Abstract: Autism Spectrum Disorder (ASD) is a neurodevelopmental disorder affecting in US 1 in 110 individuals.[1] As increasingly younger children are receiving ASD diagnoses, many pediatricians are now faced with the unique needs of parents and other caregivers of newly diagnosed toddlers and young children. This chapter provides an overview of ASD designed to offer information and resources to pediatricians that could, in turn, be provided to families of children newly diagnosed with ASD.

INTRODUCTION

Ellen and Reid have one son named Mason. Mason is 26 months old, delayed in learning to talk and has just been diagnosed with autism. Ellen and Reid have heard of autism but don't feel that they understand what it is. What does this mean for Mason? What will he be like when he's five, ten, twenty years old? These uncertainties scare Ellen and Reid. They have heard that some children with ASD never learn to talk and can have difficulties in school. However, they have also heard that there are programs designed to help children with autism to learn and they want to enroll Mason in one of these programs as soon as they can. The team who diagnosed Mason gave Ellen and Reid a few brochures but they don't feel it's enough information. They decide to visit Mason's pediatrician to see if he can help them better understand ASD and take the necessary steps to enter intervention programming.

Neurodegenerative Diseases, edited by Shamim I. Ahmad.

Ellen and Reid are one example of a family who need more information about ASD. Pediatricians, nurse practitioners and other health care staff are often the front line medical staff available to parents for consultation when they either have concerns about their child's development or when they have questions about a new diagnosis. Families may inquire about a number of different topics related to ASD including prevalence, etiology, core symptom domains, the diagnosis process and intervention programming. Medical practitioners would benefit from knowing how a diagnosis of autism impacts a family and how to provide parents with guidance to better understand ASD so that they can take steps to reduce with the symptoms and promote their child's development.

PREVALENCE OF AUTISM SPECTRUM DISORDERS

In the current Diagnostic and Statistical Manual (DSM-IV-TR),[2] five different diagnoses are included under pervasive developmental disorders or as they shall be referred to in this chapter, autism spectrum disorders. These include: (1) autistic disorder, (2) Asperger's syndrome, (3) Pervasive Developmental Disorder Not Otherwise Specified (PDD-NOS), (4) Rett's syndrome and (5) Childhood Disintegrative Disorder (CDD). Although the DSM-IV-TR describes a prevalence rate for autism, the most common pervasive developmental disorder, of 1 in 15000 several recent research reports find the prevalence much higher, occurring at a rate of 1 in 110 live births[1] or 1 in 165.[3] The other ASDs occur at lesser rates than autism. Asperger's syndrome, which includes children who do not demonstrate the language and cognitive delays found in children with autism,[2] is noted to occur at a quarter of the rate of autism.[3] Both Rett's syndrome and CDD are considerably rare and children with these diagnoses demonstrate different trajectories of development than children with autism[4,5] where children demonstrate marked regression in development culminating in severe mental retardation and autism symptoms.[4,5] Rett's syndrome is the only diagnosis under the ASD umbrella that has an identified genetic etiology.[4] Rett's syndrome is rare, occurring in 1 in 10,000-20,000 females[4] while CDD is estimated to occur in 1 or 2 children in 100,000.[3]

ETIOLOGY

Although the specific mechanisms and etiology of autism are unknown, researchers have accumulated evidence for both biological and environmental considerations. Over the past 50 years, notions regarding the etiology of ASD have changed drastically. During the height of the psychoanalytic movement, ASD was thought to be the product of detached or cold parenting, associated with the notion of the "refrigerator mother", a notion that has long since been disregarded.[6] Throughout the 1960's ASD was identified with schizophrenia and psychosis, however comparative studies have demonstrated this too is not the case.[6] Once considered a condition found only in children, researchers have demonstrated that ASD is a lifelong neurodevelopmental disorder.[6] Through early twin studies and later, the examination of the infant siblings and close relations of individuals diagnosed with autism, researchers found that ASD is highly heritable (90%) with a recurrence risk of 5-6% when one child in the family has been diagnosed with ASD.[7] Although some genetic markers associated for increased risk of ASD are being examined (for more information on genetics and ASD see Abrahams and Geschwind[8]), less than 10% of ASD diagnoses may be associated with a known medical condition.[9] Clear links between one or several biological or environmental mechanisms across the population of individuals with ASDs have yet to be identified. The notion of a single gene being responsible for ASD in individuals across the spectrum has been more or less abandoned by the research community.[7] Rather, it is suggested that a combination of biological factors and environmental triggers leads to the expression of the behavioural profile associated with ASD, which is referred to as the gene dosage model.[7] This model suggests that the impact of these factors is cumulative where once a certain threshold has been reached, behavioural symptoms can arise—a notion currently being explored by researchers.[7]

Resources on ASD Prevalence and Etiology

There are several resources that medical personnel can use to keep up to date on the changing prevalence rates of autism.For example, the **National Centers for Disease Control and Prevention (CDC)** is an American organization with a large database of online resource available, specific to the prevalence and etiology of autism: http://www.cdc.gov/ncbddd/autism/addm.html. Further, an information booklet suitable for parents can be found at the **National Institute for Mental Health (NIMH)**. It includes brief research summaries and information regarding the prevalence and etiology of ASD in the United States. Hard copies of the booklet can be ordered from NIMH. Online it can be viewed at: http://www.nimh.nih.gov/health/publications/autism/complete-index.shtml

EARLY DIAGNOSIS: SCREENING OF 'RED FLAGS'

One of the most intriguing features of autism is the range of ways it can present initially and across the lifespan. Despite this variability there are several key 'red flags' that have commonly been observed during the screening process.

Red Flags Noted by Parents

Retrospective report using home video tapes and now prospective systematic study of infant siblings of children with ASD, have revealed that autism symptoms in the three core domains can typically be recognized in children between the first and second years of life.[10] The presence of autism symptomatology can emerge in different patterns with some children demonstrating delays in early life in domains such as language, communication, social, cognitive and motor skills while others may demonstrate a marked regression in their skills.[7] Delayed development and achievement of early milestones (e.g., early language including verbal, nonverbal communication and early social skills including joint engagement and imitation) is the more common presentation. A lesser reported 20-30% of children demonstrate patterns of regression or skill loss.[10]

Delays or marked losses in early expressive language are usually what brings parents into their pediatrician's office. Most families voice concerns about language development by the time the child is 18 months old.[11] Children are now being reliably diagnosed with austism as young as 24 months of age.[12] Diagnoses at age 2 and 3 have been demonstrated to be stable over time, especially if clinicians are using standardized tools[13] including the Autism Diagnostic Observation Scale Generic (ADOS-G)[14] and the Autism Diagnostic Interview Revised (ADI-R).[15] However, some diagnostic changes may still occur up to seven years of age.[16]

Autism Screening Resources

Several reliable screening tools have been developed to assist physicians and clinicians in assessing risk for autism. The Modified—Checklist for Autism in Toddlers (M-CHAT) is available for free download for medical, clinical and educational purposes (checklist and scoring available at: www2.gsu.edu/~wwwpsy/faculty/robins.htm). This screener has been validated for toddlers 16- 30 months of age and can be administered and scored in less than two minutes during a medical appointment. Further understanding of the 'red flags' for autism can be obtained by examining a set of instructive videos contrasting typical and atypical development that have been posted online by **First Signs**, a nonprofit organization in the United States under the direction of a Scientific Advisory Board including prominent autism researchers. The site provides resources and information for both parents and clinicians regarding the behaviours of autism and their presentation in very young children with autism: http://www.firstsigns.org/

DIAGNOSIS: GETTING IT RIGHT

There are three core behavioural domains that are examined when a diagnosis of autism is being considered. In accordance with the current Diagnostic and Statistical Manual of Mental Disorders (DSM-IV-TR), in order to receive the diagnosis, a child must demonstrate: (a) impairments in social skills; (b) deficits in the domain of communication and (c) the presence of stereotypic and repetitive behaviours.[2] However, it should be noted that the criteria for an ASD diagnosis is under review and revisions will be published in the upcoming DSM-V.[17] Although the diagnosis is now often made in children prior to school entry, in each of the three domains the challenges and atypical behaviours can present differently across the lifespan and across each individual diagnosed. In order to address the immediate needs of most families with a new diagnosis, the information provided here is relevant to the behavioural presentation of these domains in toddlerhood.

Core Symptom Domains: Expression in Toddlers

Below is a description of the diagnostic criteria associated with each of the three behavioural domains in accordance with criteria set out in the DSM-IV-TR[2] as well as a description of their expression in young children with ASD.

Social Skills and Communication

For an autism diagnosis, children must demonstrate at least two of: (1) difficulties in social interaction including challenges with nonverbal social behaviours (e.g., gestures, eye contact), (2) lack of age appropriate relationships with peers, (3) lack of sharing of social experiences with another (e.g., sharing, showing behaviours) and/or (4) a lack of social reciprocity in an interaction.[2] In terms of the domain of communication and language, children must demonstrate challenges through either delayed verbal language, lack of ability to start or maintain a conversation, the presence of repetitive language (e.g., echolalia, stereotyped language) or a lack of symbolic play.[2]

A notable deficit in joint engagement occurs in many young children with ASD and is considered an early warning sign.[18] Joint attention or engagement, involves the coordination of one's attention between another individual and an object or activity and have been found to be foundational for not only language skills but also skills in the social and academic domains.[18] Early joint attention behaviours can include sharing and directing attention as well as following another person's eye gaze.[19] A lack of imitation skills is also seen as an early sign of autism.[19] Immediate imitation, copying what others sounds or actions, is another preverbal skill that is foundational to later language development.[19] These early social communicative skills have been identified as predictors of early language development.[18,20]

The American Academy of Neurology and Child Neurology Society practice recommendations, for autism assessment four different diversions in communication development that indicate assessment taking place with a young child should include: (1) a lack of babbling or gestures in the first year, (2) lack of single words by 16 months, (3) lack of generative two word phrases by two years of age and, (4) any loss of language.[21] In terms of a lack of babbling or gestures, early symbolic gestures are also often either delayed in children with ASD.[22] Alternatively, when children with ASD do demonstrate gestures, they are less often coordinated with vocalizations than those of typically developing children.[23] Criteria 2 through 4 relate to verbal language. Children with ASD often develop verbal language at a different rate and later in life than typically developing children.[24] Delays in verbal language or less frequent and varied use of language than typically developing children often encourages assessment.

Stereotypic and Repetitive Behaviours

The domain of stereotypic and repetitive behaviours (SRBs) and restricted or circumscribed interests includes a wide range and topography of behaviours (for an examination of the forms and functions of SRBs see Turner, 1999[25]). In accordance with DSM-IV-TR criteria, children must demonstrate one or more of an intense preoccupation with a particular interest, an intense lack of flexibility (e.g., scheduling, routines, rituals), repetitive motor movements (e.g., hand flapping, spinning) or intense focus on components of an object (e.g., the wheels of a toy car).[2]

In some ways SRBs are not unique to children with autism or other developmental delays. Very young typically developing children can demonstrate in the first 12 months of life including repetitive motor behaviours such as waving and rocking which can serve to reduce anxiety and create repetition in learning.[26] However, for typically developing children these behaviours, most often, fade after the first year while in the population of children with ASD, the frequency and intensity of the behaviours tends to increase. Mixed results have provoked discussion in the literature as to whether or not in very young children, there are consistently observable differences in SRBs between ASD population compared to typically developing children and children with developmental delays.[26,27] Watt et al[27] explored differences between these populations and found that the sample of children with ASD demonstrated repetitive object exploration (e.g., banging, tapping, spinning or rolling of objects), repetitive motor movements (e.g., hand movements, body rigidity/stiffening) and repetitive sensory behaviours with significantly greater frequency and for significantly longer period than children who were typically developing and those with developmental delays.

Resources for Parents: ASD Core Symptom Domains

Although parents of newly diagnosed children with autism may have some understanding of what a diagnosis of autism means, it is common for them to feel confused about the core symptoms domains and how autism may impact not only their child's development but also their family functioning. **Autism Speaks,** an American autism science and advocacy group, has created a package called the "*100 Day Kit*" designed to support families through the first 100 days after the diagnosis. The kit contains basic information regarding ASD, information regarding many different kinds of intervention programming and steps to help parents take the necessary steps forward to obtain support: http://www.autismspeaks.org/community/family_services/100_day_kit.php.

FIRST STEPS INTO EARLY INTERVENTION

Research is accumulating to indicate that early intervention provided before the age of three and a half years may be of greater impact than intervention provided after five years of age.[11] A variety of intervention programs are available for children with ASD and recently, a number of programs have been developed specifically for very young children with ASD such as Mason who was introduced in the beginning of this chapter. However, research examining the efficacy of individual programs and the details of the programs' impact on children's developmental trajectories over time varies by intervention and in some cases, is just beginning to emerge. Available services and funding policies will differ based on a family's geographic location, however, for many families with a newly diagnosed young child such as Mason's family, a number of intervention programs may be available and there are several considerations to be made when choosing and implementing an intervention.

Intervention Components

In 2001, the National Research Council[28] published recommendations for evidence based intervention practices for individuals with ASD. Based upon these recommendations, the intervention needs to be individualized to the child's needs and using positive and empirically supported methods, address the development of the child's functional communication, facilitate the teaching of social skills in natural situations with peers, address the child's cognitive development (including functional academic skills) and strategically address challenging behaviours that can negatively impact the child and his or her family. Ongoing evaluation of the child's progress is necessary to re-assess the goals targeted within the intervention and the appropriateness of the intervention program.

Intensity

Research examining intervention 'dosage' is emerging. Precedent for intensity was set in early studies using applied behaviour analytic interventions conducted by Ivar Lovaas in the 1980s and 1990s. These studies recommend at least 40 hours of week of this behavioural mass trialed intervention approach in order to obtain optimal child outcomes.[29] More recently, recommendations from the NRC[28] indicated that children should receive at least 25 hours a week of intervention. Yet, these levels of intervention intensity delivered by highly trained clinicians can be difficult to manage in terms of the financial costs, the human resources involved and the time required from the family. Dawson and colleagues[30] recently reported in a 2010 randomized control trial that families of young children enrolled in their examination of the Early Start Denver Model were not choosing to utilize all 20 hours of "clinician implemented intervention" offered to them with the mean accepting just over 15 hours a week. Families may vary in their willingness and their ability to accept high level of intrusion into their homes amidst the many other demands inherent in these parents' lives. This should be a consideration when contemplating the implementation of parent mediated vs clinician directed intervention services.

Parent vs Clinician Implemented Intervention

The active inclusion of parents in intervention is a recommended practice.[28] Several interventions that target the communication and language development of toddlers utilize a parent-mediated method of "intervention delivery" including the Early Start Denver Model,[30] Hanen More Than Words Program,[31] Joint Attention Based Intervention[32] and Pivotal Response Training.[33] Although the content and delivery of the programs vary, each of these programs engages in parent training or education, which Mahoney et al[34] call the "process of providing parents and other primary caregivers with specific knowledge and childrearing skills with the goal of promoting the development and competence of their children". Evidence is now accumulating which supports the use of parent implemented interventions with young children with ASD including positive effects for children (e.g., increased communication skills,[35] social initiations,[36] joint attention[32]) and positive effects for parents (e.g., increased positive affect[37] and mastery of intervention content.[33] However, this body of research is relatively young and it is unclear what level and forms of supports are required to assist parents in implementing high quality intervention over time as their children grow and develop.

Resources for Parents and Caregivers: Intervention

Knowing where to begin accessing services for a child with autism can be a complicated road to navigate. Understanding the range of services available for comprehensive early intervention programs is necessary before parents make important decisions regarding their child's plan of intervention. Physicians should feel confident in directing parents to reputable sources and organizations as a good place to begin looking for intervention services in their area. **Autism Speaks** provides a comprehensive overview of the different treatment options that are available in the United States, many of which are also available in Canada: http://www.autismspeaks.org/treatment/index.php. **The Association for Science in Autism Treatment** is a nonprofit organization which aims to provide a range of evidence-based resources for caregivers and professionals seeking information regarding autism: http://www.asatonline.org/. More information regarding interventions that have an evidence base can be found on the **National Autism Center (NAC)** website (www.nationalautismcenter.org). The NAC is a nonprofit organization that recently published a comprehensive report, the National Standards Report, that summarizes effective research validated educational and behavioural interventions for children with autism. This report can serve as good guide for clinicians to guide parents in selecting appropriate interventions.

THE IMPACT OF ASD ON THE FAMILY

Parental Stress and Functioning

Research has demonstrated that parents of young children with autism such as Ellen and Reid undergo an enormous amount of stress during the diagnostic process[38] and in particular, during the six-month period following diagnosis.[39] Further research has demonstrated that overall, parents of children with ASD report higher levels of stress than parents of children with other developmental disabilities[40] and that the process of developing care and education plans for children with ASD is also uniquely stressful.[38]

Importance of Reliable Sources

Several examinations of the information sources that parents use to obtain information about ASD have been published recently. Surveys exploring the choice parents make in regard to treatment[41,42] are beginning to emerge however, how parents make sense of the conflicting reports found online from professionals and those familiar with autism, or reported in the popular media to make decisions regarding their children's health, education and supports is not well understood. These surveys demonstrate that the information sources most often used by parents are other parents of children with ASD[43] while information and support were also frequently sought out from books, web pages and newsletters than medical and educational professionals.[43] Similarly, when a small sample of parents of young toddlers were asked directly, what information supports they found useful, they reported that information gleaned from other parents is most useful but desired to have more information provided by medical and clinical staff in the time immediately after the child's diagnosis and the transition into early intervention.[44] The importance of social supports have been documented within the autism literature[45,46] where social support may be a predictor of successful adaptation and a coping mechanism for the increased stress that accompanies having a child with exceptional needs. However, with the sheer number of behavioural, educational, biomedical and alternative treatments available to families of individuals with ASD and their toll on resources including human, family and financial, it is important that families receive up to date, relevant and accurate information about the evidence around these interventions. It is difficult to examine the quality of information received by parents from nonprofessional sources. Medical and clinical front line professionals have the potential to provide and direct parents to reliable sources of information. These efforts will ensure that vulnerable and stressed parents of young children with ASD do not fall prey to inaccurate information that may prevent optimal care for their youngster.

Resources for Parents and Caregivers: Supports

A wide range of support groups are available for caregivers of individuals with ASD. These are accessible through countrywide societies that can be accessed online. Caregivers and practitioners worldwide can seek out local groups through autism societies and advocacy organizations at regional and national levels.

- *The United Kingdom*
  The National Autistic Society is a broad resource which links caregivers with local support groups and provides resources to a range of professionals working with children with autism. It has a comprehensive directory for local organizations

and services throughout the UK: http://www.autism.org.uk/en-gb/our-services/ get-help-in-your-area.aspx

• *The United States of America and Canada*

   The Autism Society of America (ASA) is a general ASD information resource suitable for both caregivers and professionals. Local organizations throughout the country, affiliated with the ASA, can be accessed at the national website: http://www.autismsource.org/

   The Autism Society of Canada (ASC) is a nationwide organization providing general ASD information and facilitates connections with local organizations within Canada: http://www.autismsocietycanada.ca/provincial_territorial_ societies/overview/index_e.html

CONCLUSION

Families who are in the process of obtaining or have just received a diagnosis of autism for their young child such as Ellen and Reid, can feel enormous amounts of stress and anxiety. Obtaining a diagnosis for a child can be both relieving and frightening for parents. In any case, families are forced to navigate a wide array and large volume of information in order to effectively navigate this new terrain for their child. The internet is a common place to search for autism information but it can be difficult to identify good sources of information. Pediatricians can assist families by first identifying evidence based accurate sources of information and then helping families parcel through and unpack the dense information in order to support parents in their learning and to facilitate families' uptake of appropriate services and supports.

REFERENCES

1. Kogan, MD, Blumberg SJ, Schieve LA et al. Prevalence of parent-reported diagnosis of autism spectrum disorder among children in the US, 2007. Pediatrics 2009; 124:1395-1403.
2. American Psychiatric Association. Diagnostic and statistical manual of mental disorders. 4th edn. Washington, DC: American Psychiatric Association 1994.
3. Fombonne E. Epidemiology of autistic disorder and other pervasive developmental disorders. J Clin Psychiatry 2005; 66:3-8.
4. Kirby RS, Lane JB, Skinner SA et al. Longevity in Rett Syndrome: Analysis of the North American database. J Pediatr 2010; 156:135-138.
5. Malhotra S, Gupta N. Childhood disintegrative disorder. J Autism Dev Disord 1999; 29:491-498.
6. Fombonne E. Modern views of autism. Can J Psychiatry 2003; 48:503-505.
7. Elsabbagh M, Johnson MH. Getting answers from babies about autism. Trends Cogn Sci 2010; 14:81-87.
8. Abrahams BS, Geschwind DH. Advances in autism genetics: On the threshold of a new neurobiology. Nat Rev 2008; 9:341-355.
9. Johnson CP, Myers SM. American academy of pediatrics council on children with disabilities. Identification and evaluation of children with autism spectrum disorders. Pediatrics 2007; 120:1183-1215.
10. Rogers S. What are infant siblings teaching us about autism in infancy? Autism Res 2009; 2:125-137.
11. Wetherby AM, Woods J, Allen L et al. Early indicators of autism spectrum disorders in the second year of life. J Autism Dev Disord 2004; 34:473-493.
12. Lord C. Follow-up of two-year-olds referred for possible autism. J Child Psychol Psychiatry 1995; 36:1365-1382.
13. Charman T, Baird G. Practitioner review: Diagnosis of autism spectrum disorder in 2 and 3 year old children. J Child Psychol Psychiatry 2002; 43:289-305.
14. Lord C, Risi S, Lambrecht L et al. The autism diagnostic observation schedule-generic: A standard measure of social and communication deficits associated with the spectrum of autism. J Autism Dev Disord 2000; 30:205-233.

15. LeCouteur A, Lord C, Rutter M. The Autism Diagnostic Interview-Revised (ADI-R). Los Angeles: Western Psychological Services 2003.

16. Zwaigenbaum L, Thurm A, Stone W et al. Studying the emergence of autism spectrum disorders in high-risk infants: Methodological and practical issues. J Autism Dev Disord 2007; 37:466-480.

17. American Psychiatric Association. Proposed revisions: Autistic disorder. Retrieved online September 9, 2010 from: http://www.dsm5.org/ProposedRevisions/Pages/proposedrevision.aspx?rid = 94.

18. Mundy P, Crowson M. Joint attention and early social communication: Implications for research on intervention with autism. J Autism Dev Disord 1997; 27:653-676.

19. Toth K, Munson J, Meltzoff A et al. Early predictors of communication development in young children with autism spectrum disorder: Joint attention, imitation and toy play. J Autism Dev Disord 2006; 36:993-1005.

20. Sigman M, Ruskin E. Continuity and change in the social competence of children with autism, Down syndrome and developmental delays. Monogr Soc Res Child Dev 1999; 64.

21. Filipek PA, Accardo PJ, Ashwal S et al. Practice parameter: Screening and diagnosis of autism: Report of the quality standards subcommittee of the American Academy of Neurology and the Child Neurology Society. Neurology 2007; 55:468-479.

22. Woods JJ, Wetherby AM. Early identification of and intervention for infants and toddlers who are at risk for autism spectrum disorder. Lang Speech Hear Serv Sch 2003; 34:180-193.

23. Landa R. Early communication development and intervention for children with autism. Ment Retard Dev Disabil Rese Rev 2007; 13:16-25.

24. Siller M, Sigman M. The behaviors of parents of children with autism predict the subsequent development of their children's communication. J Autism Dev Disord 2002; 32:77-89.

25. Turner MA. Annotation: Repetitive behaviour in autism: A review of psychological research. J Child Psychol Psychiatry 1999; 40:839-849.

26. Honey E, McConachie H, Randle V et al. One-year change in repetitive behaviors in young children with communication disorders including autism. J Autism Dev Disord 2008; 38:1439-1450.

27. Watt N, Wetherby AM, Barber A et al. Repetitive and stereotyped behaviours in children with autism spectrum disorders in the second year of life. J Autism Dev Disord 2008; 38:1518-1533.

28. National Research Council. Educating children with autism. Washington, DC: National Academy Press 2001.

29. Lovaas OI. The development of a treatment-research project for developmentally disabled and autistic children. J Appl Behav Anal 1993; 26:617-630.

30. Dawson G, Rogers S, Munson J et al. Randomized, controlled trial of an intervention for toddlers with autism: The early start Denver model. Pediatrics 2010; 125:17-23.

31. Sussman F. More than words: Helping parents promote communication and social skills in children with autism spectrum disorders. Toronto, ON: The Hanen Centre 1999.

32. Kasari C, Gulsrud AC, Wong C et al. Randomized controlled caregiver mediated joint engagement intervention for toddlers with autism. J Autism Dev Disord 2010 Doi: 10.1007/s10803-010-0955-5

33. Minjarez MB, Williams SE, Mercier EM et al. Pivotal response group treatment program for parents of children with autism. J Autism Dev Disord 2010. Advance online publication. doi: 10.1007/s10803-010-1027-6

34. Mahoney G, Kaiser A, Girolametto L et al. Parent education in early intervention: A call for renewed focus. Topics Early Child Spec Ed 1999; 19:131-140.

35. Stahmer AC, Gist K. The effects of an accelerated parent education program on technique mastery and child outcome. Journal of Positive Behavior Interventions 2001; 3:75-82.

36. Girolametto L, Sussman F, Weitzman E. Using case study methods to investigate the effects of interactive intervention for children with autism spectrum disorders. J Commun Disord 2007; 40:470-492.

37. Koegel RL, Bimbela A, Schreibman L. Collateral effects of two parent training programs on family interactions. J Autism Dev Disord 1996; 26:347-359.

38. Keenan M, Dillenburger K, Doherty A et al. The experiences of parents during diagnosis and forward planning for children with autism spectrum disorder. J Appl Res in Intellect Disabil 2010; 23:390-397

39. Stuart M, McGrew JH. Caregiver burden after receiving a diagnosis of an autism spectrum disorder. Res Autism Spectr Disord 2009; 3:86-97.

40. Schieve LA, Blumberg SJ, Rice C et al. The relationship between autism and parenting stress. Pediatrics 2007; 119:S114-S121.

41. Mackintosh VH, Myers BJ, Goin-Kochel RP. Sources of information and support used by parents of children with autism spectrum disorders. Journal on Developmental Disabilities 2005; 12:41-52.

42. Patterson SY, Smith V. Patterns of parent-child interaction: The varied impact of the More Than Words program on the natural language learning environment of toddlers with autism spectrum disorder (ASD). Submitted paper.

43. Dunn ME, Burbine T, Bowers CA et al. Moderators of stress in parents of children with autism. Community Ment Health J 2001; 37:39-52.

44. Weiss MJ. Hardiness and social support as predictors of stress in mothers of typical children, children with autism and children with mental retardation. Autism 2002; 6:115-130.

CHAPTER 5

AUTOSOMAL RECESSIVE
CHARCOT-MARIE-TOOTH NEUROPATHY

Carmen Espinós,*[1] Eduardo Calpena,[1,2] Dolores Martínez-Rubio[1,2]
and Vincenzo Lupo[1,2]

[1] Unit 732, Centro de Investigación Biomédica en Red de Enfermedades Raras (CIBERER), Valencia, Spain;
[2] Unit of Genetics and Molecular Medicine, Instituto de Biomedicina de Valencia (IBV), CSIC, Valencia, Spain
*Corresponding Author: Carmen Espinós—Email: cespinos@ibv.csic.es

Abstract: Charcot-Marie-Tooth (CMT) disease, a hereditary motor and sensory neuropathy that comprises a complex group of more than 50 diseases, is the most common inherited neuropathy. CMT is generally divided into demyelinating forms, axonal forms and intermediate forms. CMT is also characterized by a wide genetic heterogeneity with 29 genes and more than 30 loci involved. The most common pattern of inheritance is autosomal dominant (AD), although autosomal recessive (AR) forms are more frequent in Mediterranean countries. In this chapter we give an overview of the associated genes, mechanisms and epidemiology of AR-CMT forms and their associated phenotypes.

INTRODUCTION

Peripheral neuropathy is one of the most common referrals to neurologists and within this group of disorders, genetic cause is common. Inherited peripheral neuropathies present with varied symptoms and temporal course and represent one of the groups of Mendelian neurological disorders with higher prevalence. They are clinically and genetically heterogeneous disorders and lead to a progressive degeneration of the peripheral nerves. Hereditary neuropathies are classified into hereditary motor neuropathies (HMN), hereditary sensory neuropathies (HSN), hereditary motor and sensory neuropathies (HMSN) and hereditary sensory and autonomic neuropathies (HSAN).

Charcot-Marie-Tooth (CMT) disease is an HMSN that comprises a complex group of more than 50 disease entities and it is the most common inherited neuropathy with

---

Neurodegenerative Diseases, edited by Shamim I. Ahmad.

Table 1. Classification of CMT forms

	Inheritance	Hystological Features	CMT Types
CMT1	Autosomal dominant	Demyelinating	CMT1A-D, CMT1F
CMT2	Autosomal dominant	Axonal	CMT2A-N
DI-CMT	Autosomal dominant	Intermediate	DI-CMTA-D
CMTX	X linked	Demyelinating or axonal	CMTX1-5
CMT4	Autosomal recessive	Demyelinating	CMT4A-H, CMT4J
CMT4C/ AR-CMT2	Autosomal recessive	Axonal	CMT4C1-4

an estimated prevalence of 28/100 000.[1,2] Patients usually present with distal muscle atrophy in the legs, areflexia, foot deformity and steppage gait in the first or second decade of life. In most cases, hands are also involved as the disease progresses. CMT can be classified according to electrophysiological and nerve biopsy findings into three types (*Neuromuscular Disease Center*, http://neuromuscular.wustl.edu/index.html; Table 1): (i) Demyelinating CMT characterized by nerve conduction velocities (NCVs) of <38m/s and demyelinating traits on sural nerve biopsies with proliferation of Schawnn cells forming onion bulbs; (ii) Axonal CMT showing normal or slightly reduced NCV and loss of myelinated axons; and (iii) intermediate CMT in which NCV lies between 30-40 m/s and nerve pathology shows axonal and demyelinating features.[3,4]

CMT is characterized by a wide genetic heterogeneity not just because of the large number of genes/loci involved (29 genes and more than 30 loci) (*Inherited Peripheral Neuropathies Mutation Database*, http://www.molgen.ua.ac.be/CMTmutations), but also because the disease may segregate with different Mendelian patterns and even, mutations in the same gene could be inherited as a recessive and a dominant trait. The most common pattern of inheritance is autosomal dominant (AD), which can be demyelinating (CMT1) or axonal (CMT2). The main CMT form is Type 1A (CMT1A; MIM 118220) caused by a duplication of 1.4-Mb on chromosome 17 that affects the *PMP22* gene and represents approximately 70% of CMT1 cases.[3] Mutations in the *PMP22* (CMT1A) and *MPZ* (CMT1B; MIM 118220) genes are relatively frequent: 2.9% and 1.5%, respectively.[4] The remaining genes associated with CMT1 are responsible for less than 1% of the cases. Regarding CMT2, the most common form is CMT2A (MIM 118210) due to mutations in the *MFN2* gene (approximately 20%)[5,6] and to a lesser extent, CMT2I (MIM 607677) and CMT2J (MIM 607736) both caused by mutations in the *MPZ* gene.[7] The X-linked type caused by mutations in the *GJB1* gene (CMTX1; MIM 302800) is also a frequent cause of CMT and can lead to both demyelinating and axonal forms. Finally, autosomal recessive (AR) forms are frequent in countries around the Mediterranean basin,[8] and as other diseases of recessive transmission, are more severe and present with an earlier onset than the dominant ones. AR-CMT forms are named CMT4 (demyelinating, Table 2) and AR-CMT2 (axonal, Table 3). The most prevalent forms are CMT4A (MIM 214400) and their allelic variants, AR-CMT2K/CMT4C4 (MIM 606598) and CMT2K (MIM 607831), due to mutations in the *GDAP1* gene and CMT4C (MIM 601596) caused by mutations in the *SH3TC2* gene. Another point to be considered is that mutations in some CMT genes, as *MFN2, NEFL* or *HSP27*, with an autosomal dominant inheritance can also show an autosomal recessive inheritance.[9-11]

Table 2. Genes involved in demyelinating CMT with an autosomal recessive inheritance (CMT4)

Gene	CMT Type	MIM	Clinical Features	Putative Function of Protein
GDAP1	CMT4A	214400	Very early onset (<2 yo); vocal cord palsy	Maintenance of the mitochondrial network
MTMR2	CMT4B1	601382	Focally folded myelin sheaths; cranial nerve involvement	A phosphatase involved in transcription and cell proliferation
SBF2/ MTMR13	CMT4B2	604563	Focally folded myelin sheaths; early-onset glaucoma	MTMR13 interacts with MTMR2 and is involved in transcription and cell proliferation
SH3TC2	CMT4C	601596	Severe scoliosis; ataxia	Involved in vesicular transport and membrane trafficking
NDRG1	CMT4D	601455	Hearing loss; possible CNS involvement	Maintenance of the myelin sheaths, involved in cell growth arrest and differentiation
EGR2	CMT4E	605253	Congenital hypomyelinating neuropathy	Transcription factor involved in the myelination process
PRX	CMT4F	145900	Hypertrophic neuropathy of Dejerine-Sottas	Maintenance of the peripheral nerve myelin
HK1	CMT4G	605285	Prominent sensory loss	Major regulator of the cell's energy metabolism involved mainly in generation of ATP
FGD4	CMT4H	609311	Variable onset (almost congenital/ during childhood); slow progression	FGD4 is a member of the Rho family and plays a role in early stages of myelination
FIG4	CMT4J	611228	Asymmetric, rapidly progressive paralysis	A phosphatase that desphosphorylates $PI(3,5)P_2$, involved in trafficking of endosomal vesicles.

In this chapter we give an overview of the AR-CMT forms. The large number of genes involved in CMT makes for a complex genetic analysis which would be almost impossible to carry out in a single specialized laboratory. Mutations in some genes have been associated with specific CMT phenotypes, although it is usual to find variability among and within families. Moreover, if these private clinical manifestations are absent, the frequency of a CMT form or the patient ethnicity gains relevance for genetics analysis. Here, we try to emphasize that there are wide, defining features that can be common to several forms of AR-CMT.

Table 3. Genes involved in axonal CMT with an autosomal recessive inheritance (AR-CMT2)

Gene	CMT Type	MIM	Clinical Features	Putative Function of Protein
LMNA	CMT4C1/ AR-CMT2A/ CMT2B1	605588	Rapid evolution involvement of proximal muscles	Nuclear lamina component, transcription factor
MED25/ ARC 92/ ACID1	CMT4C3/ AR-CMT2B/ CMT2B2	605589	Typical CMT2 phenotype	Coactivator involved in regulating transcription of RNA polymerase II-dependent genes
GDAP1	CMT4C4/ AR-CMT2K	606598	Very early onset (<2 yo); vocal cord paresis	Maintenance of the mitochondrial network
Unknown	CMT4C2/ ARCMT2H/ ARCMT2C	607731	Brisk patellar and upper limbs reflexes absent, plantar anattainable	

THE MOST PREVALENT AR-CMT FORMS

CMT4A, CMT2K and AR-CMT2K: the *GDAP1*-Associated Neuropathies

Mutations in the *GDAP1 (ganglioside-induced differentiation-associated protein-1)* gene are responsible for both the demyelinating and axonal autosomal recessive forms, CMT4A (MIM 214400) and AR-CMT2K/CMT4C4 (MIM 606598), respectively and for CMT2K (MIM 607831) an autosomal dominant axonal form. Among them, the most prevalent form is AR-CMT2K.[12] Moreover, in a number of *GDAP1* cases nerve pathology shows an intermediate phenotype with axonal as well demyelinating findings.[13-17]

Different mutations have been described in the *GDAP1* gene. Inheritance in most of *GDAP1* mutations is autosomal recessive, although in a minor number of cases mutations are transmitted in a dominant mode of inheritance.[18-21] So far, three founder mutations have been described in *GDAP1*: p.Q163X in Spanish families and in North American Hispanic families,[18] p.S194X in families from Maghreb countries and in families with Moroccan ancestry,[15,18] and p.L239F in Central and Eastern European population.[22]

The *GDAP1* gene encodes for a protein of 40 kDa, localized in the outer mitochondrial membrane and is expressed in both the peripheral nervous system and central nervous system (brain and cerebellum).[23] GDAP1 participates in the dynamics of mitochondrial network, probably in the fission pathway of mitochondria without increasing the risk of apoptosis.[24-27] According to the inheritance pattern, different possible functions for GDAP1 have been postulated. Recessive mutations seem to cause a loss of fission activity and dominant mutations may affect to mitochondrial fusion because they lead to mitochondrial aggregation.[27]

Clinical Features of the GDAP1-Associated Neuropathies

Clinically, patients with mutations in the *GDAP1* gene show a relatively uniform phenotype in terms of age at onset and pattern of muscle involvement. Autosomal recessive

GDAP1 mutations are associated with a severe phenotype that is characterized by disability due to weakness of limb muscles that usually begins before the age of 3 years and patients are wheelchair-bound in the second decade of life. In some cases, however, disease evolved at a slower pace even among patients belonging to the same family.[28,29] Affected subjects with inherited dominant *GDAP1* mutations show much milder phenotypes and onset during the second decade of life.[18,19] Moreover, in *GDAP1*-associated neuropathy vocal cord palsy and diaphragmatic dysfunction are frequent clinical features.[28,30] In fact, respiratory function should be evaluated in these patients because life span could be compromised due to respiratory failure.

CMT4C, a Common Neuropathy in Several Populations

Mutations in the *SH3TC2* gene have been reported to be responsible for a demyelinating CMT neuropathy (CMT4C, MIM 601596) and also for intermediate phenotype.[31,32] In both forms, mutations are transmitted as an autosomal recessive trait. CMT4C has been described as one of the most frequent autosomal recessive demyelinating CMT forms in several populations as in England and in countries from North Africa.[33,34]

The *SH3TC2* gene encodes for a protein that contains multiple SH3 and TPR domains therefore may be a constituent of multiprotein complexes.[31] SH3TC2 protein is highly conserved in vertebrates. Northern blot and RT-PCR analysis showed that SH3TC2 protein is strongly expressed in neural tissues, including brain, spinal cord and sciatic nerve.[31] SH3TC2 participates in the endocytic pathway of cellular traffic and is also anchored to plasma membrane.[32] Recently a small GTPase, Rab11, has been reported as an effector of SH3TC2, which mediates its localization.[35,36] A *Sh3tc2* knockout mouse study revealed that *Sh3tc2* is specifically expressed in Schwann cells and the pathology of the peripheral nerve has revealed lengthened nodes of Ranvier.[37]

To date more than 20 different *SH3TC2* mutations have been reported in Caucasian non-Gypsy families from Turkey, Germany, Italy, Greece, Iran, UK, Czech Republic and Spain.[31-33,38-40] Moreover, the *SH3TC2* p.R954X mutation, a recurrent change detected in different studies,[31,33] has been postulated as a founder event in a French-Canadian cohort,[38] and other two *SH3TC2* mutations, p.C737_P738delinsX and p.R1109X, have been exclusively associated with Gypsy population (see *HMSN-Lom, HMSN-Russe and CMT4C, three CMT forms in Gypsy population*).[41]

Clinical Features of CMT4C

CMT4C presents with an early onset although is less severe than other autosomal recessive demyelinating CMT forms as CMT4A in which there is often an early loss of ambulation. Spine deformities are a hallmark of CMT4C.[33,34] Almost all patients develop severe scoliosis and foot deformities (pes cavus or planus). Moreover, hypoacusia and facial paresis have been observed in some patients, although they have also been found in other CMT patients as HMSN-Lom ones (MIM 601455).[42] Sural nerve neuropathology show private features: very thin myelin sheaths with extensive Schawnn cell proliferation with multiple small onion bulbs.[34,43] As in the *Sh3tc2* knockout mouse, biopsies from patients have also revealed that the nodes of Ranvier show an abnormal organization, providing a new marker to diagnose CMT4C.[37]

THE MOST SEVERE AR-CMT FORMS

CMT4B1 and CMT4B2, the Myotubularin Disorders

CMT4B is genetically heterogeneous: mutations in *MTMR2 (myotubularin-related2)* and *MTMR13/SBF2 (myotubularin-related 13/set-binding factor 2)* genes can lead to CMT4B1 (MIM 601382)[44,45] and CMT4B2 (MIM 604563),[46] respectively.

MTMR2 is synthesized in fetal liver and brain and in several adult tissues, mainly in brain, spinal cord and corpus callosum. MTMR13 is mainly detected in cerebellum, placenta, testis, fetal brain and sciatic nerve. Homodimeric MTMR2 interacts with homodimeric MTMR13 to form a tetrameric complex.[47] This interaction increases the activity of the MTMR2, a lipid phosphatase that dephosphorylate PI3P and PI(3,5)P2 and that could be implicated in vacuolar fusion.[48] When MTMR13, a catalytically inactive protein is mutated, the MTMR2 PI3P activity could be misregulated leading to anomalous levels of PI3P and/or PI(3,5)P2 and subsequent membrane trafficking defects.

Clinical Features of CMT4B1 and CMT4B2

Clinically both CMT forms are similar, characterized by severe disability and cranial nerve involvement. The disease manifests in infancy. Some patients present with a very severe phenotype and are diagnosed as congenital hypomelinating neuropathy (CHN). Histopathology shows demyelination with outfoldings of the myelin sheaths. CMT4B2 also presents juvenile glaucoma in some patients.[44,45,49]

CMT4E, a Congenital Hypomelinating Neuropathy

The Charcot-Marie-Tooth disease Type 4E (CMT4E) is a recessive form of Congenital Hypomelinating Neuropathy (CHN) caused by mutations in the *EGR2 (early growth response gene-2)* gene (MIM 129010). Mutations in this gene are also associated with autosomal dominant CMT1D (MIM 607678) and Déjèrine-Sottas neuropathy (MIM 145900).

The *EGR2* gene is a member of the early growth response gene family, which encodes a Cys$_2$His$_2$ zinc-finger transcription factor.[50,51] EGR2 target genes include myelin proteins and enzymes required for synthesis of normal myelin lipids, some of them as *MPZ*, *PMP22*, *CX32* and *PRX (periaxin),* associated with several forms of CMT.[50-52] Studies of *Erg2-null*, *Egr2 hipomorphic* and *Eg2r* mutants confirmed that EGR2 is absolutely required for both development and maintenance of proper peripheral nerve myelin.[50,51,53]

Clinical Features of CMT4E

CMT4E is characterized clinically by early onset of hypotonia, areflexia, distal muscle weakness and very slow NCV. Respiratory compromise and cranial nerve dysfunction are commonly associated with *EGR2* mutations and a lesser number of patients suffer from scoliosis.[54] To date the *EGR2* p.I268N mutation is the only one identified with an autosomal recessive inheritance pattern in three affected siblings from a consanguineous marriage.[55] They were floppy at birth, had delayed motor milestones and walk with the aid of crutches. The nerve conduction velocities were 3m/s and the sural nerve biopsies showed the absence of myelin in virtually all fibers.[56] The *Egr2$^{I268N/I268N}$* mutant mouse initially grows normally, but develops rapidly progressive weakness and finally dies in a few days, due to conduction block or neuromuscular junction failure.[57]

CMT4F, a Déjèrine-Sottas Neuropathy-Like

Mutations in the *PRX* (*periaxin*) gene lead to a broad spectrum of severe demyelinating neuropathy. Patients manifesting CMT4F or hypertrophic neuropathy of Déjèrine-Sottas (MIM 145900) have been reported with nonsense and frameshift mutations in the *PRX* gene. Curiously one Gypsy patient with a large deletion in homozygosis in the *PRX* gene has been reported.[58]

The *PRX* (*periaxin*) gene encodes two proteins, L- and S- periaxin, proteins of myelinating Schawnn cells, which are required for the maintenance of peripheral nerve system.[59] The protein has four domains: PDZ, nuclear localization signal, repeat and acidic domains.[60,61] The PDZ domain interacts with plasma membrane proteins and with the cortical cytoskeleton and has been associated with the stabilization of myelin in the peripheral nervous system.[62] Thus, periaxins are thought to play a role in stabilizating the Schwann cell-axon unit.[63]

Clinical Features of CMT4F

CMT4F phenotype is characterized by an early-onset and slowly-progressive distal motor and sensory neuropathies.[64] Histopathological examination of nerve biopsy specimens showed severe loss of myelinated fibers, onion bulb formations and folded myelin.[65] Other clinical features as areflexia, sensory ataxia and foot deformities have also been associated with CMT4F patients.[66]

CMT4H, a Neuropathy with a Slow Progression

CMT4H is a disease associated with Rho GTPase signaling. The causative gene is *FRABIN (FGD1-related F-actin binding protein) or FGD4 (FYVE, RhoGEF and PH domain-containing protein 4)*, which encodes for a ubiquitously expressed Rho GTPase.[67,68] Rho-GTPases play a role in regulating signal transduction pathways in eukaryotes. FGD4 contains a FYVE domain and two PH domains, which are known to bind to phosphoinositides.[67,68] Levels of this protein in rat are lower in postnatal and adult tissues suggesting that FGD4 may have a role in early stages of myelination.

Clinical Features of CMT4H

The hallmark of CMT4H is the slow progression of disease: patients remain ambulant into middle age.[69,70] CMT4H is an infantile neuropathy (onset from almost congenital to childhood). Skeletal deformities are not always observed. Nerve biopsy shows numerous outfoldings of the myelin sheath and redundant myelin loops.[70,71]

CMT4J, a Type Allelic with an Amyotrophic Lateral Sclerosis Form

Charcot-Marie-Tooth disease Type 4J (CMT4J; MIM 611228) is caused by mutations in the *FIG4* gene.[72] This gene is also involved in amyotrophic lateral sclerosis 11 (ALS11;

MIM 612577), an autosomal dominant ALS form.[73] To date four CMT4J families have been reported.[72]

The FIG4 is a vacuolar phosphatase, localized to plasma membrane and is involved in regulating phosphoinositides content and vesicular trafficking: this protein desphosphorylates PI(3,5)P2 and interacts with FAB1 and VAC14 in a protein complex that regulates the overall concentration of PI(3,5)P2.[74]

Clinical Features of CMT4J

The phenotype is characterized by childhood-onset and by coordination disorder and severe disability. Patients develop rapidly progressive, asymmetric motor neuron degeneration and minimal symptoms of sensory loss.[75] PI(3,5)P2 mediates retrograde trafficking of endosomal vesicles to the trans-Golgi network.[76,77]

FOUNDER MUTATIONS RELATED TO AR-CMT FORMS

HMSN-Lom, HMSN-Russe and CMT4C, Three CMT Forms in Gypsy Population

The Gypsy (Roma) is a transnational founder population of around 8-10 million people in Europe whose current genetic profile is the result of profound bottlenecks, genetic drift and differential admixture.[78-80] In Gypsy population, a number of confined disease-causing mutations inherited in an autosomal recessive way and evidence of a founder effect have been reported.[81] To date three CMT4 forms have been exclusively associated with Gypsies: (i) CMT4D/HMSN-Lom (MIM 601455) due to the p.R148X mutation in the *NDRG1* gene;[82,83] (ii) CMT4C (MIM 601596), which could be caused by the p.C737_P738delinsX and/or p.R1109X mutations in the *SH3TC2* gene; (iii) and CMT4G/HMSN-Russe (MIM 605285), which has recently been associated with a G > C change in a novel alternative untranslated exon (AltT2) in the *HK1* gene.[84]

An ancestral founder mutation was postulated as causative of HMSN-Lom, since a common haplotype on chromosome 8q24, cosegregating with the disease, was found in Gypsy patients.[82] Later the p.R148X change was characterized in the *NDRG1* (*N-myc downstreamregulated gene 1*) gene which is inherited as an autosomal recessive trait.[42] The carrier rate for this mutation has been estimated in 4.5/100 and even in some Gypsy communities it is up to 16/100.[79] The *NDRG1* p.R148X mutation, currently distributed throughout Europe, probably occurred before of Gypsy diaspora from India.[85] With the exception of this mutation, defects in the *NDRG1* gene are an extremely rare cause of CMT disease. In fact only one more mutation, g.2290787G > A, in the *NDRG1* gene has been identified in a patient affected by severe demyelinating neuropathy.[86] NDRG1 is highly expressed in the Schwann cell and appears to play a role in growth arrest and cell differentiation, therefore, these findings pointed to NDRG1 having a role in the peripheral nervous system, possibly in Schawnn cell signaling between cytoplasm and the nucleus, necessary for axonal survival.

Two mutations causative of CMT4C have been described in the *SH3TC2* gene and exclusively associated with Gypsy population: p.R1109X and p.C737_P738delinsX, being the p.R1109X mutation much more common.[41] The *SH3TC2* p.R1109X mutation is probably an ancestral founder mutation because it has been detected in Turkish families and probably is distributed across Europe.[87] In Spain, this mutation would have arrived around the end of the 18th century due to a split from an original group.[41] What is known so far about SH3TC2 protein has been described in a previous section (see *CMT4C, a Common Neuropathy in Several Populations*).

The mutation responsible for HMSN-Russe has been recently reported: a G > C change in a novel alternative untranslated exon (AltT2) in the *HK1* (*hexokinase 1*) gene.[84] In a similar way to other confined mutations in Gypsy population, an ancestral founder event has been postulated for this mutation since all HMSN-Russe patients from several European countries share a common haplotype on chromosome 10q232-q23.[41,88,89] The mutational mechanism and functional effects could lead to disrupt translational regulation causing an increased antiapoptotic activity or an impairment of a novel *HK1* function in the peripheral nervous system.[84]

Clinical Features of HMSN-Lom, HMSN-Russe and CMT4C

The three CMT neuropathies associated with Gypsy population are demyelinating forms. The clinical aspects of CMT4C have already been described in a previous section (see *CMT4C, a Common Neuropathy in Several Populations*). Both neuropathies, HMSN-Lom and HMSN-Russe, have an onset in the first decade or early in the second decade of life.[83,89,90] Motor involvement is greater than sensory. Skeletal deformities are frequent, mainly foot deformities. The disease is steadily progressive, with later involvement of the upper limbs, which leads to disability. The clinical manifestations of both neuropathies are similar although they tend to me more severe in HMSN-Lom.[91] Sensorineural deafness usually developed during the third decade, is an invariable hallmark of HMSN-Lom.

CMT2B1 in North-Western African Population

CMT2B1/AR-CMT2A (MIM 605588), an autosomal recessive axonal type of CMT, is caused by a unique homozygous p.R298C mutation in the *lamin A/C* (*LMNA*) gene.[92] Mutations in this gene can cause more than ten different clinical syndromes (MIM 150330), many of which show overlapping features and could be classified into 4 major types: diseases of striated and cardiac muscle; lipodystrophy syndromes; peripheral neuropathy; and premature aging.[93] The p.R298C is the only known *LMNA* mutation to be responsible for a pure peripheral nerve phenotype. Two different dominant mutations in *LMNA* have been associated with peripheral neuropathy,[94,95] but always in combination with several clinical signs (such as muscular dystrophy, cardiac disease, or leuconychia). All CMT2B1 patients are from a restricted region of Northwest Algeria and Eastern Morocco and carry a homozygous common ancestral haplotype at the *LMNA locus*, which is suggestive of a founder event.[96,97]

LMNA gene encodes lamins, which are intermediate filaments of the nuclear lamina. The main isoforms in somatic cells are lamin A and lamin C.[98] The phenotypic heterogeneity of diseases resulting from a mutation in the *LMNA* gene can be explained by the numerous roles of the nuclear lamina, including a role in maintaining nuclear structure, regulating transcription, controlling differentiation and chromatin organization.[99] *Lmna* null mice present an axonal pathological phenotype that is highly similar to that presented by patients with AR-CMT2.[92] Since *LMNA* is ubiquitously expressed, the finding of site-specific amino acid substitutions indicates the existence of distinct functional domains in lamin A/C that are essential for the maintenance and integrity of different cell lineages. The p.R298C substitution, located in the lamin A/C central rod domain, has been predicted to impair protein-protein interactions which are essential for the maintenance of cellular function.[92]

Clinical Features of CMT2B1

CMT2B1 is characterized by a variable age of onset, although the disease usually begins in the second decade. Median NCVs are either preserved or slightly reduced.[96] There is a severe rarefaction of myelinated fibers with no evidence of demyelination or remyelination processes, or of onion bulb formations.[92,100] The severity and course of the disease is highly variable. Patients can present a severe CMT phenotype with distal wasting and weakness of all four extremities and areflexia and these features can coexist with proximal muscles affection.[96,100] The impairment can become severe in patients with the longest disease duration, affecting also the scapular muscles which might be a hallmark of CMT2B1.[96]

Charcot-Marie-Tooth Disease Type 2B2 in Costa Rican Population

The Charcot-Marie-Tooth 2B2 (CMT2B2, MIM 605589) is an autosomal recessive form of CMT mapped to chromosome 19q13.3 by linkage analysis, in an extended consanguineous family of Spanish ancestry in Costa Rica.[101] The p.A335V mutation in the *MED25 (mediator complex subunit 25)* gene is the unique homozygous change identified in the mapped region and it has been identified to be the responsible for CMT2B2.[102] To date, this unique mutation is suggested to be a founder effect.

The *MED25* gene encodes a component of the transcriptional coactivator complexes related to the yeast Mediator. RT-PCR expression analysis demonstrated that MED25 is ubiquitously expressed and underlines the sciatic nerve and dorsal root ganglia expression, which are affected in peripheral neuropathies.[102,103] Furthermore, *Med25* expression correlates with *Pmp22* gene dosage and expression in both transgenic rats and mice. After permanent sciatic nerve transsection, without allowing nerve degeneration, *Pmp22* transcrit levels were strongly reduced while *Med25* expression was only moderately decreased.[102] The p.A335V mutation resides in a proline-rich region and causes a decrease in binding specificity for SH3 domains resulting in an interaction with an extended range of SH3 domain proteins.[102]

Clinical Features of CMT2B2

According to clinical features, CMT2B2 patients present an adult onset phenotype (range, 26 to 42 years) with symmetrical weakness and atrophy in the ankles and showed sensory deficit in a symmetrical "stocking-glove" pattern. Motor-nerve conduction velocity (MCV) of the median and ulnar nerves is normal or slightly reduced, indicative of an axonal degenerative process.[101]

CONCLUSION

CMT is the most prevalent hereditary neurological condition and during the last two decades many causative genes and loci associated with CMT have been identified. CMT diseases are the clinical manifestations of peripheral nerve dysfunction resulting from abnormalities in Schwann cells and their myelin sheath, with an intimate contribution of the axon-glia communication. In this chapter we have focused on the autosomal recessive forms (AR-CMT). The list of AR-CMT loci and genes is predicted to be higher in the forthcoming years, since the number of patients is increasing and the relationship between phenotypes and genotypes is difficult to be established. For better finding of this relationship and improve the diagnostic criteria, histological aspects may be indicative of which is the disease-causative gene (i.e.,abnormal organization of nodes of Ranvier in CMT4C patients). Thus, after the recording of clinical and electrophysiological data, an examination of one nerve biopsy per family may determine which gene is most likely mutated. This screening step would facilitate and accelerate diagnosis by molecular biological analysis, which may be impossible in non-specialised laboratories. However, this screening criteria cannot be sufficient because some phenomena previously reserved exclusively for one form of CMT disease, have later been observed for other ones. Some clinical features firstly considered hallmarks for some CMT forms, as hoarseness for CMT4A or early glaucoma for CMT4B2, have rarely been described in some patients. To date, several AR-CMT forms are related to mutation with founder effect and are group-specific, as in the case of the Gypsy population (HMSN-Russe, HMSN-Lom). Nowadays, the pathogenic roles of AR-CMT genes are only partially known. Although the functions of several proteins, in particular GDAP1, PRX, LMNA, MTMR2, MTMR13 and to certain extent SH3TC2, have been investigated extensively, it is still unknown the relationship between their respective genes coding and the phenotypes. To answer why and how the axonal degeneration occurs it is necessary to understand the mechanism underlying the Schwann cells/axons interactions. When the molecular biology of peripheral nerves will be better understood, the development of novel therapeutic strategies that would result in effective treatment for these diseases and the discovery of more specific markers for diagnosis will become apparent.

ACKNOWLEDGEMENTS

This work was supported by the Fondo de Investigación Sanitaria [grants numbers PI08/90857, CP08/00053 and PS09/00095]. C.E. has a "Miguel Servet" contract funded by the Fondo de Investigación Sanitaria. The CIBERER is an initiative of the Instituto de Salud Carlos III.

REFERENCES

1. Combarros O, Calleja J, Polo JM et al. Prevalence of hereditary motor and sensory neuropathy in Cantabria. Acta Neurol Scand 1987; 75:9-12.
2. Skre H. Genetic and clinical aspects of Charcot-Marie-Tooth's disease. Clin Genet 1974; 6:98-118.
3. Reilly MM. Axonal Charcot-Marie-Tooth disease: the fog is slowly lifting! Neurology 2005; 65:186-187.
4. Boerkoel CF, Takashima H, Garcia CA et al. Charcot-Marie-Tooth disease and related neuropathies: mutation distribution and genotype-phenotype correlation. Ann Neurol 2002; 51:190-201.
5. Zuchner S, Mersiyanova IV, Muglia M et al. Mutations in the mitochondrial GTPase mitofusin 2 cause Charcot-Marie-Tooth neuropathy type 2A. Nat Genet 2004; 36:449-451.
6. Lawson VH, Graham BV, Flanigan KM. Clinical and electrophysiologic features of CMT2A with mutations in the mitofusin 2 gene. Neurology 2005; 65:197-204.
7. Szigeti K, Nelis E, Lupski JR. Molecular diagnostics of Charcot-Marie-Tooth disease and related peripheral neuropathies. Neuromolecular Med 2006; 8:243-254.

8. Vallat JM, Grid D, Magdelaine C et al. Autosomal recessive forms of Charcot-Marie-Tooth disease. Curr Neurol Neurosci Rep 2004; 4:413-419.
9. Nicholson GA, Magdelaine C, Zhu D et al. Severe early-onset axonal neuropathy with homozygous and compound heterozygous MFN2 mutations. Neurology 2008; 70:1678-1681.
10. Yum SW, Zhang J, Mo K et al. A novel recessive Nefl mutation causes a severe, early-onset axonal neuropathy. Ann Neurol 2009; 66:759-770.
11. Houlden H, Laura M, Wavrant-De Vrieze F et al. Mutations in the HSP27 (HSPB1) gene cause dominant, recessive and sporadic distal HMN/CMT type 2. Neurology 2008; 71:1660-1668.
12. Martínez-Rubio MD, Jaijo T, Sevilla T et al. Rationalisation of molecular diagnosis of the Charcot-Marie-Tooth neuropathy. Third International Charcot-Marie-Tooth Consortium Meeting. Antwerpen (Belgium) 2009.
13. Cuesta A, Pedrola L, Sevilla T et al. The gene encoding ganglioside-induced differentiation-associated protein 1 is mutated in axonal Charcot-Marie-Tooth type 4A disease. Nat Genet 2002; 30:22-25.
14. Baxter RV, Ben Othmane K, Rochelle JM et al. Ganglioside-induced differentiation-associated protein-1 is mutant in Charcot-Marie-Tooth disease type 4A/8q21. Nat Genet 2002; 30:21-22.
15. Nelis E, Erdem S, Van Den Bergh PY et al. Mutations in GDAP1: autosomal recessive CMT with demyelination and axonopathy. Neurology 2002; 59:1865-1872.
16. Birouk N, Azzedine H, Dubourg O et al. Phenotypical features of a Moroccan family with autosomal recessive Charcot-Marie-Tooth disease associated with the S194X mutation in the GDAP1 gene. Archives of neurology 2003; 60:598-604.
17. Senderek J, Bergmann C, Ramaekers VT et al. Mutations in the ganglioside-induced differentiation-associated protein-1 (GDAP1) gene in intermediate type autosomal recessive Charcot-Marie-Tooth neuropathy. Brain 2003; 126:642-649.
18. Claramunt R, Pedrola L, Sevilla T et al. Genetics of Charcot-Marie-Tooth disease type 4A: mutations, inheritance, phenotypic variability and founder effect. Journal of medical genetics 2005; 42:358-365.
19. Chung KW, Kim SM, Sunwoo IN et al. A novel GDAP1 Q218E mutation in autosomal dominant Charcot-Marie-Tooth disease. J Hum Genet 2008; 53:360-364.
20. Cassereau J, Chevrollier A, Gueguen N et al. Mitochondrial complex I deficiency in GDAP1-related autosomal dominant Charcot-Marie-Tooth disease (CMT2K). Neurogenetics 2009; 10:145-150.
21. Cavallaro T, Ferrarini M, Taioli F et al. Autosomal dominant Charcot-Marie-Tooth disease type 2 associated with GDAP1 gene. Third International Charcot-Marie-Tooth Consortium Meeting. Antwerpen (Belgium) 2009.
22. Kabzinska D, Strugalska-Cynowska H, Kostera-Pruszczyk A et al. L239F founder mutation in GDAP1 is associated with a mild Charcot-Marie-Tooth type 4C4 (CMT4C4) phenotype. Neurogenetics 2010.
23. Pedrola L, Espert A, Valdes-Sanchez T et al. Cell expression of GDAP1 in the nervous system and pathogenesis of Charcot-Marie-Tooth type 4A disease. J Cell Mol Med 2008; 12:679-689.
24. Pedrola L, Espert A, Wu X et al. GDAP1, the protein causing Charcot-Marie-Tooth disease type 4A, is expressed in neurons and is associated with mitochondria. Hum Mol Genet 2005; 14:1087-1094.
25. Niemann A, Ruegg M, La Padula V et al. Ganglioside-induced differentiation associated protein 1 is a regulator of the mitochondrial network: new implications for Charcot-Marie-Tooth disease. The Journal of Cell Biology 2005.
26. Niemann A, Wagner KM, Ruegg M et al. GDAP1 mutations differ in their effects on mitochondrial dynamics and apoptosis depending on the mode of inheritance. Neurobiol Dis 2009; 36:509-520.
27. Wagner KM, Ruegg M, Niemann A et al. Targeting and function of the mitochondrial fission factor GDAP1 are dependent on its tail-anchor. PLoS One 2009; 4:e5160.
28. Sevilla T, Cuesta A, Chumillas MJ et al. Clinical, electrophysiological and morphological findings of Charcot-Marie-Tooth neuropathy with vocal cord palsy and mutations in the GDAP1 gene. Brain 2003; 126:2023-2033.
29. Azzedine H, Ruberg M, Ente D et al. Variability of disease progression in a family with autosomal recessive CMT associated with a S194X and new R310Q mutation in the GDAP1 gene. Neuromuscul Disord 2003; 13:341-346.
30. Sevilla T, Jaijo T, Nauffal D et al. Vocal cord paresis and diaphragmatic dysfunction are severe and frequent symptoms of GDAP1-associated neuropathy. Brain 2008; 131:3051-3061.
31. Senderek J, Bergmann C, Stendel C et al. Mutations in a gene encoding a novel SH3/TPR domain protein cause autosomal recessive Charcot-Marie-Tooth type 4C neuropathy. American journal of human genetics 2003; 73:1106-1119.
32. Lupo V, Galindo MI, Martinez-Rubio D et al. Missense mutations in the SH3TC2 protein causing Charcot-Marie-Tooth disease type 4C affect its localization in the plasma membrane and endocytic pathway. Hum Mol Genet 2009; 18:4603-4614.
33. Azzedine H, Ravise N, Verny C et al. Spine deformities in Charcot-Marie-Tooth 4C caused by SH3TC2 gene mutations. Neurology 2006; 67:602-606.

34. Houlden H, Laura M, Ginsberg L et al. The phenotype of Charcot-Marie-Tooth disease type 4C due to SH3TC2 mutations and possible predisposition to an inflammatory neuropathy. Neuromuscul Disord 2009; 19:264-269.
35. Roberts RC, Peden AA, Buss F et al. Mistargeting of SH3TC2 away from the recycling endosome causes Charcot-Marie-Tooth disease type 4C. Hum Mol Genet 2010; 19:1009-1018.
36. Stendel C, Roos A, Kleine H et al. SH3TC2, a protein mutant in Charcot-Marie-Tooth neuropathy, links peripheral nerve myelination to endosomal recycling. Brain 2010; 133:2462-2474.
37. Arnaud E, Zenker J, de Preux Charles AS et al. SH3TC2/KIAA1985 protein is required for proper myelination and the integrity of the node of Ranvier in the peripheral nervous system. Proc Natl Acad Sci USA 2009; 106:17528-17533.
38. Gosselin I, Thiffault I, Tetreault M et al. Founder SH3TC2 mutations are responsible for a CMT4C French-Canadians cluster. Neuromuscul Disord 2008; 18:483-492.
39. Laura M, Houlden H, Blake J et al. Charcot-Marie-Tooth tyoe 4C caused by mutation of KIAA1985 gene: report of 5 families with variable phenotype. Third International Charcot-Marie-Tooth Consortium Meeting Antwerpen (Belgium) 2009.
40. Lassuthová P, Mazanec R, Haberlová J et al. High frequency of SH3TC2 (KIAA1985) mutations in Czech HMSN I patients. Third International Charcot-Marie-Tooth Consortium Meeting. Antwerpen (Belgium) 2009.
41. Claramunt R, Sevilla T, Lupo V et al. The p.R1109X mutation in SH3TC2 gene is predominant in Spanish Gypsies with Charcot-Marie-Tooth disease type 4. Clinical Genetics 2007; 71:343-349.
42. Kalaydjieva L, Gresham D, Gooding R et al. N-myc downstream-regulated gene 1 is mutated in hereditary motor and sensory neuropathy-Lom. American Journal of Human Genetics 2000; 67:47-58.
43. LeGuern E, Guilbot A, Kessali M et al. Homozygosity mapping of an autosomal recessive form of demyelinating Charcot-Marie-Tooth disease to chromosome 5q23-q33. Hum Mol Genet 1996; 5:1685-1688.
44. Azzedine H, Bolino A, Taieb T et al. Mutations in MTMR13, a new pseudophosphatase homologue of MTMR2 and Sbf1, in two families with an autosomal recessive demyelinating form of Charcot-Marie-Tooth disease associated with early-onset glaucoma. American Journal of Human Genetics 2003; 72:1141-1153.
45. Senderek J, Bergmann C, Weber S et al. Mutation of the SBF2 gene, encoding a novel member of the myotubularin family, in Charcot-Marie-Tooth neuropathy type 4B2/11p15. Hum Mol Genet 2003; 12:349-356.
46. Bolino A, Muglia M, Conforti FL et al. Charcot-Marie-Tooth type 4B is caused by mutations in the gene encoding myotubularin-related protein-2. Nat Genet 2000; 25:17-19.
47. Previtali SC, Zerega B, Sherman DL et al. Myotubularin-related 2 protein phosphatase and neurofilament light chain protein, both mutated in CMT neuropathies, interact in peripheral nerve. Hum Mol Genet 2003; 12:1713-1723.
48. Berger P, Berger I, Schaffitzel C et al. Multi-level regulation of myotubularin-related protein-2 phosphatase activity by myotubularin-related protein-13/set-binding factor-2. Hum Mol Genet 2006; 15:569-579.
49. Hirano R, Takashima H, Umehara F et al. SET binding factor 2 (SBF2) mutation causes CMT4B with juvenile onset glaucoma. Neurology 2004; 63:577-580.
50. Topilko P, Schneider-Maunoury S, Levi G et al. Krox-20 controls myelination in the peripheral nervous system. Nature 1994; 371:796-799.
51. Le N, Nagarajan R, Wang JY et al. Analysis of congenital hypomyelinating Egr2Lo/Lo nerves identifies Sox2 as an inhibitor of Schwann cell differentiation and myelination. Proc Natl Acad Sci USA 2005; 102:2596-2601.
52. LeBlanc SE, Ward RM, Svaren J. Neuropathy-associated Egr2 mutants disrupt cooperative activation of myelin protein zero by Egr2 and Sox10. Mol Cell Biol 2007; 27:3521-3529.
53. Decker L, Desmarquet-Trin-Dinh C, Taillebourg E et al. Peripheral myelin maintenance is a dynamic process requiring constant Krox20 expression. J Neurosci 2006; 26:9771-9779.
54. Szigeti K, Wiszniewski W, Saifi GM et al. Functional, histopathologic and natural history study of neuropathy associated with EGR2 mutations. Neurogenetics 2007; 8:257-262.
55. Warner LE, Mancias P, Butler IJ et al. Mutations in the early growth response 2 (EGR2) gene are associated with hereditary myelinopathies. Nat Genet 1998; 18:382-384.
56. Harati Y, Butler IJ. Congenital hypomyelinating neuropathy. J Neurol Neurosurg Psychiatry 1985; 48:1269-1276.
57. Baloh RH, Strickland A, Ryu E et al. Congenital hypomyelinating neuropathy with lethal conduction failure in mice carrying the Egr2 I268N mutation. J Neurosci 2009; 29:2312-2321.
58. Barankova L, Siskova D, Huhne K et al. A 71-nucleotide deletion in the periaxin gene in a Romani patient with early-onset slowly progressive demyelinating CMT. Eur J Neurol 2008; 15:548-551.
59. Dytrych L, Sherman DL, Gillespie CS et al. Two PDZ domain proteins encoded by the murine periaxin gene are the result of alternative intron retention and are differentially targeted in Schwann cells. J Biol Chem 1998; 273:5794-5800.

60. Sherman DL, Brophy PJ. A tripartite nuclear localization signal in the PDZ-domain protein L-periaxin. J Biol Chem 2000; 275:4537-4540.

61. Sherman DL, Fabrizi C, Gillespie CS et al. Specific disruption of a schwann cell dystrophin-related protein complex in a demyelinating neuropathy. Neuron 2001; 30:677-687.

62. Boerkoel CF, Takashima H, Stankiewicz P et al. Periaxin mutations cause recessive Dejerine-Sottas neuropathy. Am J Hum Genet 2001; 68:325-333.

63. Gillespie CS, Sherman DL, Fleetwood-Walker SM et al. Peripheral demyelination and neuropathic pain behavior in periaxin-deficient mice. Neuron 2000; 26:523-531.

64. Kijima K, Numakura C, Shirahata E et al. Periaxin mutation causes early-onset but slow-progressive Charcot-Marie-Tooth disease. J Hum Genet 2004; 49:376-379.

65. Guilbot A, Williams A, Ravise N et al. A mutation in periaxin is responsible for CMT4F, an autosomal recessive form of Charcot-Marie-Tooth disease. Hum Mol Genet 2001; 10:415-421.

66. Kabzinska D, Kochanski A, Drac H et al. A novel Met116Thr mutation in the GDAP1 gene in a Polish family with the axonal recessive Charcot-Marie-Tooth type 4 disease. J Neurol Sci 2006; 241:7-11.

67. Stendel C, Roos A, Deconinck T et al. Peripheral nerve demyelination caused by a mutant Rho GTPase guanine nucleotide exchange factor, frabin/FGD4. Am J Hum Genet 2007; 81:158-164.

68. Delague V, Jacquier A, Hamadouche T et al. Mutations in FGD4 encoding the Rho GDP/GTP exchange factor FRABIN cause autosomal recessive Charcot-Marie-Tooth type 4H. Am J Hum Genet 2007; 81:1-16.

69. Houlden H, Hammans S, Katifi H et al. A novel Frabin (FGD4) nonsense mutation p.R275X associated with phenotypic variability in CMT4H. Neurology 2009; 72:617-620.

70. Fabrizi GM, Taioli F, Cavallaro T et al. Further evidence that mutations in FGD4/frabin cause Charcot-Marie-Tooth disease type 4H. Neurology 2009; 72:1160-1164.

71. De Sandre-Giovannoli A, Delague V, Hamadouche T et al. Homozygosity mapping of autosomal recessive demyelinating Charcot-Marie-Tooth neuropathy (CMT4H) to a novel locus on chromosome 12p11.21-q13.11. J Med Genet 2005; 42:260-265.

72. Chow CY, Zhang Y, Dowling JJ et al. Mutation of FIG4 causes neurodegeneration in the pale tremor mouse and patients with CMT4J. Nature 2007; 448:68-72.

73. Chow CY, Landers JE, Bergren SK et al. Deleterious variants of FIG4, a phosphoinositide phosphatase, in patients with ALS. Am J Hum Genet 2009; 84:85-88.

74. Volpicelli-Daley L, De Camilli P. Phosphoinositides' link to neurodegeneration. Nat Med 2007; 13:784-786.

75. Zhang X, Chow CY, Sahenk Z et al. Mutation of FIG4 causes a rapidly progressive, asymmetric neuronal degeneration. Brain 2008; 131:1990-2001.

76. Rutherford AC, Traer C, Wassmer T et al. The mammalian phosphatidylinositol 3-phosphate 5-kinase (PIKfyve) regulates endosome-to-TGN retrograde transport. J Cell Sci 2006; 119:3944-3957.

77. Zhang Y, Zolov SN, Chow CY et al. Loss of Vac14, a regulator of the signaling lipid phosphatidylinositol 3,5-bisphosphate, results in neurodegeneration in mice. Proc Natl Acad Sci USA 2007; 104:17518-17523.

78. Gresham D, Morar B, Underhill PA et al. Origins and divergence of the Roma (gypsies). Am J Hum Genet 2001; 69:1314-1331.

79. Morar B, Gresham D, Angelicheva D et al. Mutation history of the roma/gypsies. Am J Hum Genet 2004; 75:596-609.

80. Kalaydjieva L, Morar B, Chaix R et al. A newly discovered founder population: the Roma/Gypsies. Bioessays 2005; 27:1084-1094.

81. Kalaydjieva L, Lochmuller H, Tournev I et al 125th ENMC International Workshop: Neuromuscular disorders in the Roma (Gypsy) population, 23-25 April 2004, Naarden, The Netherlands. Neuromuscul Disord 2005; 15:65-71.

82. Kalaydjieva L, Hallmayer J, Chandler D et al. Gene mapping in Gypsies identifies a novel demyelinating neuropathy on chromosome 8q24. Nat Genet 1996; 14:214-217.

83. Kalaydjieva L, Nikolova A, Turnev I et al. Hereditary motor and sensory neuropathy—Lom, a novel demyelinating neuropathy associated with deafness in gypsies. Clinical, electrophysiological and nerve biopsy findings. Brain 1998; 121 (Pt 3):399-408.

84. Hantke J, Chandler D, King R et al. A mutation in an alternative untranslated exon of hexokinase 1 associated with hereditary motor and sensory neuropathy—Russe (HMSNR). Eur J Hum Genet 2009; 17:1606-1614.

85. Kalaydjieva L, Gresham D, Calafell F. Genetic studies of the Roma (Gypsies): a review. BMC Med Genet 2001; 2:5.

86. Hunter M, Bernard R, Freitas E et al. Mutation screening of the N-myc downstream-regulated gene 1 (NDRG1) in patients with Charcot-Marie-Tooth Disease. Hum Mutat 2003; 22:129-135.

87. Gooding R, Colomer J, King R et al. A novel Gypsy founder mutation, p.Arg1109X in the CMT4C gene, causes variable peripheral neuropathy phenotypes. J Med Genet 2005; 42:e69.

88. Rogers T, Chandler D, Angelicheva D et al. A novel locus for autosomal recessive peripheral neuropathy in the EGR2 region on 10q23. Am J Hum Genet 2000; 67:664-671.

89. Thomas PK, Kalaydjieva L, Youl B et al. Hereditary motor and sensory neuropathy-russe: new autosomal recessive neuropathy in Balkan Gypsies. Ann Neurol 2001; 50:452-457.

90. Tournev I, Kalaydjieva L, Youl B et al. Congenital cataracts facial dysmorphism neuropathy syndrome, a novel complex genetic disease in Balkan Gypsies: clinical and electrophysiological observations. Ann Neurol 1999; 45:742-750.

91. Guergueltcheva V, Tournev I, Bojinova V et al. Early clinical and electrophysiologic features of the two most common autosomal recessive forms of Charcot-Marie-Tooth disease in the Roma (Gypsies). J Child Neurol 2006; 21:20-25.

92. De Sandre-Giovannoli A, Chaouch M, Kozlov S et al. Homozygous defects in LMNA, encoding lamin A/C nuclear-envelope proteins, cause autosomal recessive axonal neuropathy in human (Charcot-Marie-Tooth disorder type 2) and mouse. Am J Hum Genet 2002; 70:726-736.

93. Worman HJ, Bonne G. "Laminopathies": a wide spectrum of human diseases. Exp Cell Res 2007; 313:2121-2133.

94. Goizet C, Yaou RB, Demay L et al. A new mutation of the lamin A/C gene leading to autosomal dominant axonal neuropathy, muscular dystrophy, cardiac disease and leuconychia. J Med Genet 2004; 41:e29.

95. Benedetti S, Bertini E, Iannaccone S et al. Dominant LMNA mutations can cause combined muscular dystrophy and peripheral neuropathy. J Neurol Neurosurg Psychiatry 2005; 76:1019-1021.

96. Bouhouche A, Birouk N, Azzedine H et al. Autosomal recessive axonal Charcot-Marie-Tooth disease (ARCMT2): phenotype-genotype correlations in 13 Moroccan families. Brain 2007; 130:1062-1075.

97. Hamadouche T, Poitelon Y, Genin E et al. Founder effect and estimation of the age of the c.892C > T (p.Arg298Cys) mutation in LMNA associated to Charcot-Marie-Tooth subtype CMT2B1 in families from North Western Africa. Ann Hum Genet 2008; 72:590-597.

98. Lin F, Worman HJ. Structural organization of the human gene encoding nuclear lamin A and nuclear lamin C. J Biol Chem 1993; 268:16321-16326.

99. Capell BC, Collins FS. Human laminopathies: nuclei gone genetically awry. Nat Rev Genet 2006; 7:940-952.

100. Tazir M, Azzedine H, Assami S et al. Phenotypic variability in autosomal recessive axonal Charcot-Marie-Tooth disease due to the R298C mutation in lamin A/C. Brain 2004; 127:154-163.

101. Leal A, Morera B, Del Valle G et al. A second locus for an axonal form of autosomal recessive Charcot-Marie-Tooth disease maps to chromosome 19q13.3. Am J Hum Genet 2001; 68:269-274.

102. Leal A, Huehne K, Bauer F et al. Identification of the variant Ala335Val of MED25 as responsible for CMT2B2: molecular data, functional studies of the SH3 recognition motif and correlation between wild-type MED25 and PMP22 RNA levels in CMT1A animal models. Neurogenetics. 2009.

103. Mittler G, Stuhler T, Santolin L et al. A novel docking site on Mediator is critical for activation by VP16 in mammalian cells. EMBO J 2003; 22:6494-6504.

CHAPTER 6

CREUTZFELDT-JAKOB DISEASE

Beata Sikorska,[1] Richard Knight,[2] James W. Ironside[2]
and Paweł P. Liberski*[,1]

[1]Department of Molecular Pathology and Neuropathology, Medical University Lodz, Lodz, Poland;
[2]National CJD Surveillance Unit, Western General Hospital, Edinburgh, UK
*Corresponding Author: Pawel P. Liberski—Email: ppliber@csk.umed.lodz.pl

Abstract: Creutzfeldt-Jakob disease (CJD), a neurodegenerative disorder that is the commonest form of human prion disease or transmissible spongiform encephalopathies (TSEs). Four types of CJD are known: Sporadic (sCJD), familial or genetic (gCJD); iatrogenic (iCJD) and variant CJD (vCJD). The latter results from transmission of bovine spongiform encephalopathy (BSE) from cattle to humans. The combination of PrP^{Sc} peptide (either 21 kDa or 19 kDa) and the status of the codon 129 of the gene (*PRNP*) encoding for PrP (either Methionine or Valine) is used to classify sCJD into 6 types: MM1 and MV1, the most common; VV2; MV2 (Brownell/Oppenheimer syndrome); MM2; VV1 and sporadic fatal insomnia, in that order of prevalence. Genetic CJD is caused by diverse mutations in the *PRNP* gene. The neuropathology of CJD consists of spongiform change, astro- and microgliosis and poorly defined neuronal loss. In a proportion of cases, amyloid plaques, like those of kuru, are seen. PrP immunohistochemistry reveals different types of PrP^{Sc} deposits—the most common is the synaptic-type, but perivacuolar, perineuronal and plaque-like deposits may be also detected.

INTRODUCTION

Creutzfeldt-Jakob disease (CJD) was first described by two German researchers in twenties of the last century.[1-3] Modern evaluation of historic Jakob's sections revealed that only 3 of 5 cases met recent criteria of CJD.[4] The Creutzfeldt's case did not belong to the category of CJD as currently understood. CJD was transmitted to chimpanzee by Gibbs, Gajdusek and others in 1968.[5]

CJD is the most common prion disease. It occurs as sporadic, familial, iatrogenic (infectious) and genetic forms. It is thus unique as a disease, both infectious and genetic.

Neurodegenerative Diseases, edited by Shamim I. Ahmad.
©2012 Landes Bioscience and Springer Science+Business Media.

CLINICAL FEATURES

Different prion diseases have somewhat different clinical profiles, but the overall typical picture is that of a uniformly progressive, fatal, encephalopathic illness with dementia, cerebellar ataxia and myoclonus being common. With no simple, absolute clinical diagnostic test for prion disease, definitive diagnosis requires neuropathology i.e., autopsy or brain biopsy. However, on the basis of internationally agreed clinical diagnostic criteria[6,7] relatively reliable clinical diagnoses can often be made.

There are no systemic abnormalities due to prion disease: No pyrexia and other index of infection and routine hematology, biochemistry and immunological tests are normal. Also there is no cerebrospinal fluid pleocytosis (although the total protein may be mildly to moderately raised). The overall diagnostic process is essentially the suspicion of prion disease based on clinical pattern, the exclusion of other diagnoses and supportive investigation findings (EEG, CSF and MRI findings which are not prion-disease specific and tonsil biopsy in the case of variant CJD).[7-11] These tests are essentially supportive tests when prion disease is already reasonably suspected; positive results must be obtained in the light of the overall clinical situation.

SPORADIC CJD (sCJD)

This is commonest form of CJD existing worldwide. Typically, it affects the middle-aged and elderly (median age at death around 65 years) and consists of a rapidly progressive dementia with cerebellar ataxia, visual disturbances and myoclonus terminating in an akinetic mute state. The median survival is only around 4 months in most countries with survival of more than two years being rare.[12] There are less common or atypical forms with different presentations and overall clinical pictures. The commonest variations in presentation are the so-called Brownell-Oppenheimer CJD (a pure cerebellar syndrome) and Heidenhain's syndrome (pure visual impairment leading to cortical blindness); they progress to a similar preterminal state as typical sCJD.[13,14] Other specific presentations exist but are exceptionally rare. Young age disease onsets (even in the second decade of life) occur but are rare and late age disease are more common. The duration of illness depends on age at onset, gender, *PRNP*-129 genotype and prion protein-type.[15] In some cases, typical features, such as myoclonus, never develop. These clinical variations are associated with pathological variations and the different clinico-pathological profiles correlate to some degree with *PRNP* codon 129 polymorphism and the prion protein-type (with the most typical forms associated with MMI and MVI characteristics).[16] Recent reports concerning the co-occurrence of more than one abnormal prion protein within a single brain have prompted to revise the clinco-pathological-molecular classification of sCJD.[17,18]

The recently described 'Protease Sensitive' prion disease is of uncertain nosological status but is a form of sporadic prion disease with distinctive neuropathological and prion protein characteristics.[19,20] Its clinical profile is not yet well delineated but it appears to have a relatively longer duration than typical sCJD.

In around two-thirds of sCJD cases, the electroencephelogram (EEG) shows characteristic periodic bi- or tri-phasic complexes at some stage of the illness.[8] The cerebrospinal fluid (CSF) 14-3-3 protein test is positive in the majority of cases. Cerebral magnetic resonance imaging (MRI) frequently shows basal ganglia (putamen and caudate) and cortical hyperperintesity, especially on FLAIR and DWI sequences.[7,9] These investigation results vary with respect to *PRNP*-129 genotype and prion protein-type.[21]

Neuropathology of Sporadic Creutzfeldt-Jakob Disease

sCJD comprises broad spectra of clinico-pathological variants. In nongenetic cases of prion diseases, neuropathological examination is the only way to establish the diagnosis of definite disease. The brain biopsy in all transmissible spongiform encephalopathies (TSEs), sCJD included, is extremely rarely undertaken so definite cases in most instances are recognized during autopsy.

The gross pathology of the brain in sCJD is not characteristic, most often a various degree of focal or diffuse brain atrophy is observed. On the contrary, the microscopic changes are very distinctive. The classic triad of histopathological findings for CJD consists of marked neuronal loss, spongiform change and astrogliosis. Additionally, various forms of deposits of PrP^{Sc} are observed.

Spongiform change (Fig. 1) is the most specific microscopic alteration in CJD and is characterized by a fine vacuolation of the neuropil of the gray matter with round or oval vacuoles varying from 20-200 microns in diameter. Ultrastructurally, the vacuoles are localized in cell processes, mainly dendrites and contain curled membrane fragments and amorphous material.[22] Infrequently, vacuolation may also be seen within the cytoplasm of larger neurons within the cortex. The vacuoles may appear in any layer of the cerebral

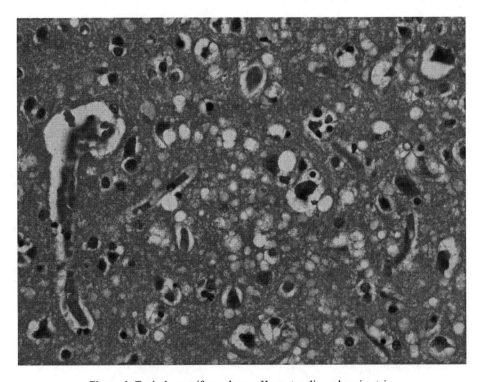

Figure 1. Typical spongiform change. Haematoxylin and eosin stain.

cortex. Distribution of spongiform change varies from case to case and between subtypes of sCJD but usually it is also observed in the basal ganglia, thalamus and the cerebellar cortex. In some cases, spongiform change may become confluent, resulting in large vacuoles which substantially distort the cortical architecture (status spongiosus).[23] Spongiform change must be distinguished from nonspecific "spongiosis" which may be observed in Alzheimer's disease, dementia with Lewy bodies and also in hypoxic damage or brain oedema.

The neuronal loss seems to be selective in prion diseases. GABAergic neurons are considered the main target of neuronal loss in experimental prion diseases and in CJD.[23,24] Interestingly, the hippocampus and dentate gyrus, which are most vulnerable areas in many neurodegenerative diseases, are relatively well preserved in most cases of CJD.[25]

Immunohistochemical reaction for the prion protein (PrP) is the gold standard for diagnosis of human prion diseases. Since most of anti-PrP antibodies do not discriminate between normal (PrP^c) and pathological (PrP^{Sc}) isoforms of prion protein, special pretreatments are required to eliminate reaction with PrP^C.[25] There are several patterns of PrP-immunoreactivity in human brains, these include: Synaptic pattern (fine dot-like deposits) (Fig. 2); pericellular/perineuronal pattern (dot-like deposits around neuronal perikarya) (Fig. 3); coarse deposits (granular or patchy/perivacuolar deposits) (Fig. 4); plaque-like deposits (not visible without PrP-immunohistochemistry and not showing congophilia or other tinctorial characteristics of amyloid) (Fig. 5); Kuru-plaques (visible in routine staining, congophilic and exhibiting tinctorial features of amyloid).

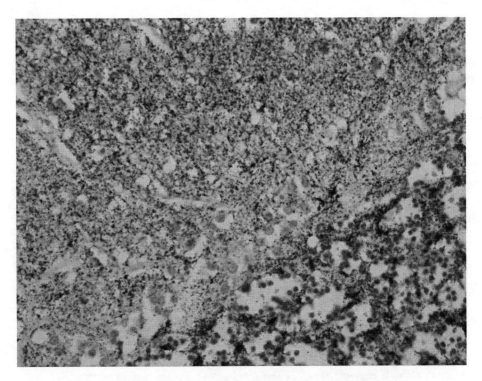

Figure 2. Synaptic pattern (fine dot-like deposits). PrP-immunohistochemistry.

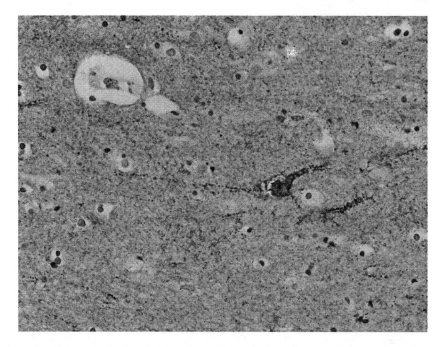

Figure 3. Pericellular/perineuronal pattern (dot-like deposits around neuronal perikarya). PrP-immunohistochemistry.

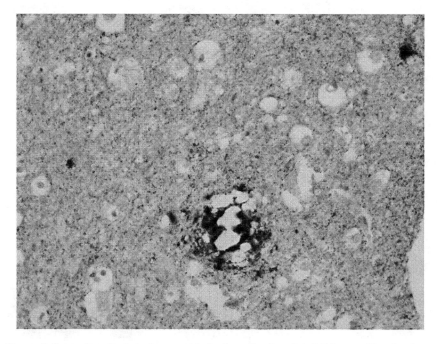

Figure 4. Coarse deposits (granular or patchy/perivacuolar deposits). PrP-immunohistochemistry.

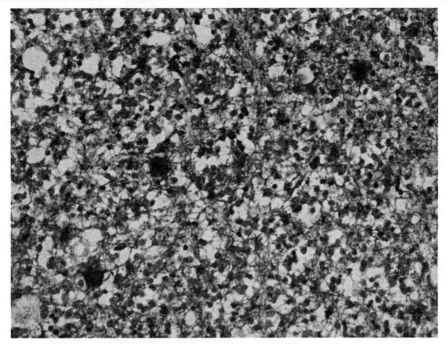

Figure 5. Plaque-like deposits (not visible without PrP-immunohistochemistry and not showing congophilia or other tinctorial characteristics of amyloid). PrP-immunohistochemistry.

It should be noted that both spongiform change and PrP-immunoreactivity may be focal and in some cases staining of several blocks is required.[25] In addition to the abovementioned patterns, PrP-immunoreactivity may be observed in neuronal perikarya as diffuse immunoreactivity, dot-like immunoreactivity or intracytoplasmic inclusions. Discrete PrP deposits were detected in the posterior root nerve fibres,[26] skeletal muscle[27] and in vessel walls.[28]

Six subtypes of sporadic Creutzfeldt-Jakob disease with distinctive clinico-pathological features have been identified largely based on two types of PrP^{Sc} and the status of polymorphic codon 129 of the prion protein gene (*PRNP*) that encodes either for methionine (M) or for valine (V).[16] Two major human PrP^{Sc} types were identified by Parchi et al;[16] Type 1 with a relative molecular mass of 21 kDa and the primary cleavage site at residue 82 and Type 2 with relative molecular mass of 19 kDa and the primary cleavage site at residue 97. The two PrP^{Sc} types, in conjunction with the codon 129 of *PRNP* genotype provided a molecular basis for the disease classification.[16,29] More recently, cases presenting more than one PrP^{Sc} type and mixed histopathological features were described leading to new classifications.[30,31] Currently, in addition to predominant pure subtypes of sCJD, mixed subtypes are also recognized.[18] In our opinion, for a routine neuropathological examination the classic Parchi and Gambetti classification[16] is the most useful. The six subtypes differ in distribution of pathological changes, clinical features and molecular characteristics.

Subtype 1: (sCJDMM1 and sCJDMV1)

This subtype is observed in patients who are MM homozygous or MV heterozygous at codon 129 of the PrP gene (*PRNP*) and carry PrP^{Sc} Type 1. Clinical duration is short, 3-4 months.[32] The most common presentation in sCJDMM1 patients is cognitive impairment leading to frank dementia, gait or limb ataxia, myoclonic jerks and visual signs leading to cortical blindness (Heidenhain's syndrome).

Neuropathologically, subtype 1 sCJD is characterized by the presence of fine spongiform change, astrogliosis and neuronal loss. Spongiform change affects all layers of the cerebral cortex except for the first; in the cerebellum, it affects the molecular layer. The cerebral neocortex, is more severely affected than basal ganglia, thalamus and the cerebellum while the brain stem is virtually spared. The hippocampal cortex is spared. PrP-immunoexoression shows synaptic pattern of staining. Most intensive immunolabelling is observed in the areas of most severe histopathological changes. The immunostaining is often not uniform and some large regions may remain unstained.[16,32]

Subtype 2: (sCJDVV2)

This is the second most common subtype of sCJD. It accounts for about 16% of all cases[32] and corresponds to the previously described cerebellar/ataxic variant. The mean age at onset is about 60 years and the clinical duration is approximately 6 months with a 3-18 month range. Ataxia is often among the presenting signs. Cognitive impairment, myoclonus and pyramidal signs affected the majority of the patients with progression of the disease.

The spongiform change often shows a laminar distribution and it is more severe in the frontal than in the occipital cortex which is relatively spared. In this subtype, the limbic structures including the hippocampal gyrus shows spongiform degeneration, although the lesions in the entorhinal cortex are more severe. The topography of the lesions shows that the caudal brain regions are more severely affected than the rostral regions. The PrP-immunohistochemical hallmark of this subtype is the presence of focal PrP^{Sc} aggregates that look-like plaques (plaque-like) but do not contain PrP amyloid. These plaque-like deposits are Congo red and thioflavine S negative and are not visible in a routine H and E staining. Other distinctive features are strong perineuronal pattern of immunoreactivity, with intense labeling along neuronal processes, especially prominent in the basal ganglia and thalamus. A diagnostic feature of this subtype is the immunostaining pattern of the cerebellum that shows intense immunostaining in the Purkinje and upper granule cell layers with the presence of numerous plaque-like formations.[16,32]

Subtype 3: (sCJDMV2)

This subtype accounts for 9-10% cases of sCJD and also corresponds to the ataxic variant (Betty Brownell/Oppenheimer syndrome). The duration of disease differs significantly from that of previous subtypes, with the average of 17 months. Ataxia also is the most common presenting sign, but cognitive and mental signs and symptoms are more common than in the sCJDVV2 patients.[32] Neuropathologically, this subtype is similar to sCJDVV2. The main difference is the presence of the kuru plaques in the Purkinje cell layer and superficial granule cell region of the cerebellum; and the lack of significant cerebellar cortical atrophy. Kuru plaques are typical amyloid, congophilic plaques similar to those of kuru. Furthermore, coarse spongiform change, which may be either focal or widespread, is occasionally present and brain stem lesions are less severe.[16,32]

Subtype 4: (sCJDMM2)

The fourth subtype is found in 2-8% of cases. The mean age at onset is 65 years and the average disease duration is 16 months. Cognitive impairment and aphasia are most often observed. Myoclonic jerks and pyramidal signs are also common in later stages of the disease. The characteristic feature of this type is spongiform change with large vacuoles (coarse spongiosis). The vacuoles are several times larger than the typical vacuoles of Type 1 sCJD and are widespread in the cerebral cortex, basal ganglia and thalamus. PrP-immunohistochemistry shows intense staining at the rim of the large vacuoles additionally, synaptic pattern may also be observed and some plaque-like deposits are not uncommon.[16,32]

Subtype 5: (sCJDVV1)

The most uncommon subtype affecting about 1% of all cases of sCJD. Average age at onset is 39 years (range, 24-49 years) and the disease duration is approximately 15 months. This subtype is frequently referred to as "early onset sCJD".

The neuropathological hallmark of this subtype is the dissociation between the severity of fine spongiform change, gliosis and occasionally neuronal loss and the PrP-immunostaining that is faint and of a synaptic pattern. Interestingly, the hippocampal cortex is more affected and the thalamus and the cerebellum are less affected brain regions.[16,32]

Subtype 6: Sporadic Fatal Insomnia (sFI)

Another rare subtype accounts for about 2% of all cases of sporadic prion disease. Only a few proven cases of this subtype have been reported. Its phenotype is indistinguishable from that of fatal familial insomnia (FFI), hence the name "sporadic fatal insomnia" (sFI) has been coined.[33,34] The major pathology is observed in the thalamus, especially the medial dorsal and anterior ventral nuclei, that show a profound neuronal loss and astrogliosis but generally without spongiform change. Other brain regions are less affected and PrP-immunoreactivity is minimal or even absent. Astrogliosis and neuronal loss are also minimal in basal ganglia and cerebellum.[16,32]

GENETICS OF PRION DISEASE (gPD)

This has a very varied clinico-pathological phenotype, partly depending on the underlying *PRNP* mutation but other factors are obviously relevant, including the *PRNP* 129 genotype.[35] Historically, it comprises three main clinico-pathological phenotypes: Genetic CJD (gCJD), Gerstmann-Sträussler-Scheinker disease (GSS) and fatal familial insomnia (FFI). Classical GSS presents as a progressive cerebellar ataxia with relatively late cognitive features and FFI begins with prominent sleep and autonomic disturbances.

In gCJD, the mean age at onset is slightly younger and illness duration is longer than in sCJD. The inheritance of gPD is autosomal dominant with variable (generally high) penetrance. However, a family history may be absent in around 40% of cases.[36] The commonest mutation disease is E200K-CJD which typically clinically resembles sCJD, although a polyneuropathy may be present. Given the variable clinical phenotypes of gPD, the potential similarity to sCJD and the frequently absent family history, *PRNP* mutation testing (which can be undertaken on a blood sample) is necessary to definitively distinguish gPD and sCJD; such testing should be considered in a wide variety of neuropsychiatric illnesses.

IATROGENIC CJD (iCJD)

This appears to have most commonly arisen from cadaveric-derived human growth hormone (hGH-CJD) (mostly in France, UK and USA) and dura mater grafts (mostly in Japan). The incubation period is potentially long but varies with cause (mean of 11 years in dura mater-related cases and 15 yrs in hGH-CJD). In general, it resembles sCJD with the age at onset reflecting the age of exposure and the incubation period. However, hGH-CJD tends to have a relatively young age at onset, reflecting the age at which hormone treatment is given and presents differently—as a progressive cerebellar syndrome, with relatively minor and late cognitive features.[37,38]

VARIANT CREUTZFELDT-JAKOB DISEASE (vCJD)

It affects significantly younger age group and progresses more slowly (in the UK: Mean age at onset, 28 years; mean duration of illness, 14 months) compared with sCJD. Notably, the initial symptoms are prominently psychiatric and behavioural with the absence of clear, specifically neurological features.[39] Painful sensory symptoms may be present.[40] As the illness progresses, cerebellar ataxia is usually prominent with cognitive impairment and involuntary movements, including chorea, dystonia and myoclonus.[41] All cases of definite and probable vCJD (as defined by the current internationally accepted diagnostic criteria) have been *PRNP*-129 MM individuals. There is a report of one possible case in an MV individual.[42] The incubation period of vCJD is expected to vary with *PRNP*-129 genotype (any effect on clinico-pathological expression is unknown; the reported MV case was clinically indistinguishable from the MM cases). Variant CJD has occurred as a secondary, iatrogenic illness, resulting from blood transfusion; these cases have been clinically identical to primary dietary cases.[43] The EEG usually does not show the periodic activity seen in sCJD; being reported in two cases but only in the very late disease stages.[44] The CSF 14-3-3 is positive in less than half the cases.[45] The cerebral MRI shows hyper intensity in the posterior thalamus (the 'Pulvinar Sign') in over 90% of cases, especially on FLAIR and DWI sequences.[10] In vCJD, in contradistinction to other prion diseases, a tonsil biopsy may show the disease-specific abnormality of PrP[Sc].[11]

vCJD was reported as a novel form of human prion disease in a series of 10 patients in the United Kingdom (UK) by Will et al in 1996.[46] Since then, additional cases have been identified both in the UK (currently 173 cases) and 47 cases in 10 other countries, with France having the second largest total (25 cases at time of writing). The incidence of vCJD has declined in the UK in recent years, although small numbers of new cases are still being identified.

Surveillance of CJD was re-instated in the UK in 1990 following the identification of an epidemic of a novel prion disease in cattle, bovine spongiform encephalopathy (BSE). The evidence for a link between BSE and vCJD was first suggested on the basis of epidemiology; however, subsequent laboratory and biological studies have indicated that the infectious agent responsible for vCJD is identical to the BSE agent.[47] vCJD is therefore unique amongst human prion diseases since it represents an acquired infection across a species barrier, namely from bovines to humans. The most likely route of primary infection in vCJD is by the oral route, via meat products contaminated with the BSE agent.[48]

Neuropathology of vCJD

The brain in vCJD is often unremarkable on macroscopic inspection, but in cases with a long disease duration (over 20 months) there may be both cerebellar and cerebral cortical atrophy.[49] The microscopic features of vCJD are distinct from other forms of human prion disease (Table 1): Large fibrillary amyloid plaques surrounded by a corona or halo of spongiform change, known as florid plaques, are the most striking feature (Fig. 6A). These are widespread throughout the cerebral cortex, particularly in the occipital cortex and are also present in the cerebellar molecular layer. Smaller fibrillary plaques without surrounding spongiform change are present in the basal ganglia, thalamus and cerebellar granular layer. Florid plaques are not specific for vCJD, as similar lesions have been identified in smaller numbers in dura mater-associated cases of iatrogenic CJD in Japan.[50]

Spongiform change in vCJD is most severe in the putamen and caudate nucleus, while neuronal loss and gliosis is most severe in the posterior thalamus, particularly the pulvinar (Fig. 6B).[49] The distribution of thalamic neuronal loss and gliosis correlates with the areas of hyper intensity seen on MRI scans (particularly in T2-weighted or FLAIR sequences) within the pulvinar in vCJD patients.[10] Immunohistochemistry for PrP shows a characteristic pattern of accumulation in the brain in vCJD. The florid plaques label intensely and this technique also demonstrates large numbers of smaller plaque-like lesions in clusters, which are not evident on routine stains. Pericellular accumulation of

Table 1. Neuropathological diagnostic criteria for vCJD

Cerebral and cerebellar cortices	Numerous florid plaques in routine stained sections
	Multiple small cluster plaques in PrP stained sections
	"Feathery" pericellular and pericapillary PrP accumulation
Caudate nucleus and putamen	Severe spongiform change
	Perineuronal and focal periaxonal PrP accumulation
Posterior thalamic nuclei	Marked neuronal loss and astrocytosis
	Scanty spongiform change and amyloid plaques
Brainstem and spinal cord	Perineuronal and granular PrP accumulation
Biochemistry	Type 2B PrP^{Sc} on Western blot analysis

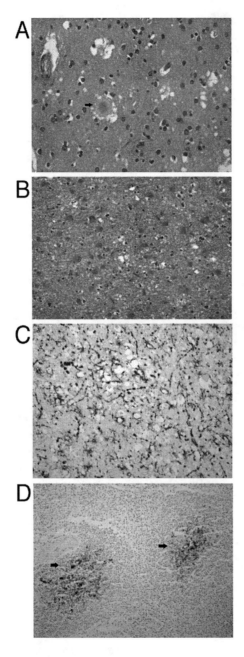

Figure 6. A) The florid plaque (arrow) in the occipital cortex in vCJD consists of a fibrillary amyloid core surrounded by a corona of spongiform change. Haematoxylin and eosin stain. B) The pulvinar nucleus in vCJD shows sever neuronal loss and astrocytosis, with little spongiform change and no amyloid plaque formation. Haematoxylin and eosin stain. C) Prion protein accumulation in the putamen in vCJD is perineuronal and periaxonal, with a beaded deposists in linear profiles. PrP-immunohistochemistry, KG9 anti-prion protein antibody. D) Prion protein accumulation in the tonsil in vCJD is concentrated within germinal centres (arrows), in follicular dendritic cells and macrophages. PrP-immunohistochemistry, KG9 anti-prion protein antibody.

PrP is also detected in an amorphous or feathery distribution around small neurones and astrocytes in the cerebral and cerebellar cortex. Occasional perivascular accumulation of PrP is also present around capillaries in the cerebral and cerebellar cortex, but there is no evidence of an amyloid angiopathy. The basal ganglia shows a perineuronal and periaxonal pattern of PrP positivity, often in a linear distribution (Fig. 6C).[49]

Peripheral Pathology in vCJD

In contrast to other forms of human prion disease, immunohistochemistry for PrP shows positivity in a variety of peripheral neural tissues, including sensory and autonomic ganglia.[51] There is also intense positivity for PrP within germinal centres in lymphoid tissues that colocalises to follicular dendritic cells (Fig. 6D).[11,51] Infectivity in both lymphoid and peripheral nervous tissues in vCJD has been demonstrated experimentally.[52] The accumulation of PrP positivity in lymphoid follicles allows tonsil biopsy to be used as an aid to premortem diagnosis in some cases of vCJD.[11] Furthermore, it has been shown that PrP positivity could be detected in lymphoid follicles in the appendix prior to the onset of clinical disease.[53] This finding has been explored in a retrospective prevalence study of PrP accumulation in appendix and tonsil tissues in the UK.[54,55] These studies suggest that the number of clinical cases of vCJD to date may not reflect accurately the true number of infections with BSE in the UK population and it is interesting to note that two of the positive appendix cases were subsequently shown to be *PRNP* codon 129 valine homozygotes.[54,56]

vCJD exhibits a characteristic PrP[res] isoform in the brain—Type 2 PrP[Sc] with predominance of the diglycosylated form, designated Type 2B (Fig. 7).[51] This has not

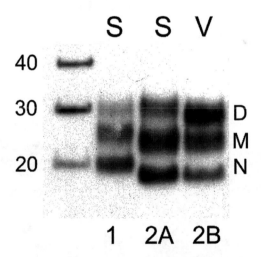

Figure 7. Western blot analysis of protease-resistant prion protein (PrP[res]) in frontal cortex samples from two cases of sporadic CJD (S), contrasted with a case of variant CJD (V). The sizes of the molecular weight markers are given in kDa and the three PrP[res] glycoforms are labelled as D (diglycosylated), M (monoclycosylated) and N (nonglycosylated). PrP[res] Type 1 (nonglycosylated band ~21kDa) and Type 2 (nonglycosylated band ~19kDa) are typical of different forms of sporadic CJD. The PrP[res] found in variant CJD has a nonglycosylated band of ~19kDa, but is characterised by a predominance of the diglycosylated form and termed Type 2B to distinguish it from the 2A isoform of sporadic CJD in which the mono- or nonglycosylated forms predominate.

Table 2. *PRNP* codon 129 genotype in prion diseases (probable and definite cases)

Codon 129 Genotype (%)	MM	MV	VV
Normal population	39	50	11
Sporadic CJD	63	19	18
Variant CJD	100	-	-

Abbreviations: M, methionine; V, valine.

been detected in sporadic CJD and is therefore useful as a biochemical marker of vCJD infection.[31] The widespread distribution of PrP[res] and infectivity in vCJD tissues indicates that the disease might be transmitted accidentally by medical or surgical procedures.[51,52] In support of this concern, there have been four incidences of apparent transmission of vCJD infectivity by blood transfusion, three of which resulted in the clinical onset of vCJD 6.5-8 years after transfusion in recipients who were *PRNP* codon 129 methionine homozygotes.[43] The fourth recipient died around 5 years after transfusion of an unrelated cause, but without clinical features of vCJD or any other neurological disease.[57] However, PrP[Sc] accumulation was detected biochemically and by immunohistochemistry in the spleen and a lymph node (but not in the brain or spinal cord), indicating asymptomatic vCJD infection in the recipient, who was a *PRNP* codon 129 heterozygote.[57] A similar case in an elderly adult haemophiliac patient in the UK has recently been described, where PrP[res] was detected biochemically in the spleen of a PRNP codon 129 heterozygote patient with no clinical or neuropathological evidence of vCJD.[58] The most likely source of vCJD infection in this case appears to be vCJD—contaminated plasma products.[58]

All patients with the histologically proven vCJD have been methionine homozygotes at codon 129 in the *PRNP* (Table 2), but one recent case of possible vCJD in a *PRNP* codon 129 heterozygous patient has been reported.[42] It remains to be established whether BSE infection in other human *PRNP* genotypes will result in a further increase in vCJD cases in the UK; continuing surveillance is required to address this important tissue.

CONCLUSION

CJD has been the first prion disease of worldwide occurrence and is unique. Causative agents are various forms of prion proteins which are mostly in misfolded form to generate the disease. It occurs either as sporadic, iatrogenic and variant (all are infectious) or a familial (genetic but also transmissible) forms. The diseases is strongly influence by the genetic backround.

REFERENCES

1. Jakob A. Uber eigenartige Erkrankungen des Zentralnervensystems mit bemerkenswertem anatomischem Befunde (spastische Pseudosclerose-Encephalopathie mit disseminierten Degenerationsherden). Deutsch Z Nervenheilk 1921(a); 70:132-146.
2. Jakob A. Uber eine der multiplen Sklero se klinisch nahestehende Erkrankung des Centralnervensystems (spastische Pseudosklerose) mit bemerkenswertem anatomischen Befunde. Med Klin 1921(b); 13:372-376.

3. Creutzfeldt HG. Uber eine egenartige herdformige Erkrankung des Zentralnervensystems. Z ges Neurol Psychiat 1920; 57:1-18.
4. Masters CL, Gajdusek DC. The spectrum of Creutzfeldt-Jakob disease and the virus induced subacute spongiform encephalopathies. In: Smith TJ, Cavanagh JB, eds. Recent Advances in Neuropathology. Edinburgh: Churchill Livingstone, 1982:139-163.
5. Gibbs CJ Jr, Gajdusek DC, Asher DM et al. Creutzfeldt-Jakob disease (spongiform encephalopathy): transmission to chimpanzee. Science 1968; 161:388-389.
6. The Revision of the Surveillance Case Definition for Variant Creutzfeldt-Jakob Disease (vCJD). World Health Organisation, 2002:30.
7. Zerr I, Kallenberg K, Summers DM et al. Updated clinical diagnostic criteria for sporadic Creutzfeldt-Jackob disease. Brain 2009; 132(10):2659-2668.
8. Zerr I, Pocchiari M, Collins S et al. Analysis of EEG and CSF 14-3-3 proteins as aids to the diagnosis of Creutzfeldt-Jakob disease. Neurology 2000; 55(6):811-815.
9. Collie DA, Sellar RJ, Zeidler M et al. MRI of Creutzfeldt-Jakob disease: imaging features and recommended MRI protocol. Clinical Radiology 2001; 56:726-739.
10. Collie DA, Summers DM, Sellar RJ et al. Diagnosing variant Creutzfeldt-Jakob disease with the pulvinar sign: MR imaging findings in 86 neuropathologically confirmed cases. Am J Neuroradiol 2003; 24:1560-1569.
11. Hill AF, Butterworth RJ, Joiner S et al. Investigation of variant Creutzfeldt-Jakob disease and other human prion diseases with tonsil biopsy samples. Lancet 1999; 353:183-184.
12. Knight RSG, Will RG. Prion diseases. JNNP 2004; 75(1):i36-i42.
13. Cooper SA, Murray KL, Heath CA et al. Sporadic Creutzfeldt-Jakob disease with cerebellar ataxia at onset in the United Kingdom. JNNP 2006; 77:1273-1275.
14. Cooper SA, Murray KL, Heath CA et al. Isolated visual symptoms at onset in sporadic Creutzfeldt-Jakob disease: the clinical phenotype of the "Heidenhain variant". Br J Opthalmol 2005; 89:1341-1342.
15. Pocchiari M, Puopolo M, Croes EA et al. Predictors of survival in sporadic Creutzfeldt-Jakob disease and other human transmissible spongiform encephalopathies. Brain 2004; 127:2348-2359.
16. Parchi P, Giese A, Capellari S et al. Classification of sporadic Creutzfeldt-Jakob disease based on molecular and phenotypic analysis of 300 subjects. Ann Neurol 1999; 46:224-233.
17. Cali I, Castellani R, Alshekhlee A et al. Co-existence of scrapie prion protein types 1 and 2 in sporadic Creutzfeldt-Jakob disease: its effect on the phenotype and prion-type characteristics. Brain 2009; 132:2643-2568.
18. Parchi P, Stranniello R, Notari S et al. Incidence and spectrum of sporadic Creutzfeldt-Jakob disease variants with mixed phenotypes and co-occurrence of PrP^{sc} types: an updated classification. Acta Neuropathol 2009; 118(5):659-671.
19. Head MW, Knight R, Zeidler M et al. A case of protease sensitive prionpathy in a patient in the UK. Neuropathol and App Neurobiol 2009; 35:628-632.
20. Gambetti P, Dong X, Yuan J et al. A novel human disease with abnormal prion protein sensitive to protease. Ann Neurol 2008; 63(6):697-708.
21. Meissner B, Kallenberg K, Sanchez-Juan P et al. MRI lesion profiles in sporadic Creutzfeldt-Jakob disease. Neurology 2009; 72:1994-2001.
22. Liberski PP YR, Gajdusek DC. The spongiform vacuole—the hallmark of slow virus diseases. In: Liberski PP, ed. Neuropathology of Slow Virus Disorders. Boca Raton: CRC Press, 1993:155-180.
23. Kovacs GG, Kalev O, Budka H. Contribution of neuropathology to the understanding of human prion disease. Folia Neuropathol 2004; 42 Suppl A:69-76.
24. Guentchev M, Hainfellner JA, Trabattoni GR et al. Distribution of parvalbumin-immunoreactive neurons in brain correlates with hippocampal and temporal cortical pathology in Creutzfeldt-Jakob disease. J Neuropathol Exp Neurol 1997; 56(10):1119-1124.
25. Budka H. Neuropathology of prion diseases. Br Med Bull 2003; 66:121-130.
26. Hainfellner JA, Budka H. Disease associated prion protein may deposit in the peripheral nervous system in human transmissible spongiform encephalopathies. Acta Neuropathol 1999; 98(5):458-460.
27. Glatzel M, Abela E, Maissen M et al. Extraneural pathologic prion protein in sporadic Creutzfeldt-Jakob disease. N Engl J Med 2003; 349(19):1812-1820.
28. Koperek O, Kovacs GG, Ritchie D et al. Disease-associated prion protein in vessel walls. Am J Pathol 2002; 161(6):1979-1984.
29. Parchi P, Castellani R, Capellari S et al. Molecular basis of phenotypic variability in sporadic Creutzfeldt-Jakob disease. Ann Neurol 1996; 39(6):767-778.
30. Hill AF, Joiner S, Wadsworth JD et al. Molecular classification of sporadic Creutzfeldt-Jakob disease. Brain 2003; 126:1333-1346.
31. Head MW, Bunn TJR, Bishop MT et al. Prion protein geterogeneity in sporadic but not variant Creutzfeldt-Jakob disease: UK Cases 1991-2002. Ann Neurol 2004; 55:851-859.

32. Gambetti P, Kong Q, Zou W et al. Sporadic and familial CJD: classification and characterisation. Br Med Bull 2003; 66:213-239.
33. Mastrianni JA, Nixon R, Layzer R et al. Prion protein conformation in a patient with sporadic fatal insomnia. N Engl J Med 1999; 340(21):1630-1638.
34. Parchi P, Capellari S, Chin S et al. A subtype of sporadic prion disease mimicking fatal familial insomnia. Neurology 1999; 52(9):1757-1763.
35. Kovacs GG, Trabattoni G, Hainfellner JA et al. Mutations of the prion protein gene: phenotypic spectrum. J Neurol 2002; 249:1567-1582.
36. Kovacs GG, Puopolo M, Ladogana A et al. Genetic prion disease: the EUROCJD experience. Hum Gen 2005; 118:166-174.
37. Brown P, Preece M, Brandel J-P et al. Iatrogenic Creutzfeldt-Jakob disease at the millennium. Neurology 2000; 55:1075-1081.
38. Brown P, Brandel J-P, Preece M et al. Iatrogenic Creutzfeldt-Jakob disease: the waning of an era. Neurology 2006; 67:389-393.
39. Spencer MD, Knight RSG, Will RG. First hundred cases of variant Creutzfeldt-Jakob disease: retrospective case note review of early psychiatric and neurological features. BMJ 2002; 324:1479-1482.
40. Macleod J, Stewart G, Zeidler M et al. Sensory features of variant Creutzfeldt-Jakob disease. J Neurol 2002; 249:706-711.
41. Zeidler M, Stewart GE, Barraclough CR et al. New variant Creutzfeldt-Jakob disease: neurological features and diagnostic tests. Lancet 1997; 350:903-907.
42. Kaski D, Mead S, Hyare H et al. Variant CJD in an individual heterozygous for PRNP codon 129. Lancet 2009; 374:2128.
43. Hewitt PE, Llewelyn CA, Mackenzie J et al. Creutzfeldt-Jakob disease and blood transfusion: results of the UK Transfusion Medicine Epidemiology Review study. Vox Sang 2006; 91:221-230.
44. Yamada M. The first Japanese case of variant Creutzfeldt-Jakob disease showing periodic electroencephalogram. Lancet 2006; 367:874.
45. Green AJE, Ramljak S, Muller WEG et al. 14-3-3 in the cerebrospinal fluid of patients with variant and sporadic Creutzfeldt-Jakob disease measured using capture assay able to detect low levels of 14-3-3 protein. Neuro Sci Lett 2002; 324:57-60.
46. Will RG, Ironside JW, Zeidler M et al. A new variant of Creutzfeldt-Jakob disease in the UK. Lancet 1996; 347:921-925.
47. Bruce E, Will RG, Ironside JW et al. Transmissions to mice indicate that "new variant" CJD is caused by the BSE agent. Nature 1997; 389:498-501.
48. Ward HJ, Everington D, Cousens SN et al. Risk factors for variant Creutzfeldt-Jakob disease: a case-control study. Ann Neurol 2006; 59:111-120.
49. Ironside JW, McCardle L, Horsburgh A et al. Pathological diagnosis of variant Creutzfeldt-Jakob disease. APMIS 2002; 11:79-87.
50. Shimizu S, Hoshi T, Homma M et al. Creutzfeldt-Jakob disease with florid-type plaques after cadaveric dura mater grafting. Arch Neurol 1999; 56:357-363.
51. Head MW, Ritchie D, Smith N et al. Peripheral tissue involvement in sporadic, iatrogenic and variant Creutzfeldt-Jakob disease: an immunohistochemical, quantitative and biochemical study. Am J Pathol 2004; 164:143-153.
52. Bruce ME, McConnell I, Will RG et al. Detection of variant Creutzfeldt-Jakob disease infectivity in extraneural tissues. Lancet 2001; 358:208-209.
53. Hilton DA, Fathers E, Edwards P et al. Prion immunoreactivity in appendix before clinical onset of variant Creutzfeldt-Jakob disease. Lancet 1998; 352:703-704.
54. Hilton DA, Ghani AC, Conyers L et al. Prevalence of lymphoreticular prion protein accumulation in UK tissue samples. J Pathol 2004; 203:733-739.
55. Clewley JP, Kelly CM, Andrews N et al. Prevalence of disease related prion protein in anonymous tonsil specimens in Britain: cross sectional opportunistic survey. Brit Med J 2009; 338:b1442.
56. Ironside JW, Bishop MT, Connolly K et al. Variant Creutzfeldt-Jakob disease: prion protein genotype analysis of positive appendix tissue samples from a retrospective prevalence study. Brit Med J 2006; 332:1186-1188.
57. Peden AH, Head MW, Ritchie DL et al. Preclinical vCJD after blood transfusion in a PRNP codon 129 heterozygous patient. Lancet 2004; 364:527-529.
58. Peden A, McCardle L, Head MW et al. Variant CJD infection in the spleen of a neurologically asymptomatic UK adult patient with haemophilia. Haemophilia 2010; 16:296-304.

CHAPTER 7

EPIGENETICS IN AUTISM AND OTHER NEURODEVELOPMENTAL DISEASES

Kunio Miyake,[1] Takae Hirasawa,[1] Tsuyoshi Koide[2] and Takeo Kubota*,[1]

[1]Department of Epigenetics Medicine, Faculty of Medicine, Interdisciplinary Graduate School of Medicine and Engineering, University of Yamanashi, Yamanashi, Japan; [2]Department of Mouse Genomics Resource Laboratory, National Institute of Genetics, Shizuoka, Japan
*Corresponding Author: Takeo Kubota—Email: takeot@yamanashi.ac.jp

Abstract: Autism was previously thought to be caused by environmental factors. However, genetic factors are now considered to be more contributory to the pathogenesis of autism, based on the recent findings of mutations in the genes which encode synaptic molecules associated with the communication between neurons. Epigenetic is a mechanism that controls gene expression without changing DNA sequence but by changing chromosomal histone modifications and its abnormality is associated with several neurodevelopmental diseases. Since epigenetic modifications are known to be affected by environmental factors such as nutrition, drugs and mental stress, autistic diseases are not only caused by congenital genetic defects, but may also be caused by environmental factors via epigenetic mechanism. In this chapter, we introduce autistic diseases caused by epigenetic failures and discuss epigenetic changes by environmental factors and discuss new treatments for neurodevelopmental diseases based on the recent epigenetic findings.

INTRODUCTION

Autism is a neuroimpairmental disease affecting approximately 30-60 per 10,000 children worldwide including Japan, with the male to female ratio of 2.5:1.[1] Autism has been classified into these three categories: (1) abnormal reciprocal social interactions including reduced interest in peers and difficulty in maintaining social interaction and failure to use eye gaze and facial expressions to communicate efficiently, (2) impaired communication including language delays, deficits in language comprehension and response to voices,

---

Neurodegenerative Diseases, edited by Shamim I. Ahmad.
©2012 Landes Bioscience and Springer Science+Business Media.

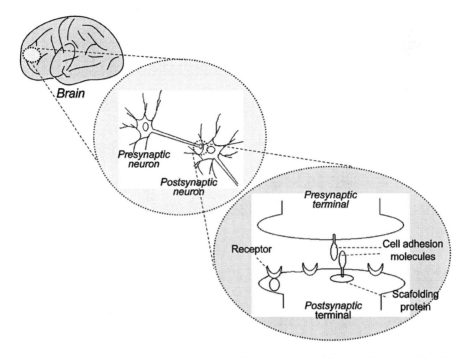

Figure 1. Location of the molecules in the synapse, which are associated with the pathogenesis of autism.

stereotyped use of words and phrases, poor pragmatics and lack of prosody, resulting in monotone and exaggerated speech patterns and (3) repetitive behaviors including motor stereotypes, repetitive use of objects, compulsions and rituals, insistence on sameness, upset to change and unusual or very narrow restricted interests.[2] Autism is diagnosed based purely on behavioral criteria such as DSM-IV99 (the diagnostic manual of the American Psychiatric Association) and ICD-10100 (the diagnostic manual of the World Health Organization.[3,4]

While various environmental factors are thought to contribute to the pathogenesis of autism,[5] recent genetic studies have revealed the involvement of genes in autism, in which mutations in the genes encoding neuronal molecules have been identified in a subset of autistic children. These molecules have been associated with synapse for neuronal connection, including synaptic scaffolding proteins, receptors and transporters on synapses and neuronal cell adhesion molecules.[6] These findings indicate that autism is a synapse disease[7] (Fig. 1).

The number of autistic children has been increasing recent years in Japan, USA and other countries.[8-11] This increase cannot be solely attributed to genetic factors, because it is unlikely that the mutation rate has suddenly increased in recent years. Therefore, environmental factors are more likely to be involved in this increase.[12]

Epigenetic is a DNA and histone-modification based genetic mechanism that is involved in neuronal gene(s) control in cells which are affected by environmental factors.[13] Thus, it is important to understand epigenetic mechanism when to study pathogenesis of autism.

In this chapter we show examples of neurodevelopmental diseases associated with epigenetics and environmental factors that affect epigenetic gene regulation and discuss future medical directions for autistic children based on recent epigenetic understanding.

CONGENITAL NEURODEVELOPMENTAL DISEASES ASSOCIATED WITH EPIGENETIC MECHANISM

Epigenetic gene control is an intrinsic mechanism for normal brain development[14] and abnormalities in the molecules associated with this mechanism are known to cause congenital neurodevelopmental diseases including autistism.[15-23] Genomic imprinting is the epigenetic phenomenon initially discovered in human diseases. In an imprinted gene, out of the two parental alleles, one allele is active and the other is inactive due to epigenetic mechanism such as DNA methylation. Therefore, defect in the active allele of the imprinted gene results in the loss of expression. This has been found in autistic diseases, Angelman syndrome and Prader-Willi syndrome.[15]

Since the number of genes in X chromosome is higher than those in Y chromosome, females (XX) have more genes than males (XY). To minimize this sex imbalance, one of the two X chromosomes in females is inactivated by epigenetic mechanism.[16] If X-inactivation does not properly occur in a female, such fetuses are believed to be aborted. This hypothesis is supported by the recent findings in cloned animals produced by somatic nuclear transfer in which failure of X-chromosome inactivation induces embryonic abortion.[17,18] Even if one of the X chromosomes is extremely small due to a large terminal deletion, so that over dosage effect of X-linked genes is small, the female show a severe congenital neurodevelopmental delay,[19] indicating that proper gene suppression by epigenetic mechanism is essential for the development.

DNA methylation is a fundamental step in epigenetic gene control and it is achieved by an addition of the methyl group (CH_3) to CpG dinucleotides mediated by DNA methyltransferase. Defect in this enzyme, DNMT3B, causes an ICF syndrome characterized by Immunodeficiency, centromere instability, facial abnormalities and mild mental retardation.[20-22]

Methyl-CpG binding proteins, which bind to the methylated DNA regions, are also important to control the gene expression. Mutations in one of the Methyl-CpG binding proteins, methyl CpG binding protein 2 (MeCP2), causes Rett syndrome, which is characterized by seizures, ataxic gait, language dysfunction and autistic behavior.[23-24] Therefore, it has been suggested that MeCP2 dysfunction, due to a mutation, leads to aberrant gene expression in the brain, which are associated with features in Rett syndrome including autism. Recent studies have shown that MeCP2 controls some neuronal genes, such as brain derived neurotrophic factor (BDNF), distal-less homeobox 5 (DLX5) and insulin-like growth factor binding protein 3 (IFGBP3).[25-28] These findings suggest that not only mutations,[6] but also epigenetic deregulation of genes that encode synaptic molecules may be attributed to autism.[29-31] Our recent findings, in which MeCP2 regulate a synaptic scaffolding protein and neuronal cell adhesion molecules and deficiency of MeCP2 leading aberrant expression of these genes, may support this hypothesis (our unpublished data).

ENVIRONMENTAL FACTORS WHICH AFFECT BRAIN FUNCTION VIA EPIGENETIC MECHANISM

In autism, both environmental factors (e.g., toxins, infections, ways of child rearing) and genetic factors (e.g., mutations in genes involved in synaptic system) have been discussed.[32-34] However, the biological mechanism that links these two factors has not been identified. Epigenetics may bridge these two[13] (see Fig. 2). Besides the intrinsic

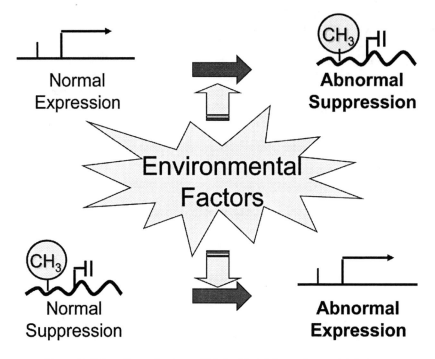

Figure 2. Epigenetic mechanism, which is affected by environmental factors.

(congenital) epigenetic defects (associated diseases described above), several lines of evidence suggest that extrinsic (environmental) factors, such as malnutrition, drugs, mental stress, maternal care and neuronal stimulation can alter the epigenetic status thereby affecting brain functions.[35-39] Therefore, it is intriguing to note that acquired neurodevelopmental diseases, including autistic diseases, may be the result of epigenetic deregulation caused by environmental factors. One example is drug addiction in which the gene expression in the dopaminergic and glutamatergic systems is mediated by epigenetic mechanism and, cocaine and alcohol alter the epigenetic state (chromatin structure) on a subset of genes, inducing drug addiction state[40,41] (Fig. 2).

Environmental factors via epigenetic mechanism are not always harmful. Imipramine, a major antidepressant, has recently been found to have the effect of restoring depressive state by alteration of epigenetic state (histone modification) of the *Bdnf* gene, leading to up-regulation of BDNF (brain-derived nutrition factor) in the hippocampus.[38] Valproic acid (VPA), a histone deacetylase (HDAC) inhibitor, is another example that alters the epigenetic state. VPA normalizes histone acetylation in the genes in the hippocampus, which lead to suppress seizure-induced cognitive impairment by blocking seizure-induced aberrant neurogenesis.[36] These indicate that chemicals that alter epigenetic gene expression, such as HDAC inhibitors, may become the candidates as new drugs for neurodevelopmental diseases including autism.[42]

The findings above are mainly obtained from animal experiments and there is little indication of its existence in humans. However, epigenetic difference is larger in monozygotic twins with older age than those with younger age,[43] suggesting that in humans epigenetic status may be altering during aging by environmental factors in humans.

EPIGENETIC TREATMENTS FOR PATIENTS
WITH NEURODEVELOPMENTAL DISEASES

Because epigenetic mechanism is reversible and variable unlike most mutational changes, epigenetic information can be potentially used as a disease-condition marker and a target for the treatment of the diseases.

Folic acid is the substrate for supplying methyl-residues to cytosine when is converted into methylated cytosine in DNA. Therefore, in order to maintain DNA methylation, proper intake of folic acid is important. In Japan young women who do not have sufficient amount of folic acid during pregnancy are now increasing and this increases the risk of having babies with neural tube defects.[44] Inappropriate supply of nutrients from mother to the fetus also increases the susceptibility of the fetus to develop diabetes mellitus by changing their epigenetic status.[45] These hypotheses are supported by experiments on rats, that protein restriction during pregnancy induces mal-nutrition and hyperlipedemia in the fetus and that supplementation of folic acid during pregnancy relieves these abnormal conditions, increases DNA methylation of the promoter regions of PPAR-alpha and glucocorticoid receptor genes in liver and leads to proper suppression of the genes.[35] These findings indicate that specific nutrient intakes may alter the phenotype of the offspring through epigenetic changes.

Royal jelly is known to have an effect of changing the phenotype from genetically identical female honeybees to a fertile queen; thus an effect of epigenetic changes. A recent study has revealed that royal jelly has effect of removing global DNA methylation, because silencing the expression of Dnmt3, a DNA methyltransferase, lead to a royal jelly-like effect on the larval development.[46]

Since 1980s folic acid has empirically been used for the treatment of autism and these studies have shown that it is effective only to a subset of the patients.[47-49] However, it is unknown whether folic acid leads epigenetic changes in autistic patients. Therefore, it is important to find if DNA methylation is altered by the treatment of folic acid. From the recent advancement in genome analysis technology, it has been possible to identify "folate-responsible genomic region (gene)" where DNA methylation is altered by administration of folic acid in the autistic children.[50,51] Such identified gene will be a therapeutic marker for folic acid treatment and the folic acid-sensitive (treatable) patients can be distinguished from the nonsensitive (nontreatable) patients (Fig. 3).

Folic acid-based treatment is relatively safe, since it is a nutrient. However, its effect is global, not gene specific. One ideal therapeutic way would be to target the specific gene(s) that can alter epigenetic status only in the genomic region (gene) associated with autistic pathogenesis. This treatment is now under development using PI polyamide that is designed on DNA sequence of a target gene.[52]

It has recently been discovered that DNA sequence is different in each neuron,[53] and epigenetic change underlies the somatic change.[54] This phenomenon is based on retrotransposition, in which a repetitive L1 sequence is inserted into various genomic regions when it is hypomethylated, which could alter the adjacent gene expression. Interestingly, the retrotransposition is activated by voluntary exercise (for example running).[55] Taken together, adequate intake of folic acid will ensure proper DNA methylation status and prevent from aberrant genomic change in neurons and mal-nutrition possibly induce hypomethylation, which leads to genomic changes in neurons, resulting in mental diseases including autism.

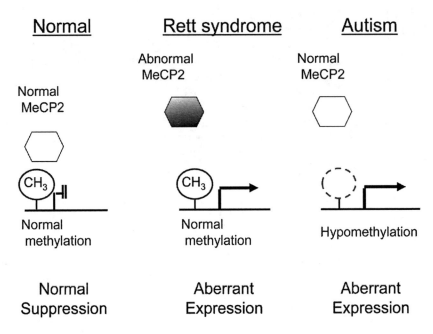

Figure 3. Pathogenesis of Rett syndrome and putative pathogenesis of autistic children to whom folic acid is effective. The gene with hypomethylation may be changed into normal hypermethylated status by administration of folic acid.

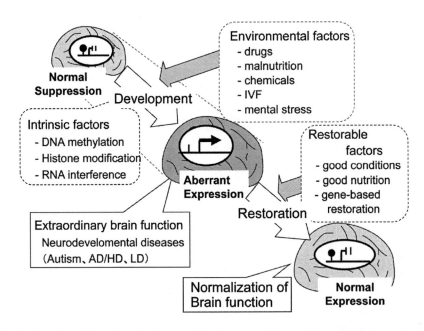

Figure 4. Overview of the epigenetic change and environmental factors. Note: epigenetic mechanism is reversible and restrable.

CONCLUSION

In summary, various environmental factors potentially rewrite epigenetic codes. However, epigenetic mechanism is reversible which is different from that of changes in genetic codes. Thus epigenetics are treatable and preventable when we will understand further as to which environmental factor(s) induce epigenetic changes and which gene is affected in this case (Fig. 4). In this context, epigenetic research is now important.

REFERENCES

1. Honda H, Shimizu Y, Imai MM. Cumulative incidence of childhood autism: a total population study of better accuracy and precision. Dev Med Child Neurol 2005; 47:10-18.
2. Silverman JL, Yang M, Lord C. Behavioural phenotyping assays for mouse models of autism. Nature Review Neuroscience 2010; 490:490-502.
3. American Psychiatric Association. Diagnostic and Statistical Manual of Mental Diseases. 4th ed, APA, Washington DC, 1994.
4. World Health Organization. The ICD-10 Classification of Mental and Behavioural Diseases. Geneva, Switzerland, 1992.
5. Herbert MR. Contributions of the environment and environmentally vulnerable physiology to autism spectrum diseases. Curr Opin Neurol 2010; 23:103-110.
6. Persico AM, Bourgeron T, Persico AM. Searching for ways out of the autism maze: genetic, epigenetic and environmental clues. Trends Neurosci 2006; 29:349-358.
7. Zoghbi HY. Postnatal neurodevelopmental diseases: meeting at the synapse?. Science 2003; 302:826-830.
8. Basic investigation report for handicapped children. report by Ministry of Health, Welfare and Labor, Japan. 2005 (in Japanese).
9. Yeargin-Allsopp M, Rice C, Karapurkar T. Prevalence of Autism in a US Metropolitan Area. JAMA 2003; 289:49-55.
10. Holden C. Autism Now. Science 2009; 323:565.
11. Fombonne E. Epidemiology of pervasive developmental diseases. Pediatric Research 2009; 65:591-598.
12. Hoekstra RA, Bartels M, Hudziak JJ. Genetics and environmental covariation between autistic traits and behavioral problems. Twin Res Hum Genet 2007; 10:853-860.
13. Qiu J. Epigenetics: unfinished symphony. Nature 2006; 441:143-145.
14. Guy J, Herndrich B, Hormes M. A mouse Mecp2-null mutation causes neurological symptoms that mimic Rett syndrome. Nat Genet 2001; 27:322-326.
15. Kubota T, Das S, Christian SL. Methylation-specific PCR symplifies imprinting analysis. Nat Genet 1997; 16:16-17.
16. Kubota T, Nonoyama S, Tonoki H. A new assay for the analysis of X-chromosome inactivation based on methylation-specific PCR. Hum Genet 1999; 104:49-55.
17. Xue F, Tian XC, Du F. Aberrant patterns of X chromosome inactivation in bovine clones. Nat Genet 2002; 31:216-220.
18. Nolen LD, Gao S, Han Z. X chromosome reactivation and regulation in cloned embryos. Developmental Biology 2005; 279:525-540.
19. Kubota T, Wakui K, Nakamura T. Proportion of the cells with functional X disomy is associated with the severity of mental retardation in mosaic ring X Turner syndrome females. Cytogenet Genome Res 2002; 99:276-284.
20. Okano M, Bell DW, Habe DA. DNA methyltransferases Dnmt3a and Dnmt3b are essential for de novo methylation and mammalian development. Cell 1999; 99:247-257.
21. Shirohzu H, Kubota T, Kumazawa A. Three novel DNMT3B mutations in Japanese patients with ICF syndrome. Am J Med Genet 2002; 112:31-37.
22. Kubota T, Furuumi H, Kamoda T. ICF syndrome in a girl with DNA hypomethylation but without detectable DNMT3B mutation. Am J Med Genet A 2004; 129:290-293.
23. Amir RE, Van den Veyver IB, Wan M. Rett syndrome is caused by mutations in X-linked MECP2, encoding methyl-CpG-binding protein 2. Nat Genet 1999; 23:185-188.
24. Chunshu Y, Endoh K, Soutome M. A patient with classic Rett syndrome with a novel mutation in MECP2 exon 1. Clin Genet 2006; 70:530-531.
25. Chen WG, Chang Q, Lin Y. Derepression of BDNF Transcription Involves Calcium-Dependent Phosphorylation of MeCP2. Science; 2003; 302:885-889.

26. Martinowich K, Hattori D, Wu H. DNA methylation-related chromatin remodeling in activity-dependent BDNF gene regulation. Science 2003; 302:890-893.
27. Horike S, Cai S, Miyano M. Loss of silent-chromatin looping and impaired imprinting of DLX5 in Rett syndrome. Nat Genet 2005; 37:31-40.
28. Itoh M, Ide S, Takashima S. Methyl CpG-binding protein 2, whose mutation causes Rett syndrome, directly regulates Insulin-like Growth Factor Binding Protein 3 in mouse and human Brains. J Neuropathol Exp Neurol 2007; 66:117-123.
29. Jiang YH, Sahoo T, Michaelis RC. A mixed epigenetic/genetic model for oligogenic inheritance of autism with a limited role for UBE3A. Am J Med Genet A 2004; 131:1-10.
30. Beaudet AL. Autism: highly heritable but not inherited. Nat Med 2007; 13:534-536.
31. Beaudet AL. Allan Award lecture: Rare patients leading to epigenetics and back to genetics. Am J Hum Genet 2008; 82:1034-1038.
32. Fombonne E. Epidemiology of pervasive developmental diseases. Pediatr Res 2009; 65:591-598.
33. Bailey A, Le Couteur A, Gottesman I. Autism as a strongly genetic disease: evidence from a British twin study. Psychol Med 1995; 25:63-77.
34. Zafeiriou DI, Ververi A, Vargiami E. Childhood autism and associated comorbidities. Brain Dev 2007; 29:257-272.
35. Burdge GC, Lillycrop KA, Phillips ES. Folic acid supplementation during the juvenile-pubertal period in rats modifies the phenotype and epigenotype induced by prenatal nutrition. J Nutr 2009; 139:1054-1060.
36. Jessberger S, Nakashima K, Clemenson GD. Epigenetic modulation of seizure-induced neurogenesis and cognitive decline. J Neurosci 2007; 27:5967-5975.
37. Weaver IC, Cervoni N, Champagne FA. Epigentic programming by maternal behavior. Nat Neurosci 2004; 9:847-554.
38. Tsankova NM, Berton O, Renthal W. Sustained hippocampal chromatin regulation in a mouse model of depression and antidepressant action. Nat Neurosci 2006; 9:519-525.
39. Ma DK, Jang MH, Guo JU. Neuronal activity-induced Gadd45b promotes epigenetic DNA demethylation and adult neurogenesis. Science 2009; 323:1074-1077.
40. Kumar A, Choi KH, Renthal W. Chromatin remodeling is a key mechanism underlying cocaine-induced plasticity in striatum. Neuron 2005; 48:303-314.
41. Pascual M, Boix J, Felipo V. Repeated alcohol administration during adolescence causes changes in the mesolimbic dopaminergic and glutamatergic systems and promotes alcohol intake in the adult rat. J Neurochem 2009; 108:920-931.
42. Renthal W, Nestler EJ. Epigenetic mechanisms in drug addiction. Trends Mol Med 2008; 14:341-350.
43. Fraga MF, Ballestar E, Paz MF. Epigenetic differences arise during the lifetime of monozygotic twins. Proc Natl Acad Sci USA 2005; 102:10604-10609.
44. Watanabe H, Fukuoka H, Sugiyama T. Dietary folate intake during pregnancy and birth weight in Japan. European J Nutr 2008; 47:341-347.
45. Park JH, Stroffers DA, Nicholles RD. Development of type 2 diabetes following intrauterine growth retardation in rats is associated with progressive epigenetic silencing of Pdx1. J Clin Invest 2008; 118:2316-2324.
46. Kucharski R, Maleszka J, Foret S. Nutritional control of reproductive status in haneybees via DNA methylation. Science 2008; 319:1827-1830.
47. Rimland B. Controversies in the treatment of autistic children: vitamin and drug therapy. J Child Neurol 1998; 3:68-72.
48. James SJ, Cutler P, Melnyk S. Metabolic biomarkers of increased oxidative stress and impaired methylation capacity in children with autism. Am J Clin Nutr 2004; 20:1611-1617.
49. Moretti P, Sahoo T, Hyland K. Cerebral folate deficiency with developmental delay, autism and response to folinic acid. Neurology 2005; 64:1088-1090.
50. Laird PW. Principles and challenges of genome-wide DNA methylation analysis. Nat Rev Genet 2005; 11:191-203.
51. Varley KE, Mitra RD. Bisulfite Patch PCR enables multiplexed sequencing of promoter methylation across cancer samples. Genome Res 2010. [Epub ahead of print].
52. Ohtsuki A, Kimura MT, Minoshima M. Synthesis and properties of PI polyamide–SAHA conjugate. Tetrahedron Lett 2009; 50:7288-7292.
53. Coufal NG, Garcia-Perez JL, Peng GE. L1 retrotransposition in human neural progenitor cells.. Nature 2009; 460:1127-1131.
54. Muotri AR, Chu VT, Marchetto MCN. Somatic mosaicism in neuronal precursor cells mediated by L1 retrotransposition. Nature 2005; 435:903-910.
55. Muotri AR, Zhao C, Marchetto MCN. Environmental Influence on L1 Retrotransposons in the Adult Hippocampus. Hypoocumpus 2009; 19:1002-1007.

CHAPTER 8

EPILEPSY AND EPILEPTIC SYNDROME

Tomonori Ono[1] and Aristea S. Galanopoulou*[,1,2]

[1]Saul R. Korey Department of Neurology, Albert Einstein College of Medicine, Bronx, New York, USA;
[2]Dominick P. Purpura Department of Neuroscience, Albert Einstein College of Medicine, Bronx, New York, USA
*Corresponding Author: Aristea S. Galanopoulou—Email: aristea.galanopoulou@einstein.yu.edu

Abstract: Epilepsy is one of the most common neurological disorders. In most patients with epilepsy, seizures respond to available medications. However, a significant number of patients, especially in the setting of medically-intractable epilepsies, may experience different degrees of memory or cognitive impairment, behavioral abnormalities or psychiatric symptoms, which may limit their daily functioning. As a result, in many patients, epilepsy may resemble a neurodegenerative disease. Epileptic seizures and their potential impact on brain development, the progressive nature of epileptogenesis that may functionally alter brain regions involved in cognitive processing, neurodegenerative processes that relate to the underlying etiology, comorbid conditions or epigenetic factors, such as stress, medications, social factors, may all contribute to the progressive nature of epilepsy. Clinical and experimental studies have addressed the pathogenetic mechanisms underlying epileptogenesis and neurodegeneration.

We will primarily focus on the findings derived from studies on one of the most common causes of focal onset epilepsy, the temporal lobe epilepsy, which indicate that both processes are progressive and utilize common or interacting pathways. In this chapter we will discuss some of these studies, the potential candidate targets for neuroprotective therapies as well as the attempts to identify early biomarkers of progression and epileptogenesis, so as to implement therapies with early-onset disease-modifying effects.

INTRODUCTION

Epilepsy is one of the most common neurological disorders affecting 50 million people worldwide.[1,2] It is a chronic neurological disorder characterized by a predisposition to generate recurrent unprovoked epileptic seizures.[3] An epileptic seizure is a transient

Neurodegenerative Diseases, edited by Shamim I. Ahmad.
©2012 Landes Bioscience and Springer Science+Business Media.

abnormal synchronization of neurons in the brain that disrupts normal patterns of neuronal activity (*electrographic seizure*) and may manifest with a variety of signs and symptoms (*electroclinical seizure*). These may include focal or generalized convulsive or atonic behaviors (i.e., tonic-clonic, myoclonic, tonic, atonic), paroxysmal abnormal sensory or autonomic symptoms, impaired consciousness or alertness (absence seizures, complex partial seizures). Epileptic syndrome, on the other hand, is used to denote "a complex of signs and symptoms that define a unique epilepsy condition".[4]

Epilepsy is a general term encompassing a variety of "*epilepsy diseases*", each of which is attributed to a single etiology.[4] Epilepsy can be associated with a constellation of neurobiologic, cognitive, psychological and social sequelae, which may greatly impact on the quality of life, especially of patients who do not respond to available therapies.[3] This has prompted the proposal to revisit the terminology of epilepsy and refer to it as a "disease" rather than a "disorder", so as to raise the level of awareness and urgency to find better ways to address these issues and alleviate or cure epilepsy.[5]

The progressive course of epilepsy (e.g., increase in frequency, duration and severity of seizures) and associated neurological dysfunction (e.g., physical, cognitive and behavioral impairment) can, in certain patients, mimic neurodegenerative diseases. The underlying pathogenetic mechanisms may relate to the progressive nature of epileptogenesis and its impact on the function and physiology of brain regions involved in cognition, the cumulative effect of the seizures and their therapies, epigenetic factors such as stress, changes in life style and environment.

Epileptogenesis is the process of forming a focus capable of generating spontaneous seizures.[6] Epileptogenesis evolves and progresses over several years in humans or months in rodents and may disrupt normal neuronal development and differentiation. In combination with the ongoing effects of seizures or epileptic discharges, epileptogenesis may result in developmental disabilities and cognitive decline in epilepsy patients. Given the chronicity and progressive nature of these processes, a key question in epilepsy research is to identify neuroprotective therapies that halt or reverse epilepsy and its sequelae. In this chapter, we will review the clinical and experimental evidence for the neurodegenerative aspects of epilepsy addressing the following questions:

1. Is epilepsy a progressive disorder with neurodegenerative features?
2. Which mechanisms underlie neurodegeneration in epilepsy?
3. Is it possible to diagnose and prevent these neurodegenerative aspects of epilepsy?

CLINICAL FEATURES OF EPILEPSY: IS EPILEPSY A NEURODEGENERATIVE DISORDER/SYNDROME?

Overview of Epilepsy and Epileptic Syndrome

Seizures are usually described as having focal (or partial) or generalized onset; however in certain cases the onset cannot be readily determined. Focal-onset seizures are generated by abnormal activity stemming from one brain region. They are further classified as simple (with intact consciousness) and complex partial seizures (with altered level of consciousness). If seizures arise or engage bilaterally distributed networks they are described as generalized, such as generalized tonic-clonic, absence, atonic or myoclonic seizures.[4,7]

The etiologies of epilepsy and epileptic syndromes are diverse. Brain lesions, including malformation of cortical development, tuberous sclerosis complex, neoplasms and hypothalamic hamartomas are major causes of drug-resistant epilepsy and often required surgical interventions. Genetic mutations, such as in sodium channels or γ-aminobutyric acid (GABA) receptor subunits, have been linked to epileptic syndromes, like juvenile myoclonic epilepsy or Dravet syndrome.[8-10] Other etiologies include infections of central nervous system (e.g., meningitis and encephalitis), vascular disease (e.g., cerebral infarction and hemorrhage), traumatic brain injury and other neurodegenerative disease (e.g., dementia and multiple sclerosis). When the etiology is known, epilepsy is categorized as symptomatic or of structural/metabolic etiology if it is unknown but suspected as cryptogenic. The term idiopathic has been used so far to denote epilepsies of presumed genetic etiology.[7,11] Epileptic syndromes are also categorized by age of onset, i.e., in neonatal, infantile, juvenile, adult, or seizure type i.e., myoclonic, sensory, absence.

The course of epileptic syndromes and their response to the available therapies differ. Seizures can be controlled with appropriate drugs in approximately 70% of cases.[12,13] However, some of them cannot discontinue their drugs as seizures may recur. The remaining 30% of patients are intractable with standard medical treatment and sometimes require surgical interventions. This medically intractable subpopulation of patients is also more likely to manifest the adverse cognitive sequelae. The choice and efficacy of available therapies differ among the various epilepsy types, as they exhibit distinct pathogenetic mechanisms. In the following sections we will focus more on the research stemming from temporal lobe epilepsy (TLE), one of the most common focal-onset epilepsies, as it has been more widely studied both clinically and experimentally and we will sporadically refer to other epileptic syndromes, as needed.

Progressive and Neurodegenerative Aspects in Temporal Lobe Epilepsy (TLE)

TLE is one of the common focal (localization-related) types of epilepsy in adult patients. Hippocampal sclerosis is the characteristic pathological finding in TLE (then called mesial temporal lobe epilepsy (MTLE)) and particularly so in the medically-refractory subpopulation of TLE patients. Characteristic semiologies of the observed seizures in TLE include aura of unusual sensations, fear, feeling of prior familiar experience (déjà vu), peculiar odor and abdominal sensation (epigastric aura), alteration of consciousness, motionless staring, automatism (unusual and purposeless movement) of mouth and extremities, post-ictal confusion.[14-16] Spontaneous seizures in MTLE begin usually in childhood and adolescence between 4 and 16 years of age.[17] A history of febrile convulsions or status epilepticus within the early years of life is retrospectively obtained in majority of patients.[14,16] Once habitual seizures begin, they often initially respond to antiepileptics and may even remit for many years. However, after this "silent" period, seizures recur and may then become drug-resistant.[14,16] Therefore, it is assumed that epileptogenesis is maintained and progressing during this period.

Cognitive performance, learning and long-term memory are often affected in TLE, since the epileptogenic focus is located in the vicinity of one of the primary brain regions involved in memory. Verbal memory deficit has been more commonly associated with language-dominant sided TLE and nonverbal memory is affected more in nonlanguage-dominant sided TLE.[18-20] Importantly, age at epilepsy onset and longer duration of illness also influence the cognitive impairment of TLE patients.[19] Patients with childhood-onset TLE have reduced total white matter volume which is associated with poorer cognitive status.[21] Children with seizure onset before 5 years of age have lower IQ regardless of the type of seizures.[22] Development of hippocampal sclerosis and atrophy and psychometric intelligence in patients with chronic TLE have been correlated with epilepsy duration[19] implying a progressive and neurodegenerative course of TLE.

Progressive and Neurodegenerative Aspects of Pediatric Epilepsies: Evolution of Seizures and Developmental Factors

Epilepsies with onset in the early stages of life have multivariate etiologies, presentations and prognosis. Of particular interest are the early life epileptic encephalopathies, which may significantly impair cognitive development, leading to language or more global regression. If not appropriately treated, residual intellectual disabilities may persist.[23-26] Among these catastrophic types of epilepsies, infantile spasms or West syndrome has attracted particular interest due to its characteristic presentation, modes of treatment and impact on the infant's development, should it not respond to therapy. West syndrome presents with a unique seizure type, infantile spasms, which are clusters of flexor, extensor, or mixed flexor-extensor spasms; a distinct electroencephalographic pattern called hypsarrhythmia and psychomotor delay/arrest.[11,27] In the majority of patients (approximately 60-70% in various cohorts), a specific underlying pathology can be found, rendering them into the symptomatic or structural/metabolic group.[27-31] Control of spasms may be obtained with some medications such as adrenocorticotrophic hormone and vigabatrin, but recurrence of spasms, evolution to other type of seizure and psychomotor retardation are frequently seen. Infantile spasms sometimes develop to Lennox-Gastaut syndrome, which has also been associated with intellectual and developmental disability. However, early cessation of spasms and normalization of the electroencephalogram (EEG) with treatment as well as good neuropsychological status prior to the onset of spasms correlated with favorable prognosis.[28,32-37] These emphasize that ongoing unremitting seizures and epileptic activity as well as the underlying pathologies may contribute to the ongoing cognitive deterioration. They also underline the feasibility of identifying new, more potent disease-modifying therapies for this condition.

NEURODEGENERATIVE CHANGES IN EPILEPSIES: CELLULAR AND MOLECULAR MECHANISMS

Cellular and Molecular Mechanisms of Epileptogenesis

As epileptogenesis develops, numerous changes at the cellular or synaptic level occur that ultimately lead to the formation of abnormal neuronal circuits with increased excitability. These include neuronal cell loss, gliosis, increased expression of immediate-early genes like c-fos and c-jun, neurogenesis, synaptogenesis, alterations in excitatory and inhibitory cell signaling, inflammatory mediators, changes in voltage-gated ion channels and autoimmune processes.[38]

Neuronal cell loss commonly observed in human epileptic tissue obtained from resective surgeries or in animal models of status epilepticus and epilepsy. Neuronal damage can occur due to excitotoxic effects of glutamate. Glutamate is an excitatory neurotransmitter in the brain that is released during prolonged seizures or other pathological conditions generating a surfeit of neuronal activity.[39] Glutamate-induced excitotoxicity is mediated by intense depolarization, calcium influx leading to activation of Ca2$^+$ dependent intracellular signals, oxidative stress/free radical damage and activation of apoptotic pathways such as caspase activity, P53 stress response and the proapoptotic Bcl protein.[39] The degree and distribution of neuronal loss in the brain is not uniform and depends upon the type or model of epilepsy, age, sex, hormonal milieu among other modifiers.[40-44]

Gliosis is often observed in the human tissue of epileptic focus. Animal experiments using chemically-induced status epilepticus demonstrated an activation and increased presence of astrocytes and microglia.[45] These glial cells are suggested to play roles in the development of epilepsy. Through releasing cytokines, such as interleukin 1β (IL-1β) and tumor necrosis factor-alpha (TNFα), glial cells cause chronic inflammatory state in the epileptic focus which may further influence neuronal excitability and survival.[46,47] In addition, glial cells may increase extracellular K$^+$ and glutamate,[48] neurogenesis and tissue remodeling[49] further promoting epileptogenesis.

Gene expression profiling experiments have associated different classes of genes with epileptogenesis: Immediate early genes, genes involved in calcium homeostasis, intra/extracellular signaling, neuronal/synaptic transmission, morphology, cell cycle/ fate, injury/survival, metabolism as well as genes of yet unknown function.[50] More than 2000 genes have been linked with TLE.[50]

A classical example of synaptic remodeling and aberrant circuitry formation is the mossy fiber sprouting observed in both human and experimental models of TLE. The mossy fibers, axons of the dentate granule cells, normally terminate their synaptic terminals into the dentate hilus and CA3 subfield (stratum lucidum). After precipitating insults or repeated seizures, the mossy fibers are reorganized and sprout into the inner molecular layer of the dentate gyrus to form new synaptic terminals with dendrites of interneurons (like basket cells) and primarily spine and dendrites of granule cells.[51] This synaptic reorganization may be triggered by neuronal cell loss and, at least during the early period after seizure, it may physiologically restore the inhibitory input to the granule cells, by forming excitatory connections upon the basket inhibitory interneurons.[51] However, in later periods, it may result into recurrent excitation of granule cells, promoting excitability. It has been estimated that a sprouted granule cell develops about 500 newly formed synaptic contacts with granule cells and less than 25 contacts upon interneurons[52] suggesting that this formation ultimately results into a primarily recurrent excitatory collateral circuit. The mossy fiber sprouting after the precipitating injury or recurrent seizures is found in not only the inner molecular layer of the dentate gyrus but also at other hippocampal regions, including dentate hilus, stratum oriens of CA3 subfield.[53,54] Thus, this synaptic reorganization is hypothesized to render hippocampus hyperexcitable.

The balance between excitatory (i.e., glutamatergic) and inhibitory (i.e., GABAergic) neuronal functions plays a crucial role in the development of epilepsy. Repeated seizures, especially if prolonged and certain types of initial precipitating insults may lead to a loss of GABAergic inhibitory interneurons in the hippocampus.[51] Functional changes may occur however even in the absence of cell loss and with seizures of lesser severity. Only a few seizure-like ictal episodes can be sufficient to cause fast and lasting enhancement in synaptic signal transmission in the hippocampal networks (long-term potentiation) through shifting the excitation/inhibition balance towards a more excitatory state.[55] This shift was attributed to both an enhancement of non-NMDA glutamatergic transmission but also a positive shift in the reversal potential of GABA to less hyperpolarizing inhibitory potentials. Similar shift to more excitatory state of neurotransmission is also found in malformations of cortical development and tuberous sclerosis complex, which are common causes of medically intractable epilepsy. Specifically, a decrease in GABAergic interneurons, decrease in GABA$_A$ receptors and their subunits, a reduction in L-glutamic acid decarboxylase isoforms (enzyme synthesizing GABA), or vesicular GABA transporter, increase in postsynaptic glutamate receptors and their subunits have been described.[56-58] Additionally, regulation of inhibitory or excitatory receptors can be effected by transcriptional factors, neuropeptides or secreted molecules that become activated in seizures. An example is the regulation of GABA$_A$ signaling through brain derived neurotrophic factor (BDNF), both at the level of GABA$_A$ receptor expression as well as the regulation of cation-chloride cotransporters that control their activity.[59,60]

Functional changes play a larger role in early life epileptogenesis, when the brain is not only more resistant to seizure-induced injury but also normally operates under a different excitation-inhibition setpoint.[61-63] Epileptogenesis is developmentally regulated. There are "critical periods" when an insult may have more disruptive effects upon neuronal differentiation and plasticity that may ultimately lead to epileptogenesis or cognitive deficits.[64,65] These do not always result in hyperexcitability, just as early life seizures do not always lead to subsequent epilepsy. Changes, however, in the quality of synaptic input may alter the way neurons communicate and networks operate, contributing to the observed cognitive impairment that follows early life seizures. The observed changes in inhibitory and excitatory neurotransmission following early life seizures are age, model, sex and cell-type specific but can also be modified by epigenetic factors and other stressors.[9,66-69]

The complexity of these processes becomes more evident when one considers the dynamic changes that normally occur in the developing brain. One example is the substantia nigra, which is involved in the control of seizures but also in cognition, as it acts as Go/No Go gatekeeper in networks that determine both the function of the prefrontal cortex as well as of seizure control.[70] During development, the substantia nigra undergoes functional changes that depend upon age and sex but also experience, i.e., early life seizures or stress.[71,72] As a result, the communication protocols that control the activity of these seizure-control and Go/NoGo decision centers may become dysfunctional, contributing to the increased susceptibility to subsequent epilepsy and cognitive deficits that are observed in patients with early life epilepsies.

An important signaling pathway that controls development is the GABA$_A$ signaling cascade. Early in life, GABA$_A$ergic currents are depolarizing and promote calcium entry into the cell which activates calcium-sensitive differentiating processes like cellular proliferation, migration, neuronal differentiation and synaptic integration.[61,62] Precocious termination of these depolarizing GABA$_A$ currents may occur after seizures or stress, in a sex, age and cell type dependent manner.[9] This may disrupt the normal developmental processes, resulting in morphological abnormalities, such as defective arborization and dendritic spine formation of the neurons.[73,74] Such changes can affect information reception and processing, further contributing to cognitive changes.

Structural and Functional Alterations in Hippocampal Sclerosis and MTLE

Hippocampal sclerosis is the pathological signature of MTLE. Its diagnostic pathologic features include loss of neurons at specific hippocampal regions (dentate hilus, CA1 pyramidal neurons, although CA3 and CA4 regions may also be affected), reactive gliosis in the hippocampus, reorganization of synaptic connections, including but not limited at the mossy fibers and dentate granule cell dispersion.[75,76] Extrahippocampal lesions or atrophy may also be noted as well as dysplastic neurons.[75-77]

A fundamental question has been whether hippocampal neuronal loss and sclerosis are the "cause" or "effect" of seizures. Clinical, neuroimaging and pathological studies favor the hypothesis that hippocampal sclerosis can be an acquired event, following prolonged seizures. Quantitative magnetic resonance imaging (MRI) have shown a consistent relationship between the degree of hippocampal sclerosis and the duration/severity of epilepsy or the total number of generalized seizures in some studies[78-81] but not in others.[82,83] In the study by Mathern et al[84] hippocampal sclerosis was strongly linked with initial precipitating injuries in both TLE and extra-temporal seizure patients. Early life frequent seizures were associated with abnormal postnatal granule cell development and aberrant axon sprouting, rather than with neuronal loss. However, in TLE patients, longer seizure durations were independently associated with decreased neuronal densities in all hippocampal subfields, but this occurred over several decades.[84] The hypothesis put forth was that the initial precipitating insult, during a critical period of development, may act as a "first hit". This primes the hippocampus and when the first seizure comes, it responds with progressive neuronal loss, gliosis and reorganization, eventually leading to epilepsy.[84]

Animal models of epilepsy have also provided ample evidence that hippocampal sclerosis results after prolonged seizures[85-89] and that this effect is age, sex and experience dependent.[40,41,43,44] However, prospective studies are needed to determine the exact temporal sequence of events and clarify whether, in some patients, pre-existing hippocampal pathologies may predispose patients to develop seizures and subsequent epilepsy.

Development of Post-Traumatic Epilepsy

Post-traumatic epilepsy is a major morbidity representing 5-6% of all epilepsy types.[90] Immediate (<24 hours) or early (<1 week) seizures are thought to be a nonspecific response to the injury. They may however raise the risk for subsequent epilepsy.[90,91] Recurrent seizures may appear after a latent period of at least several months. Post-traumatic lesions (contusions, intracerebral hematoma or subarachnoid hemorrhage) are risk factors of post-traumatic epilepsy.[91] Damage by free radicals caused by iron deposition from extravasated blood as well as excitotoxicity due to accumulation of glutamate have been postulated as basic mechanisms of epileptogenesis in post-traumatic epilepsy.[92] Indeed, intracortical injection of iron (ferric chlorides) in the rat brain has been used as a post-traumatic epilepsy model.[93,94] Other animal models of post-traumatic epilepsy include the lateral fluid percussion and the controlled cortical impact models. These demonstrate neurodegeneration, neurogenesis, astrocytosis, microgliosis, axonal, myelin injury, axonal sprouting, vascular damage and angiogenesis in the injured cortex, perifocal area, underlying hippocampus and/or thalamus.[95-98] Although the hippocampus is not the direct focus of injury, hippocampal hyperexcitability or disinhibition has also been reported in the lateral fluid percussion model.[99,100] Similar to humans, epilepsy can progressively develop after an initial latent phase and may follow the physical recovery of the animals from the post-traumatic somatomotor deficits.[96,98]

Development of Epilepsy in Neurodegenerative Disorders

Dementia is a risk factor for seizures and epilepsy. At least 10-22% of patients diagnosed with Alzheimer's disease may have one unprovoked seizure during the course of their disease.[101,102] The incidence of epilepsy in Alzheimer's disease varies between 3.6-7%.[102] Accumulation of β-amyloid may increase excitability and cause seizures.[103]

Palop et al recently demonstrated epileptic phenotype in an animal model of Alzheimer's disease.[104] These provide further proof of principle that the pathways primarily responsible for neurodegeneration and epileptogenesis interact or converge.

CURRENT STATE OF DIAGNOSIS AND TREATMENT OF NEURODEGENERATION IN EPILEPSY

Clinical Biomarkers of Epileptogenesis: Can We Clinically Detect Neurodegenerative Changes?

To cure epilepsy, both epileptogenesis and the associated neurodegeneration have to be stopped and if possible reversed. This will require early detection through biomarkers that can reliably predict progression to each of these two processes. Increasing interest has been placed upon the utilization of neuroimaging abnormalities as biomarkers of progression. In the lithium-pilocarpine model of status epilepticus, abnormal T2-weighted signal in the hippocampus correlated with the ongoing edema and injury during the early phase of epileptogenesis, prior to the appearance of spontaneous seizures. Furthermore, the same authors also showed a good predictive value for the increased T2 relaxation time in the piriform and entorhinal cortices.[105,106] In humans, early detection of MRI findings suggestive of hippocampal sclerosis have been sporadically reported after prolonged seizures and prior to the onset of spontaneous seizures.[107-109] Prospective randomized studies are needed to investigate the temporal relationship of first appearance of hippocampal sclerosis versus the onset of epilepsy. The availability of more advanced and potentially more sensitive neuroimaging methods, such as MR spectroscopy,[110-115] diffusion tensor imaging [105,110,116-121] or functional neuroimaging such as positron emission tomography[122-125] may enhance our ability to detect epilepsy-relevant pathology in MRI-negative epileptogenic foci.

The role of the EEG as a biomarker has been long investigated. Although EEG abnormalities that increase the risk for subsequent epilepsy after an initial precipitating event have been identified,[126-130] they do not yet carry the diagnostic sensitivity needed to initiate therapeutic or neuroprotective interventions. To identify and further investigate the validity of candidate biomarkers in predicting progression and outcome, a multi-center prospective study is ongoing, in which children with febrile status epilepticus are being followed up (FEBSTAT study).[131]

Do Current Treatments Improve Neurodegenerative Consequence in Epilepsy?

At present the existing antiepileptic therapies aim to stop seizures, as there are no documented pure neuroprotective therapies. Given the substantial evidence that cognitive impairment correlates with seizure severity, it is hoped that seizure control will also improve or halt the progression of cognitive decline.[132] Indeed, the clinical experience with the medical treatment of epileptic syndromes, even of the more devastating pediatric ones, shows that early cessation of spasms and seizures with appropriate drugs also improves neurodevelopmental outcome.[23-26,28,32-37] This may also be partially due to the fact that antiepileptic drugs directly inhibit signaling pathways that promote neurodegeneration, such as the glutamatergic pathway. Seizure control can therefore be an indirect neuroprotective therapy.[133]

Even in patients with medically-intractable forms of epilepsy who undergo surgical resection procedures to control their seizures, surgery may improve cognitive functions. This has been reported in both adults and children after temporal lobectomies[20,134-136] or even functional hemispherectomy for more dramatic forms of epilepsy, like Rasmussen's encephalitis.[137,138]

However, many of these patients may still be faced with residual deficits. In others, especially the very young patients, the benefit of stopping seizures may need to be weighted against the potentially adverse effects of certain antiepileptic therapies on brain development.[139] It is therefore urgent to identify new and safer neuroprotective therapies. These may involve benign lifestyle modifications, such as environmental enrichment,[140] or more specialized therapies targeting the signaling pathways and metabolic processes implicated in neurodegeneration and cognitive dysfunction.

Potential Treatments Suppressing Neurodegeneration and Epileptogenesis

A shift in the focus of experimental studies has been made, redirecting efforts to identify new neuroprotective and anti-epileptogenic therapies rather than simply antiepileptic ones. The current approaches target the known cellular and molecular mechanisms of epileptogenesis and/or eurodegeneration and are listed below.[6,141]

1. Among the antiepileptic drugs, compounds with potential as anti-epileptogenic therapies include phenobarbital,[142] valproic acid,[143] levetiracetam,[144] and ethosuximide.[145] However, some of these observations are model-specific. Ketogenic diet also has shown neuroprotective properties.[146] Documentation of anti-epileptogenic and disease-modifying effect in these treatments needs more confirmation with clinical trials.

2. Drugs with antioxidant properties or actions as free radical scavengers include lipoic acid,[147] adenosine,[148] melatonin,[149] edaravone,[150,151] the antiepileptic drug zonisamide,[152] and vitamins C and E.[153]

3. Strategies targeting molecular pathways involved in neurodegeneration[141] include: (a) stabilization of mitochondrial membrane with neuro-immunophilins (like FK506)[154] and NMDA receptor antagonists,[155] (b) exogenous administration of neurotrophic factors such as fibroblast growth factor 2, brain derived neurotrophic factor[156] and neuropeptide Y;[157] (c) molecular manipulation of glutamate receptors (blockade of group I and activation of group II and/or III metabotropic glutamate receptor subunit);[158] (d) anti-apoptotic drugs: e.g., corticotrophin releasing hormone.[159]

4. Anti-inflammatory drugs, such as inhibitors of cyclooxygenase-2, reduce the activation of prostanoid pathways (prostaglandin E2 production), microglial activation, leukocyte infiltration, cytokine release, oxidative stress and neurodegeneration.[6]

5. Inhibition of the mTOR pathway with rapamycin has an antiepileptogenic potential.[6] Rapamycin inhibits the serine threonine protein kinase *mammalian target of rapamycin* (mTOR).[6] Under normal conditions, mTOR activity is inhibited by hamartin (TSC1) and tuberin (TSC2).[6] In tuberous sclerosis complex, a loss of function mutation in either TSC1 or TSC2 results in increased mTOR activity and, therefore, activation of its downstream pathways, which ultimately leads to increased cell growth and tumor formation.[160] Epilepsy occurs in 80-90%

of TSC patients and seizures often originate within or around hamartomatous lesions or tubers, but the specific mechanisms of ictogenesis are unknown.[161,162] In the previous experiments with mice models of tuberous sclerosis, in which the mTOR pathway is activated, rapamycin can suppress epileptogenesis and improve the underlying pathology.[163,164]

6. Transplantation of neuronal precursor cells, embryonic stem cells, induced pluripotent stem cells and mesenchymal stem cells are under investigation in experimental models of epilepsy due to their potential to replenish neuronal populations that have been lost in the epileptic focus.[165]

CONCLUSION

Epilepsy can be associated with progressive cognitive decline, resembling, at times, a neurodegenerative disease. Factors that predispose to such bad outcome include early age of onset and long duration of epilepsy, underlying etiology and the type of epilepsy. Significant progress has been done in promoting our understanding of the pathophysiology of epileptogenesis and epilepsy-related neurodegeneration, especially from clinical studies and experimental models of temporal lobe epilepsy. Interventions that may potentially increase the ability of the brain to withstand seizure-induced injury (neuroprotective treatments) have been identified. However more effective and safer neuroprotective and disease-modifying therapies are needed. The model-specific and syndrome-specific differences in etiology, pathogenesis, course and treatment of the various epileptic syndromes will require however more intense research and validation in appropriate animal models. Early predictors of epileptogenesis and neurodegeneration are warranted so as to initiate early treatment to cure epilepsy and its sequelae.

ACKNOWLEDGEMENTS

We would like to acknowledge the funding by NINDS/NICHD grant NS62947, NIH NINDS grant NS20253, as well as the Heffer Family Foundation.

REFERENCES

1. Hirtz D, Thurman DJ, Gwinn-Hardy K et al. How common are the "common" neurologic disorders? Neurology 2007; 68:326-337.
2. WHO. Epilepsy. Fact Sheet 2009. http://www.who.int/mediacentre/factsheets/fs999/en/index.html
3. Fisher RS, van Emde Boas W, Blume W et al. Epileptic seizures and epilepsy: definitions proposed by the International League Against Epilepsy (ILAE) and the International Bureau for Epilepsy (IBE). Epilepsia 2005; 46:470-472.
4. Engel J Jr. A proposed diagnostic scheme for people with epileptic seizures and with epilepsy: report of the ILAE Task Force on Classification and Terminology. Epilepsia 2001; 42:796-803.
5. Engel J Jr. Do we belittle epilepsy by calling it a disorder rather than a disease? Epilepsia 2010; 51:2363-2364.
6. Pitkanen A. Therapeutic approaches to epileptogenesis—hope on the horizon. Epilepsia 2010; 51 Suppl 3:2-17.
7. Berg AT, Berkovic SF, Brodie MJ et al. Revised terminology and concepts for organization of seizures and epilepsies: report of the ILAE Commission on Classification and Terminology, 2005-2009. Epilepsia 2010; 51:676-685.
8. Escayg A, Goldin AL. Sodium channel SCN1A and epilepsy: mutations and mechanisms. Epilepsia 2010; 51:1650-1658.

9. Galanopoulou AS. Dissociated gender-specific effects of recurrent seizures on GABA signaling in CA1 pyramidal neurons: role of GABA(A) receptors. J Neurosci 2008; 28:1557-1567.
10. Macdonald RL, Kang JQ, Gallagher MJ. Mutations in GABAA receptor subunits associated with genetic epilepsies. J Physiol 2010; 588:1861-1869.
11. Commission on Classification and Terminology of the International League Against Epilepsy. Proposal for revised classification of epilepsies and epileptic syndromes. Epilepsia 1989; 30:389-399.
12. Kwan P, Brodie MJ. Early identification of refractory epilepsy. N Engl J Med 2000; 342:314-319.
13. Brodie MJ. Diagnosing and predicting refractory epilepsy. Acta Neurol Scand Suppl 2005; 181:36-39.
14. French JA, Williamson PD, Thadani VM et al. Characteristics of medial temporal lobe epilepsy: I. Results of history and physical examination. Ann Neurol 1993; 34:774-780.
15. Engel J Jr. Introduction to temporal lobe epilepsy. Epilepsy Res 1996; 26:141-150.
16. Engel J Jr. Mesial temporal lobe epilepsy: what have we learned? Neuroscientist 2001; 7:340-352.
17. Engel J. Natural history of mesial temporal lobe epilepsy with hippocampal sclerosis: how does kindling compare with other commonly used animal models? In: Corcoran ME, Moshé SL, eds. Kindling 6. New York: Springer, 2005:371-384.
18. Kwan P, Brodie MJ. Neuropsychological effects of epilepsy and antiepileptic drugs. Lancet 2001; 357:216-222.
19. Motamedi G, Meador K. Epilepsy and cognition. Epilepsy Behav 2003; 4 Suppl 2:S25-S38.
20. Andersson-Roswall L, Engman E, Samuelsson H et al. Cognitive outcome 10 years after temporal lobe epilepsy surgery: a prospective controlled study. Neurology 2010; 74:1977-1985.
21. Hermann B, Seidenberg M, Bell B et al. The neurodevelopmental impact of childhood-onset temporal lobe epilepsy on brain structure and function. Epilepsia 2002; 43:1062-1071.
22. O'Leary DS, Lovell MR, Sackellares JC et al. Effects of age of onset of partial and generalized seizures on neuropsychological performance in children. J Nerv Ment Dis 1983; 171:624-629.
23. Tassinari CA, Michelucci R, Forti A et al. The electrical status epilepticus syndrome. Epilepsy Res Suppl 1992; 6:111-115.
24. Galanopoulou AS, Bojko A, Lado F et al. The spectrum of neuropsychiatric abnormalities associated with electrical status epilepticus in sleep. Brain Dev 2000; 22:279-295.
25. McVicar KA, Shinnar S. Landau-Kleffner syndrome, electrical status epilepticus in slow wave sleep and language regression in children. Ment Retard Dev Disabil Res Rev 2004; 10:144-149.
26. Tassinari CA, Rubboli G. Cognition and paroxysmal EEG activities: from a single spike to electrical status epilepticus during sleep. Epilepsia 2006; 47 Suppl 2:40-43.
27. Hrachovy RA, Frost JD Jr. Infantile epileptic encephalopathy with hypsarrhythmia (infantile spasms/West syndrome). J Clin Neurophysiol 2003; 20:408-425.
28. Lombroso CT. A prospective study of infantile spasms: clinical and therapeutic correlations. Epilepsia 1983; 24:135-158.
29. Berg AT, Cross JH. Towards a modern classification of the epilepsies? Lancet Neurol 2010:9.
30. Partikian A, Mitchell WG. Neurodevelopmental and epilepsy outcomes in a North American cohort of patients with infantile spasms. J Child Neurol 2010; 25:423-428.
31. Pellock JM, Hrachovy R, Shinnar S et al. Infantile spasms: a US consensus report. Epilepsia 2010; 51:2175-2189.
32. Koo B, Hwang PA, Logan WJ. Infantile spasms: outcome and prognostic factors of cryptogenic and symptomatic groups. Neurology 1993; 43:2322-2327.
33. Lux AL, Edwards SW, Hancock E et al. The United Kingdom Infantile Spasms Study (UKISS) comparing hormone treatment with vigabatrin on developmental and epilepsy outcomes to age 14 months: a multicentre randomised trial. Lancet Neurol 2005; 4:712-717.
34. Primec ZR, Stare J, Neubauer D. The risk of lower mental outcome in infantile spasms increases after three weeks of hypsarrhythmia duration. Epilepsia 2006; 47:2202-2205.
35. Kivity S, Lerman P, Ariel R et al. Long-term cognitive outcomes of a cohort of children with cryptogenic infantile spasms treated with high-dose adrenocorticotropic hormone. Epilepsia 2004; 45:255-262.
36. Darke K, Edwards SW, Hancock E et al. Developmental and epilepsy outcomes at age 4 years in the UKISS trial comparing hormonal treatments to vigabatrin for infantile spasms: a multi-centre randomised trial. Arch Dis Child 2010; 95:382-386.
37. Riikonen RS. Favourable prognostic factors with infantile spasms. Eur J Paediatr Neurol 2010; 14:13-18.
38. Giblin KA, Blumenfeld H. Is epilepsy a preventable disorder? New evidence from animal models. Neuroscientist 2010; 16:253-275.
39. Naegele JR. Neuroprotective strategies to avert seizure-induced neurodegeneration in epilepsy. Epilepsia 2007; 48 Suppl 2:107-117.
40. Albala BJ, Moshe SL, Okada R. Kainic-acid-induced seizures: a developmental study. Brain Res 1984; 315:139-148.
41. Stafstrom CE, Thompson JL, Holmes GL. Kainic acid seizures in the developing brain: status epilepticus and spontaneous recurrent seizures. Brain Res Dev Brain Res 1992; 65:227-236.

42. Galanopoulou AS, Vidaurre J, Moshe SL. Under what circumstances can seizures produce hippocampal injury: evidence for age-specific effects. Dev Neurosci 2002; 24:355-363.
43. Galanopoulou AS, Alm EM, Veliskova J. Estradiol reduces seizure-induced hippocampal injury in ovariectomized female but not in male rats. Neurosci Lett 2003; 342:201-205.
44. Raol YS, Budreck EC, Brooks-Kayal AR. Epilepsy after early-life seizures can be independent of hippocampal injury. Ann Neurol 2003; 53:503-511.
45. Shapiro LA, Wang L, Ribak CE. Rapid astrocyte and microglial activation following pilocarpine-induced seizures in rats. Epilepsia 2008; 49 Suppl 2:33-41.
46. Pitkanen A, Sutula TP. Is epilepsy a progressive disorder? Prospects for new therapeutic approaches in temporal-lobe epilepsy. Lancet Neurol 2002; 1:173-181.
47. Vezzani A, Ravizza T, Balosso S et al. Glia as a source of cytokines: implications for neuronal excitability and survival. Epilepsia 2008; 49 Suppl 2:24-32.
48. de Lanerolle NC, Lee TS, Spencer DD. Astrocytes and epilepsy. Neurotherapeutics 2010; 7:424-438.
49. Yang T, Zhou D, Stefan H. Why mesial temporal lobe epilepsy with hippocampal sclerosis is progressive: uncontrolled inflammation drives disease progression? J Neurol Sci 2010; 296:1-6.
50. Wang YY, Smith P, Murphy M et al. Global expression profiling in epileptogenesis: does it add to the confusion? Brain Pathol 2010; 20:1-16.
51. Cavazos JE, Cross DJ. The role of synaptic reorganization in mesial temporal lobe epilepsy. Epilepsy Behav 2006; 8:483-493.
52. Boyett JM, Buckmaster PS. Somatostatin-immunoreactive interneurons contribute to lateral inhibitory circuits in the dentate gyrus of control and epileptic rats. Hippocampus 2001; 11:418-422.
53. Buckmaster PS, Dudek FE. In vivo intracellular analysis of granule cell axon reorganization in epileptic rats. J Neurophysiol 1999; 81:712-721.
54. Sutula T. Seizure-induced axonal sprouting: assessing connections between injury, local circuits and epileptogenesis. Epilepsy Curr 2002; 2:86-91.
55. Lopantsev V, Both M, Draguhn A. Rapid plasticity at inhibitory and excitatory synapses in the hippocampus induced by ictal epileptiform discharges. Eur J Neurosci 2009; 29:1153-1164.
56. Badawy RA, Harvey AS, Macdonell RA. Cortical hyperexcitability and epileptogenesis: understanding the mechanisms of epilepsy—part 2. J Clin Neurosci 2009; 16:485-500.
57. André VM, Cepeda C, Vinters HV et al. Interneurons, GABAA currents and subunit composition of the GABAA receptor in type I and type II cortical dysplasia. Epilepsia 2010; 51(Suppl 3):166-170.
58. Cepeda C, André VM, Yamazaki I et al. Comparative study of cellular and synaptic abnormalities in brain tissue samples from pediatric tuberous sclerosis complex and cortical dysplasia type II. Epilepsia 2010; 51(Suppl 3):160-165.
59. Rivera C, Voipio J, Thomas-Crusells J et al. Mechanism of activity-dependent downregulation of the neuron-specific K-Cl cotransporter KCC2. J Neurosci 2004; 24:4683-4691.
60. Brooks-Kayal AR, Raol YH, Russek SJ. Alteration of epileptogenesis genes. Neurotherapeutics 2009; 6:312-318.
61. Ben-Ari Y. Excitatory actions of gaba during development: the nature of the nurture. Nat Rev Neurosci 2002; 3:728-739.
62. Galanopoulou AS. GABA(A) receptors in normal development and seizures: friends or foes? Curr Neuropharmacol 2008; 6:1-20.
63. Rakhade SN, Jensen FE. Epileptogenesis in the immature brain: emerging mechanisms. Nat Rev Neurol 2009; 5:380-391.
64. Knudsen EI. Sensitive periods in the development of the brain and behavior. J Cogn Neurosci 2004; 16:1412-1425.
65. Ben-Ari Y, Holmes GL. Effects of seizures on developmental processes in the immature brain. Lancet Neurol 2006; 5:1055-1063.
66. Sanchez RM, Koh S, Rio C et al. Decreased glutamate receptor 2 expression and enhanced epileptogenesis in immature rat hippocampus after perinatal hypoxia-induced seizures. J Neurosci 2001; 21:8154-8163.
67. Zhang G, Raol YH, Hsu FC et al. Effects of status epilepticus on hippocampal GABAA receptors are age-dependent. Neuroscience 2004; 125:299-303.
68. Isaeva E, Isaev D, Khazipov R et al. Selective impairment of GABAergic synaptic transmission in the flurothyl model of neonatal seizures. Eur J Neurosci 2006; 23:1559-1566.
69. Swann JW, Le JT, Lee CL. Recurrent seizures and the molecular maturation of hippocampal and neocortical glutamatergic synapses. Dev Neurosci 2007; 29:168-178.
70. O'Reilly RC. Biologically based computational models of high-level cognition. Science 2006; 314:91-94.
71. Veliskova J, Moshe SL. Sexual dimorphism and developmental regulation of substantia nigra function. Ann Neurol 2001; 50:596-601.
72. Galanopoulou AS, Moshe SL. The epileptic hypothesis: developmentally related arguments based on animal models. Epilepsia 2009; 50 Suppl 7:37-42.
73. Cancedda L, Fiumelli H, Chen K et al. Excitatory GABA action is essential for morphological maturation of cortical neurons in vivo. J Neurosci 2007; 27:5224-5235.

74. Wang DD, Kriegstein AR. GABA regulates excitatory synapse formation in the neocortex via NMDA receptor activation. J Neurosci 2008; 28:5547-5558.
75. Mathern GW, Babb TL, Armstrong TM. Hippocampal sclerosis. Philadelphia: Lippincott-Raven, 1997:133-155.
76. Thom M, Sisodiya SM, Beckett A et al. Cytoarchitectural abnormalities in hippocampal sclerosis. J Neuropathol Exp Neurol 2002; 61:510-519.
77. Blumcke I, Zuschratter W, Schewe JC et al. Cellular pathology of hilar neurons in ammon's horn sclerosis. J Comp Neurol 1999; 414:437-453.
78. Kalviainen R, Salmenpera T, Partanen K et al. Recurrent seizures may cause hippocampal damage in temporal lobe epilepsy. Neurology 1998; 50:1377-1382.
79. Theodore WH, Bhatia S, Hatta J et al. Hippocampal atrophy, epilepsy duration and febrile seizures in patients with partial seizures. Neurology 1999; 52:132-136.
80. Tasch E, Cendes F, Li LM et al. Neuroimaging evidence of progressive neuronal loss and dysfunction in temporal lobe epilepsy. Ann Neurol 1999; 45:568-576.
81. Salmenpera T, Kalviainen R, Partanen K et al. Hippocampal and amygdaloid damage in partial epilepsy: a cross-sectional MRI study of 241 patients. Epilepsy Res 2001; 46:69-82.
82. Cendes F, Andermann F, Gloor P et al. Atrophy of mesial structures in patients with temporal lobe epilepsy: cause or consequence of repeated seizures? Ann Neurol 1993; 34:795-801.
83. Spanaki MV, Kopylev L, Liow K et al. Relationship of seizure frequency to hippocampus volume and metabolism in temporal lobe epilepsy. Epilepsia 2000; 41:1227-1229.
84. Mathern GW, Adelson PD, Cahan LD et al. Hippocampal neuron damage in human epilepsy: Meyer's hypothesis revisited. Prog Brain Res 2002; 135:237-251.
85. Cavazos JE, Sutula TP. Progressive neuronal loss induced by kindling: a possible mechanism for mossy fiber synaptic reorganization and hippocampal sclerosis. Brain Res 1990; 527:1-6.
86. Cavazos JE, Das I, Sutula TP. Neuronal loss induced in limbic pathways by kindling: evidence for induction of hippocampal sclerosis by repeated brief seizures. J Neurosci 1994; 14:3106-3121.
87. Bengzon J, Kokaia Z, Elmer E et al. Apoptosis and proliferation of dentate gyrus neurons after single and intermittent limbic seizures. Proc Natl Acad Sci USA 1997; 94:10432-10437.
88. Pretel S, Applegate CD, Piekut D. Apoptotic and necrotic cell death following kindling induced seizures. Acta Histochem 1997; 99:71-79.
89. Kotloski R, Lynch M, Lauersdorf S et al. Repeated brief seizures induce progressive hippocampal neuron loss and memory deficits. Prog Brain Res 2002; 135:95-110.
90. Lowenstein DH. Epilepsy after head injury: an overview. Epilepsia 2009; 50 Suppl 2:4-9.
91. Agrawal A, Timothy J, Pandit L et al. Post-traumatic epilepsy: an overview. Clin Neurol Neurosurg 2006; 108:433-439.
92. Willmore LJ, Ueda Y. Posttraumatic epilepsy: hemorrhage, free radicals and the molecular regulation of glutamate. Neurochem Res 2009; 34:688-697.
93. Willmore LJ, Sypert GW, Munson JB. Recurrent seizures induced by cortical iron injection: a model of posttraumatic epilepsy. Ann Neurol 1978; 4:329-336.
94. Sharma V, Babu PP, Singh A et al. Iron-induced experimental cortical seizures: electroencephalographic mapping of seizure spread in the subcortical brain areas. Seizure 2007; 16:680-690.
95. Reilly PL. Brain injury: the pathophysiology of the first hours. 'Talk and die revisited'. J Clin Neurosci 2001; 8:398-403.
96. Thompson HJ, Lifshitz J, Marklund N et al. Lateral fluid percussion brain injury: a 15-year review and evaluation. J Neurotrauma 2005; 22:42-75.
97. Pitkanen A, McIntosh TK. Animal models of posttraumatic epilepsy. J Neurotrauma 2006; 23:241-261.
98. Pitkanen A, Immonen RJ, Grohn OH et al. From traumatic brain injury to posttraumatic epilepsy: what animal models tell us about the process and treatment options. Epilepsia 2009; 50 Suppl 2:21-29.
99. Coulter DA, Rafiq A, Shumate M et al. Brain injury-induced enhanced limbic epileptogenesis: anatomical and physiological parallels to an animal model of temporal lobe epilepsy. Epilepsy Res 1996; 26:81-91.
100. Reeves TM, Lyeth BG, Phillips LL et al. The effects of traumatic brain injury on inhibition in the hippocampus and dentate gyrus. Brain Res 1997; 757:119-132.
101. Mendez M, Lim G. Seizures in elderly patients with dementia: epidemiology and management. Drugs Aging 2003; 20:791-803.
102. Larner AJ. Epileptic seizures in AD patients. Neuromolecular Med 2010; 2010:71-77.
103. Minkeviciene R, Rheims S, Dobszay MB et al. Amyloid beta-induced neuronal hyperexcitability triggers progressive epilepsy. J Neurosci 2009; 29:3453-3462.
104. Palop JJ, Chin J, Roberson ED et al. Aberrant excitatory neuronal activity and compensatory remodeling of inhibitory hippocampal circuits in mouse models of Alzheimer's disease. Neuron 2007; 55:697-711.
105. Roch C, Leroy C, Nehlig A et al. Magnetic resonance imaging in the study of the lithium-pilocarpine model of temporal lobe epilepsy in adult rats. Epilepsia 2002; 43:325-335.

106. Roch C, Leroy C, Nehlig A et al. Predictive value of cortical injury for the development of temporal lobe epilepsy in 21-day-old rats: an MRI approach using the lithium-pilocarpine model. Epilepsia 2002; 43:1129-1136.
107. Wieshmann UC, Woermann FG, Lemieux L et al. Development of hippocampal atrophy: a serial magnetic resonance imaging study in a patient who developed epilepsy after generalized status epilepticus. Epilepsia 1997; 38:1238-1241.
108. VanLandingham KE, Heinz ER, Cavazos JE et al. Magnetic resonance imaging evidence of hippocampal injury after prolonged focal febrile convulsions. Ann Neurol 1998; 43:413-426.
109. Kuster GW, Braga-Neto P, Santos-Neto D et al. Hippocampal sclerosis and status epilepticus: cause or consequence? A MRI study. Arq Neuropsiquiatr 2007; 65:1101-1104.
110. Tokumitsu T, Mancuso A, Weinstein PR et al. Metabolic and pathological effects of temporal lobe epilepsy in rat brain detected by proton spectroscopy and imaging. Brain Res 1997; 744:57-67.
111. Connelly A, Van Paesschen W, Porter DA et al. Proton magnetic resonance spectroscopy in MRI-negative temporal lobe epilepsy. Neurology 1998; 51:61-66.
112. Woermann FG, McLean MA, Bartlett PA et al. Short echo time single-voxel 1H magnetic resonance spectroscopy in magnetic resonance imaging-negative temporal lobe epilepsy: different biochemical profile compared with hippocampal sclerosis. Ann Neurol 1999; 45:369-376.
113. Shih JJ, Weisend MP, Lewine J et al. Areas of interictal spiking are associated with metabolic dysfunction in MRI-negative temporal lobe epilepsy. Epilepsia 2004; 45:223-229.
114. Gomes WA, Lado FA, de Lanerolle NC et al. Spectroscopic imaging of the pilocarpine model of human epilepsy suggests that early NAA reduction predicts epilepsy. Magn Reson Med 2007; 58:230-235.
115. Shen J, Zhang L, Tian X et al. Use of short echo time two-dimensional 1H-magnetic resonance spectroscopy in temporal lobe epilepsy with negative magnetic resonance imaging findings. J Int Med Res 2009; 37:1211-1219.
116. Rugg-Gunn FJ, Eriksson SH, Symms MR et al. Diffusion tensor imaging of cryptogenic and acquired partial epilepsies. Brain 2001; 124:627-636.
117. Dube C, Yu H, Nalcioglu O et al. Serial MRI after experimental febrile seizures: altered T2 signal without neuronal death. Ann Neurol 2004; 56:709-714.
118. Nairismagi J, Grohn OH, Kettunen MI et al. Progression of brain damage after status epilepticus and its association with epileptogenesis: a quantitative MRI study in a rat model of temporal lobe epilepsy. Epilepsia 2004; 45:1024-1034.
119. Kim CH, Koo BB, Chung CK et al. Thalamic changes in temporal lobe epilepsy with and without hippocampal sclerosis: a diffusion tensor imaging study. Epilepsy Res 2010; 90:21-27.
120. Liacu D, de Marco G, Ducreux D et al. Diffusion tensor changes in epileptogenic hippocampus of TLE patients. Neurophysiol Clin 2010; 40:151-157.
121. Shon YM, Kim YI, Koo BB et al. Group-specific regional white matter abnormality revealed in diffusion tensor imaging of medial temporal lobe epilepsy without hippocampal sclerosis. Epilepsia 2010; 51:529-535.
122. Mirrione MM, Schiffer WK, Siddiq M et al. PET imaging of glucose metabolism in a mouse model of temporal lobe epilepsy. Synapse 2006; 59:119-121.
123. Jupp B, O'Brien TJ. Application of coregistration for imaging of animal models of epilepsy. Epilepsia 2007; 48 Suppl 4:82-89.
124. Goffin K, Van Paesschen W, Dupont P et al. Longitudinal microPET imaging of brain glucose metabolism in rat lithium-pilocarpine model of epilepsy. Exp Neurol 2009; 217:205-209.
125. Guo Y, Gao F, Wang S et al. In vivo mapping of temporospatial changes in glucose utilization in rat brain during epileptogenesis: an 18F-fluorodeoxyglucose-small animal positron emission tomography study. Neuroscience 2009; 162:972-979.
126. Shinnar S, Kang H, Berg AT et al. EEG abnormalities in children with a first unprovoked seizure. Epilepsia 1994; 35:471-476.
127. Shinnar S, Berg AT, Moshe SL et al. The risk of seizure recurrence after a first unprovoked afebrile seizure in childhood: an extended follow-up. Pediatrics 1996; 98:216-225.
128. Berg AT, Shinnar S, Levy SR et al. Early development of intractable epilepsy in children: a prospective study. Neurology 2001; 56:1445-1452.
129. Spooner CG, Berkovic SF, Mitchell LA et al. New-onset temporal lobe epilepsy in children: lesion on MRI predicts poor seizure outcome. Neurology 2006; 67:2147-2153.
130. Nordli DR, Moshe SL, Shinnar S. The role of EEG in febrile status epilepticus (FSE). Brain Dev 2010; 32:37-41.
131. Shinnar S, Hesdorffer DC, Nordli DR Jr et al. Phenomenology of prolonged febrile seizures: results of the FEBSTAT study. Neurology 2008; 71:170-176.
132. Beume LA, Steinhoff BJ. Long-term outcome of difficult-to-treat epilepsy in childhood. Neuropediatrics 2010; 41:135-139.
133. Sutula TP, Hagen J, Pitkanen A. Do epileptic seizures damage the brain? Curr Opin Neurol 2003; 16:189-195.
134. Gleissner U, Sassen R, Schramm J et al. Greater functional recovery after temporal lobe epilepsy surgery in children. Brain 2005; 128:2822-2829.

135. Jambaque I, Dellatolas G, Dulac O et al. Verbal and visual memory impairment in children with epilepsy. Neuropsychologia 1993; 31:1321-1337.
136. Elsharkawy AE, May T, Thorbecke R et al. Long-term outcome and determinants of quality of life after temporal lobe epilepsy surgery in adults. Epilepsy Res 2009; 86:191-199.
137. Jonas R, Nguyen S, Hu B et al. Cerebral hemispherectomy: hospital course, seizure, developmental, language and motor outcomes. Neurology 2004; 62:1712-1721.
138. Tubbs RS, Nimjee SM, Oakes WJ. Long-term follow-up in children with functional hemispherectomy for Rasmussen's encephalitis. Childs Nerv Syst 2005; 21:461-465.
139. Bittigau P, Sifringer M, Ikonomidou C. Antiepileptic drugs and apoptosis in the developing brain. Ann N Y Acad Sci 2003; 993:103-114.
140. Dhanushkodi A, Shetty AK. Is exposure to enriched environment beneficial for functional post-lesional recovery in temporal lobe epilepsy? Neurosci Biobehav Rev 2008; 32:657-674.
141. Hamed SA. The multimodal prospects for neuroprotection and disease modification in epilepsy: relationship to its challenging neurobiology. Restor Neurol Neurosci 2010; 28:323-348.
142. Sutula T, Cavazos J, Golarai G. Alteration of long-lasting structural and functional effects of kainic acid in the hippocampus by brief treatment with phenobarbital. J Neurosci 1992; 12:4173-4187.
143. Pitkanen A, Kubova H. Antiepileptic drugs in neuroprotection. Expert Opin Pharmacother 2004; 5:777-798.
144. Yan HD, Ji-qun C, Ishihara K et al. Separation of antiepileptogenic and antiseizure effects of levetiracetam in the spontaneously epileptic rat (SER). Epilepsia 2005; 46:1170-1177.
145. Blumenfeld H, Klein JP, Schridde U et al. Early treatment suppresses the development of spike-wave epilepsy in a rat model. Epilepsia 2008; 49:400-409.
146. Muller-Schwarze AB, Tandon P, Liu Z et al. Ketogenic diet reduces spontaneous seizures and mossy fiber sprouting in the kainic acid model. Neuroreport 1999; 10:1517-1522.
147. Meyerhoff JL, Lee JK, Rittase BW et al. Lipoic acid pretreatment attenuates ferric chloride-induced seizures in the rat. Brain Res 2004; 1016:139-144.
148. Yokoi I, Toma J, Liu J et al. Adenosines scavenged hydroxyl radicals and prevented posttraumatic epilepsy. Free Radic Biol Med 1995; 19:473-479.
149. Kabuto H, Yokoi I, Ogawa N. Melatonin inhibits iron-induced epileptic discharges in rats by suppressing peroxidation. Epilepsia 1998; 39:237-243.
150. Miyamoto R, Shimakawa S, Suzuki S et al. Edaravone prevents kainic acid-induced neuronal death. Brain Res 2008; 1209:85-91.
151. Kamida T, Fujiki M, Ooba H et al. Neuroprotective effects of edaravone, a free radical scavenger, on the rat hippocampus after pilocarpine-induced status epilepticus. Seizure 2009; 18:71-75.
152. Komatsu M, Hiramatsu M, Willmore LJ. Zonisamide reduces the increase in 8-hydroxy-2'-deoxyguanosine levels formed during iron-induced epileptogenesis in the brains of rats. Epilepsia 2000; 41:1091-1094.
153. Mori A, Yokoi I, Noda Y et al. Natural antioxidants may prevent posttraumatic epilepsy: a proposal based on experimental animal studies. Acta Med Okayama 2004; 58:111-118.
154. Guo X, Dawson VL, Dawson TM. Neuroimmunophilin ligands exert neuroregeneration and neuroprotection in midbrain dopaminergic neurons. Eur J Neurosci 2001; 13:1683-1693.
155. Chen CM, Lin JK, Liu SH et al. Novel regimen through combination of memantine and tea polyphenol for neuroprotection against brain excitotoxicity. J Neurosci Res 2008; 86:2696-2704.
156. Paradiso B, Marconi P, Zucchini S et al. Localized delivery of fibroblast growth factor-2 and brain-derived neurotrophic factor reduces spontaneous seizures in an epilepsy model. Proc Natl Acad Sci USA 2009; 106:7191-7196.
157. Noe F, Nissinen J, Pitkanen A et al. Gene therapy in epilepsy: the focus on NPY. Peptides 2007; 28:377-383.
158. Moldrich RX, Chapman AG, De Sarro G et al. Glutamate metabotropic receptors as targets for drug therapy in epilepsy. Eur J Pharmacol 2003; 476:3-16.
159. Elliott-Hunt CR, Kazlauskaite J, Wilde GJ et al. Potential signalling pathways underlying corticotrophin-releasing hormone-mediated neuroprotection from excitotoxicity in rat hippocampus. J Neurochem 2002; 80:416-425.
160. Napolioni V, Moavero R, Curatolo P. Recent advances in neurobiology of Tuberous Sclerosis Complex. Brain Dev 2009; 31:104-113.
161. Holmes GL, Stafstrom CE. Tuberous sclerosis complex and epilepsy: recent developments and future challenges. Epilepsia 2007; 48:617-630.
162. Wong M. Mechanisms of epileptogenesis in tuberous sclerosis complex and related malformations of cortical development with abnormal glioneuronal proliferation. Epilepsia 2008; 49:8-21.
163. Zeng LH, Rensing NR, Wong M. The mammalian target of rapamycin signaling pathway mediates epileptogenesis in a model of temporal lobe epilepsy. J Neurosci 2009; 29:6964-6972.
164. Ljungberg MC, Sunnen CN, Lugo JN et al. Rapamycin suppresses seizures and neuronal hypertrophy in a mouse model of cortical dysplasia. Dis Model Mech 2009; 2:389-398.
165. Naegele JR, Maisano X, Yang J et al. Recent advancements in stem cell and gene therapies for neurological disorders and intractable epilepsy. Neuropharmacology 2010; 58:855-864.

CHAPTER 9

FRONTOTEMPORAL LOBAR DEGENERATION

Enrico Premi, Alessandro Padovani and Barbara Borroni*

Centre for Ageing Brain and Neurodegenerative Disorders, Neurology Unit, University of Brescia, Brescia, Italy
Corresponding Author: Barbara Borroni—Email: bborroni@inwind.it

Abstract: Frontotemporal Lobar Degeneration (FTLD) is an heterogeneous neurodegenerative disorder characterized by behaviour and language disturbances, associated with degeneration of the frontal and temporal lobes. Three different clinical presentations have been described, namely behavioural variant Frontotemporal Dementia (bvFTD), Semantic Dementia (SD) and Progressive Non-Fluent Aphasia (PNFA). The associated histopathology includes different neuropathological hallmarks, the most frequent being tau-positive inclusions (FTLD-TAU) or tau-negative and TDP-43 positive inclusions (FTLD-TDP). The majority of familial FTLD cases are caused by mutations within *Microtubule-Associated Protein Tau (MAPT)* gene, leading to FTLD-TAU, or *Progranulin (PGRN)* gene, leading to FTLD-TDP. In the last few years, imaging, biological and genetic biomarkers have been developed, helping in clinical evaluation and diagnostic accuracy. Though current pharmacologic interventions are only symptomatic, recent research argues for possible disease-modifying strategies in the near future.

INTRODUCTION

At the end of the XIX century, Arnold Pick described a series of patients with progressive aphasia, apraxia and behavior changes, associated with severe frontotemporal atrophy.[1,2] Initially, these cases were grouped under the same label of Alzheimer's disease pathology. Thereafter, the identification of the round silver staining inclusions (Pick's bodies) allowed the researchers to characterize the specific hallmark of this entity.[3-5]

By a clinical point of view, patients with behavioural disturbances, deficits of executive functions and language impairment were labeled under the term of Frontotemporal Lobar Degeneration (FTLD).[6] However, at autopsy, Pick's bodies were found only in a subgroup

Neurodegenerative Diseases, edited by Shamim I. Ahmad.
©2012 Landes Bioscience and Springer Science+Business Media.

of these patients, suggesting that FTLD is characterized by a more heterogeneous histopathology than previously thought.

Different diagnostic criteria have been developed over years, with the attempt to better define the clinical correlates of FTLD neuropathology.[6,7] The first criteria were developed by the Lund and Manchester group in 1994.[8] These criteria permitted a good discrimination between FTLD and Alzheimer's disease,[9] but no hints on the number of clinical features necessary for FTLD diagnosis were defined. In 1998, new diagnostic criteria were published by Neary et al[10] and the three major clinical syndromes were described. Indeed, behavioural variant Frontotemporal Dementia (bvFTD) and two language variants (Progressive Non-Fluent Aphasia, PNFA and Semantic Dementia, SD) were considered.[11] Finally, in 2001 revised clinical criteria were published by McKhann and colleagues.[12] In these criteria, Progressive Supranuclear Palsy (PSP) and Corticobasal syndrome (CBS) were considered under the label of FTLD, as overlapping both clinically and neuropathologically.

EPIDEMIOLOGY

FTLD is considered the second cause of presenile dementia, accounting for 20% of all the cases under the age of 65 years.[13] The disease onset is usually in the sixth decade, but it can be from the third to the ninth decade. In FTLD, the mean age at onset is lower than in Alzheimer's disease and other neurodegenerative dementias.[14,15]

The prevalence of FTLD in population-based studies has varied between 2.7/100,000 inhabitants in the Netherlands,[16] to 15.1/100,000 in subjects aged <65 years in Cambridgeshire, UK.[14] In early-onset cases, the incidence was 3.5/100,000 person-year in the Cambridge study[17] and 3.3/100,000 person-year in the Rochester study.[18]

A population-based study in Northern Italy reported an overall prevalence of 17.6/100,000 inhabitants, with a higher prevalence in patients aged 66-75 (78/100,000 inhabitants).[19] Taken together these data suggest that FTLD is a common cause of early-onset dementia, but it is frequent in advanced age as well.

The estimated median survival is approximately of 6-11 years from symptom onset and 3-4 years from diagnosis.[20] Some studies suggested that in FTLD survival is shorter than that found in Alzheimer's disease.[21] The worse predictor of survival in FTLD is the presence of associated motor neuron disease.

CLINICAL PRESENTATION

bvFTD is the most common clinical presentation, characterized by changes in personality and social conduct (including breaches of interpersonal etiquette and tactlessness). Disturbances could range from inertia, loss of interest in personal affairs and responsibility, to social disinhibition and socially inappropriate behaviors.[9] Distractibility, overactivity, pacing and wandering are common,[22] with relative preservation of memory functions. Perseverations, stereotyped and compulsive behaviors are present,[23] as emotional blunting and loss of insight. Dietary changes typically take the form of overeating and a preference for sweet foods.[24] Sometimes, personality changes have been defined as a "change in self".[25] At onset, language deficits are less common than in other clinical subtypes, but these can arise during disease course. Cognitive deficits occur in the domains of attention, abstraction, planning, problem solving and judgment. Attention and working memory

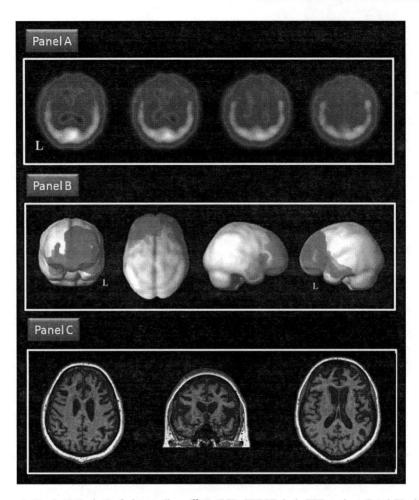

Figure 1. Panel A) Brain Perfusion analisys (^{99m}Tc-ECD $SPECT$) in bvFTD patient. Panel B) single subject Statistical Parametric Mapping analysis on SPECT image (patient vs healthy control group, $P < 0.005$); hypoperfusion pattern in bvFTD patient is highlighted in red. Panel C) MRI atrophy in bvFTD patient. As described in the text, there is a deep and selective fronto-temporal atrophy/hypoperfusion.

may be involved, with a variable preservation of episodic memory. However, memory test performances may be affected, because of patient's frontal disturbances rather than a primary amnesia. In contrast to Alzheimer's disease, at onset visuospatial functions are well preserved.[26] Behavioral disturbances may help to distinguish FTLD from Alzheimer's disease, as in the former lack of concern and insight, presence of repetitive stereotyped behaviors and confabulations are more common.[27] Neuroimaging studies have demonstrated structural (gray matter atrophy) and functional (hypoperfusion and hypometabolism) frontal involvement[28-31] (see Fig. 1). More recently, it has been reported that white matter is affected as well and damage in frontal tracts (superior longitudinal fasciculus) is specific for bvFTD and correlates with behavioral disturbances.[32] The heterogeneous clinical presentation reflects specific brain damage: dorsomedial frontal atrophy correlated with apathy and orbitofrontal atrophy is associated with disinhibited syndrome.[33] Atrophy in medial paralimbic region

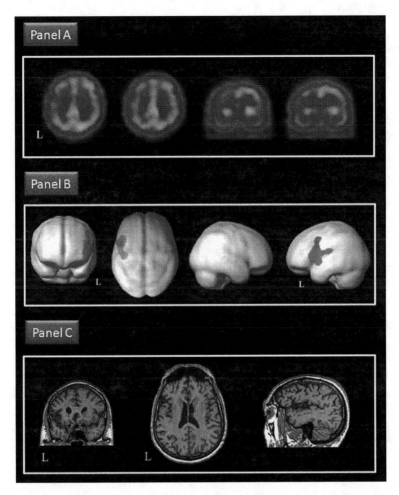

Figure 2. Panel A) Brain Perfusion analisys (*⁹⁹ᵐTc-ECD SPECT*) in PNFA patient. Panel B) single subject Statistical Parametric Mapping analysis on SPECT image (patient vs healthy control group, *P* < 0.005); hypoperfusion pattern in PNFA patient is highlighted in red. Panel C) MRI atrophy in PNFA patient. As described in the text, there is a selective left inferior frontal atrophy/hypoperfusion.

(anterior cingulated, orbitofrontal and frontoinsular cortices) help to differentiate FTLD from Alzheimer's disease.[34] Motor neuron disease (MND) can co-occur with any of the FTLD clinical variants, but is more commonly associated with bvFTD.[35]

The second prototypic syndrome, namely PNFA, is a disorder of expressive language with difficulty in initiating speech, slow rate of speech, with phonologic and grammatical errors, word retrieval difficulties, as well as difficulties in reading and writing. The disorder of language occurs in the absence of impairment in other cognitive domains, although behavioral changes may emerge later in the disease course. Neuropsychological evaluation could show impairment in working memory and executive functions, with a substantial preservation of episodic memory and visuospatial ability.[36] Behavioural abnormalities are less severe than in bvFTD and SD.[37] PNFA is characterized by atrophy in left frontal operculum, premotor and supplementary premotor areas and anterior insula[38] (Fig. 2).

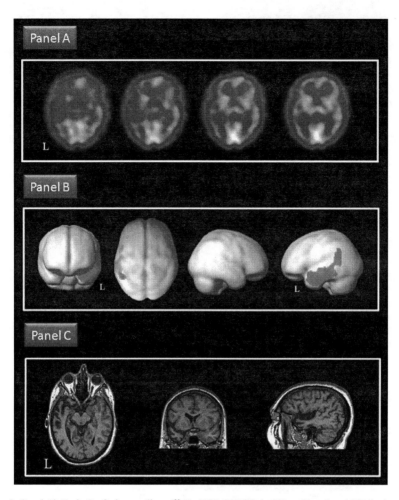

Figure 3. Panel A) Brain Perfusion analisys (^{99m}Tc-*ECD SPECT*) in SD patient. Panel B) single subject Statistical Parametric Mapping analysis on SPECT image (patient vs healthy control group, $P < 0.005$); hypoperfusion pattern in SD patient is highlighted in red. Panel C) MRI atrophy in SD patient. As described in the text, there is a selective left temporal atrophy/hypoperfusion.

SD is characterized by fluent, anomic aphasia and behavioural changes, with asymmetric degeneration of the temporal poles and white matter damage (inferior longitudinal fasciculus)[32] (Fig. 3). Cognitive testing denotes profound semantic loss, with deficit in word comprehension and naming, deficits in face and object recognition; visuospatial ability and episodic memory are well preserved.[10] Primary left-side involvement is associated with progressive loss of semantic knowledge of words, objects and concepts. On the other side, patients with right involvement present with prosopagnosia and behavioral abnormalities.[39,40]

HISTOPATHOLOGY

At autopsy, FTLD patients are characterized by gross atrophy of the frontal and anterior temporal lobes,[41] with relative sparing of posterior brain regions until the most advanced stages of disease.[42] Microscopic examination shows loss of pyramidal neurons and microvacuolar degeneration in layers II and III of the frontal and temporal cortex, with a variable degree of cortical gliosis. White matter shows both axonal and myelin loss and gliosis.[43] Recent advances in neuropathology have defined a new histopathological classification of FTLD, barely overlapping with clinical diagnoses. Mackenzie et al have recently published an update on nosology of FTLD.[44] Virtually all cases of FTLD can be subclassified into the following four major categories, which are based on the presence or the absence of specific inclusion bodies: (i) FTLD with tau inclusions (FTLD-TAU), (ii) FTLD with tau-negative and TDP-43-positive inclusions (FTLD-TDP) and (iii) FTLD with tau/TDP-43 negative and FUS-positive inclusions (FTLD-FUS), (iv) FTLD with positive immunohistochemistry against proteins of the ubiquitin proteasome system (FTLD-UPS).

Tau-Positive FTLD (FTLD-TAU)

FTLD-related tauopathies are classified based on both the morphologic features and the biochemical composition of tau inclusions. Pick's disease, the prototypical tauopathy of FTLD, is characterized by the presence of Pick's bodies, which are solitary, round or oval, argyrophilic inclusions found in the cytoplasm of neurons. In Pick's disease, hyperphosphorylated and aggregated forms of tau accumulate in neurons and, in certain cases, glia.[45] Tau is a complex protein that regulates microtubules dynamics,[46] and six different isoforms resulting from alternative splicing of exons 2, 3 and 10 exist.[47] The splicing-in of exon 10 introduces an extra microtubule-binding domain leading to the so-called four-repeat tau (4R), whereas the splicing-out of exon 10 leads to three-repeat tau (3R). The composition of the different isoforms of tau can be identical to those observed in Alzheimer's disease, with all six alternatively spliced forms present as neurofibrillary tangles. Alternatively, a predominance of three- or four-repeat forms may be present. The histological profile of Pick's disease is defined by predominant 3R tau neuropathology.[48] Anatomically, Pick's bodies are most commonly localized in the dentate gyrus of hippocampus, amygdala and frontal and temporal neocortex.[43] They are detected by Bielschowsky staining.

Less common FTLD-related tauopathies include the argyrophilic grain disease, the sporadic multiple system tauopathy with dementia, the neurofibrillary tangle dementia and the Amyotrophic Lateral Sclerosis (ALS) parkinsonism-dementia complex of Guam.[49]

TAR DNA-Binding Protein 43 (TDP-43)-Positive FTLD (FTLD-TDP)

Tau-negative FTLD pathology has been termed as FTLD tau-negative ubiquitin positive (FTLD-U) for ages.[50] In 2006, the hallmark of a subgroup of FTLD-U cases was identified and TDP-43 protein was found in the vast majority of FTLD-U brains (FTLD-TDP). Up to now, FTLD-TDP represents the most common histological profile observed in FTLD.[51] This protein accumulates in cytoplasm, neuritis and occasionally it forms intranuclear lentiform inclusions.[52] Phosphorylation of S409/410 residue is a consistent feature of FTLD-TDP.[53] Antibodies against this residue do not stain nuclear TDP-43 in physiological conditions, thus suggesting a pathological role of phosphorylation of this site. TDP-43 is usually ubiquitinated and cleaved (probably by caspase-3[54] to produce C-terminal fragments.[55] The overexpression of these fragments appears to evoke splicing abnormalities, leading to the pathological process.[56] Protein inclusions are found in the dentate gyrus of hippocampus, in layer II of the frontotemporal cortex, in cranial nerve nuclei and in the anterior horn cells of the spinal cord.[49] FTLD-TDP can present with any of the major clinical phenotypes of the FTLD syndrome[57] and, rarely it may lead to clinical diagnosis of Parkinson's disease or Corticobasal syndrome.[58]

TDP-43 pathology is characterized by four different patterns with clinical and genetic correlations, based on the anatomical distribution, morphology and inclusion type. Clinical presentation of Type 1 is primarily bvFTD or PNFA. Type 1 histopathology is associated with progranulin mutations. Type 2 usually presents with sporadic SD, while Type 3 is correlated with FTD-ALS presentation (including familial cases linked to chromosome 9p). Type 4 is correlated with *Valosin Containing protein (VCP)* mutations.[57,59-61]

However, TDP-43 inclusions are also found in patients with sporadic ALS and the mutations have been correlated to inherited and sporadic ALS.[62]

Little is known about the role of TDP-43 in neurodegeneration. The abnormal redistribution of TDP-43 to cytoplasm correlated with a loss-of-function mechanism.[63] In support of this hypothesis, reduction of TDP-43 in mouse models correlated with motor impairment and structural abnormalities of motor neurons.[64] However, many other studies postulated a toxic gain of function mechanism induced by pathological cellular inclusions.[65,66] The pathogenic mechanism associated with TDP-43 proteinopathy is still under investigation, but multiple disease mechanism remains the most probable working hypothesis.

FUS-Positive FTLD (FTLD-FUS)

Five to 20% of FTLD-U cases present a negative staining for TDP-43.[67] Recent advances have shown that the majority of these cases consist of positive staining for the protein fused in sarcoma (FUS).[68] Furthermore, abundant FUS-positive pathology was found in cases previously termed as neuronal cytoplasmic inclusions of basophilic inclusions body disease (BIBD) and in cases of neuronal intermediate filament inclusion disease (NIFID).[44,69] This ubiquitously expressed DNA/RNA binding protein regulates gene expression and translocation of FUS and it leads to sarcoma and hematological malignancies.[70] Mutations within *FUS* gene are associated with familial ALS, while the majority of FTLD-FUS cases do not show any genetic mutation within *FUS*. Clinically, patients with FUS pathology are characterized by very early age at onset, behavioural syndrome and MRI caudate atrophy.[68] The latter has been proposed as a useful clinical predictor of this FTLD subtype.[71] As for TDP-43 proteinopathies, complex pathogenic mechanism (gain and loss of function) is plausible. Several works have suggested that altered nuclear import is a key event in the pathogenesis.[72] Modifications of FUS include phosphorylation and arginine methylation. As in other RNA binding proteins, these interactions are crucial for nuclear-cytoplasmic shuttling.[73]

FTLD with Positive Immunohistochemistry against Proteins of the Ubiquitin Proteasome System (FTLD-UPS)

FTLD-UPS encompasses cases with tau/TDP-43/FUS negative inclusions.[74] FTLD-UPS remains appropriate for familial FTLD linked to chromosome 3, caused by mutation in *CHMP2B* gene.[75,76] It is possible that the inclusions herein detected have a more heterogeneous composition resulting from a defect of endosomal function.[77]

GENETICS

FTLD patients present a high rate of positive family history (up to 40%), with approximately 10% of patients showing an autosomal dominant inheritance pattern.[78] Familial FTLD is more common in patients with bvFTD and FTLD-ALS than in patients with PNFA and SD. In 1994, FTLD was firstly linked to chromosome 17q21-22, defining frontotemporal dementia with parkinsonism linked to chromosome 17 (FTDP-17).[79] In 1998, *Microtuble Associated Protein Tau (MAPT)* gene, encoding for protein tau, was identified as the causal gene in FTDP-17 families with tau-positive histopathology.[80] In 2006, *Progranulin (PGRN)* gene mutations were identified in FTDP-17 families with tau-negative histopathology.[81] *MAPT* and *PGRN* mutations account for the majority of familial FTLD (http://www.molgen.ua.ac.be/FTDMutations).

FTLD Associated with *MAPT* Gene Mutations

The microtubule-associated protein tau in neurons binds to axonal microtubules, promoting microtubule assembly and it is involved in stabilization and in signal transduction. Two different types of mutations have been described, either exonic (which lead to aminoacid change) or intronic (which disturb the regulation of the alternative splicing). As in other proteinopathies, mutations may lead to loss (decrease) of function, with decreased binding affinity to microtubules (and consequent deficit in axonal transport). On the other side, a toxic gain-of function mechanism can be represented by an increased self aggregation into toxic filamentous inclusions. These mechanisms are not mutually exclusive and often act synergistically. As stated, many *MAPT* mutations are located in or adjacent to the alternatively spliced exon 10, altering the normal 1:1 ratio between 3R and 4R isoforms in favour to 4R.[82] *MAPT* mutations are characterised by a disease onset between the ages of 40 and 60 years (mean 55) and by a wide spectrum of clinical phenotypes, i.e., bvFTD, SD, PNFA, CBD, PSP and even ALS.[84]

Furthermore, *MAPT* haplotype, i.e., H1 and H2, can modify the risk for developing a tauopathy and can modulate disease phenotype. H1/H1 haplotype has been associated with PSP and CBS.[85-89]

FTLD Associated with *PGRN* Gene Mutations

Progranulin is a secreted growth factor with a wide range of functions (inflammation, tumor growth in nonbrain tissue, promotion of neural survival, stimulation of neuritic outgrowth).[90-92] It is secreted as a precursor protein.[93] Mutations (nonsense or missense) are found in all *PGRN* exons and lead to the loss of at least 50% of functional *PGRN*, suggesting an haploinsufficiency mechanism.[94,95] In patients with *PGRN* mutations, expression pattern of many genes in the frontal cortex are altered, compared to controls.[96] Compared with *MAPT*, *PGRN* mutations are associated with more variable age at onset (range 35-89 years, mean 60 years) and penetrance (estimated at only 50% by age 60 years and 90% by age 70 years).[97] *PGRN* gene mutations accounted for 5%-10% of sporadic FTLD cases and for 20%-25% of the familial ones.[98-100] However, in Europe a lower prevalence was reported.[101] Clinical presentation is heterogeneous: 20-25% of patients present with PNFA, but also bvFTD and CBS are common clinical phenotypes. In some cases, clinical diagnosis of Alzheimer's disease has been postulated.[102] Parkinsonism may be the leading feature, whereas motor neuron disease is uncommon.[103]

Rare Mutations

Mutations in the *CHMP2B* gene were found only in a single Danish family. CHMP2B is part of the endosomal complex (ESCRTIII) involved in endosomal/lysosomal system.[103] Cases of myopathy with succeeding development of FTLD were found to be associated with mutations in the *VCP* gene.[104] *VCP* is an adenosine triphosphatase that is involved in protein degradation in the endoplasmic reticulum. *VCP* mutation carriers show variable penetrance and phenotypic heterogeneity, with myopathy (frequently inclusion body myositis) in the most of cases, while FTLD and Paget's disease occur in fewer than 50%.[105] Mutations in *TARDP* gene and *FUS* gene are associated with familial ALS and seldom with FTLD.[106-111]

International genome-wide association studies are underway with the attempt to identify additional causative and risk-modifying genes.[112] As initial finding, an association with locus 9p13.2-21.3 has been consistently identified in patients with clinical presentation ranging from FTLD to ALS or a combined phenotype.[113] Another genome wide-scan has shown an association between FTLD-TDP and variants in *TMEM106B*.[114]

TREATMENT

No specific treatment is currently available.[115] Neurochemical[116] and functional PET[117] studies have demonstrated impairment of serotonin metabolism. Furthermore, many of the behavioural symptoms of FTLD (depression, compulsions, disinhibition) respond to selective serotonin reuptake inhibitors (SSRI). Different clinical trials have been performed with SSRIs, but different biases limited the obtained results, i.e., small sample size, absence of placebo group, short-term follow-up. However, Huey and collaborators have published a meta-analisys suggesting that the use of serotoninergic drugs in FTLD is associated with a significant reduction of behavioral disturbances.[118]

In clinical practice SSRI are often prescribed as first-line drugs for behavioural symptoms in FTLD. Second-line therapy (in patients that are refractory to SSRI and only in selected cases as first choice) may be represented by atypical antipsychotics.[119] However, extrapyramidal adverse effects and the reported increase mortality risk in the elderly represent important red flags to keep in mind.[120] From this perspective, quetiapine is preferred for its relatively low dopamine D_2 receptor antagonism. Acetylcholinesterase inhibitors, commonly used in Alzheimer's disease, have been tested in FTLD patients; however, no study has shown a cholinergic deficit in FTLD.[121,122] Memantine is a noncompetitive NMDA antagonist, used in Alzheimer's disease. A case series has shown improvement of behavioral disturbances in FTLD as well.[123] Several clinical trials are currently in progress in FTLD (www.clinicaltrials.gov).

CONCLUSION

FTLD is challenging not only for clinicians, that face up every day with the management of the disease, but also for researchers, as actors in disentangling phatogenic and molecular basis leading to neurodegeneration. As in other neurodegenerative diseases, clinical and research perspectives represent the two faces of the same coin. The progressive knowledge of the genetic and pathological bases of FTLD has allowed us to define new type of biomarkers to trace the disease[124,125] and to speculate about new therapeutic targets.[126] Nevertheless, the precise definition of the underlying phatogenetic mechanisms, for setting pharmacological targets and disease-modifying drugs and a world-wide collaborative strategy to recruit a large cohort of patients to demonstrate treatment effect in FTLD, are mandatory.[127]

REFERENCES

1. Pick A. Uber die Beziehungen der senilen Hirnatrophie zur Aphasie. Prager Med Wochenschr 1882; 17:165-167.
2. Kertesz A, McMonagle P, Blair M et al. The evolution and pathology of frontotemporal dementia. Brain 2005; 128:1996-2005.
3. Gans A. Betrachtungen uber Art und Ausbreitung des Krankhaften Prozessess in einem Fall von Pickscher Atrophie des Sternhirns. Z Gesamte Neurol Psychiatr 1922; 80:10-28.
4. Onari K, Spatz H. Anatomische Beitaege zur lehre von Pickschen umschriebenen Grosshirnrinden-Atrophie ("Picksche Krankheit"). Z Gesamte Neurol Psychiatr 1926; 101:470-511.
5. Schneider C. Uber Picksche Krankheit. Monatschr Psychiatr Neurol 1927; 65:230-275.
6. Neary D, Snowden JS, Goulding P et al. Dementia of frontal lobe type. J Neurol Neurosurg Psychiatry 1988; 51:353-361.
7. Gustafson L. Frontal lobe degeneration of non-Alzheimer type II. Clinical picture and differential diagnosis. Arch Gerontol Geriatr 1987; 6:209-223.
8. Lund and Manchester Groups. Consensus statement. Clinical and neuropathological criteria for frontotemporal dementia. J Neurol Neurosurg Psychiatry 1994; 57:416-418.
9. Miller BL, Ikonte BS, Ponton M et al. A study of the Lund- Manchester research criteria for frontotemporal dementia: clinical and single photon emission CT correlations. Neurology 1997; 48:937-942.
10. Neary D, Snowden JS, Gustafson L et al. Frontotemporal lobar degeneration: a consensus on clinical diagnostic criteria. Neurology 1998; 51:1546-1554.
11. Edwards-Lee T, Miller BL, Benson DF et al. The temporal variant of frontotemporal dementia. Brain. 1997; 120(Pt 6):1027-1040.
12. McKhann GM, Dickson D, Trojanowski JQ et al. Clinical and Pathological Diagnosis of Frontotemporal Dementia. Report of the work group on frontotemporal dementia and Pick's disease. Arch Neurol 2001; 58:1803-1809.
13. Skibinski G, Parkinson NJ, Brown JM et al. Mutations in the endosomal ESCRTIII-complex subunit CHMP2B in frontotemporal dementia. Nat Genet 2005; 37(8):806-808
14. Ratnavalli E, Brayne C, Dawson K et al. The prevalence of frontotemporal dementia. Neurology 2002; 58:1615-1621.
15. Ikeda M, Ishikawa T, Tanabe H. Epidemiology of Frontotemporal Lobar Degeneration. Dement Geriatr Cogn Disord 2004; 17:265-268.
16. Rosso SM, Donker Kaat L, Baks T et al. Frontotemporal dementia in the Netherlands: patient characteristics and prevalence estimates from a population-based study. Brain 2003; 126:2016-2022.
17. Mercy L, Hodges JR, Dawson K et al. Incidence of early-onset dementias in Cambridgeshire, United Kingdom. Neurology 2008; 71:1496-1499.
18. Knopman DS, Petersen RC, Edland SD et al. The incidence of frontotemporal lobar degeneration in Rochester, Minnesota, 1990 through 1994. Neurology 2004; 62:506-508.
19. Borroni B, Alberici A, Grassi M et al. Is frontotemporal lobar degeneration a rare disorder? Evidence from a preliminary study in Brescia county, Italy. J Alzheimers Dis 2010; 19(1):111-116.
20. Kertesz A, Blair M, McMonagle P et al. The diagnosis and course of frontotemporal dementia. Alzheimer Dis Assoc Disord 2007; 21:155-163.
21. Roberson ED, Hesse JH, Rose KD et al. Frontotemporal dementia progresses to death faster than Alzheimer disease. Neurology 2005; 65:719-725.

22. Liu W, Miller BL, Kramer JH et al. Behavioral disorders in the frontal and temporal variants of frontotemporal dementia. Neurology 2004; 62:742-748.
23. Mendez MF, Shapira JS, Miller BL. Stereotypical movements and frontotemporal dementia. Mov Disord 2005; 20:742-745.
24. Ikeda M, Brown J, Holland AJ et al. Changes in appetite, food preference and eating habits in frontotemporal dementia and Alzheimer's disease. J Neurol Neurosurg Psychiatry 2002; 73:371-376
25. Miller BL, Seeley WW, Mychack P et al. Neuroanatomy of the self: evidence from patients with frontotemporal dementia. Neurology 2001; 57:817-821.
26. Kramer JH, Jurik J, Sha SJ et al. Distinctive neuropsychological patterns in frontotemporal dementia, semantic dementia and Alzheimer disease. Cogn Behav Neurol 2003; 16:211-218.
27. Levy ML, Miller BL, Cummings JL et al. Alzheimer disease and frontotemporal dementias: behavioral distinctions. Arch Neurol 1996; 53:687-690.
28. Rosen HJ, Gorno-Tempini ML, Goldman WP et al. Patterns of brain atrophy in frontotemporal dementia and semantic dementia. Neurology 2002; 58:198-208.
29. McNeill R, Sare GM, Manoharan M, Testa et al. J Neurol Neurosurg Psychiatry 2007; 78(4):350-355.
30. Grimmer T, Diehl J, Drzezga A et al. Region-specific decline of cerebral glucose metabolism in patients with frontotemporal dementia: a prospective 18F-FDG-PET study. Dement Geriatr Cogn Disord 2004.
31. Whitwell JL, Avula R, Senjem ML et al. Gray and white matter water diffusion in the syndromic variants of frontotemporal dementia. Neurology 2010; 74(16):1279-1287.
32. Borroni B, Brambati SM, Agosti C et al. Evidence of white matter changes on diffusion tensor imaging in frontotemporal dementia. Arch Neurol 2007; 64(2):246-251.
33. Rosen HJ, Allison SC, Schauer GF et al. Neuroanatomical correlates of behavioural disorders in dementia. Brain 2005; 128:2612-2625.
34. Rabinovici GD, Seeley WW, Kim EJ et al. Distinct MRI atrophy patterns in autopsy-proven Alzheimer's disease and frontotemporal lobar degeneration. Am J Alzheimers Dis Other Demen 2007; 22:474-488.
35. Lomen-Hoerth C. Characterization of amyotrophic lateral sclerosis and frontotemporal dementia. Dement Geriatr Cogn Disord 2004; 17:337-341.
36. Gorno-Tempini ML, Dronkers NF, Rankin KP et al. Cognition and anatomy in three variants of primary progressive aphasia. Ann Neurol 2004; 55:335-346.
37. Rosen HJ, Allison SC, Ogar JM et al. Behavioral features in semantic dementia vs other forms of progressive aphasias. Neurology 2006; 67:1752-1756.
38. Josephs KA, Duffy JR, Strand EA et al. Clinicopathological and imaging correlates of progressive aphasia and apraxia of speech. Brain 2006; 129:1385-1398.
39. Seeley WW, Bauer AM, Miller BL et al. The natural history of temporal variant frontotemporal dementia. Neurology 2005; 64:1384-1390.
40. Gorno-Tempini ML, Rankin KP, Woolley JD et al. Cognitive and behavioral profile in a case of right anterior temporal lobe neurodegeneration. Cortex 2004; 40:631-644.
41. Brun A. Frontal lobe degeneration of the non-Alzheimer type: I. Neuropathology Arch Gerontol Geriatr 1987; 6:193-208.
42. Short RA, Broderick DF, Patton A. Different patterns of magnetic resonance imaging atrophy for frontotemporal lobar degeneration syndromes. Arch Neurol 2005; 62(7):1106-1110.
43. Dickson DW. Neuropathology of Pick's disease. Neurology 2001; 56:S16-S20.
44. Mackenzie IR, Neumann M, Bigio EH. Nomenclature and nosology for neuropathologic subtypes of frontotemporal lobar degeneration: an update. Acta Neuropathol 2010; 119:1-4.
45. Spillantini MG, Goedert M. Tau protein pathology in neurodegenerative diseases. Trends Neurosci 1998; 21:428-433.
46. Lee G, Neve RL, Kosik KS. The microtubule binding domain of tau protein. Neuron 1989; 2:1615-1624.
47. Andreadis A, Brown WM, Kosik KS. Structure and novel exons of the human tau gene. Biochemistry 1992; 31:10626-10633.
48. Cooper PN, Jackson M, Lennox G et al. Tau, ubiquitin and alpha B-crystallin immunohistochemistry define the principal causes of degenerative frontotemporal dementia. Arch Neurol 1995; 52:1011-1015.
49. Cairns NJ, Bigio EH, Mackenzie IR et al. Neuropathologic diagnostic and nosologic criteria for frontotemporal lobar degeneration: consensus of the Consortium for Frontotemporal Lobar Degeneration. Acta Neuropathol 2007; 114:5-22.
50. Knopman DS. Overview of dementia lacking distinctive histology: pathological designation of a progressive dementia. Dementia 1993; 4:132-136. Neumann M, Sampathu DM, Kwong LK et al. Ubiquitinated TDP-43 in frontotemporal lobar degeneration and amyotrophic lateral sclerosis. Science 2006; 314:130-133.
51. Neumann M, Sampathu DM, Kwong LK et al. Ubiquitinated TDP-43 in frontotemporal lobar degeneration and amyotrophic lateral sclerosis. Science 2006; 314:130-133.

52. Arai T, Hasegawa M, Akiyama H et al. TDP-43 is a component of ubiquitin-positive tau-negative inclusions in frontotemporal lobar degeneration and amyotrophic lateral sclerosis. Biochem Biophys Res Commun 2006; 351:602-611.
53. Hasegawa M, Arai T, Nonaka T et al. Phosphorylated TDP-43 in frontotemporal lobar degeneration and amyotrophic lateral sclerosis. Ann Neurol 2008; 64(1):60-70.
54. Zhang YJ, Xu YF, Dickey CA et al. Progranulin mediates caspase-dependent cleavage of TAR DNA binding protein-43. J Neurosci 2007; 27:10530-10534.
55. Igaz LM, Kwong LK, Xu Y et al. Enrichment of C-terminal fragments in TAR DNA-binding protein-43 cytoplasmic inclusions in brain but not in spinal cord of frontotemporal lobar degeneration and amyotrophic lateral sclerosis. Am J Pathol 2008; 173:182-194.
56. Igaz LM, Kwong LK, Chen-Plotkin A et al. Expression of TDP-43 C-terminal fragments in vitro recapitulates pathological features of TDP-43 proteinopathies. J Biol Chem 2009; 284:8516-8524.
57. Davidson Y, Kelley T, Mackenzie IR et al. Ubiquitinated pathological lesions in frontotemporal lobar degeneration contain the TAR DNA-binding protein, TDP-43. Acta Neuropathol (Berl) 2007; 113:521-533.
58. Masellis M, Momeni P, Meschino W et al. Novel splicing mutation in the progranulin gene causing familial corticobasal syndrome. Brain 2006; 129:3115-3123.
59. Josephs KA, Stroh A, Dugger B et al. Evaluation of subcortical pathology and clinical correlations in FTLD-U subtypes. Acta Neuropathol 2009; 118:349-358.
60. Mackenzie IR, Shi J, Shaw CL et al. Dementia lacking distinctive histology (DLDH) revisited. Acta Neuropathol 2006; 112:551-559.
61. Forman MS, Mackenzie IR, Cairns NJ et al. Novel ubiquitin neuropathology in frontotemporal dementia with valosin-containing protein gene mutations. J Neuropathol Exp Neurol 2006; 65:571-581.
62. Sreedharan J, Blair IP, Tripathi VB et al. TDP-43 mutations in familial and sporadic amyotrophic lateral sclerosis. Science 2008; 319:1668-1672.
63. Buratti E, Baralle FE. Multiple roles of TDP-43 in gene expression, splicing regulation and human disease. Front Biosci 2008; 13:867-878.
64. Kabashi E, Lin L, Tradewell ML et al. Gain and loss of function of ALS-related mutations of TARDBP (TDP-43) cause motor deficits in vivo. Hum Mol Genet 2010; 19:671-683.
65. Johnson BS, McCaffery JM, Lindquist S et al. A yeast TDP-43 proteinopathy model: exploring the molecular determinants of TDP-43 aggregation and cellular toxicity. Proc Natl Acad Sci USA 2008; 105:6439-6444.
66. Zhang YJ, Xu YF, Cook C et al. Aberrant cleavage of TDP-43 enhances aggregation and cellular toxicity. Proc Natl Acad Sci USA 2009; 106:7607-7612.
67. Mackenzie IR, Foti D, Woulfe J et al. A typical frontotemporal lobar degeneration with ubiquitin positive, TDP-43-negative neuronal inclusions. Brain 2008; 131:1282-1293.
68. Neumann M, Rademakers R, Roeber S et al. A new subtype of frontotemporal lobar degeneration with FUS pathology. Brain 2009; 132(Pt 11):2922-2931.
69. Urwin H, Josephs KA, Rohrer JD et al. FUS pathology defines the majority of tau- and TDP-43 negative frontotemporal lobar degeneration. Acta Neuropathol 2010; 120(1):33-41.
70. Lagier-Tourenne C, Cleveland DW. Rethinking ALS: the FUS about TDP-43. Cell 2009; 136(6):1001-1004.
71. Josephs KA, Whitwell JL, Parisi JE et al. Caudate atrophy on MRI is a characteristic feature of FTLD-FUS. Eur J Neurol 2010 17:969-975.
72. Dormann D, Rodde R, Edbauer D et al. ALS-associated fused in sarcoma (FUS) mutations disrupt transportin-mediated nuclear import. EMBO J 2010; 29:2841-2857.
73. Bedford MT, Clarke SG. Protein arginine methylation in mammals: who, what and why. Mol Cell 2009; 33:1-13.
74. Mackenzie IR, Neumann M, Bigio EH et al. Nomenclature for neuropathologic subtypes of frontotemporal lobar degeneration: consensus recommendations. Acta Neuropathol 2009; 117(1):15-18.
75. Holm IE, Englund E, Mackenzie IRA et al. A reassessment of the neuropathology of frontotemporal dementia linked to chromosome 3 (FTD-3). J Neuropathol Exp Neurol 2007; 66:884-889.
76. Holm IE, Isaacs A, Mackenzie IR. Absence of FUS-immunoreactive pathology in frontotemporal dementia linked to chromosome 3 (FTD-3) caused by mutation in the CHMP2B gene. Acta Neuropathol 2009; 118:719-720.
77. Urwin H, Ghazi-Noori S, Collinge J et al. The role of CHMP2B in frontotemporal dementia. Biochem Soc Trans 2009; 37:208-212.
78. Rohrer JD, Guerreiro R, Vandrovcova J et al. Neurology 2009; 73(18):1451-1456
79. Lynch T, Sano M, Marder KS et al. Clinical characteristics of a family with chromosome 17-linked disinhibition-dementia-parkinsonism-amyotrophy complex. Neurology 1994; 44:1878-1884.
80. Spillantini MG, Murrell JR, Goedert M et al. Mutation in the tau gene in familial multiple system tauopathy with presenile dementia. Proc Natl Acad Sci USA 1998; 95:7737-7741.

81. Baker M, Mackenzie IR, Pickering-Brown SM et al. Mutations in progranulin cause tau-negative frontotemporal dementia linked to chromosome 17. Nature 2006; 442:916-919.

82. D'Souza I, Poorkaj P, Hong M et al. Missense and silent tau gene mutations cause frontotemporal dementia with parkinsonism-chromosome 17 type, by affecting multiple alternative RNA splicing regulatory elements. Proc Natl Acad Sci USA 1999; 96:5598-5603.

83. Seelaar H, Kamphorst W, Rosso SM et al. Distinct genetic forms of frontotemporal dementia. Neurology 2008; 71:1220-1226.

84. Bird T, Nochlin D, Poorkaj P et al. A clinical pathological comparison of three families with frontotemporal dementia and identical mutations in the tau gene (P301L). Brain 1999; 122 (Pt 4):741-756.

85. Baker M, Litvan I, Houlden H et al. Association of an extended haplotype in the tau gene with progressive supranuclear palsy. Hum Mol Genet 1999; 8:711-715

86. Borroni B, Yancopoulou D, Tsutsui M et al. Association between tau H2 haplotype and age at onset in frontotemporal dementia. Arch Neurol 2005; 62(9):1419-1422.

87. Borroni B, Perani D, Agosti C et al. Tau haplotype influences cerebral perfusion pattern in frontotemporal lobar degeneration and related disorders. Acta Neurol Scand 2008; 117(5):359-366.

88. Baba Y, Tsuboi Y, Baker MC et al. The effect of tau genotype on clinical features in FTDP-17. Parkinsonism Relat Disord 2005; 11(4):205-208.

89. Boeve BF, Tremont-Lukats I, Waclawik A et al. Longitudinal characterization of two siblings with frontotemporal dementia and parkinsonism linked to chromosome 17 associated with the S305N tau mutation. Brain 2005; 128 (pt 4):752-772.

90. He Z, Bateman A. Progranulin (granulin-epithelin precursor, PC-cell-derived growth factor, acrogranin) mediates tissue repair and tumorigenesis. J Mol Med 2003; 81:600-612.

91. Ahmed Z, Mackenzie IR, Hutton ML et al. Progranulin in frontotemporal lobar degeneration and neuroinflammation. J Neuroinflammation 2007; 4:7.

92. Van Damme P, Van Hoecke A, Lambrechts D et al. Progranulin functions as a neurotrophic factor to regulate neurite outgrowth and enhance neuronal survival. J Cell Biol 2008; 181:37-41.

93. Hrabal R et al. The hairpin stack fold, a novel protein architecture for a new family of protein growth factors. Nat Struct Biol 1996; 3:747-752.

94. Cruts M, Van Broeckhoven C. Loss of progranulin function in frontotemporal lobar degeneration. Trends Genet 2008; 24:186-194.

95. Yu CE, Bird TD, Bekris LM et al. The spectrum of mutations in progranulin. Arch Neurol 2010; 67(2):161-170.

96. Chen-Plotkin AS, Geser F, Plotkin JB et al. Variations in the progranulin gene affect global gene expression in frontotemporal lobar degeneration. Hum Mol Genet 2008; 17:1349-1362.

97. van Swieten JC, Heutink P. Mutations in progranulin (GRN) within the spectrum of clinical and pathological phenotypes of frontotemporal dementia. Lancet Neurol 2008; 7:965-974.

98. Le Ber I, van der Zee J, Hannequin D et al. Progranulin null mutations in both sporadic and familial frontotemporal dementia. Hum Mutat 2007; 28:846-855.

99. Lo´pez de Munain A, Alzualde A, Gorostidi A et al. Mutations in progranulin gene: clinical, pathological and ribonucleic acid expression findings. Biol Psychiatry 2008; 63:946-952.

100. Gijselinck I, Van Broeckhoven C, Cruts M. Granulin mutations associated with frontotemporal lobar degeneration and related disorders: an update. Hum Mutat 2008; 29(12):1373-1386.

101. Borroni B, Archetti S, Alberici A et al. Progranulin genetic variations in frontotemporal lobar degeneration: evidence for low mutation frequency in an Italian clinical series. Neurogenetics 2008; 9:197-205.

102. Le Ber I, Camuzat A, Hannequin D et al. Phenotype variability in progranulin mutation carriers: a clinical, neuropsychological, imaging and genetic study. Brain 2008; 131:732-746.

103. Beck J, Rohrer JD, Campbell T et al. A distinct clinical, neuropsychological and radiological phenotype is associated with progranulin gene mutations in a large UK series. Brain 2008; 131:706-720.

104. Watts GD, Wymer J, Kovach MJ et al. Inclusion body myopathy associated with Paget disease of bone and frontotemporal dementia is caused by mutant valosin-containing protein. Nat Genet 2004; 36:377-381.

105. Kimonis VE, Mehta SG, Fulchiero EC et al. Clinical studies in familial VCP myopathy associated with Paget disease of bone and frontotemporal dementia. Am J Med Genet A 2008; 146A:745-757.

106. Borroni B, Archetti S, Del Bo R et al. TARDBP Mutations in Frontotemporal Lobar Degeneration: Frequency, Clinical Features and Disease Course. Rejuvenation Res 2010; (ahead of print).

107. Borroni B, Bonvicini C, Alberici A et al. Mutation within TARDBP leads to frontotemporal dementia without motor neuron disease. Hum Mutat 2009; 30:E974-E983.

108. Benajiba L, Le Ber I, Camuzat A et al. TARDBP mutations in motoneuron disease with frontotemporal lobar degeneration. Ann Neurol 2009; 65:470-473.

109. Kovacs GG, Murrell JR, Horvath S et al. TARDBP variation associated with frontotemporal dementia, supranuclear gaze palsy and chorea. Mov Disord 2009; 24:1843-1847.

110. Blair IP, Williams KL, Warraich ST et al. FUS mutations in amyotrophic lateral sclerosis: clinical, pathological, neurophysiological and genetic analysis. J Neurol Neurosurg Psychiatry 2009; 81:639-641.

111. Ticozzi N, Silani V, LeClerc AL et al. Analysis of FUS gene mutation in familial amyotrophic lateral sclerosis within an Italian cohort. Neurology 2009; 73:1180-1185.
112. Van Deerlin V, Martinez-Lage M, Hakonarson H et al. Genome-wide association study of frontotemporal lobar degeneration with or without concomitant motor neuron disease and TDP-43 neuropathology. 6th International Conference on Frontotemporal Dementias 2008; Rotterdam.
113. Vance C, Al-Chalabi A, Ruddy D et al. Familial amyotrophic lateral sclerosis with frontotemporal dementia is linked to a locus on chromosome 9p13.2-21.3. Brain 2006; 129:868-876.
114. Van Deerlin VM, Sleiman PM, Martinez-Lage M et al. Common variants at 7p21 are associated with frontotemporal lobar degeneration with TDP-43 inclusions. Nat Genet 2010; 42(3):234-239.
115. Perry RJ, Miller BL. Behavior and treatment in frontotemporal dementia. Neurology 2001; 56 (Suppl 4):S46-S51.
116. Procter AW, Qume M, Francis PT. Neurochemical features of frontotemporal dementia. Dement Geriatr Cog Disord 1999; 10 (suppl 1): 80-84.
117. Franceschi M, Anchisi D, Pelati O et al. Glucose metabolism and serotonin receptors in the frontotemporal lobe degeneration. Ann Neurol 2005; 57:216-225.
118. Huey ED, Putnam KT, Grafman J. A systematic review of neurotransmitter deficits and treatments in frontotemporal dementia. Neurology 2006; 66:17-22.
119. Moretti R, Torre P, Antonello RM et al. Olanzapine as a treatment of neuropsychiatric disorders of Alzheimer's disease and other dementias: a 24-month follow-up of 68 patients. Am J Alzheimers Dis Other Demen 2003; 18:205-214.
120. Pijnenburg YA, Sampson EL, Harvey RJ et al. Vulnerability to neuroleptic side effects in frontotemporal lobar degeneration. Int J Geriatr Psychiatry 2003; 18:67-72.
121. Hansen LA, Deteresa R, Tobias H et al. Neocortical morphometry and cholinergic neurochemistry in Pick's disease. Am J Pathol 1988; 131:507-518.
122. Mendez MF, Shapira JS, McMurtray A et al. Preliminary findings: behavioral worsening on donepezil in patients with frontotemporal dementia. Am J Geriatr Psychiatry 2007; 15(1):84-87.
123. Swanberg MM. Memantine for behavioral disturbances in frontotemporal dementia: a case series. Alzheimer Dis Assoc Disord 2007; 21:164-166.
124. Borroni B, Alberici A, Archetti S et al. New insights into biological markers of frontotemporal lobar degeneration spectrum. Curr Med Chem 2010; 17(10):1002-1009.
125. Mioshi E, Hsieh S, Savage S et al. Clinical staging and disease progression in frontotemporal dementia. Neurology 2010; 74(20):1591-1597.
126. Trojanowski JQ, Duff K, Fillit H et al. New directions for frontotemporal dementia drug discovery. Frontotemporal Dementia (FTD) Working Group on FTD Drug Discovery. Alzheimers Dement 2008; 4(2):89-93.
127. Knopman DS, Kramer JH, Boeve BF et al. Development of methodology for conducting clinical trials in frontotemporal lobar degeneration. Brain 2008; 131:2957-2968.

CHAPTER 10

GERSTMANN-STRÄUSSLER-SCHEINKER DISEASE

Paweł P. Liberski

Department of Molecular Pathology and Neuropathology, Medical University Lodz, Lodz, Poland
Email: ppliber@csk.am.lodz.pl

Abstract: Gerstmann-Sträussler-Scheinker (GSS) is a slowly progressive hereditary autosomal dominant disease (OMIM: 137440) and the first human transmissible spongiform encephalopathy (TSE) in which a mutation in a gene encoding for prion protein (PrP) was discovered. The first "H" family had been known by the Viennese neuropsychiatrists since the XXth century and was reported by Gerstmann, Sträussler and Scheinker in 1936. In this chapter we present the clinical, neuropathological and molecular data on GSS with the mutations in the PRNP gene: at codons 102, 105, 117, 131, 145, 187, 198, 202, 212, 217 and 232. In several families with GSS the responsible mutations are unknown.

INTRODUCTION

Gerstmann-Sträussler-Scheinker (GSS) is a slowly progressive hereditary autosomal dominant neurodegenerative disease of the Central Nervous System (CNS) (OMIM: 137440) and the first human transmissible spongiform encephalopathy (TSE) in which a mutation in a gene encoding for prion protein (PrP) was discovered. The true prevalence is difficult to estimate but numbers in a range of 1-10/100 000 000 are quoted.[1]

According to Budka et al,[2] GSS is defined as a neurodegenerative disease "in family with dominantly inherited progressive ataxia and/or dementia): encephalo(myelo)pathy with multicentric PrP plaques" (Figs. 1,2).

The first "H" family had been known by the Viennese neuropsychiatrists since the 20th century and was reported by Dimitz in 1913,[3] then by Gerstmann in 1928[4] and again by Gerstmann, Sträussler and Scheinker in 1936.[5] Later on Scheinker became an established neuropathologist and published, among other works, *Neuropathology in Its Clinicopathologic Aspects.*[6]

Neurodegenerative Diseases, edited by Shamim I. Ahmad.
©2012 Landes Bioscience and Springer Science+Business Media.

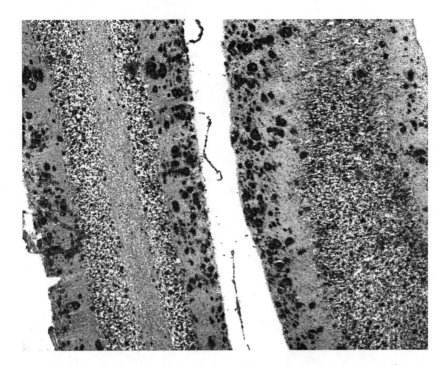

Figure 1. Cerebellum loaded with plaques from a case of the 232 mutation.[87]

Gerstmann in 1928 described a peculiar reflex, when a patient extended both arms in front of the chest and then the head was turned to one side, both arms crossed moving to the midline.[4] The arm contralateral to direction of turning was placed above the other arm. Subsequent members of the same family were described by von Braunmühl[7] and Franz Seitelberger, then director of the Obersteiner Institute, Vienna, Austria.[8,9] Seitelberger, four years before the discovery of the transmissible nature of kuru by Gajdusek et al[10] stressed the close neuropathological similarity (amyloid plaques—Figs. 1,2) of these two diseases and in a sense preconceived the transmissible nature of GSS.[11] The history of original GSS "H" family from Vienna is interesting. This family stems from a little rural town in the lower Austria (Niederoestereich) and had been diagnosed by local physicians as suffering from a form of hereditary neurosyphilis. As this diagnosis would stigmatize them, they decided to hide from doctors. In 1990, one of us consulted on a female case suspected of CJD whose father died with a diagnosis of "Friedreich ataxia". The maiden name of this case "H" was the name of the GSS family.[12,13] This discovery enabled modern studies of that fascinating kindred.

In 1981 a seminal paper by Masters et al[11] was published. In this paper, several GSS cases were proved transmissible to nonhuman primates. The Fujisaki strain of GSS (codon 102 mutation) first isolated by Tateishi et al[14] was passaged to mice, rats, guinea pigs and Squirrel monkeys. Another case with the same mutation[15] was passaged to Spider monkeys and to Marmosets.[16] To date, only inocula derived from 5 brains with 102[Leu] could be transmitted.[17,18]

Of note, in the 1981 paper, Masters et al mentioned the "CG" family described by Worster-Drought et al.[19-21] While phenotypically similar to GSS in regard to amyloid plaques, this family had been later proved to represent Familial British Dementia with genetic alteration obviously different from that in GSS.[22]

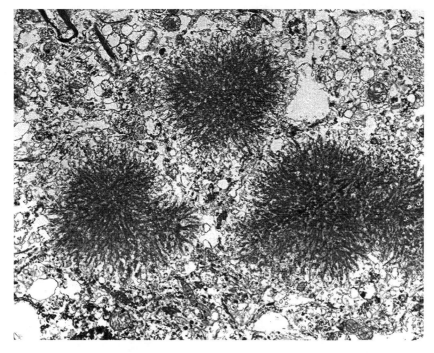

Figure 2. Electron microscopy of numerous plaques from a case of GSS from the "H" family.[12,13]

CODON 102 MUTATION (102^{Leu} 129^{Met})

GSS with mutation at codon 102 of the prion protein gene (*PRNP*) was the first target of molecular-genetic approach in 1989.[23] At codon 102, a mutation leading to a substitution of Pro (CCG) by Leu (CTG) was found. This mutation was subsequently found in several families from Japan including, "I" family;[24,25] Germany[26,27]—in the well characterized Sch. family,[28-31] Israel,[32] Hungary;[33] Poland,[34] UK,[35,36] Italy,[37-39] and in the original Viennese "H" family.[40]

The original family from which 4 cases were described by Seitelberger[9] numbered then 81 members; currently expanded to 221 member including 20 definitive GSS cases.[12] The disease manifests as slowly progressive cerebellar ataxia with dementia appearing late. The last case of GSS from this family (children of this female were tested for a mutation and proved negative for the codon 102 mutation) exhibited, however, features of otherwise typical CJD—i.e. early symptoms of dementia and a characteristic periodic EEG.

For some GSS families harboring the 102 codon mutation, a typical feature is heterogeneity of neurological signs and symptoms. The classical ataxic type starts in second to sixth decade and the duration of the disease ranges from a few months to a few years. Dysartria and alterations of saccadic eye movements, pyramidal and extrapyramidal signs and symptoms, cognitive changes leading to frank dementia are typical features. In a proportion of cases, a CJD-like disease type with myoclonic jerks and periodic EEG pattern is observed. MRI demonstrated mild atrophy of the cerebellum and the brain.

A separate problem is status of the codon 129 in a coupling with a mutated codon 102. In almost all GSS cases with this mutation, it is coupled with 129^{Met}. Cases coupled with 129[Val] are rare. A case described by Young et al[41] was a 33-year-old male, clinically significantly different from those of 129[Met], with seizures as a first sign, lower limb paraesthesias and bilateral deafness. Dementia was not observed. This male died at the age of 45, some 12 years after the first sings and symptoms were noticed.[42]

GSS with a mutation of codon 102 is transmissible to non human primates[11,17] as well as to rodents.[14] It seems that GSS cases with 102 mutations transmitted thus far.

Antibodies raised against different segments of prion protein (PrP) sequence help to elucidate the composition of peptides forming plaques.[43] Plaques were labeled with Abs raised against PrP 90-102 and, in much smaller proportion, with Abs raised against peptide PrP 58-71. Plaque cores were also strongly stained with Abs raised against residues 95-108, 127-147 and 151-165. Abs raised against PrP residues 23-40 (N-terminus) and 220-231 (C-terminus) stained peripheries of plaques as ring-shaped structures. Some plaque cores are labeled, however, with all Abs irrespective whether raised against either midportion of PrP or its N- and C-terimini. The latter findings indicate that both truncated peptides and full-length PrP may form amyloid fibrils but the truncated fibrils predominated.

MUTATION OF CODON 105 (105[Leu] 129[Val])

This mutation was found in 5 GSS families, all from Japan.[44-52] The disease manifests as spastic paraparesis with brisk reflexes and the presence of Babinski sign; in terminal stages, a patient becomes teraplegic, demented, with tremor and limb rigidity. Illness starts around 40-50 year of age and lasts 6-12 years. PrP deposits are encountered mainly in the cerebral cortex and less frequently in striatum. The cerebellum is affected only minimally. In two cases, sparse neurofibrillary tangles composed of paired helical filaments were seen.[44] Numerous neurofibrillary tangles [NFTs] were found in a case of 57-year-old female with dementia but not spastic paraparesis.[52] Of note, in a case described by Amano et al,[44] another type of plaques were seen—localized in the fifth and sixth layers of cerebral cortex, weakly PAS-positive, confluent and of laminar distribution.

MUTATION OF CODON 117 (117[Val] 129[Val])

This mutation was discovered in families characterized by dementia but not by otherwise typical for GSS cerebellar ataxia ("telencephalic type").[53-62] Mastrianni et al[58] described the cerebellar syndrome. In an Alsatian family, in earlier generations, only "pure" dementia was observed. In more recent generations, more complex pattern of signs and symptoms, including dementia, were noticed. Amyloid plaques were reactive with Abs raised against the central region of PrP while Abs to the C- and N-termini of the molecule stained the peripheries of plaques.[63] A 7 kDa peptide of PrP[d] was found by Western blot,[63-65] but the presence of PrP[d] varied; in some samples no PrP[d] was found.[64,65] Also in the brain of an asymptomatic carrier, no PrP[d] was seen.[64,65]

MUTATION OF CODON 131 (131[Met] 129[Val])

This mutation was found in only one family characterized by dementia, apraxia, cerebellar ataxia, extrapyramidal signs and brisk tendom reflexes.[66] The disease started in the 5th decade and lasted for 9 years. MRI demonstrated cerebral and cerebellar atrophies. The family history was negative. Numerous PrP-amyloid plaques and diffuse deposits were seen in cerebral cortex, basal ganglia and cerebellum.

MUTATION OF CODON 145 (145[Stop])

This mutation was discovered by Kitamoto et al[67] in a case with spastic paraparesis and progressive severe dementia. Neuropathological examination revealed numerous PrP plaques and PrP deposits in the wall of brain vessels as well as meningeal vessels (PrP-congophilic angiopathy). NFT were seen in the neocortex.

MUTATION OF CODON 187 (187[Arg] 129[Val])

This mutation was found in one American GSS family from the USA.[68] Nine affected cases were characterized by dementia, cerebellar ataxia, myoclonic jerks and seizures. The median age at onset is 42 years (range 33-50 years) and the duration of illness ranges from 8 to 19 years. Neuropathological examination revealed PrP[d] deposits in cerebral cortex of distinct "curly" appearance and laminar pattern. PrP plaques were absent and spongiform change was not seen.

MUTATION OF CODON 198 (198[Ser] 129[Val])

This mutation was discovered in a family from Indiana ("Indiana kindred", IK)[69] and in another unrelated family.[70] Patients are homozygous or heterozygous in respect to Val at codon 129.

The IK is characterized by pyramidal and cerebellar signs, dementia, dysarthria and progressive clumsiness and difficulties of walking, prominent parkinsonian features, bradykinesia, cogwheel rigidity but no tremors, optokinetic nystagmus and sleep disturbances.[71,72] Alterations of saccadic eye movements[73] may be detected before other signs and symptoms appear. The disease starts between 40 and 70 years of age and in patients homozygous for 129[Val Val] the beginning is approximately 10 years earlier than in heterozygous cases 129[Val Met] patients. The disease lasts approximately 5 years (from 2 to 12 years) but may be accelerated to 1-2 years.

Neuropathological examination revealed changes otherwise typical for GSS.[74-82] Neurites around plaques contained NFT composed, not unlike those of Alzheimer's disease, of hyperphosphorylated MAP-τ. Spongiform change was occasionally visible around plaques.

Antibodies raised against different segments of PrP peptides helped to resolve the question of plaque composition.[43,83] Plaques are composed of two species of PrP—7 and 11 kDa spanning PrP residues 81-150 and 58-150 respectively. In contrast, nonfibrillar (pre-amyloid) PrP is immunolabeleld with antibodies raised against residues 23-40 and 220-231.[76] Abs raised against PrP residues 23-40 (N-terminus) and 220-231 (C-terminus) stained peripheries of plaques as ring-shaped structures.[84]

MUTATION OF CODON 202 (202^{Asn} 129^{Val}) AND MUTATION OF CODON 212 (212^{Pro} 129^{Met})[75,80,82]

The duration of illness of a case with 202^{Asn} was 6 years, the disease started in the 8th decade of life and manifested as dementia with cerebellar signs. PrP plaques were seen in both brain and the cerebellum. Numerous NFT were visible in the cerebral cortex. The patient with mutation 212^{Pro} became ill at the year 60 and the disease lasted for 8 years.[1] Slurred speech, cerebellar ataxia leading to total incapacitation but not dementia were seen. PrP plaques were visible in both brain and the cerebellum but their density was the lowest among all GSS families.

MUTATION OF CODON 217 (217^{Arg} 129^{Val})

This mutation was described by Karen Hsiao et al[56] in 2 patients from a Swedish-American family[74,76,85] with psychotic manic-depression disturbances, dementia, ataxia and parkinsonian features. The neuropathological picture is similar to that of IK; numerous PrP plaques and NFTs composed of paired helical filaments. PrP^d in plaques coexists with Aβ peptide.

MUTATION OF CODON 232 (232^{Thr})

This mutation was found by Liberski et al[86,87] in a case diagnosed earlier as olivo-ponto-cerebellar degeneration with spastic paraparesis and dementia. The disease started in the 5th decade of life and lasted for 6 years. Numerous PrP plaques were visible in the cerebral and cerebellar cortex and subcortical nuclei; in substantia, nigra Lewy bodies were seen occasionally.

UNKNOWN MUTATIONS

A few families (Italian-Canadian family;[88] "N" family[89]) and some others[90-94] were reported as GSS but the mutations were not known. The disease described by de Courten-Myers and Mandybur[95] was Alzheimer as later proved by immunohistochemistry.[96]

BIOLOGY OF GSS

Nomenclature

PrP^{Sc} is the pathological misfolded protein, insoluble in denaturing detergent; however, some pathological isoforms of PrP^{Sc} have recently been found not to be PK-resistant.[97] The neutral term "PrP^{d}" denotes the misfolded species of PrP which is disease-associated. PrP 27-30 is the proteolytic cleavage product of PrP^{Sc}.[98,99]

PrP PEPTIDES IN GSS

Two types of the unglycosylated isoforms of PrP^d may be present in human TSEs—Type 1 (21 kDa) and Type 2 (19 kDa). In GSS smaller peptides in the range of 7 to 8 kDa have been found. In GSS 102^{Leu}, two patterns of PrP^d were observed on the Western blot: a single 8 kDa band or 3 bands of 29, 27 and 21 kDa.[100] The 21 kDa peptide is N-truncated while the 8 kDa band is both N- and C-truncated. N-terminal sequencing revealed N-terminal cleavage site at residues 78, 80 and 82 for the 8 kDa peptide and residues 78 and 82 for the 21 kDa peptide. An additional cleavage site at residue 74 was found by mass spectroscopy for the 8 kDa peptide. It is interesting, that the 21 kDa PrP^d was purified only from those GSS brains where spongiform change was also seen.[100] A peptide of 7 kDa is associated with GSS 117^{Val}.[63] Microsequencing and mass spectroscopy revealed several N-terminal cleavage sites at residues 85-95. The most frequent were residues 88, 90 and 92 of PrP. The C-terminus was also ragged and ended with residues 148, 152 or 153. Of note, the 7 kDa peptide was derived only from the mutated allele.

CONCLUSION

In conclusion, GSS is a neurodegenerative diverse of diverse clinicopathological picture. The hallmark of microscopic picture is a multicentric plaque and thus, GSS is a prion disease similar to kuru.

REFERENCES

1. Kong Q, Surewicz WK, Petersen RB et al. Inherited prion diseases. In: Prusiner SB, ed. Prion Biology and Diseases. New York: Cold Spring Harbor Laboratory Press, 2004:673-775.
2. Budka H, Aguzzi A, Brown P et al. Neuropathological diagnostic criteria for Creutzfeldt-Jakob disease (CJD) and other human spongiform encephalopathies (prion diseases). Brain Pathol 1995; 4:459-466.
3. Dimitz L. Bericht der Vereines fur Psychiatrie und Neurologie in Wien (Vereinsjahr 1912/1913), Sitzung vom 11 Juni 1912. Jahrb Psychiatr Neurol 1913; 34:384.
4. Gerstmann J. über ein noch nicht beschriebenes Reflexphanomen beieiner Erkrankung des zerebellaren Systems. Wien Medizin Wochenschr 1928; 78:906-908.
5. Gerstmann J, Sträussler E, Scheinker I. Uber eine eigenartige hereditär-familiäre Erkrankung des Zetralnervensystems. Zugleich ein Beitrag zur Frage des vorzeitigen lokalen Alterns. Z. ges. Neurol Psychiat 1936; 154:736-762.
6. Scheinker I. Neuropathology in its Clinicopathologic Aspects. Charles C Thomas:Springfield, 1947.
7. Braunmühl von A. Uber eine eigenartige hereditaer-familiaere Erkrankung des Zentralnervensystems. Arch Psychiatr Z Neurol 1954; 191:419-449.
8. Seitelberger F. Eigenartige familiar-hereditare Krankheit des Zetralnervensystems in einer niederosterreichischen Sippe. Wien Klin Wochen 1962; 74:687-691.
9. Seitelberger F. Neuropathological conditions related to neuroaxonal dystrophy. Acta Neuropathol (Berl) 1971; 7:17-29.
10. Gajdusek DC, Gibbs CJ, Alpers MP. Experimental transmission of a kuru-like syndrome to chimpanzees. Nature 1966; 209:794-796.
11. Masters CL, Gajdusek DC, Gibbs CJ Jr. Creutzfeldt-Jakob disease virus isolations from the Gerstmann-Sträussler syndrome. With an analysis of the various forms of amyloid plaque deposition in the virus induced spongiform encephalopathies. Brain 1981; 104:559-588.
12. Hainfellner J, Brantner-Inhaler S, Cervenakova L et al. The original Gerstmann-Sträussler-Scheinker family of Austria: divergent clinicopathological phenotypes but constant PrP genotype. Brain Pathol 1995; 5:201-213.
13. Liberski PP, Budka H. Ultrastructural pathology of Gerstmann-Sträussler-Scheinker disease. Ultrastr Pathol 1995; 19:23-36.
14. Tateishi J, Kitamoto T, Hashiguchi H et al. Gerstmann-Sträussler-Scheinker disease: immunohistological and experimental studies. Ann Neurol 1988; 24:35-40.

15. Rosenthal NP, Keesy J, Crandall B et al. Familial neurological disease associated with spongiform encephalopathy. Arch Neurol 1976; 33:252-259.

16. Baker HF, Duchen LW, Jacobs JM et al. Spongiform encephalopathy transmitted experimentally from Creutzfeldt-Jakob and familial Gerstmann-Sträussler-Scheinker diseases. Brain 1990; 113:1891-1909.

17. Brown P, Gibbs C Jr, Rodgers Johnson P et al. Human spongiform encephalopathy: The National Institutes of Health series of 300 cases of experimentally transmitted disease. Ann Neurol 1994; 35:513-529.

18. Brown—personal communication, 2005.

19. Worster-Drought C, Greenfield JG, McMenemey WH. A form of familial presenile dementia with spastic paralysis (including the pathological examination of a case). Brain 1940; 63:237-254.

20. Worster-Drought C, Greenfield JG, McMenemey WH. A form of familial presenile dementia with spastic paralysis. Brain 1944; 67:38-43.

21. Worster-Drought C, Hill TR, McMenemey WH. Familial presenile dementia with spastic paralysis. J Neurol Psychopathol 1933; 14:27-34.

22. Masters CL, Beyreuther K. The Worster-Drought syndrome and other syndromes of dementia with spastic paraparesis: the paradox of molecular pathology. J Neuropathol Exp Neurol 2001; 60:317-319.

23. Hsiao K, Baker HF, Crow TJ et al. Linkage of a prion protein missense variant to Gerstmann-Sträussler syndrome. Nature 1989; 338:342-345.

24. Kuzuhara S, Kanazawa I, Sasaki H et al. Gerstmann-Sträussler-Scheinker disease. Ann Neurol 1983; 14:216-225.

25. Yamada M, Tomimotsu H, Yokota T et al. Involvement of the spinal posterior horn in Gerstmann-Sträussler-Scheinker disease (PrP P102L). Neurology 1999; 52:260-265.

26. Brown P, Goldfarb LG, Brown WT et al. Clinical and molecular genetic study of a large German kindred with Gerstmann-Sträussler-Scheinker syndrome. Neurology 1991; 41:375-379.

27. Goldgaber D, Goldfarb L, Brown P et al. Mutations in familial Creutzfeldt-Jakob disease and Gerstmann-Sträussler-Scheinker syndrome. Exp Neurol 1989; 106:204-206.

28. Boellaard JW, Schlote W. Subakute spongiforme Encephalopathie mit multiformer Plaquebildung. "Eigenartige familiar-hereditare Kranknheit des Zentralnervensystems [spino-cerebellare Atrophie mit Demenz, Plaques and plaqueähnlichen im Klein- and Grosshirn" (Gerstmann, Sträussler, Scheinker)]. Acta Neuropathol (Berl) 1980; 49:205-212.

29. Doerr-Schott J, Kitamoto T, Tateishi J et al. Technical communication. Immunogold light and electron microscopic detection of amyloid plaques in transmissible spongiform encephalopathies. Neuropathol Appl Neurobiol 1990; 16:85-89.

30. Schlote W, Boellaard JW, Schumm F et al. Gerstmann-Sträussler-Scheinker's disease. Electron-microscopic observations on a brain biopsy. Acta Neuropathol (Berl) 1980; 52:203-211.

31. Schumm F, Boellaard JW, Schlote W et al. Morbus Gerstmann-Sträussler-Scheinker. Familie SCh.—Ein Bericht uber drei Kranke. Arch Psychiatr Nervenkr 1981; 230:179-196.

32. Goldhammer Y, Gabizon R, Meiner Z et al. An Israeli family with Gerstmann-Sträussler-Scheinker disease manifesting the codon 102 mutation in the prion protein gene. Neurology 1993; 43:2718-2719.

33. Majtenyi C, Brown P, Cervenakova L et al. A three-sister sibship of Gerstmann-Sträussler-Scheinker disease with a CJD phenotype. Neurology 2000; 54:2133-2137.

34. Kulczycki J, Collinge J, Lojkowska W et al. Report on the first Polish case of the Gerstmann-Sträussler-Scheinker syndrome. Folia Neuropathol 2001; 39:27-31.

35. Cameron E, Crawford AD. A familial neurological disease complex in a Bedfordshire community. JR Coll Gen Pract 1974; 24:435-436.

36. Collinge J, Harding AE, Owen F et al. Diagnosis of Gerstmann-Sträussler syndrome in familial dementia with prion protein gene analysis. Lancet 1989; 2:15-17.

37. Barbanti P, Fabbrini G, Salvatore M et al. Polymorphism at codon 129 or codon 219 of PRNP and clinical heterogeneity in a previously unreported family with Gerstmann-Sträussler-Scheinker disease (PrP-P102L mutation). Neurology 1996; 47:734-741.

38. Bianca M, Bianca S, Vecchio I et al. Gerstmann-Sträussler-Scheinker disease with P102L-V129 mutation: a case with psychiatric manifestations at onset. Ann Genet 2003; 46:467-469.

39. De Michele G, Pocchiari M, Petraroli R et al. Variable phenotype in a P102L Gerstmann-Sträussler-Scheinker Italian family. Can J Neurol Sci 2003; 30:233-236.

40. Kretzschmar HA, Honold G, Seitelberger F et al. Prion protein mutation in family first reported by Gerstmann, Sträussler and Scheinker. Lancet 1991; 337:1160.

41. Young K, Clark HB, Piccardo P et al. Gerstmann-Sträussler-Scheinker disease with the PRNP P102L mutation and valine at codon 129. Molec Brain Res 1997; 44:147-150.

42. Goodbrand IA, Ironside JW, Nicolson D et al. Prion protein accumulations in the spinal cords of patients with sporadic and growth hormone-associated Creutzfeldt-Jakob disease. Neurosci Lett 1995; 183:127-130.

43. Piccardo P, Ghetti B, Dickson DW et al. Gerstmann-Sträussler-Scheinker disease (PRNP P102L): Amyloid deposits are best recognized by antibodies directed to epitopes in PrP region 90-165. J Neuropathol Exp Neurol 1995; 54:790-801.

44. Amano N, Yagishita S, Yokoi S. Gerstmann-Sträussler-Scheinker syndrome—a variant type: amyloid plaques and Alzheimer's neurofibrillary tangles in cerebral cortex. Acta Neuropathol 1992; 84:15-23.

45. Itoh Y, Yamada M, Hayakawa M et al. A variant of Gerstmann-Sträussler-Scheinker disease carrying codon 105 mutation with codon 129 polymorphism of the prion protein gene: a clinicopathological study. J Neurol Sci 1994; 127:77-86.
46. Kitamoto M, Amano N, Terao Y et al. A new inherited prion disease (PrP P105L mutation) showing spastic paraparesis. Ann Neurol 1993; 34:808-813.
47. Kitamoto T, Ohta M, Doh-Ura K et al. Novel missense variants of prion protein in Creutzfeldt-Jakob disease or Gerstmann-Sträussler-Scheinker syndrome. Bioch Bioph Res Comm 1993; 191:709-714.
48. Kubo M, Nishimura T, Shikata E et al. A case of variant Gerstmann-Sträussler-Scheinker disease with the mutation of codon P105L. Rinsho Shinkeigaku 1995; 35:873-877.
49. Nakazato Y, Ohno R, Negishi T et al. An autopy case of Gerstmann-Sträussler-Scheinker's disease with spastic paraplegia as its principal feature. Clin Neuropathol 1991; 31:987-992.
50. Yamada M, Itoh Y, Fujigasaki H et al. A missense mutation at codon 105 with codon 129 polymorphism of the prion protein gene in a new variant of Gerstmann-Sträussler-Scheinker disease. Neurology 1993; 43:2723-2724.
51. Yamada M, Itoh Y, Inaba A et al. An inherited prion disease with a PrP P105L mutation: clinicopathologic and PrP heterogeneity. Neurology 1999; 53:181-188.
52. Yamazaki M, Oyanagi K, Mori O et al. Variant Gerstmann-Sträussler syndrome with the P105L prion gene mutation: an unusual case with nigral degeneration and widespread neurofibrillary tangles. Acta Neuropathol (Berl) 1999; 98:506-511.
53. Doh-Ura K, Tateishi J, Sakaki Y et al. Pro → Leu change at position 102 of prion protein is the most common but not the sole mutation related to Gerstmann-Sträussler-Scheinker syndrome. Bioch Biophys Res Comm 1989; 163:974-979.
54. Heldt N, Boellaard JW, Brown P et al. Gerstmann-Sträussler-Scheinker disease with A117V mutation in a second French-Alsatian family. Clin Neuropathol 1998; 17:229-234.
55. Heldt N, Floquet J, Warter JM et al. Syndrome de Gerstmann-Sträussler-Scheinker: Neuropathologie de trois cas dans une famille alsacienne. In: Court LA, Cathala F, eds. Virus Non Conventionnels et Affections du Systeme Nerveux Central. Paris: Masson, 1983:290-297.
56. Hsiao K, Dlouhy SR, Farlow MR et al. Mutant prion proteins in Gerstmann-Sträussler-Scheinker disease with neurofibrillary tangles. Nat Genet 1992; 1:68-71.
57. Mallucci GR, Campbell TA, Dickinson A et al. Inherited prion disease with an alanine to valine mutation at codon 117 in the prion protein gene. Brain 1999; 122:1823-1837.
58. Mastrianni JA, Curtis MT, Oberholtzer JC et al. Prion disease (PrP—A117V) presenting with ataxia instead of dementia. Neurology 1996; 45:2042-2050.
59. Mohr M, Tranchant C, Steinmetz G et al. Gerstmann-Sträussler-Scheinker disease and the French-Alsatian A117V variant. Clin Exp Pathol 1999; 47:161-175.
60. Tranchant C, Doh-Ura K, Steinmetz G et al. Mutation of codon 117 of the prion gene in Gerstmann-Sträussler-Scheinker disease. Rev Neurol (Paris) 1991; 147:274-278.
61. Tranchant C, Doh-Ura K, Warter JM et al. Gerstmann-Sträussler-Scheinker disease in an Alsatian family: clinical and genetic studies. J Neurol Neurosurg Psychiatry 1992; 55:185-187.
62. Tranchant C, Sergeant N, Wattez A et al. Neurofibrilllary tangles in Gerstmann-Sträussler-Scheinker syndrome with the A117V prion gene mutation. J Neurol Neurosurg Psychiatry 1997; 63:240-246.
63. Tagliavini F, Lievens PM-J, Tranchant C et al. A 7-kDa prion protein (PrP) fragment, an integral component of the PrP region required for infectivity, is the major amyloid protein in Gerstmann-Sträussler-Scheinker disease A117V. J Biol Chem 2001; 276:6009-6015.
64. Ghetti B, Bugiani O, Tagliavini F et al. Gerstmann-Sträussler-Scheinker disease. In: Dickson D, ed. Neurodegeneration: The Molecular Pathology of Dementia and Movement Disorders. Basel: ISN Neuropath Press, 2003:318-325.
65. Piccardo P, Liepnieks JJ, William A et al. Prion proteins with different conformations accumulate in Gerstmann-Sträussler-Scheinker disease caused by A117V and F198S mutations. Am J Pathol 2001; 158:2201-2207.
66. Panegyres PK, Toufexis K, Kakulas BA et al. A new PRNP mutation (G131V) associated with Gerstmann-Sträussler-Scheinker disease. Arch Neurol 2001; 58:1899-1902.
67. Kitamoto T, Iizuka R, Tateishi J. An amber mutation of prion protein in Gerstmann-Sträussler syndrome with mutant PrP plaques. Bioch Biophys Res Comm 1993; 192:525-531.
68. Butefisch CM, Gambetti P, Cervenakova L et al. Inherited prion encephalopathy associated with the novel PRNP H187R mutation: a clinical study. Neurology 2000; 55:517-522.
69. Dlouhy SR, Hsiao K, Farlow MR et al. Linkeage of the Indiana kindred of Gerstmann-Sträussler-Scheinker disease to the prion protein gene. Nat Genet 1992; 1:64-67.
70. Mirra SS, Young K, Gearing M et al. Coexistence of prion protein (PrP) amyloid, neurofibrillary tangles and Lewy bodies in Gerstmann-Sträussler-Scheinker disease with prion gene (PRNP) mutation F198S. Brain Pathol 1997; 7:1378.
71. Farlow MR, Tagliavini F, Bugiani O et al. Gerstmann-Sträussler-Scheinker disease. In: Vinken PJ, Bruyn GW, Klawans HL, eds. Hereditary Neuropathies and Spinocerebellar Atrophies. Amsterdam: Elsevier Science Publishers, 1991:619-633.

72. Farlow MR, Yee RD, Dlouhy SR et al. Gerstmann-Sträussler-Scheinker disease. I. Extending the clinical spectrum. Neurology 1989; 39:1446-1452.
73. Yee RD, Farlow MR, Suzuki DA et al. Abnormal eye movements in Gerstmann-Sträussler-Scheinker disease. Arch Ophthalmol 1992; 110:68-74.
74. Ghetti B, Dlouhy SR, Giaccone G et al. Gerstmann-Sträussler-Scheinker diseases and the Indiana kindred. Brain Pathol 1995; 5:61-95.
75. Ghetti B, Piccardo P, Lievens PMJ et al. Phenotypic and prion protein (PrP) heterogeneity in Gerstmann-Sträussler-Scheinker disease (GSS) with a proline to a leucine mutation at PRNP residue 102. In: The 6th International Conference on Alzheimer's Disease and Related Disorders, Amsterdam. Neurobiol Aging 1998; 19:298.
76. Ghetti B, Tagliavini F, Giaccone G et al. Familial Gerstmann-Sträussler-Scheinker disease with neurofibrillary tangles. Mol Neurobiol 1994; 8:41-48.
77. Ghetti B, Tagliavini F, Masters CL et al. Gerstmann-Sträussler-Scheinker disease. II Neurofibrillary tangles and plaques with PrP-amyloid coexist in an affected family. Neurology 1989; 39:1453-1461.
78. Nochlin D, Sumi SM, Bird TD et al. Familial dementia with PrP positive amyloid palques: a variant of Gerstmann-Sträussler syndrome. Neurology 1989; 39:910-918.
79. Pearlman RL, Towfighi J, Pezeshkpour GH et al. Clinical significance of types of cerebellar amyloid plaques in human spongiform encephalopathies. Neurology 1988; 38:1249-1254.
80. Young K, Piccardo P, Kish SJ et al. Gerstmann-Sträussler-Scheinker disease (GSS) with a mutation at prion protein (PrP) residue 212. W: The 74th Annual Meeting of the American Association of Neuropathologists Inc, Minneapolis, Minnesota. J Neuropathol Exp Neurol 1998; 57:518.
81. Bugiani O, Giaccone G, Verga L et al βPP participates in PrP-amyloid plaques of Gerstmann-Sträussler-Scheinker disease, Indiana kindred. J Neuropathol Exp Neurol 1993; 52:64-70.
82. Piccardo P, Dlouhy SR, Lievens PMJ et al. Phenotypic variability of Gerstmann-Sträussler-Scheinker disease is associated with prion protein heterogeneity. J Neuropathol Exp Neurol 1998; 57:979-988.
83. Tagliavini F, Prelli F, Porro M et al. Amyloid fibrils in Gerstmann-Sträussler-Scheinker disease (Indiana and Swedish kindreds) express only PrP peptides encoded by the mutant allele. Cell 1994; 79:695-703.
84. Giaccone G, Verga L, Bugiani O et al. Prion protein preamyloid and amyloid deposits in Gerstmann-Sträussler-Scheinker disease, Indiana kindred. Proc Natl Acad Sci USA 1992; 89:9349-9353.
85. Ikeda S, Yanagisawa N, Glenner GG et al. Gerstmann-Sträussler-Scheinker disease showing β-protein amyloid deposits in the peripheral regions of PrP-immunoreactive amyloid plaques. Neurodegeneration 1992; 1:281-288.
86. Bratosiewicz-Wasik J, Wasik T, Liberski PP. Molecular approaches to mechanisms of prion diseases. Folia Neuropathol 2004; 42:33-46.
87. Liberski PP, Bratosiewicz J, Barcikowska M et al. A case of sporadic Creutzfeldt-Jakob disease with Gerstmann-Sträussler-Scheinker phenotype but no alterations in the PRNP gene. Acta Neuropathol (Berl) 2000; 100:233-234.
88. Hudson AJ, Farrell MA, Kalnins R et al. Gerstmann-Sträussler-Scheinker disease with coincidental familial onset. Ann Neurol 1983; 14:670-678.
89. Peiffer J. Gerstmann-Sträussle's disease, atypical multiple sclerosis and carcinomas in family of sheepbreeders. Acta Neuropathol (Berl) 1982; 56:87-92.
90. Ceballos AC, Baringo FT, Pelegrin VC. Gerstmann-Sträussler syndrome clinical and neuromorphofunctional disgnosis: a case report. Actas Luso Esp Neurol Psiquiatr Cienc Afines 1996; 24:156-160.
91. Galatioto S, Ruggeri D, Gullotta F. Gerstmann-Sträussler-Scheinker syndrome in a Sicilian patient. Neuropathological aspects. Pathologica 1995; 87:659-665.
92. Ikeda S, Yanagisawa N, Allsop D et al. Gerstmann-Sträussler-Scheinker disease showing β-protein type cerebellar and cerebral amyloid angiopathy. Acta Neuropathol 1994; 88:262-266.
93. Mighelli A, Attanasio A, Claudia M et al. Dystrophic neurites around amyloid plaques of human patients with Gerstmann-Sträussler-Scheinker disease contain ubiquitinated inclusions. Neurosci Lett 1991; 121:55-58.
94. Vinters HV, Hudson AJ, Kaufmann JCE. Gerstmann-Sträussler-Scheinker disease: autopsy study of a familial case. Ann Neurol 1986; 20:540-543.
95. de Courten-Myers G, Mandybur TI. Atypical Gerstmann-Sträussler syndrome or familial spinocerebellar ataxia and Alzheimer's disease? Neurology 1987; 37:269-275.
96. de Courten-Myers G,—personal communication, 2005.
97. Gabizon R, Telling G, Meiner Z et al. Insoluble wild-type and protease-resistant mutant prion protein in brains of patients with inherited prion diseases. Nat Med 1996; 2,59-64.
98. Bolton DC, McKinley MP, Prusiner SB. Identification of a protein that purifies with the scrapie prion. Science 1982; 218:1309-1311.
99. McKinley MP, Bolton DC, Prusiner SB. A protease resistant protein is a structural component of the scrapie prion. Cell 1983; 35:57-62.
100. Parchi P, Chen SG, Brown P et al. Different patterns of truncated prion protein fragments correlate with distinct phenotypes in P102L Gerstmann-Sträussler-Scheinker disease. Proc Natl Acad Sci USA 1998; 95:8322-8327.

CHAPTER 11

JUVENILE NEURONAL CEROID LIPOFUSCINOSES

Shiyao Wang

Peking University, People's Hospital, Beijing, China
Email: samuelwsy@163.com

Abstract   Juvenile neuronal ceroid lipofuscinoses (JNCL) is the most common type of the neuronal ceroid lipofuscinoses (NCLs), a group of pediatric neurodegenerative diseases. In this chapter the genetic and biochemical basis, pathogenesis, clinical features, histopathological features, diagnosis and therapeutic strategies of the JNCL are reviewed. The premature death of the patients and subnormal life quality are inevitable due to the lack of understanding of pathogenesis and limitation in treatment. Hence we are still a long way to conquer the disease.

INTRODUCTION

Juvenile neuronal ceroid lipofuscinoses is one type of the neuronal ceroid lipofuscinoses (NCLs), a group of inherited, autosomal recessive, progressive pediatric neurodegenerative diseases.[1] The incidence of NCLs in USA and Scandinavian countries was estimated in 1988 to be 1:12500,[2] while there is no accurate statistics in China or other part of the world.

The NCLs were originally classified into four groups based on storage materials and the clinical onset of symptoms: infantile (INCL), late-infantile (LINCL), juvenile (JNCL) and adult (ANCL);[1] but another classification based on molecular genetics is adopted more widely now.[3] The NCLs have common histopathological characteristics: these include progressive neuronal loss and the accumulation of lipofuscin-like autofluorescent storage material and common clinical features such as rapidly progressing vision loss, epileptic seizures, loss of cognitive function, motor dysfunction, and premature death.[4] In NCLs group JNCL (also known as Batten disease, Spielmeyer-Vogt-Sjogren disease and CLN3) is most common caused by mutations in *CLN3*[5] and this has been focused in this chapter.

Neurodegenerative Diseases, edited by Shamim I. Ahmad.
©2012 Landes Bioscience and Springer Science+Business Media.

GENETIC AND BIOCHEMICAL BASIS

The gene *CLN3* (OMIM 607042), mutated in JNCL, was identified in 1995 by the International Batten Disease Consortium. It maps at chromosome region16p12.1[6] and contains at least 15 exons and spans 15 kb.[7] Subsequently Lee mapped the mouse gene at chromosome 7 in a region of syntenic homology with human 16p12.[8] The International Batten Disease Consortium demonstrated that a deletion of 1.02 kb in the *CLN3* gene is responsible for 73% of Batten disease.[5] *CLN3* encodes a protein, CLN3P, containing 438 amino acids and an estimated molecular weight of 48 kDa.[9] Despite much research has been carried out to understand the CLN3 protein function and *CLN3* gene's biological roles,[10-13] we are still some way away.

PATHOGENESIS

Recently our group has reviewed the pathogenesis of JNCL,[14] including apoptosis, autophagy, dysfunction in the structure associated with plasmalemma, oxidative stress and disruption of nitric oxide signaling, dysfunction of mitochondria and lysosome, imbalanced intracellular pH, and some other related mechanisms.

In apoptosis the defect in *CLN3* results in ceramide accumulation and increase of mitochondrial membrane permeabilization, which eventually induce caspase-dependent and caspase-independent cell death. In JNCL autophagy exists but is disrupted because of the immaturity of autophagic vacuoles which leads to failure of autophagy circulation.

CLINICAL FEATURES

The clinical features of JNCL have been reviewed by Hofmann and Peltonen,[15] and these have been summarized in Table 1.

Table 1. Clinical features of typical JNCL

Features	Details
Onset age	Between ages 4 and 8 typically.
Rapidly progressing vision loss	The first clinical sign of the disease: patients become totally blind within few years.
Epileptic seizures	Epileptic seizures of generalized, complex-partial or myoclonic type may occur by early second decade. Speech disturbances (festinating stuttering, often mislabeled as echolalia) and slow decline in cognition occur around the time of onset of seizures.
Behavioral problems and motor dysfunction	Behavioral problems, occur in the second decade. Motor dysfunction is a combination of extrapyramidal, pyramidal and cerebellar disturbances.
Psychiatric problems	Some individuals with JNCL experience multiple psychiatric problems such as: disturbed thoughts, attention problems, somatic complaints and aggressive behavior.[16]
Premature death	Most individuals live until the late teens or early 20s; some may live into their 30s.

Figure 1. Fingerprint patterns are the predominant-type of intraneuronal inclusions in the juvenile form. Reproduced from: Haltia M. Biochim Biophys Acta 2006; 1762:850-856;[1] ©2006 with permission from Elsevier.

HISTOPATHOLOGY

The histopathological features of the JNCL have also been reviewed by Hofmann and Peltonen[15] and summarized by Dinesh Rakheja.[17] The autopsy of the brain showed that the grey and white matters were both, grossly atrophic and diffused loss of neurons showing changes known as the Shaffer-Speilmayer process. In the eyes, the retinal neuroepithelium shows degenerative changes at first and later extensive loss of retinal neurons in every layers.

Under the electron microscopy, JNCL storage material has the characteristic inclusions: the compactly bound laminated bodies resembling fingerprints and hence called fingerprint profiles (Fig. 1).[1] Vacuolated lymphocytes can be seen in the peripheral blood smear (Fig. 2).[18]

DIAGNOSIS

The diagnosis of JNCL has been presented elsewhere,[18] and is summarized below. The diagnosis needs to combine the clinical features, ophthalmological testing,

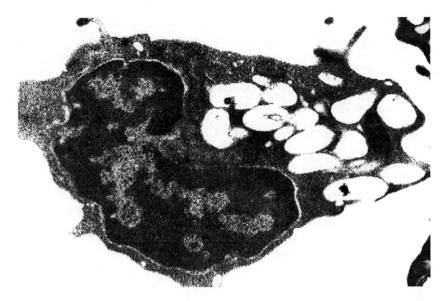

Figure 2. Membrane-bound vacuoles within a blood lymphocyte, juvenile-type. Reproduced from: Williams RE et al. Biochim Biophys Acta 2006; 1762:865-872;[18] ©2006 with permission from Elsevier.

neuro-radiological testing, molecular genetic testing and pathological testing. The clinical features include onset age, rapidly progressing vision loss, epilepsy, behavioral problems and motor dysfunction, psychiatric problems and premature death. Visual failure is usually the first symptom in JNCL, so rapidly progressive visual failure at the young age should be paid attention to. The first epileptic seizure is noticed at the early second decade. The seizures tend to increase in frequency and severity as the patients grow up. Progressive abnormality and an increase in paroxysmal activity can be seen in the electroencephalogram (EEG). The extrapyramidal signs include impaired balance, rigidity, hypokinesia, stooped posture and shuffling gait and tremor. As the gradual psychomotor deterioration becomes evident, a specific verbal Wechsler Intelligence Scale for Children (WISC-R) profile can be applied.

MRI is not that useful for diagnosis as it is usually normal in the first decade. Several European and North American centers are now screening for mutation in *CLN3* as the diagnostic tool.[19] Vacuolated lymphocytes can be observed in the blood (Fig. 2) and Fingerprint profiles should be found under the electron microscopy (Fig. 1) as the pathological diagnosis.

THERAPEUTIC STRATEGY

There is no specific treatment for this disease, hence symptomatic treatment such as the drugs for seizures and psychiatric problems can be used for a limited success. Also counseling and prenatal care are also very crucial.

The treatment has been reviewed by Hobert,[20] which is summarized below.

Generalized care is focused on minimizing symptoms such as seizures, behavioral problems, and depression. Anticonvulsant medications are a mainstay of therapy, and diazepam, haloperidol, and lithium are used to treat depression. Dietary supplements such as food with antioxidant, fish oil extracts have been studied but without any positive reports.

Several other therapies have been tried for other types of NCLs, such as targeted small molecules, enzyme replacement therapy, stem cell therapy, and gene therapy. We are still awaiting more research to be carried out in the direction of effective therapies for JNCLs.

CONCLUSION

Due to poor prognosis and lack of effective therapy, Juvenile neuronal ceroid lipofuscinoses is a big problem in the field of pediatric neurodegenerative diseases. Although the genetic basis of the disease has been studied and advanced diagnosis methods are available, the treatments are still limited and hence the life quality remains sub-normal and the early death is inevitable. High hope exists that the future research will remove more curtains in the direction of investigation and therapies for this complicated juvenile neurodegenerative disease.

REFERENCES

1. Haltia M. The neuronal ceroid-lipofuscinoses: from past to present. Biochim Biophys Acta 2006; 1762:850-856.
2. Santavuori P, Lauronen L, Kirveskari K et al. Neuronal ceroid lipofuscinoses in childhood. Suppl Clin Neurophysiol 2000; 53:443-451.
3. Jalanko A, Braulke T. Neuronal ceroid lipofuscinoses. Biochim Biophys Acta 2009; 1793:697-709.
4. Haltia M. The neuronal ceroid-lipofuscinoses. J Neuropathol Exp Neurol 2003; 62:1-13.
5. Isolation of a novel gene underlying Batten disease, CLN3. The International Batten Disease Consortium. Cell 1995; 82:949-957.
6. MIM ID *607042 CLN3 GENE; CLN3: http://www.ncbi.nlm.nih.gov/omim/607042
7. Mitchison HM, Munroe PB, O'Rawe AM et al. Genomic structure and complete nucleotide sequence of the Batten disease gene, CLN3. Genomics 1997;40:346-350.
8. Lee RL, Johnson KR, Lerner TJ. Isolation and chromosomal mapping of a mouse homolog of the Batten disease gene CLN3. Genomics 1996;35:617-619.
9. Narayan SB, Rakheja D, Tan L et al. CLN3P, the Batten's disease protein, is a novel palmitoyl-protein Delta-9 desaturase. Ann Neurol 2006; 60:570-577.
10. Luiro K, Yliannala K, Ahtiainen L et al. Interconnections of CLN3, Hook1 and Rab proteins link Batten disease to defects in the endocytic pathway. Hum Mol Genet 2004; 13:3017-3027.
11. Ramirez-Montealegre D, Pearce DA. Defective lysosomal arginine transport in juvenile Batten disease. Hum Mol Genet 2005; 14:3759-3773.
12. Hobert JA, Dawson G. A novel role of the Batten disease gene CLN3: association with BMP synthesis. Biochem Biophys Res Commun 2007; 358:111-116.
13. Tuxworth RI, Vivancos V, O'Hare MB et al. Interactions between the juvenile Batten disease gene, CLN3, and the Notch and JNK signalling pathways. Hum Mol Genet 2009; 18:667-678.
14. Wang SY, Jin WN, Wu D. Mechanisms of juvenile neuronal ceroid lipofuscinosis (JNCL). Yi Chuan 2009; 31:779-784.
15. Hofmann SL, Peltonen L. In: Scriver CR, Beaudet AL, Sly WS, Valle D, eds. The Metabolic and Molecular Bases of Inherited Disease. New York: McGraw-Hill, 2001:3877-3894.
16. Backman ML, Santavuori PR, Aberg LE et al. Psychiatric symptoms of children and adolescents with juvenile neuronal ceroid lipofuscinosis. J Intellect Disabil Res 2005; 49:25–32.
17. Rakheja D, Narayan SB, Bennett MJ. Juvenile neuronal ceroid-lipofuscinosis (Batten disease): a brief review and update. Curr Mol Med 2007; 7:603-608.
18. Williams RE, Aberg L, Autti T et al. Diagnosis of the neuronal ceroid lipofuscinoses: An update. Biochim Biophys Acta 2006; 1762:865-872.
19. Zhong N. Molecular genetic testing for neuronal ceroid lipofuscinoses. Adv Genet 2001; 45:141-158.
20. Hobert JA, Dawson G. Neuronal ceroid lipofuscinoses therapeutic strategies: past, present and future. Biochim Biophys Acta 2006; 1762:945-953.

CHAPTER 12

KURU:
The First Prion Disease

Paweł P. Liberski,*,1 Beata Sikorska[1] and Paul Brown[2]

[1]Laboratory of Electron Microscopy and Neuropathology, Department of Molecular Pathology and Neuropathology, Medical University Lodz, Lodz, Poland; [2]CEA/DSV/iMETI/SEPIA, Fontenay-aux-Roses, France
*Corresponding Author: Paweł P. Liberski—Email: ppliber@csk.am.lodz.pl

Abstract: Kuru disease is linked with the name of D. Carleton Gajdusek and he was the first to show that this human neurodegenerative disease can be transmitted to chimpanzees and subsequently classified as a transmissible spongiform encephalopathy (TSE), or slow unconventional virus disease. It was first reported to Western world in 1957 by Gajdusek and Vincent Zigas,[1,2] and in 1975 a complete bibliography of kuru was published by Alpers et al.[3] "Kuru" in the Fore language in Papua New Guinea means to shiver from fever and cold. The disease has been found to spread through ritualistic cannibalism and is an invariably fatal cerebellar ataxia accompanied by tremor, choreiform and athetoid movements. Neuropathologically, kuru is characterized by the presence of amyloid "kuru" plaques.

INTRODUCTION

The recognition of kuru as a neurodegenerative disease that is also transmissible and infectious and subsequent proof that Creutzfeldt-Jakob disease (CJD) belongs to the same category[4] points that kuru is not an strange and unique disease caused by cannibalism on remote island, but a representative of a novel class of diseases which made D. Carleton Gajdusek to win the Nobel Prize in 1976 and subsequently to Stanley B. Prusiner in 1997. Indirectly, kuru was also linked to a third Nobel Prize winner, Kurt Wüthrich, who determined the structure of the prion protein.[5] Interestingly, the same year transmissibility of CJD was published by Van Rossum who wrote in the "Handbook of Neurology"[6] "We consider the following facts to be in favour of the endogenous nature of the disease: the absence of changes (clinical and anatomical) pointing to an exogenous change, for

Neurodegenerative Diseases, edited by Shamim I. Ahmad.
©2012 Landes Bioscience and Springer Science+Business Media.

instance infection or intoxication; and the asymmetry of clinical and anatomical findings. Indirectly, these indications of an endogenous cause are also in favour of the view that we are dealing with a disease and not with a syndrome."

As Gajdusek stressed for the last time,[7] the solving of the kuru riddle helped to develop ideas of molecular casting and further to understand such diverse areas as dermatoglypes, osmium shadowing in electron microscopy of amyloid-enhancing factors. To this end, Gajdusek recalled a metaphore of Ice-9 by Kurt Vonnegut and industrial viruses of "The man in white suit" movie from 1951.

BACKGROUND AND ETHNOGRAPHIC SETTING

"Kuru" in the Fore language of Papua New Guinea means to shiver from fever or cold. The Fore language speakers used the nominative form of the verb to describe the fatal disease which decimated their children and adult women, but rarely men.

Kuru was and still is restricted to natives of the Foré linguistic group in Papua New Guinea's Eastern Highlands and neighboring linguistic groups that exchange women with Fore people but not the neighboring groups into which kuru-affected peoples did not settle. It seems that Kuru first appeared at or shortly after the turn of 20th century[8,9] in Uwami village of Keiagana people. Within 20 years it had spread further into the Kasokana[10] and Miarasa villages of North Fore and a decade later had reached the South Fore at the Wanikanto and Kamira villages. Kuru became endemic in villages which it entered and became hyperendemic in the South Fore region.

CANNIBALISM

Ritualistic endocannibalism (eating of relatives as part of a mourning ritual) was practiced not only in the kuru area but in many surrounding Eastern Highland groups in which kuru never developed.[10]

The first European who witnessed kuru was Ted Ubank, a gold prospector, in 1936.[10] In the late 1930s and 1940s, many gold miners, Protestant missionaries and government officials made contacts with the northern periphery of the kuru region and became thoroughly familiar with the ritual endocannibalism of Eastern Highland tribes. Gajdusek has been asked when the hypothesis of cannibalism as a vehicle to spread kuru was first envisaged. His response was that "anyone would come to the conclusion that a disease endemic among cannibals must be spread through eating corpses". The hypothesis of cannibalism was thus not widely discussed because it was taken for granted, not because of any lack of insight and it was proved in 1965 by the transmission of kuru to chimpanzees.[11]

Among the Foré, kuru was believed to result from sorcery.[10,12,13] To cause kuru, a would-be-sorcerer would need to obtain a part of the victim's body (nail clippings, hair) or excreta, particularly feces- or urine-soaked vegetation, saliva, blood, or partially consumed food such as peelings from sweet potato eaten by the victim, or clothing ("maro"). These were packed with leaves and made into a "kuru bundle" and placed partially submerged into one of numerous in the Fore regions swamps. Subsequently, the sorcerer shook the package daily until the sympathetic kuru tremor was induced in his victim. As a result, kinsmen of a kuru victim attempted to identify and subsequently kill a suspected sorcerer if they could not bribe or intimidate him to release a victim from the kuru spell. Killing of a sorcerer—tukabu—was a ritualistic form of vendetta; it included crushing with stones the bones of the neck, arm and thigh, as well as the loins, biting the trachea and grinding the genitalia with stones and clubs.

KURU ETIOLOGY—THE INSIGHT INTO A NOVEL CLASS OF PATHOGENS

Although on epidemiological grounds the etiology of kuru was thought to be infectious, patients had no obvious signs of the CNS infection and all attempts to transmit kuru to small laboratory animals were unsuccessful. In other wide ranging investigations, neither exhaustive genetic analyses nor the search for nutritional deficiencies or environmental toxins resulted in a tenable hypothesis.[14,15]

On July 21, 1959, Gajdusek received a letter from William Hadlow, at the Rocky Mountain Laboratory in Hamilton Montana,[16-19] which pointed out the analogies between kuru and scrapie, a slow neurodegenerative disease of sheep and goats known to be endemic in the United Kingdom since the 18th century and experimentally transmitted in 1936.

A similar observation was made by veterinary neuropathologist Innes[20] during his visit to the Gajdusek laboratory.[21,22] Hadlow, in his recollection of that seminal observation pointed out intracellular vacuoles as those neuropathological changes that attracted his attention some forty years ago.[23,24] In 1965, in a monograph "Slow, Latent and Temperate Virus Infections" which resulted from a meeting convened in 1964, Gibbs and Gajdusek[25] wrote that kuru could be transmitted to chimpanzee.

EPIDEMIOLOGY OF KURU—A STRONG SUPPORT OF THE CANNIBALISM THEORY

Kuru incidence increased in the 1940 and 1950s to reach the mortality rate in some villages of 35 per 1000 of 12 000 Fore people. The female to male ratio was distorted in particular in the South Fore the female: male ratio was 1:1.67 in contrast to 1:1 ratio in unaffected Kamano people. This ratio increased to 1:2 to even 1:3 in certain South Fore areas.

The absence of kuru cases in South Fore among children born after 1954 and the rising of age of kuru cases year by year suggested that transmission of kuru to children stopped in 1950s[26-28] when cannibalism ceased to be practiced among the Fore people. Also, brothers with kuru tend to die at the same age which suggested that they were infected at similar age but not at the similar time. The assumption that affected brothers were infected with kuru at the same age led to a calculation of minimal age of exposure for males to be in a range of 1-6 years with a mean incubation period of 3-6 years and the maximum incubation period of 10-14 years.[8]

The most fascinating clue toward the solving of the kuru riddle came from epidemiological studies. The almost "formal" proof that kuru was indeed transmitted by cannibalism was provided by Klitzman et al[29] who studied clusters of kuru patients who participated in a limited number of kuru feasts in 1940s and 1950s. Three clusters were identified, one of which will be recalled here. Two brothers, Ob and Kasis from Awande village, North Fore developed signs and symptoms of kuru in 1975, 21 or 27 years after the latest or the earlier exposure, respectively. They participated in 2 feasts for kuru victims. Taken together, those 3 pairs of a simultaneous development of kuru in persons infected at the same time suggested that cannibalism indeed played a role in kuru spreading and that infectivity followed almost the same course in paired individuals.

CLINICAL MANIFESTATIONS

Kuru is an invariably fatal cerebellar ataxia accompanied by tremor, choreiform and athetoid movements (Fig. 1). In contrast to the neuropathological picture, it is remarkably uniform in clinical signs, symptoms and evolution. The progressive dementia that is so characteristic of sporadic CJD, is barely noticeable in patients with kuru and then only late in the course of the illness. However, kuru patients often displayed emotional changes, including inappropriate euphoria and compulsive laughter (the journalistic "laughing death" or "laughing disease"), or apprehension and depression. Kuru is divided clinically into three stages: ambulant, sedentary and terminal (the Pidgin expressions, *wokabaut yet*; i.e.,"is still walking", *sindaun pinis*; i.e.,"is able only to sit" and *slip pinis*; i.e.,"is unable to sit up").

The duration of kuru, as measured from the onset of prodromal signs and symptoms until death was at about 12 months (3-23 months).[30-32]

There is an ill defined prodromal period (*kuru laik i-kamp nau*—i.e.,"kuru is about to begin") characterized by the presence of headache and limb pains, frequently in the joints; knees and ankles came first, followed by elbows and wrists; sometimes, interphalangeal joints were first affected, abdominal pains and loss of weight. This period lasted approximately a few months.

The prodromal period is followed by the "ambulant stage", the end of which is defined when the patient is unable to walk without a stick. As patients were well aware that kuru heralded death in about a year, they became withdrawn and quiet. A

Figure 1. Several children affected with kuru. Courtesy of late D. Carleton Gajdusek.

fine 'shivering' tremor, starting in the trunk, amplified by cold and associated with a "goose flesh", is often followed by titubation and other types of abnormal movements. Attempts to maintain balance result in clawing of the toes and curling of the feet. Plantar reflex is always flexor while clonus, in particular ankle clonus and patellar clonus, are hallmarks of the clinical picture. The gait of kuru is characteristic "hesitant, wide-based and staggering, with irregular placement of the feet and lurching to either side or with a high steppage and stamping as the raise foot was put down". Resting tremor is a cardinal sign of kuru. A horizontal convergent strabismus is a typical sign, especially in younger patients; nystagmus was common but the papillary responses were preserved.

The second "sedentary" stage begins when the patient is unable to walk without constant support and ends when he or she is unable to sit without it. Postural instability, severe ataxia, tremor and dysarthria progress endlessly through this stage.

In the third stage, the patient is bedridden and incontinent, with dysphasia and primitive reflexes and eventually succumbs in a state of advanced starvation. Extraocular movements were jerky or, to the contrary, slow and rigid. A strong grasp reflex occurrs as well as fixed dystonic postures, athetosis and chorea. Terminally, "the patient lies moribund inside her hut surrounded by a constant group of attending relatives. She barely moves and is weak and wasted. Her pressure sores may have spread widely to become huge rotting ulcers which attract a swarm of flies. She is unable to speak. The jaws are clenched and have to be forced open in order to put food or fluid in. Despite her mute and immobile state she can make clear signs of recognition with her eyes and may even attempt to smile".

NEUROPATHOLOGY

The first systematic examination of kuru neuropathology (12 cases) was published by Klatzo et al.[33,34] Of utmost interest was neuronal alterations observed in anterior motor neurons of the spinal cord and in different brain stem nuclei, the cerebellum and the cerebral cortex were totally nonspecific in nature but nonetheless sufficient for Klatzo et al to draw a parallel between kuru and Creutzfeldt-Jakob disease (CJD). It was observed that neurons were shrunken and hyperchromatic or pale with dispersion of Nissl substance or contained intracytoplasmic vacuoles similar to those already described in scrapie. In the striatum, some neurons were vacuolated to such a degree that they looked "moth-eaten". Neuronophagia was observed. A few binucleated neurons were visible and torpedo formation was noticed in the Purkinje cell layer, along with empty baskets that marked the presence of degenerated Purkinje cells. In the medulla, neurons of the vestibular nuclei and the lateral cuneatus were frequently affected; the spinal nucleus of the trigeminal nerve and nuclei of VIth, VIIth and motor nucleus of the VIth cranial nerves were affected less frequently while nuclei of the XIIth cranial nerve, the dorsal nucleus of Xth cranial nerve and nucleus ambiguous were relatively spared. In the cerebral cortex, the deeper layers were affected more than the superficial layers and the neurons in the hippocampal formation were normal. In the cerebellum the paleocerebellar structure (vermis and flocculo-nodular lobe) was most severely affected and spinal cord pathology was most severe in the corticospinal and spinocerebellar tracts. Astro- and microglial proliferation was widespread; the latter formed rosettes and appeared as rod- or amoeboid types or as macrophages (gitter cells). Interestingly, the significance of spongiform change was not noted by Klatzo et al,[33,34] but "small spongy spaces", were noted in 7 of 13 cases studied by Beck and Daniel.[35-39]

The most striking neuropathologic feature of kuru was the presence of numerous amyloid plaques (Figs. 2-3), described as "spherical bodies with a rim of radiating filaments" found in 6 of 12 cases studied by Klatzo et al,[33,34] and in "about three-quarters" of the 13 cases of Beck and Daniel;[35,39] they became known as "kuru plaques". These measured 20-60 μm in diameter, were round or oval and consisted of a dark-stained core with delicate radiating periphery surrounded by a pale "halo". Kuru plaques were most abundant in the granular cell layer of the cerebellum, basal ganglia, thalamus and cerebral cortex in that order of frequency. Kuru plaques are metachromatic and stain with PAS, Alcian blue (Fig. 2) and Congo-red and a proportion of them are weakly argentophilic when impregnated according to Belschowsky or von Braunmühl techniques. Of historical interest, another unique disease reported by Seitelberger[40] as "A peculiar hereditary disease of the central nervous system in a family from lower Austria" (germ. Eigenartige familiar-hereditare Krankenheit des Zentralnerven systems in einer niederoosterreichen Sippe) was described by Neumann et al[41] who was thus the first person to suggest a connection between GSS and kuru. Indeed, the latter was transmitted to non human primates in 1981.[42]

Renewed interest in kuru pathology has been provoked by the appearance in 1997 of a variant form of CJD characterized by numerous amyloid plaques, including "florid" or "daisy" plaques—a kuru plaque surrounded by a corona of spongiform vacuoles.[43] To this end, a few papers re-evaluating a historical material has been published. Hainfellner et al[44] studied by PrP (prion protein)-immunohistochemistry (Fig.

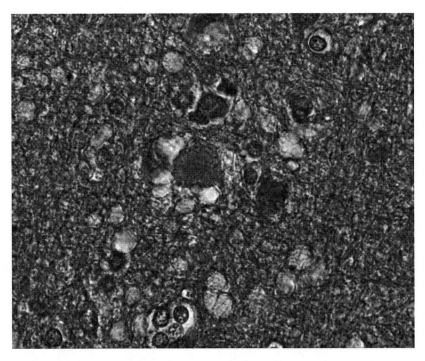

Figure 2. A kuru plaque stained with Alcian blue.[44] A color version of this figure is available at www. landesbioscience.com/curie.

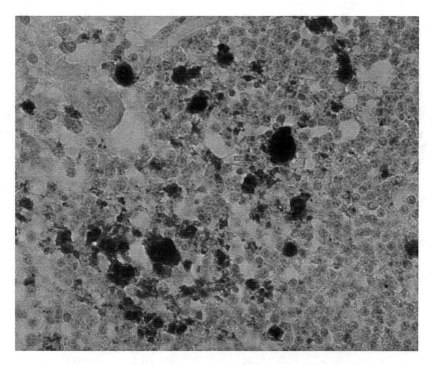

Figure 3. A cluster of kuru plaques stained with anti-PrP antibodies.[44]

3) the case of a young male kuru victim by the name Kupenota from the South Fore region whose brain tissue had transmitted disease to chimpanzees, and McLean et al examined a series of 11 cases of kuru.[45] In contrast to the classical studies described above, both papers stressed the presence of typical spongiform change present in deep layers (III-V) of the cingulate, occipital, enthorrinal and insular cortices and in the subiculum. Spongiform change was also observed in the putamen and caudate and some putaminal neurons contained intraneuronal vacuoles. Spongiform change were prominent in the molecular layer of the cerebellum, in periaqueductal gray matter, basal pontis, central tegmental area and inferior olivary nucleus. The spinal cord showed only minimal spongiform change.

There is no ultrastructural observations on kuru in humans except by a paper by Peat and Field[46] who described "intracytoplasmic dense barred structures" known as normal structures[47] and Field et al[48] reported on the typical ultrastructure of the kuru plaques and "herring-bone" structures, again the normal structure of Hirano bodies.[49] Our studies on formalin-fixed paraffin-embedded Kuru specimens seen under electron microscope revealed typical plaques composed of amyloid fibrils (Fig. 4).

Immunohistochemical studies revealed that misfolded PrP[d] (d, from "disease") was present not only as kuru plaques but also in synaptic and at perineuronal sites and in the spinal cord the *substantia gelatinosa* was particularly affected, as in iatrogenic CJD cases following peripheral inoculation.[50] The latter finding may suggest a common pathogenetic pathway. Brandner et al[51] studied one very recent case of kuru and basically confirmed the findings of Hainfellner et al.[44]

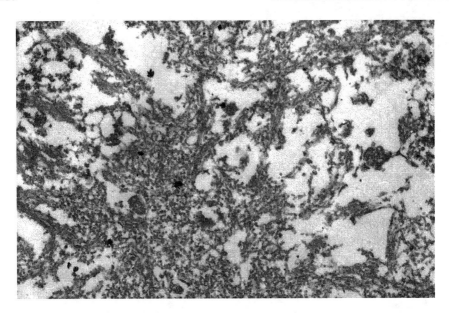

Figure 4. Amyloid fibrils of kuru plaques. Material reversed from paraffin and reprocessed to electron microscopy.

GENETICS AND MOLECULAR BIOLOGY OF KURU

Lee et al[52] first showed that in 2 kuru cases were 129^MetMet homozygotes. Subsequently, the whole ORF of the PRNP gene was sequenced.[52] None of the known (or a novel) mutation was found in all these 5 patients who were 129^ValVal homozygotes. Further studies found that individuals of 129^ValVal and 129^MetVal genotype were susceptible to kuru, but those of 129^MetMet genotype were overrepresented in the younger age group while those of 129^ValVal 129^MetVal were overrepresented in much older age group.[53-58] In contrast, those people who survived the epidemic were characterized by almost the total absence of 129^MetMet homozygotes. The more recent cases studied by Lantos et al[59] McLean et al[45] and us[44] were all 129^MetMet homozygotes. A recent genome-wide studies confirmed a strong association of kuru with a single nucleotide polymorphism (SNP) localized within the codon 129 and also with two other SNPs localized within genes *RARB* (the gene encoding retinoic acid receptor beta) and STMN2 (the gene encoding SCG10).[55]

The practice of endocannibalism underlying the kuru epidemic created a selective force on the prion protein genotype.[60,61] As in CJD, homozygosity at codon 129 (129^MetMet or 129^ValVal) is overrepresented in kuru.[53-58] Furthermore, Mead et al[56,57] found that among Fore women over fifty years of age, there is a remarkable overrepresentation of heterozygosity (129^MetVal) at codon 129, which is consistent with the interpretation that 129^MetVal makes an individual resistant to TSE agents and that such a resistance was selected by cannibalistic rites. Because of this 129^MetVal heterozygote advantage, it has been suggested that the heterozygous genotype at codon 129 has been sustained by a widespread ancient practice of human cannibalism.[62] In conclusion, Collinge et al[63] suggested that the survival advantage of the *PRNP* 129^MetMet heterozygotes provides a basis for a selection pressure not only in Fore but also in those human populations that practiced cannibalism.

The molecular strain typing of kuru cases was performed by the Collinge's group.[64,65] This typing is based on the electrophoretic mobility of de-, mono- and diglycosylated bands of PrP^d following digestion with proteinase K.[66] The kuru specimens revealed type 2 ($PrP^{Met\ Met}$) or 3 ($129^{Val\ Val}$) PrP^d patterns and the glycoform ratio was similar to that of sporadic CJD but not vCJD.[67] The latter notion is supported by the fact of a similar transmission rate of kuru to transgenic mice lacking mouse PrP gene but expressing human $PrP\ 129^{Val\ Val}$ gene.[64,65] In contrast, kuru is not transmissible to normal wild-type mice. Collectively, those data suggest that kuru is similar to sporadic CJD but not variant CJD.

CONCLUSION

Kuru, a nearly extinct horrible disease of a cannibalistic Stone Age tribe in remote areas of Papua New Guinea, still exerts an influence on many aspects of neurodegeneration research. First, it showed that a human neurodegenerative disease can result from an infection with an infectious agent and then called a "slow virus".[68] This discovery opened a window into the new class of human diseases including Creutzfeldt-Jakob disease, Gerstmann-Sträussler-Scheinker disease and, recently, fatal familial insomnia. Parenthetically CJD was pointed out as a possible analogue of kuru, based of non specific neuropathological findings but Gerstmann-Sträussler-Scheinker disease was identified as linked to kuru because of the presence of numerous amyloid plaques not unlike kuru plaques. The kuru plaque became a link to Alzheimer disease and, as Gajdusek suggested,[69] all amyloidoses share a common pathogenetic mechanism—processing of a normal protein into an amyloid deposit. This event underlies all "conformational disorders", including pathogenetically novel classes of neurodegenerations like α-synucleinopathies, tauopathies and expanded triplet disorders.

We may also speculate what would happen if kuru had not been discovered or did not exist. The infectious nature of Creutzfeldt-Jakob disease would probably not have been suspected until the identification of cases of iatrogenic CJD in human growth hormone or dura mater recipients, or in the vCJD outbreak in the UK. Creutzfeldt-Jakob disease and Gerstmann-Sträussler-Scheinker disease would have remained for decades as obscure neurodegenerations of merely academic interest. The familial forms of Creutzfeldt-Jakob disease would not have benefited from *PRNP* gene analysis, but only later would have been studied by linkage analysis and reverse genetics probably. The whole field would have probably remained of only arcane interest to veterinarians until the BSE epidemic began to exert its devastating effect. The discovery of vCJD would have been delayed, as no surveillance would have been initiated for Creutzfeldt-Jakob disease. And, perhaps most importantly, the sea-change in mentality that has led to the conception of 'protein-misfolding diseases', including not only the neurodegenerative but also an increasing number of nonneurological disorders, would have been delayed by decades.

REFERENCES

1. Gajdusek DC, Zigas V. Kuru. Clinical, pathological and epidemiological study of an acute progressive degenerative disease of the central nervous system among natives of the Eastern Highlands of New Papua. Am J Med 1959; 26:442-469.
2. Gajdusek DC, Zigas V. Studies of kuru. I. The ethnologic setting of kuru. Am J Trop Med Hyg 1961; 10:80-91.
3. Alpers MP, Gajdusek DC, Gibbs CJ Jr. Bibliography of Kuru. Bethesda:Natl Inst Health, 1975.

4. Gibbs CJ Jr, Gajdusek DC, Asher DM et al. Creutzfeldt-Jakob disease (spongiform encephalopathy): transmission to chimpanzee. Science 1968; 161:388-389.

5. Jaskolski M, Liberski PP. Kurt Wüthrich-cowinner of the Nobel Prize in Chemistry, Acta Neurobiol Exp 2002; 62:288-289.

6. Van Rossum A. Spastic pseudosclerosis (Creutzfeldt-jakob disease). In: PJ Vinken, GW Bruyn, eds. Handbook of Clinical Neurology. Vol 6. Diseases of Basal Ganglia. Amsterdam: North-Holland Publ Comp, 1968: 726-760.

7. Gajdusek DC. Kuru and its contribution to medicine. Phil Trans R Soc B 2008; 363:3697-3700.

8. Mathews JD. The changing face of kuru: a personal perspective. Phil Trans R Soc 2008; 363:3679-3684.

9. Mathews JD. The epidemiology of kuru. Papua new Guinea Med J 1967; 10:76-82.

10. Lindenbaum S. Kuru sorcery. Disease and danger in the New Guinea Highlands. Palo Alto:Mayfield Publishing Company, 1979.

11. Gajdusek DC. Unconventional viruses and the origin and disappearance of kuru. Science 1977; 197:943-960.

12. Zigas V, Gajdusek DC. Kuru. Clinical, pathological and epidemiological study of a recently discovered acute progressive degenerative disease of the central nervous system reaching "epidemic" proportions among natives of the Eastern Highlands of New Guinea. P N G Med J 1959; 3:1-31.

13. Zigas V. Origin of investigations on slow virus infections in man. In: SB Prusiner, Hadlow WJ, eds. Slow Transmissible Diseases of the Nervous System, vol 1. New York: Academic Press, 1979:3-6.

14. Sorenson ER, Gajdusek DC. Nutrition in the kuru region. I. Gardening, Food handling and diet of the Fore people. Acta Tropica 1969; 26:281-330.

15. Sorenson ER, Gajdusek DC. The study of child growth and development in primitive cultures. A research archive for ethnopediatric film investigations of styles in the patterning of the nervous system. Pediatrics 1966; 37 (suppl):149-243.

16. Hadlow WJ. Kuru likened to scrapie: the story remembered. Phil Trans R Soc B 2008; 363:3644.

17. Hadlow WJ. Neuropathology and the scrapie-kuru connection. Brain Pathol 1995; 5:27-31.

18. Hadlow WJ. Scrapie and kuru. Lancet 1959; 2:289-290.

19. Hadlow WJ. The Scrapie-kuru connection: recollections of how it came about. In: Prusiner SB, Collinge J, Powell J, Anderton B, eds. Prion Diseases of Humans and Animals. New York: Ellis Horwood, 1993: 40-46.

20. Innes JRM, Saunders LZ. Comparative neuropathology, Academic Press, New York 1962; 839.

21. Goldfarb LG, Cervenakova L, Gajdusek DC. Genetic studies in relation to kuru: an overview. Curr Mol Med 2004; 4:375-384.

22. Gajdusek, telephone conversation, 2008.

23. Zlotnik I. Significance of vacuolated neurones in the medulla of sheep infected with scrapie. Nature 1957a; 180,393-394.

24. Zlotnik I. Vacuolated neurons in sheep affected with scrapie. Nature 1957b;179,737.

25. Gibbs CJ Jr, Gajdusek DC. Attempts to demonstrate a transmissible agent in kuru, amyotrophic lateral sclerosis and other subacute and chronic progressive nervous system degenerations of man. In: Gajdusek DC, Gibbs CJ Jr, Alpers M, eds. Slow, Latent and Temperate virus infections. NINDB Monograph No. 2. US Department of Health, Education and Welfare, 1965:39-48.

26. Gajdusek DC, Zigas V, Baker J. Studies on kuru. III. Patterns of kuru incidence: demographic and geographic epidemiological analysis. Am J Trop Med Hyg 1961; 10:599-627.

27. Mathews JD. A transmission model for kuru. Lancet 1967; 285:821-825.

28. Matthews JD. Kuru as an epidemic disease. In: Hornabrook RW, ed. Essays on Kuru, Papua New Guinea Inst of Med Res Monographs no 3. Berkshire: Faringdon, 1976:83-104.

29. Klitzman RL, Alpers MP, Gajdusek DC. The natural incubation period of kuru and the episodes of transmission in three clusters of patients. Neuroepidemiology 1984; 3:3-20.

30. Alpers MP. Kuru: a clinical study. Mimeographed. US Dept Health, Education, Welfare 1964; 1-38.

31. Alpers MP. Kuru: age and duration studies. Mimeographed. Dept Med, Univ Adelaide 1964; 12.

32. Zigas V, Gajdusek DC. Kuru. Clinical study of a new syndrome resembling paralysis agitans in natives of the eastern Highlands of Australian new Guinea. Med J Australia 1957; 44:745-754.

33. Klatzo I, Gajusek DC, Zigas V. Evaluation of pathological findings in twelve cases of kuru. In: Van Boagert L, Radermecker J, Hozay J, Lowenthal A, eds. Encephalitides. Amsterdam: Elsevier Publ. Comp, 1959:172-190.

34. Klatzo I, Gajusek DC. Pathology of kuru. Lab Invest 1959; 8:799-847.

35. Beck E, Daniel PM, Alpers MP et al. Experimental kuru in chimpanzees. A pathological report. Lancet 1966; 2:1056-1059.

36. Beck E, Daniel PM, Asher DM et al. Experimental kuru in chimpanzees. A neuropathological study. Brain 1973; 96:441-462.

37. Beck E, Daniel PM, Gajdusek DC. A comparison between the neuropathological changes in kuru and scrapie, system degeneration. Proc of the VIth Int Congress Neuropathol, Zurich 1965; 213-218.

38. Beck E, Daniel PM. Kuru and scrapie compared: are they examples of system degeneration? In: Gajdusek DC, Gibbs CJ Jr., Alpers MP, eds. Slow, Latent and Temperate Virus Infections. Washington, DC: US Dept Health, Education, Welfare, 1965:85-93.

39. Beck E, Daniel PM. Prion diseases from a neuropathologist's perspective. In: Prusiner SB, Collinge J, Powell J, Anderton B, eds. Prion Diseases of Humans and Animals. New York, London, Toronto, Sydney: Singapore Ellis Horwood, 1993:63-65.

40. Seitelberger F. Eigenartige familiar-hereditare Krankheit des Zetralnervensystems in einer niederosterreichischen Sippe. Wien Klein Wochen 1962; 74:687-691.

41. Neuman MA, Gajdusek DC, Zigas V. Neuropathologic findings in exotic neurologic disorder among natives of the Highlands of New Guinea. J Neuropathol Exp Neurol 1964; 23:486-507.

42. Masters CL, Gajdusek DC, Gibbs CJ Jr. Creutzfeldt-Jakob disease virus isolations from the Gerstmann-Sträussler syndrome. With an analysis of the various forms of amyloid plaque deposition in the virus induced spongiform encephalopathies. Brain 1981; 104:559-588.

43. Sikorska B, Liberski PP, Sobów T et al. Ultrastructural study of florid plaques in variant Creutzfeldt-Jakob disease: a comparison with amyloid plaques in kuru, sporadic Creutzfeldt-Jakob disease and Gerstmann-Sträussler-Scheinker disease. Neuropathol Appl Neurobiol 2009; 35:46-59.

44. Hainfellner J, Liberski PP, Guiroy DC et al. Pathology and immunohistochemistry of a kuru brain. Brain Pathol 1997; 7:547-554.

45. McLean CA, Ironside JW, Alpers MP et al. Comparative neuropathology of kuru with the new variant of Creutzfeldt-Jakob disease: evidence for strain of agent predominating over genotype of host. Brain Pathol 1998; 8:428-437.

46. Peat A, Field EJ. An unusual structure in kuru brain. Acta Neuropathol (Berl) 1970; 15:288-292.

47. Liberski PP. The occurrence of cytoplasmic lamellar bodies in scrapie infected and normal hamster brains. Neuropatol. Polska 1988,26:79-85.

48. Field EJ, Mathews JD, Raine CS. Electron microscopic observations on the cerebellar cortex in kuru. J Neurol Sci 1969; 8:209-224

49. Liberski PP, Yanagihara R, Gibbs CJ, Jr et al. Re-evaluation of experimental Creutzfeldt-Jakob disease: serial studies of the Fujisaki strain of Creutzfeldt-Jakob disease virus in mice. Brain 1990; 113:121-137.

50. Goodbrand IA, Ironside JW, Nicolson D et al. Prion protein accumulations in the spinal cords of patients with sporadic and growth hormone-associated Creutzfeldt-Jakob disease. Neurosci Lett 1995; 183:127-130.

51. Brandner S, Whitfield J, Boone K et al. Central and peripheral pathology of kuru: pathological analysis of a recent case and comparison with other forms of human prion diseases. Phil Trans R Soc B 2008; 363:3755-3763.

52. Lee H-S, Brown P, Cervenakova L et al. Increased susceptibility to kuru of carriers of the PRNP 129 Methionine/Methionine genotype. J Inf Dis 2001; 183:192-196.

53. Cervenakova L, Goldfarb LG, Garruto R et al. Phenotype-genotype studies in kuru: implications for new variant Creutzfeldt-Jakob disease. Proc Natl Acad Sci USA 1998; 95:13239-13241.

54. Matthews JD, Glasse RM, Lindenbaum S. Kuru and cannibalism. Lancet 1968; 292:449-452.

55. Mead S, Poulter M, Uphill J et al. Genetic risk factors for variant Creutzfeldt-Jakob disease: a genome-wide association study. Lancet Neurology 2009; 8:57-66.

56. Mead S, Stumpf MP, Whitfield J et al. Balancing selection at the prion protein gene consistent with prehistoric kurulike epidemics. Science 2003; 300:640-643.

57. Mead S, Whitfield J, Poulter M et al. Genetic susceptibility, evolution and the kuru epdemic. Phil Trans R Soc B 2008; 363:3741-3746.

58. Mead S. Prion disease genetics. Eur J Hum Genet 2006; 14:273-81.

59. Lantos B, Bhata K, Doey LJ et al. Is the neuropathology of new variant Creutzfeldt-Jakob disease and kuru similar? Lancet 1997; 350:187-188.

60. Aguzzi A, Heikenwalder M. Prion diseases: cannibals and garbage piles. Nature 2003; 423:127-129.

61. Brookfield JF. Human evolution: a legacy of cannibalism in our genes? Curr Biol 2003; 13:R592-593.

62. Marlar RA, Leonard BL, Billman BR et al. Biochemical evidence of cannibalism at a prehistoric Puebloan site in southwestern Colorado. Nature 2000; 407:25-26.

63. Collinge CJ, Whitfield J, McKintosch E et al. Kuru in the 21st century-an acquired human prion disease with very long incubation periods. Lancet 2006; 367:2068-2074.

64. Wadsworth JDF, Joiner S, Linehan JM et al. The origin of the prion agent of kuru: molecular and biological strain typing. Phil Trans R Soc B 2008; 363:3747-3753.

65. Wadsworth JDF, Joiner S, Linehan JM et al. Kuru prions and sporadic Creutzfeldt-Jakob disease prions have equivalent transmission properties in transgenic and wild-type mice.Proc Natl ACad SCi USA 2007; 105:3885-3890.

66. Collinge J, Sidle KCL, Meads J et al. Molecular analysis of prion strain variation and the etiology of „new variant" CJD. Nature 1996; 383:685-670.

67. Hill AF, Desbruslais M, Joiner S et al. The same prion strain causes vCJD and BSE. Nature 1997; 389:448-450.

68. Gibbs CJ. Spongiform encephalopathies—slow, latent and temperate virus infections—in retrospect. In: Prusiner SB, Collinge J, Powell J, Anderon B, eds. Prion Diseases of Humans and Animals. New York, London, Toronto, Sydney, Singapore: Ellis Horwood, 1993:77-91.

69. Gajdusek DC. Infectious amyloids: subacute spongiform encephalopathies as transmissible cerebral amyloidoses. In: BN Fields, DM Knippe, PM Howley eds. Fields Virology, 3rd ed. Philadelphia: Lippincott-Raven, 1996:2851-2900.

CHAPTER 13

LEUKODYSTROPHIES

Seth J. Perlman and Soe Mar*

Division of Pediatric and Developmental Neurology Department of Neurology, Washington University School of Medicine, Saint Louis, Missouri, USA
*Corresponding Author: Soe Mar—Email: mars@neuro.wustl.edu

Abstract: Leukodystrophies comprise a broad group of progressive, inherited disorders affecting mainly myelin. They often present after a variable period of normalcy with a variety of neurologic problems. Though the ultimate diagnosis is not found in many patients with leukodystrophies, distinctive features unique to them aid in diagnosis, treatment and prognostication. The clinical characteristics, etiologies, diagnostic testing and treatment options are reviewed in detail for some of the major leukodystrophies: X-linked adrenoleukodystrophy, Krabbe disease, metachromatic leukodystrophy, Pelizaeus-Merzbacher disease, Alexander disease, Canavan disease, megalencephalic leukoencephalopathy with subcortical cysts and vanishing white matter disease.

INTRODUCTION

Leukodystrophies are inherited disorders primarily affecting brain myelin development and maintenance. Although myelin is most prominently found in the white matter, many leukodystrophies also affect nonwhite matter regions of the nervous system. They have also been shown to be associated with axonal injury.[1] They must be distinguished from other causes of acquired myelin disorders that are often considered as part of a differential diagnosis: inflammatory conditions such as multiple sclerosis, infections such as progressive multifocal leukoencephalopathy (PML), toxin-mediated disorders and chromosomal disorders. An alternate term that has been suggested is *leukoencephalopathy*, though this is often used as a broader term to also include these other conditions. There are a large number of different leukodystrophies with different characteristics and etiologies, ages of onset, clinical courses and prognoses.[2]

Neurodegenerative Diseases, edited by Shamim I. Ahmad.
©2012 Landes Bioscience and Springer Science+Business Media.

Leukodystrophies typically present after an initial period of normal development, though the age of clinical onset varies depending upon the specific disorder and normal development may not be seen in cases of infantile leukodystrophies. Behavioral changes and cognitive deterioration may be the initial presenting symptoms, progressing to regression of motor development and a variable array of findings including: spasticity (though hypotonia may also be seen), weakness, ataxia, nystagmus, swallowing dysfunction, enunciation difficulties, movement disorders, optic atrophy, neuropathy (if peripheral myelin is affected) and epilepsy.[3,4] Up to half of patients with leukoencephalopathies are never given a specific diagnosis.[5,6] Many leukodystrophies are well-defined and may be differentiated by distinctive characteristics in their clinical presentations and their associated biochemical, imaging and pathologic findings. While it is beyond the scope of this chapter to describe all leukodystrophies, we will highlight several of the most important and common types.

DEMYELINATING AND DYSMYELINATING DISORDERS

X-Linked Adrenoleukodystrophy

Clinical Characteristics

X-linked adrenoleukodystrophy (X-ALD) is one of the most common leukodystrophies, with a minimum incidence of 1 in 21,000 males.[7] The rapidly progressive childhood cerebral form is the most common type of X-ALD. These children typically present around 3-10 years of age with behavioral and cognitive deterioration, then follow a progressive course to profound neurodevelopmental disability within a few years. The other most common form of X-ALD presents in male adults as an adrenomyeloneuropathy (AMN), following a more slowly progressive course involving spastic paraparesis, sphincter dysfunction and sensory changes. All phenotypes of X-ALD will often involve a variable degree and timing of adrenocortical dysfunction. Some patients may present with an Addison-only form, whereas a small percentage of males—typically diagnosed after X-ALD is found in a family member—may be asymptomatic and show biochemical evidence but no clinical signs of disease.[2] Finally, a significant number of heterozygous female carriers will develop an AMN-like syndrome in adulthood, though cerebral and adrenocortical involvement in these patients is rare.[8-10]

Etiology and Pathophysiology

X-ALD is an X-linked recessive disease caused by mutations in the *ABCD1* gene. There is no clear correlation between genotype and phenotype, with a single kindred often displaying multiple manifestations of the disease.[11] The gene encodes the peroxisomal ATP-binding cassette trans-membrane transporter ALDP. The detectable biochemical abnormality is the accumulation in serum and all tissues of very long chain fatty acids (VLCFA) due to impaired peroxisomal degradation, though the primary effects are on the nervous system and adrenal cortex. Oxidative stress, inflammatory and autoimmune processes may all play a role in the various phenotypes of the disease.[8]

Diagnosis, Pathology and Imaging

The biochemical defect of elevated plasma VLCFA levels can be reliably found in untreated affected males, including during the neonatal period, and only 15% of heterozygote females may have a normal result.[12] Mutation analysis of *ABCD1* may therefore also be necessary in such patients or their family members. Neuroimaging often provides the primary information ultimately leading to diagnosis of X-ALD. The classic pattern of the cerebral form involves initial involvement of white matter spreading from the splenium of the corpus callosum to the parietooccipital lobes (Fig. 1); other described patterns (the differentiation of which can aid in prognostication) include primary involvement of the frontal lobes or genu of the corpus callosum, the corticospinal tracts (primarily in adults), cerebellar white matter and combined involvement of the parieto-occipital and frontal white matter.[13-15] U-fibers and cortex are typically spared. Proton MR spectroscopy demonstrates an abnormally reduced ratio of N-acetylaspartate (NAA) to choline that is apparent before the development of visible signal abnormalities on conventional MRI.[16,17] The pathologic findings in X-ALD comprise cytoplasmic inclusions of cholesterol esterified with VLCFA, with cerebral disease demonstrating perivascular inflammatory infiltrates behind an edge of confluent demyelination, while peripheral disease demonstrates a distal axonopathy with distal degeneration of myelin.[13]

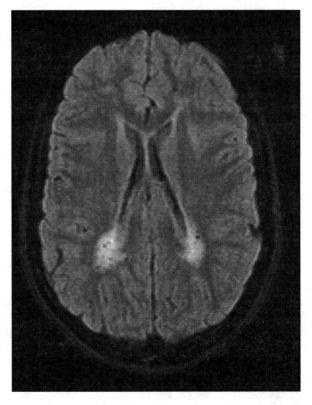

Figure 1. A 9 year-old boy with XALD. In this FLAIR sequence from a brain MRI there are marked T2/FLAIR hyperintense lesions in the white matter surrounding the occipital horns of the lateral ventricles.

Treatment

Patients with X-ALD must be carefully assessed for adrenocortical dysfunction and, if found, given hormone replacement therapy with glucocorticoids and sometimes mineralocorticoids. Treatment of patients with X-ALD with Lorenzo's oil, a 4:1 mixture of glyceryl trioleate and glyceryl trierucate and a reduced fat diet can normalize plasma levels of VLCFA. Although this does not affect endocrine function and has not been shown to alter neurologic disease progression in symptomatic boys, it may exert a preventive effect in asymptomatic boys with normal MRI findings.[8,18] Allogeneic hematopoietic stem cell transplant (HSCT), if performed at an early stage of disease when patients are free of neurologic and significant neuropsychological involvement and have only limited abnormalities on brain MRI, can arrest the cerebral inflammatory demyelination of X-ALD.[18,19] Autologous stem cell gene therapy techniques and valproic acid have also recently been investigated with initially promising results.[20,21]

Krabbe Disease

Clinical Characteristics

Krabbe disease, also known as globoid cell leukodystrophy is a progressive white matter disorder with three subtypes that have been delineated: infantile, juvenile and adolescent-adult forms. The infantile form is most frequent with an incidence of 1 in 70,000 to 100,000. The infantile can be separated into early and late forms based on the onset of the disease. In early infantile form, the clinical symptoms occur between 1 and 6 months of age and include hyperirritability, an exaggerated startle and dysphagia. These are followed by progressive spasticity, stagnation and regression of development, seizures, blindness, deafness, rapid motor and mental deterioration and death. Life expectancy varies between 5 months and 3 years of age. The late-onset infantile form usually presents before 18 months of age. The clinical symptoms of other late onset forms are heterogeneous, ranging from insidious isolated visual impairment to cognitive deterioration, gait abnormality, ataxia, hemiparesis and spasticity. The age of death of late onset forms varies from 18 months to more than 14 years.[1,22,23,15]

Etiology and Pathophysiology

Krabbe disease has autosomal recessive inheritance and is caused by a deficiency of galactocerebroside β galactosidase (GALC) activity. GALC is a lysosomal enzyme involved in the metabolism of important galactolipids found in myelin.[24,25] The accumulation of cerebrosides and psychosine results in the apoptotic death of oligodendrocytes and associated demyelination. Various mutations in the *GALC* gene, located on chromosome 14q31, have been identified.[26]

Pathology and Imaging Findings

Microscopic pathology reveals diffuse demyelination with loss of oligodendrocytes throughout the brain, but preservation of U fibers. Extensive fibrillary astrocytic gliosis replaces the lost oligodendrocytes and myelin. Infiltration of numerous macrophages, often multinucleated ("globoid cells"), is the unique feature of Krabbe disease. In areas

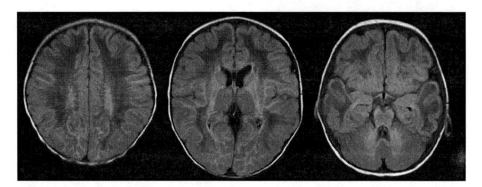

Figure 2. A 10 month old girl with infantile Krabbe disease. In this FLAIR sequence from a brain MRI there are T2/FLAIR hyperintense lesions in the predominantly parietooccipital periventricular and cerebellar white matter, the corticospinal tracts and the posterior corpus callosum. Grey matter structures in the basal ganglia are also involved. Courtesy of Jayne Ness, MD, PhD, The University of Alabama at Birmingham.

of demyelination, axonal degeneration typically is observed. Peripheral nerves also are involved in the disease process, with pathology demonstrating segmental demyelination.[27] MRI of infantile Krabbe disease (Fig. 2) typically shows T2/FLAIR hyperintense lesions of periventricular and cerebellar white matter and the pyramidal tracts. Posterior corpus callosum and parietooccipital white matter abnormalities are also seen and grey matter structures are commonly involved.[28] In the juvenile and adult forms white matter changes are characteristically confined to periventricular parietooccipital regions in the beginning of the disease. Proton MR spectroscopy has provided evidence of significant neuroaxonal loss in infantile globoid cell leukodystrophy, minor neuroaxonal damage in the juvenile form and nearly normal white matter and no evidence of neuroaxonal loss in adult globoid cell leukodystrophy, thus suggesting that axonal loss might be the marker of severe disease.[29]

Treatment and Outcomes

Umbilical cord blood transplantation has been shown to reduce the clinical severity if treatment is given in very young asymptomatic patients predicted to have a severe phenotype. However, children who underwent transplantation after the onset of symptoms had minimal neurologic improvement.[30]

Metachromatic Leukodystrophy

Clinical Characteristics

Metachromatic leukodystrophy (MLD) is an autosomal recessive lysosomal storage disease with an estimated incidence of 1 in 40,000 births.[2] It is divided into three subtypes based on age of onset after a period of initially normal development: late infantile, juvenile and adult. The late infantile form is the most severe phenotype and may present between 6 months and 2 years of age with motor developmental delay, weakness, hypotonia and ataxia. As the disorder progresses peripheral neuropathy occurs and worsens, speech and cognition deteriorate, seizures may occur, hypotonia eventually transitions into a spastic quadriplegia, vision and hearing decline and the child eventually is left in a decerebrate state.[31,32] The juvenile form may have onset as either an early (4 to 8 years old) or late (6 to 16 years old) subgroup with the early subgroup resembling the late infantile form with predominantly motor involvement initially, whereas the late juvenile subgroup typically presents with behavioral and cognitive problems. The adult form may present anytime after 16 years of age and, like the late juvenile form, typically begins with emotional and intellectual dysfunction and may appear psychiatric in nature. As it progresses motor function is affected and peripheral neuropathy, to a varying degree, may be seen.[33,34] All forms of the disease follow a relentlessly progressive course that culminates in a profoundly debilitated state, with the pace of deterioration and progression to death increasing with later onset from within a few years to several decades after initial symptoms.

Etiology and Pathophysiology

MLD is caused by mutations in the *ARSA* gene causing deficiency in the lysosomal enzyme arylsulfatase A, which leads to accumulation of sulfatide.[2] Rarely, the disorder may be caused by mutations in the *PSAP* (prosaposin) gene, leading to deficient sphingolipid activator protein SAP-B (saposin B) that has a role in stimulating the action of ARSA.[35] This accumulation leads to demyelination from oligodendroglial and Schwann cell injury and death, although the mechanism by which this occurs is not completely understood. Greater than 150 distinct mutations in *ARSA* have been described. Those causing no enzyme activity are termed I-type alleles and those causing pathologically reduced enzyme activity are termed A-type alleles; there has been described some degree of genotype-phenotype correlation of allelic makeup with clinical subtype, although this is not universal and greatly varying ages of onset have been seen among siblings with identical mutations.[36-39]

Diagnosis, Pathology and Imaging

Although measurement of ARSA activity in leukocytes or cultured skin fibroblasts may demonstrate disease-causing reduction (typically from zero to 10% activity in MLD), it cannot differentiate between pathologic deficiency and ARSA pseudodeficiency that can occur in healthy individuals.[40] Measurement of urine sulfatide, molecular genetic testing of *ARSA* (or, if urine sulfatide is elevated but enzyme activity is normal, of *PSAP*), or biopsy of nerve or brain are therefore often undertaken as confirmatory tests. Pathologic examination of affected tissues reveals the characteristic metachromatically staining sulfatide deposited throughout the white matter and in Schwann cells and macrophages of peripheral nerve, while MRI demonstrates confluent, symmetric and non-enhancing white matter hypodensity that initially spares U-fibers and may start with occipital or frontal predominance (Fig. 3).[2]

Treatment

At present, HSCT is the only therapy available that has been clearly shown to alter the course of MLD involving the central nervous system, with greatest efficacy seen in patients with juvenile or adult forms still early in the course of the disease.[41] Other therapies currently under investigation include gene therapy techniques, warfarin (which

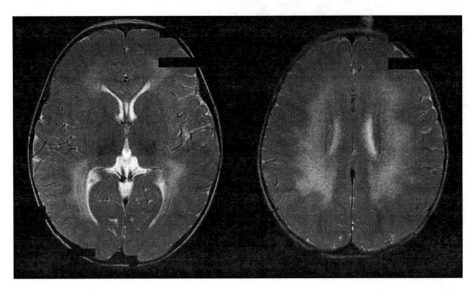

Figure 3. A 2-year, 10-month-old boy with the late-infantile form of metachromatic leukodystrophy. In this axial T2-weighted noncontrast MRI there are symmetric, confluent white matter abnormalities in all lobes that exhibit relative sparing of the subcortical U-fibers.

may affect the role vitamin K plays in sulfatide synthesis) and recombinant human ARSA (though this is not believed to cross the blood-brain barrier).[38,42,43]

HYPOMYELINATING DISORDERS

Pelizaeus-Merzbacher Disease

Clinical Characteristics

Pelizaeus-Merzbacher disease (PMD) is a rare disorder with an X-linked recessive mode of inheritance. Three forms have been delineated based on age and severity of presentation: classical, connatal and transitional. Children with the classical type (the most common form of the disease) present within the first year of life with irregular nystagmoid eye movements, hypotonia, severe developmental delay, microcephaly, growth retardation, ataxia and seizures. Spasticity, cerebellar signs and movement abnormalities including dystonia, choreoathetosis and hyperkinesis become more prominent with age. Very slow neurological deterioration occurs, usually noted after 10-12 years of age and they may survive until the 6th decade of life. Connatal PMD is the most severe form. Severe hypotonia, stridor, feeding impairment and nystagmus are noted during the neonatal period. Few, if any, developmental milestones are reached. The affected children undergo rapid progression and typically die during the first decade. Patients with the transitional form have overlapping features between the connatal and classic forms, with notable slower and later onset than the connatal form.[2,44] X-linked spastic paraplegia Type 2 (SPG2) is an allelic condition of PMD. The pure form of SPG2 may present with only a spastic gait and ambulation problems, though there also exists a complicated form with additional nystagmus, ataxia and intention tremor.[45]

Etiology and Pathophysiology

PMD and SPG2 are caused by mutation in the *PLP* gene coding for proteolipid protein located on the long arm of the X chromosome (Xq21.33-Xq22).[46-48] The *PLP* gene encodes two proteolipid proteins in oligodendrocytes, PLP and DM20. While PLP is the prominent protein in CNS myelin, DM20 may be involved in oligodendrocyte differentiation and survival.[49-51] Many different mutations have been identified in *PLP*, with gene duplications being the most common cause of PMD. Clinical severity is correlated with the specific mutation. Mutations that affect only PLP, but not DM20, cause a milder clinical syndrome.[52] The most severe presentations of PMD are usually caused by missense mutations, although a broad range of clinical phenotypes may still be seen.[53] A benign form of PMD is seen in deletions or null mutations of the *PLP* gene.[52,54]

Pathology and Imaging Findings

In PMD central white matter is reduced in volume with signs of deficient myelination (Fig. 4). In the connatal form, central myelin is completely absent with reduced number of oligodendrocytes. In the classical form, the central white matter shows a patchy distribution of dysmyelination with preserved myelin islands, giving a "tigroid" appearance. Axons are relatively well preserved, though some axonal loss

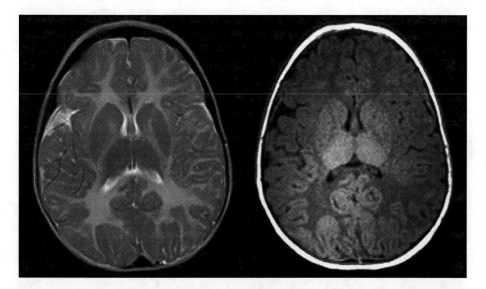

Figure 4. A 12-month-old boy with Pelizaeus-Merzbacher disease. T2 (left) and T1-MPR (right) axial sequences from a brain MRI show diffuse white matter signal intensity abnormalities with no evidence of myelination on the T1-MPR or T2 weighted images. There is bilateral, symmetric abnormal T2/FLAIR signal hyperintensity in the internal and external capsules.

may be seen in completely demyelinated areas. In SPG2, demyelination affects mainly the longitudinal spinal tracts.[55] The extent of axonal injury increases with age and may account for the progression of neurological signs and symptoms in patients with PMD.[52] It may be difficult to diagnose hypomyelination during the first few months of life, though the absence of myelin in the pons, cerebellum, posterior limb of the internal capsule, splenium of corpus callosum and optic radiations in the first 3 months in the right clinical setting may suggest PMD.[56] MRI patterns of PMD have been divided into 3 subtypes; Type I, diffusely hemispheric and corticospinal; Type II, diffusely hemispheric without brain stem lesions; and Type III, patchy in the hemispheres.[57] There is no clear relationship between MRI findings and clinical phenotype, but the degree of hypomyelination has been correlated with the severity of clinical handicap.[58] Magnetic resonance spectroscopy (MRS) results in patients with PMD are inconsistent: normal NAA/Cr ratio with decreased Ch/Cr ratio;[59,60] decreased NAA/Cr ratio with normal Cho/Cr ratio;[61] and increased NAA concentration.[62]

Treatment

There is currently no specific treatment available for PMD.

Alexander Disease

Clinical Characteristics

Alexander disease is a rare, mostly sporadic, progressive disorder of CNS, although familial cases have been reported. Based on the age of onset and severity of symptoms three subtypes have been described: infantile, juvenile and adult forms. The infantile form (birth to 2 years) is the most severe phenotype. Patients often present with macrocephaly, seizures, spasticity, motor and bulbar dysfunction. These children usually die within the first decade of life. Children with the juvenile form often present between 2 to 12 years of age with bulbar symptoms and signs followed by ataxia and spasticity. They typically survive into the 2nd to 4th decades of life. Adult-onset Alexander disease presents with gradually or episodically progressive symptoms of dysarthria, dysphonia, dysphagia, pyramidal signs, ataxia and palatal myoclonus or tremor.[63,64] The clinical course and survival of adult-onset Alexander disease are quite variable: the mean age of onset is in the late thirties, although an asymptomatic carrier over age 60 has been described.[65]

Etiology and Pathophysiology

The genetic basis of Alexander disease was established following discovery of mutations in different glial fibrillary acidic protein (GFAP) residues. *GFAP* mutations result in an over-expression of abnormal GFAP, which causes a lethal encephalopathy.[66] Another distinct pathological feature of Alexander disease is paucity of myelin, which is most pronounced and severe in the frontal white matter. In areas of myelin paucity, most axons are intact. In juvenile and adult forms, the pathology may be mainly limited to the brainstem and cerebellum.[67]

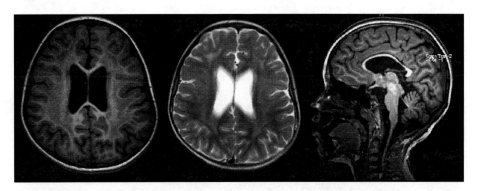

Figure 5. A 6 year-old girl with infantile Alexander disease. Axial T1-MPR (left) and T2 (center) images from a brain MRI show white matter changes with increased T2 and decreased T1 signal intensity, most prominent anteriorly and largely sparing the subcortical and periventricular white matter. There is a periventricular rim with high signal on T1-weighted images and low signal on T2-weighted images. Sagittal T1-MPR images (right) show corpus callosum, frontal and parietal cortical, cerebellar and pontine atrophy. Also noted is thickening of the inferior aspect of the tectum that may represent an incidental glioma.

Pathology and Imaging Findings

The most distinctive pathologic feature of Alexander disease is the presence and widespread deposition of cytoplasmic inclusions localized in astrocyte cytoplasm, termed Rosenthal fibers. They are mainly found in perivascular, perivenular and subpial spaces of cerebral hemispheres (mainly frontal), cerebellum and brainstem.[68] These Rosenthal fibers contain GFAP, which is now believed to modulate astrocyte motility and shape. GFAP also may be involved in glial cell adhesion, myelination and cell signaling.[69] Van der Knaap et al defined MR imaging criteria to diagnose Alexander disease (4 of 5 required): (1) extensive cerebral white matter changes with frontal predominance, (2) a periventricular rim with high signal on T1-weighted images and low signal on T2-weighted images, (3) abnormalities of basal ganglia and thalami, (4) brain stem abnormalities and (5) abnormal contrast enhancement. Although MRI (Fig. 5) is useful for diagnostic purposes and can document the extent of white matter disease, it does not always predict the severity and course of the disease.[64] The discrepancy between severe white matter changes on MRI and mild clinical features in some patients with Alexander disease may be explained by axonal preservation.[70] Decreasing NAA concentration over the time on proton MR spectroscopy may predict ongoing subclinical neuronal degeneration in these patients. This possibility is supported by research showing absent or severely reduced NAA in the oldest patients with the most severe clinical course.[70] These findings suggest that the degree and timing of the axonal degeneration may determine the phenotypic severity, rate of neurological decline and age of symptom onset in Alexander disease.

Treatment and Outcomes

There is no cure for Alexander Disease. The treatment for Alexander disease is symptomatic and supportive.

SPONGIFORM DISORDERS

Canavan Disease

Clinical Characteristics

Canavan disease is a rare autosomal recessive spongiform leukodystrophy most often seen in children of Ashkenazi Jewish descent.[71] It most commonly presents in infancy, with symptom onset between 3 and 6 months of age, although a congenital form may occur rarely and juvenile- and adult-onset forms, though controversial, may exist.[72,73] Initial symptoms include lethargy or irritability, a weak cry and suck, poor head control, hypotonia, decreased movement of the extremities and decreased visual responsiveness. Head growth is accelerated and macrocephaly develops and, as the condition progresses, spasticity and hyperreflexia develop, tonic extensor spasms occur, development arrests and regresses, blindness and optic atrophy are seen and about half of affected children develop seizures as they progress to a decerebrate state, with death typically occurring in childhood or the teenage years.[73]

Etiology and Pathophysiology

Canavan disease is caused by mutations in the *ASPA* gene leading to deficiency in aspartoacylase, the enzyme located in oligodendroglia that hydrolyzes NAA into aspartic acid and acetate.[74-77] Among the Ashkenazi population nearly all disease occurs due to only a few mutations, whereas in other patient groups different mutations of *ASPA* are found.[71,74] NAA accumulates in the CNS, causing dysfunction and injury to oligodendrocytes, vacuolation and spongiform changes and myelin destruction. The mechanism by which this occurs is not known, but there is increased water content in affected white matter and it has been suggested that, among other things, NAA may act as an osmoregulator and molecular water pump.[2,78]

Diagnosis, Pathology and Imaging

Biochemical diagnosis may be made by identifying elevated NAA in urine and serum (or amniotic fluid for prenatal testing) organic acid screening, while molecular genetic studies may be useful as confirmatory or prenatal tests. On pathologic examination, the brain is enlarged and extensive sponge-like vacuolation and demyelination is primarily seen in the subcortical and cerebellar white matter, although brainstem, globus pallidus, hypothalamus and thalamus are also affected.[2] MRI demonstrates symmetric, diffuse white matter involvement with T1 hypo- and T2 hyperintensity along with evidence of restricted diffusion in the deep white matter and brainstem (Fig. 6), whereas MR spectroscopy shows markedly increased levels of NAA.[79,80]

Treatment

No specific therapies are yet available and treatment is symptomatic and supportive. Investigational treatments include the use of gene therapy techniques and lithium, which may reduce the NAA content in affected white matter.[81,82]

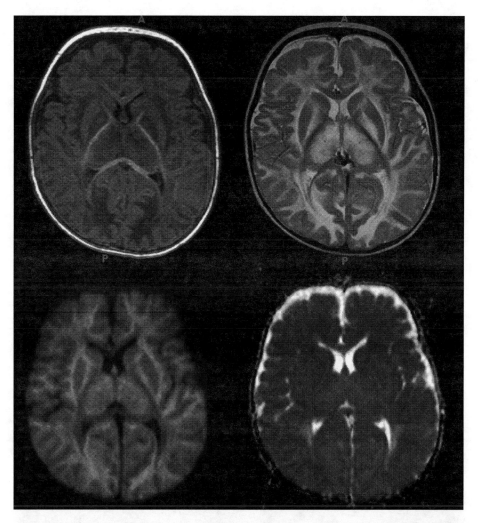

Figure 6. A 7-month-old boy with Canavan disease. Brain MRI T1-MPR (upper left) and T2 sequences (upper right) show diffuse signal abnormality throughout the bilateral cerebral white matter. Trace-diffusion (lower left) high signal and low ADC on the ADC map (lower right) indicate diffusion restriction within these regions. Note the preservation of myelin within the posterior limbs of the internal capsules and the corpus callosum.

Megalencephalic Leukoencephalopathy with Subcortical Cysts

Clinical Characteristics

Megalencephalic leukoencephalopathy with subcortical cysts (MLC) is a recently described, rare, autosomal recessive disease that occurs with greater frequency in Turkish and Agarwal (an ethnic group in India) populations.[83] It presents with macrocephaly at birth or within the first year of life, after which the growth rate normalizes, with initially normal neurologic exam and development in most cases.[84] Severity and course vary, but delayed and unsteady walking is seen early and is followed, over years, by the slow development of prominent ataxia, hyperreflexia, epilepsy, dysarthria, inability to walk and mild cognitive decline.[2]

Etiology and Pathophysiology

Mutations in *MLC1* genes are found in the majority of patients with MLC, although a significant percentage have typical MLC disease and no mutations.[85-87] The gene encodes for a transmembrane protein with unknown function.

Diagnosis, Pathology and Imaging

No measurable biochemical abnormalities have been identified and many patients do not have *MLC1* mutations (though genetic testing may be used prenatally or for confirmation in families with known mutations). Diagnosis of MLC is based on the clinical picture and imaging findings (Fig. 7), with MRI showing diffusely abnormal and mildly swollen cerebral white matter (with relative preservation of the corpus callosum, internal capsule and brainstem), mildly abnormal cerebellar white matter, increased white matter diffusivity, anterior temporal and frontoparietal cysts that may grow in size and number and eventual cerebral atrophy.[84,88,89] Brain pathologic findings were described in one patient and consisted of vacuolization and spongiform changes in the white matter, minor myelin degradation with splitting between the outer lamellae of the sheaths and normal cortex.[90]

Treatment and Outcomes

No specific treatment exists and therapy is supportive and symptomatic.

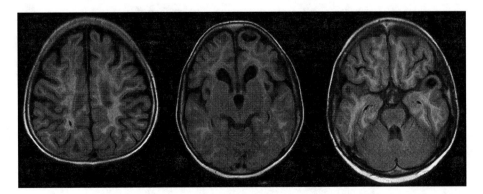

Figure 7. A 7 year-old girl with megalencephalic leukoencephalopathy with subcortical cysts. FLAIR sequence from the brain MRI shows diffusely abnormal cerebral white matter with relative preservation of the brainstem, mildly abnormal cerebellar white matter and primarily anterior temporal and frontoparietal cysts along with cerebral atrophy. Courtesy of Jayne Ness, MD, PhD, The University of Alabama at Birmingham.

CYSTIC DISORDERS

Vanishing White Matter Disease

Clinical Characteristics

Vanishing white matter disease (VWM, childhood ataxia with central nervous system hypomyelination, or Cree encephalopathy) is an autosomal recessive disease that is likely among the more common leukodystrophies.[91] VWM manifests a broad range of clinical presentations and ages of onset, from congenital forms with rapid neurological decline and death, to adult-onset forms with slow symptom progression; the classical and most common form presents between 2 and 5 years of age with progressive gait difficulties and mild cognitive impairment.[92] Characteristic of the disease are episodes of acute deterioration after minor head trauma, febrile infections, prolonged sun exposure and even fear that manifest as hypotonia, irritability, vomiting, seizures, unconsciousness that may progress to coma or death and lack of full recovery afterwards.[2] Extracranial involvement often includes ovarian dysgenesis in antenatal forms and ovarian failure in later-onset disease (termed ovarioleukodystrophy).[93]

Etiology and Pathophysiology

VWM and its varying presentations are caused by mutations in any of the five genes, *eIF2B1* through *eIF2B5*, that encode subunits (alpha through epsilon) of eukaryotic translation initiation factor 2B (eIF2B).[94,95] Regulation of the translation of mRNA to polypeptides involves eukaryotic transcription factors and mutant eIF2B in VWM may cause disease by disrupting the normal stress-elicited compensatory mechanisms that inhibit synthesis of new proteins and that induce signals promoting both cellular survival and apoptosis.[96] While members of the same affected family and individuals with the same mutation have been described to display significant phenotypic heterogeneity, some genotype-phenotype correlations have been noted.[94,97-101]

Diagnosis, Pathology and Imaging

No reliable routine laboratory abnormalities have been identified that aid in the diagnosis of VWM, though decreased CSF asialotransferrin to transferrin ratio and decreased lymphocyte eIF2B nucleotide guanine exchange (GEF) activity have both been proposed as possible markers of disease.[102,103] Molecular genetic testing of eIF2B subunits can be performed and is essential in infants with the antenatal or early infantile forms of VWM, but may not demonstrate a mutation in approximately 10% of patients with VWM.[104-106] Pathologic findings include vacuolation and "foamy" oligodendrocytes, white matter changes ranging from cystic to cavitary, variable degrees of axon loss and no significant inflammation.[2] MRI may show diffuse signal abnormalities in the cerebral white matter that may have a signal intensity near that of CSF, diffuse disappearance of the cerebral white matter, a fluid-filled space between the ependyma and cortex if all cerebral white matter has vanished, relative sparing of the temporal lobes, noncystic cerebellar white matter, no abnormal enhancement, radiating stripes within abnormal white matter on sagittal and coronal images and involvement of the inner (but sparing of the outer) rim of the corpus callosum.[97]

Treatment

There is no specific therapy for VWM, but avoidance and prompt treatment of possible stressors that may precipitate an episode is essential.

CONCLUSION

Patients with leukodystrophies, though often never given an ultimate diagnosis, can present in a wide variety of unique ways. Challenging and frustrating for both the clinician and the patient are the typically progressive natures of these diseases. The paucity of specific therapies available to treat them, and the often grim prognoses they carry, also contribute to the difficulty in caring for affected patients and families. Some promising tools for identification, evaluation and treatment of these diseases are under development, while many more are yet to be discovered and investigated. Continued research into genetic and pathophysiologic etiologies, distinctive biochemical and imaging features and potential avenues of treatment are all absolute necessities to improve the clinical care available for patients suffering from this difficult group of disorders.

REFERENCES

1. Mar S, Noetzel M. Axonal damage in leukodystrophies. Pediatr Neurol 2010; 42(4):239-242.
2. van der Knaap MS, Valk J, Barkhof F. Magnetic resonance of myelination and myelin disorders. 3rd ed. Berlin; New York: Springer; 2005.
3. Lyon G, Fattal-Valevski A, Kolodny EH. Leukodystrophies: clinical and genetic aspects. Top Magn Reson Imaging 2006; 17(4):219-242.
4. Costello DJ, Eichler AF, Eichler FS. Leukodystrophies: classification, diagnosis and treatment. Neurologist 2009; 15(6):319-328.
5. Bonkowsky JL, Nelson C, Kingston JL et al. The burden of inherited leukodystrophies in children. Neurology 2010.
6. van der Knaap MS, Breiter SN, Naidu S et al. Defining and categorizing leukoencephalopathies of unknown origin: MR imaging approach. Radiology 1999; 213(1):121-133.
7. Bezman L, Moser AB, Raymond GV et al. Adrenoleukodystrophy: incidence, new mutation rate and results of extended family screening. Ann Neurol 2001; 49(4):512-517.
8. Moser HW, Mahmood A, Raymond GV. X-linked adrenoleukodystrophy. Nat Clin Pract Neurol 2007; 3(3):140-151.
9. Restuccia D, Di Lazzaro V, Valeriani M et al. Neurophysiological abnormalities in adrenoleukodystrophy carriers. Evidence of different degrees of central nervous system involvement. Brain 1997; 120 (Pt 7):1139-1148.
10. Schmidt S, Traber F, Block W et al. Phenotype assignment in symptomatic female carriers of X-linked adrenoleukodystrophy. J Neurol 2001; 248(1):36-44.
11. Moser HW, Moser AB, Smith KD et al. Adrenoleukodystrophy: phenotypic variability and implications for therapy. J Inherit Metab Dis 1992; 15(4):645-664.
12. Moser AB, Kreiter N, Bezman L et al. Plasma very long chain fatty acids in 3,000 peroxisome disease patients and 29,000 controls. Ann Neurol 1999; 45(1):100-110.
13. Kim JH, Kim HJ. Childhood X-linked adrenoleukodystrophy: clinical-pathologic overview and MR imaging manifestations at initial evaluation and follow-up. Radiographics 2005; 25(3):619-631.
14. Kumar AJ, Rosenbaum AE, Naidu S et al. Adrenoleukodystrophy: correlating MR imaging with CT. Radiology 1987; 165(2):497-504.
15. Loes DJ, Fatemi A, Melhem ER et al. Analysis of MRI patterns aids prediction of progression in X-linked adrenoleukodystrophy. Neurology 2003; 61(3):369-374.
16. Eichler FS, Barker PB, Cox C et al. Proton MR spectroscopic imaging predicts lesion progression on MRI in X-linked adrenoleukodystrophy. Neurology 2002; 58(6):901-907.
17. Eichler FS, Itoh R, Barker PB et al. Proton MR spectroscopic and diffusion tensor brain MR imaging in X-linked adrenoleukodystrophy: initial experience. Radiology 2002; 225(1):245-252.

18. Semmler A, Kohler W, Jung HH et al. Therapy of X-linked adrenoleukodystrophy. Expert Rev Neurother 2008; 8(9):1367-1379.

19. Cartier N, Aubourg P. Hematopoietic stem cell transplantation and hematopoietic stem cell gene therapy in X-linked adrenoleukodystrophy. Brain Pathol 2010; 20(4):857-862.

20. Cartier N, Hacein-Bey-Abina S, Bartholomae CC et al. Hematopoietic stem cell gene therapy with a lentiviral vector in X-linked adrenoleukodystrophy. Science 2009; 326(5954):818-823.

21. Fourcade S, Ruiz M, Guilera C et al. Valproic acid induces antioxidant effects in X-linked adrenoleukodystrophy. Hum Mol Genet 2010; 19(10):2005-2014.

22. Crome L, Hanefeld F, Patrick D et al. Late onset globoid cell leucodystrophy. Brain 1973; 96(4):841-848.

23. Thomas PK, Halpern JP, King RH et al. Galactosylceramide lipidosis: novel presentation as a slowly progressive spinocerebellar degeneration. Ann Neurol 1984; 16(5):618-620.

24. Suzuki Y, Austin J, Armstrong D et al. Studies in globoid leukodystrophy: enzymatic and lipid findings in the canine form. Exp Neurol 1970; 29(1):65-75.

25. Suzuki K, Suzuki Y. Globoid cell leucodystrophy (Krabbe's disease): deficiency of galactocerebroside beta-galactosidase. Proc Natl Acad Sci USA 1970; 66(2):302-309.

26. Wenger DA, Rafi MA, Luzi P. Molecular genetics of Krabbe disease (globoid cell leukodystrophy): diagnostic and clinical implications. Hum Mutat 1997; 10(4):268-279.

27. Suzuki K. Twenty five years of the "psychosine hypothesis": a personal perspective of its history and present status. Neurochem Res 1998; 23(3):251-259.

28. Provenzale JM, Escolar M, Kurtzberg J. Quantitative analysis of diffusion tensor imaging data in serial assessment of Krabbe disease. Ann N Y Acad Sci 2005; 1064:220-229.

29. Brockmann K, Dechent P, Wilken B et al. Proton MRS profile of cerebral metabolic abnormalities in Krabbe disease. Neurology 2003; 60(5):819-825.

30. Escolar ML, Poe MD, Provenzale JM et al. Transplantation of umbilical-cord blood in babies with infantile Krabbe's disease. N Engl J Med 2005; 352(20):2069-2081.

31. Cameron CL, Kang PB, Burns TM et al. Multifocal slowing of nerve conduction in metachromatic leukodystrophy. Muscle Nerve 2004; 29(4):531-536.

32. Wang PJ, Hwu WL, Shen YZ. Epileptic seizures and electroencephalographic evolution in genetic leukodystrophies. J Clin Neurophysiol 2001; 18(1):25-32.

33. Felice KJ, Gomez Lira M, Natowicz M et al. Adult-onset MLD: a gene mutation with isolated polyneuropathy. Neurology 2000; 55(7):1036-1039.

34. Marcao AM, Wiest R, Schindler K et al. Adult onset metachromatic leukodystrophy without electroclinical peripheral nervous system involvement: a new mutation in the ARSA gene. Arch Neurol 2005; 62(2):309-313.

35. Wrobe D, Henseler M, Huettler S et al. A nonglycosylated and functionally deficient mutant (N215H) of the sphingolipid activator protein B (SAP-B) in a novel case of metachromatic leukodystrophy (MLD). J Inherit Metab Dis 2000; 23(1):63-76.

36. Arbour LT, Silver K, Hechtman P et al. Variable onset of metachromatic leukodystrophy in a Vietnamese family. Pediatr Neurol 2000; 23(2):173-176.

37. Berger J, Loschl B, Bernheimer H et al. Occurrence, distribution and phenotype of arylsulfatase A mutations in patients with metachromatic leukodystrophy. Am J Med Genet 1997; 69(3):335-340.

38. Fluharty AL. (Updated 2008/09/23). Arylsulfatase A Deficiency. GeneReviews at GeneTests: Medical Genetics Information Resource (database online) Copyright, University of Washington, Seattle:1997-2010. Available at http://www.genetests.org, Accessed 2010/09/15.

39. Polten A, Fluharty AL, Fluharty CB et al. Molecular basis of different forms of metachromatic leukodystrophy. N Engl J Med 1991; 324(1):18-22.

40. Barth ML, Ward C, Harris A et al. Frequency of arylsulphatase A pseudodeficiency associated mutations in a healthy population. J Med Genet 1994; 31(9):667-671.

41. Orchard PJ, Tolar J. Transplant Outcomes in Leukodystrophies. Seminars in Hematology 2010; 47(1):70-78.

42. Sevin C, Cartier-Lacave N, Aubourg P. Gene therapy in metachromatic leukodystrophy. Int J Clin Pharmacol Ther 2009; 47 Suppl 1:S128-131.

43. Sundaram KS, Lev M. Warfarin administration reduces synthesis of sulfatides and other sphingolipids in mouse brain. J Lipid Res 1988; 29(11):1475-1479.

44. Boulloche J, Aicardi J. Pelizaeus-Merzbacher disease: clinical and nosological study. J Child Neurol 1986; 1(3):233-239.

45. Hudson LD. Pelizaeus-Merzbacher disease and spastic paraplegia type 2: two faces of myelin loss from mutations in the same gene. J Child Neurol 2003; 18(9):616-624.

46. Willard HF, Riordan JR. Assignment of the gene for myelin proteolipid protein to the X chromosome: implications for X-linked myelin disorders. Science 1985; 230(4728):940-942.

47. Gencic S, Abuelo D, Ambler M et al. Pelizaeus-Merzbacher disease: an X-linked neurologic disorder of myelin metabolism with a novel mutation in the gene encoding proteolipid protein. Am J Hum Genet 1989; 45(3):435-442.

48. Hudson LD, Puckett C, Berndt J et al. Mutation of the proteolipid protein gene PLP in a human X chromosome-linked myelin disorder. Proc Natl Acad Sci USA 1989; 86(20):8128-8131.
49. Weimbs T, Stoffel W. Proteolipid protein (PLP) of CNS myelin: positions of free, disulfide-bonded and fatty acid thioester-linked cysteine residues and implications for the membrane topology of PLP. Biochemistry 1992; 31(49):12289-12296.
50. LeVine SM, Wong D, Macklin WB. Developmental expression of proteolipid protein and DM20 mRNAs and proteins in the rat brain. Dev Neurosci 1990; 12(4-5):235-250.
51. Yang X, Skoff RP. Proteolipid protein regulates the survival and differentiation of oligodendrocytes. J Neurosci 1997; 17(6):2056-2070.
52. Garbern JY. Pelizaeus-Merzbacher disease: Genetic and cellular pathogenesis. Cell Mol Life Sci 2007; 64(1):50-65.
53. Cailloux F, Gauthier-Barichard F, Mimault C et al. Genotype-phenotype correlation in inherited brain myelination defects due to proteolipid protein gene mutations. Clinical European Network on Brain Dysmyelinating Disease. Eur J Hum Genet 2000; 8(11):837-845.
54. Garbern JY, Yool DA, Moore GJ et al. Patients lacking the major CNS myelin protein, proteolipid protein 1, develop length-dependent axonal degeneration in the absence of demyelination and inflammation. Brain 2002; 125(Pt 3):551-561.
55. Seitelberger F. Neuropathology and genetics of Pelizaeus-Merzbacher disease. Brain Pathol 1995; 5(3):267-273.
56. Barkovich AJ. Magnetic resonance techniques in the assessment of myelin and myelination. J Inherit Metab Dis 2005; 28(3):311-343.
57. Nezu A, Kimura S, Takeshita S et al. An MRI and MRS study of Pelizaeus-Merzbacher disease. Pediatr Neurol 1998; 18(4):334-337.
58. Plecko B, Stockler-Ipsiroglu S, Gruber S et al. Degree of hypomyelination and magnetic resonance spectroscopy findings in patients with Pelizaeus Merzbacher phenotype. Neuropediatrics 2003; 34(3):127-136.
59. Takanashi J, Sugita K, Osaka H et al. Proton MR spectroscopy in Pelizaeus-Merzbacher disease. AJNR Am J Neuroradiol 1997; 18(3):533-535.
60. Spalice A, Popolizio T, Parisi P et al. Proton MR spectroscopy in connatal Pelizaeus-Merzbacher disease. Pediatr Radiol 2000; 30(3):171-175.
61. Bonavita S, Schiffmann R, Moore DF et al. Evidence for neuroaxonal injury in patients with proteolipid protein gene mutations. Neurology 2001; 56(6):785-788.
62. Takanashi J, Inoue K, Tomita M et al. Brain N-acetylaspartate is elevated in Pelizaeus-Merzbacher disease with PLP1 duplication. Neurology 2002; 58(2):237-241.
63. Li R, Johnson AB, Salomons G et al. Glial fibrillary acidic protein mutations in infantile, juvenile and adult forms of Alexander disease. Ann Neurol 2005; 57(3):310-326.
64. van der Knaap MS, Naidu S, Breiter SN et al. Alexander disease: diagnosis with MR imaging. AJNR Am J Neuroradiol 2001; 22(3):541-552.
65. Balbi P, Seri M, Ceccherini I et al. Adult-onset Alexander disease : report on a family. J Neurol 2008; 255(1):24-30.
66. Brenner M, Johnson AB, Boespflug-Tanguy O et al. Mutations in GFAP, encoding glial fibrillary acidic protein, are associated with Alexander disease. Nat Genet 2001; 27(1):117-120.
67. Spalke G, Mennel HD. Alexander's disease in an adult: clinicopathologic study of a case and review of the literature. Clin Neuropathol 1982; 1(3):106-112.
68. Quinlan RA, Brenner M, Goldman JE et al. GFAP and its role in Alexander disease. Exp Cell Res 2007; 313(10):2077-2087.
69. Tomokane N, Iwaki T, Tateishi J et al. Rosenthal fibers share epitopes with alpha B-crystallin, glial fibrillary acidic protein and ubiquitin, but not with vimentin. Immunoelectron microscopy with colloidal gold. Am J Pathol 1991; 138(4):875-885.
70. Dinopoulos A, Gorospe JR, Egelhoff JC et al. Discrepancy between neuroimaging findings and clinical phenotype in Alexander disease. AJNR Am J Neuroradiol 2006; 27(10):2088-2092.
71. Kaul R, Gao GP, Aloya M et al. Canavan disease: mutations among Jewish and nonJewish patients. Am J Hum Genet 1994; 55(1):34-41.
72. Adachi M, Schneck L, Cara J et al. Spongy degeneration of the central nervous system (van Bogaert and Bertrand type; Canavan's disease). A review. Hum Pathol 1973; 4(3):331-347.
73. Traeger EC, Rapin I. The clinical course of Canavan disease. Pediatr Neurol 1998; 18(3):207-212.
74. Kaul R, Gao GP, Balamurugan K et al. Cloning of the human aspartoacylase cDNA and a common missense mutation in Canavan disease. Nat Genet 1993; 5(2):118-123.
75. Kaul R, Gao GP, Balamurugan K et al. Canavan disease: molecular basis of aspartoacylase deficiency. J Inherit Metab Dis 1994; 17(3):295-297.
76. Matalon R, Kaul R, Casanova J et al. SSIEM Award. Aspartoacylase deficiency: the enzyme defect in Canavan disease. J Inherit Metab Dis 1989; 12 Suppl 2:329-331.

77. Matalon R, Michals-Matalon K. Biochemistry and molecular biology of Canavan disease. Neurochem Res 1999; 24(4):507-513.
78. Baslow MH. Brain N-acetylaspartate as a molecular water pump and its role in the etiology of Canavan disease: a mechanistic explanation. J Mol Neurosci 2003; 21(3):185-190.
79. Cakmakci H, Pekcevik Y, Yis U et al. Diagnostic value of proton MR spectroscopy and diffusion-weighted MR imaging in childhood inherited neurometabolic brain diseases and review of the literature. Eur J Radiol 2010; 74(3):e161-171.
80. Srikanth SG, Chandrashekar HS, Nagarajan K et al. Restricted diffusion in Canavan disease. Childs Nerv Syst 2007; 23(4):465-468.
81. Assadi M, Janson C, Wang DJ et al. Lithium citrate reduces excessive intra-cerebral N-acetyl aspartate in Canavan disease. Eur J Paediatr Neurol 2010; 14(4):354-359.
82. Janson CG, Assadi M, Francis J et al. Lithium citrate for Canavan disease. Pediatr Neurol 2005; 33(4):235-243.
83. Singhal BS, Gorospe JR, Naidu S. Megalencephalic Leukoencephalopathy With Subcortical Cysts. Journal of Child Neurology 2003; 18(9):646-652.
84. van der Knaap MS, Barth PG, Stroink H et al. Leukoencephalopathy with swelling and a discrepantly mild clinical course in eight children. Ann Neurol 1995; 37(3):324-334.
85. Leegwater PA, Yuan BQ, van der Steen J et al. Mutations of MLC1 (KIAA0027), encoding a putative membrane protein, cause megalencephalic leukoencephalopathy with subcortical cysts. Am J Hum Genet 2001; 68(4):831-838.
86. Topcu M, Saatci I, Topcuoglu MA et al. Megalencephaly and leukodystrophy with mild clinical course: a report on 12 new cases. Brain Dev 1998; 20(3):142-153.
87. van der Knaap MS, Lai V, Köhler W et al. Megalencephalic leukoencephalopathy with cysts without MLC1 defect: 2 phenotypes. Annals of Neurology 2010; 67(6):834-7.
88. Itoh N, Maeda M, Naito Y et al. An adult case of megalencephalic leukoencephalopathy with subcortical cysts with S93L mutation in MLC1 gene: a case report and diffusion MRI. Eur Neurol 2006; 56(4):243-245.
89. van der Voorn JP, Pouwels PJ, Hart AA et al. Childhood white matter disorders: quantitative MR imaging and spectroscopy. Radiology 2006; 241(2):510-517.
90. van der Knaap MS, Barth PG, Vrensen GF et al. Histopathology of an infantile-onset spongiform leukoencephalopathy with a discrepantly mild clinical course. Acta Neuropathol 1996; 92(2):206-212.
91. Schiffmann R, Boespflug-Tanguy O. An update on the leukodsytrophies. Curr Opin Neurol 2001; 14(6):789-794.
92. Fogli A, Boespflug-Tanguy O. The large spectrum of eIF2B-related diseases. Biochem Soc Trans 2006; 34(Pt 1):22-29.
93. van der Knaap MS, van Berkel CG, Herms J et al. eIF2B-related disorders: antenatal onset and involvement of multiple organs. Am J Hum Genet 2003; 73(5):1199-1207.
94. Fogli A, Wong K, Eymard-Pierre E et al. Cree leukoencephalopathy and CACH/VWM disease are allelic at the EIF2B5 locus. Ann Neurol 2002; 52(4):506-510.
95. Ohlenbusch A, Henneke M, Brockmann K et al. Identification of ten novel mutations in patients with eIF2B-related disorders. Hum Mutat 2005; 25(4):411.
96. van der Voorn JP, van Kollenburg B, Bertrand G et al. The unfolded protein response in vanishing white matter disease. J Neuropathol Exp Neurol 2005; 64(9):770-775.
97. van der Knaap MS, Pronk JC, Scheper GC. Vanishing white matter disease. Lancet Neurol 2006; 5(5):413-423.
98. Fogli A, Rodriguez D, Eymard-Pierre E et al. Ovarian failure related to eukaryotic initiation factor 2B mutations. Am J Hum Genet 2003; 72(6):1544-1550.
99. Fogli A, Schiffmann R, Bertini E et al. The effect of genotype on the natural history of eIF2B-related leukodystrophies. Neurology 2004; 62(9):1509-1517.
100. van der Knaap MS, Leegwater PA, van Berkel CG et al. Arg113His mutation in eIF2Bepsilon as cause of leukoencephalopathy in adults. Neurology 2004; 62(9):1598-1600.
101. Li W, Wang X, Van Der Knaap MS et al. Mutations linked to leukoencephalopathy with vanishing white matter impair the function of the eukaryotic initiation factor 2B complex in diverse ways. Mol Cell Biol 2004; 24(8):3295-3306.
102. Vanderver A, Hathout Y, Maletkovic J et al. Sensitivity and specificity of decreased CSF asialotransferrin for eIF2B-related disorder. Neurology 2008; 70(23):2226-2232.
103. Horzinski L, Huyghe A, Cardoso MC et al. Eukaryotic initiation factor 2B (eIF2B) GEF activity as a diagnostic tool for EIF2B-related disorders. PLoS One 2009; 4(12):e8318.
104. Labauge P, Gelot A, Fogli A et al. Autosomal dominant leukodystrophy and childhood ataxia with central nervous system hypomyelination syndrome. Ann Neurol 2006; 60(4):485; author reply 485-486.
105. van der Knaap MS, Scheper GC. Non-eIF2B-related cystic leukoencephalopathy of unknown origin. Ann Neurol 2006; 59(4):724.
106. Labauge P, Fogli A, Castelnovo G et al. Dominant form of vanishing white matter-like leukoencephalopathy. Ann Neurol 2005; 58(4):634-639.

CHAPTER 14

MACHADO-JOSEPH DISEASE AND OTHER RARE SPINOCEREBELLAR ATAXIAS

Antoni Matilla-Dueñas

Basic, Translational and Molecular Neurogenetics Research Unit, Department of Neurosciences, Health Sciences Research Institute Germans Trias i Pujol (IGTP), Universitat Autònoma de Barcelona, Barcelona, Spain
Email: amatilla@igtp.cat; amatilla@btnunit.org

Abstract: The spinocerebellar ataxias (SCAs) are a group of neurodegenerative diseases characterised by progressive lack of motor coordination leading to major disability. SCAs show high clinical, genetic, molecular and epidemiological variability. In the last one decade, the intensive scientific research devoted to the SCAs is resulting in clear advances and a better understanding on the genetic and nongenetic factors contributing to their pathogenesis which are facilitating the diagnosis, prognosis and development of new therapies. The scope of this chapter is to provide an updated information on Machado-Joseph disease (MJD), the most frequent SCA subtype worldwide and other rare spinocerebellar ataxias including dentatorubral-pallidoluysian atrophy (DRPLA), the X-linked fragile X tremor and ataxia syndrome (FXTAS) and the nonprogressive episodic forms of inherited ataxias (EAs). Furthermore, the different therapeutic strategies that are currently being investigated to treat the ataxia and non-ataxia symptoms in SCAs are also described.

INTRODUCTION

Ataxias are a heterogeneous group of diseases characterised by progressive lack of motor coordination due to degeneration of the cerebellum and its connections. They fall into three categories: (1) acquired ataxias with nongenetic causes (2) hereditary ataxias, comprising recessive, dominant and X-linked inherited ataxias, and (3) sporadic ataxias. They are rare disorders with a prevalence of 10-15 per 100,000 population and affect mostly young adults. Due to their progressive nature they can lead to major disability and premature death.

Neurodegenerative Diseases, edited by Shamim I. Ahmad.
©2012 Landes Bioscience and Springer Science+Business Media.

The spinocerebellar ataxias (SCAs) are highly heterogeneous neurodegenerative diseases[1,2] characterised by lack of coordination of the gait and are often associated with poor coordination of hands, speech and eye movements. The term "spinocerebellar ataxias" is commonly used for those inherited ataxias presenting an autosomal dominant inheritance. SCAs are usually slowly progressive and often associated with cerebellar and brain atrophy as seen from brain imaging studies. Up to date, there are more than 35 different SCA subtypes (see Chapter 27) where the age of onset and clinical symptoms overlap in most of them and thus it is often difficult to distinguish among them based only on clinical or neuroimaging assessment. SCAs can be diagnosed via identifying the relevant genetic deficit which enables molecular diagnosis of at risk, a/presymptomatic, prenatal or pre-implantation of the different SCA subtypes and facilitates genetic counselling. The prevalence of individual SCA subtypes varies from region to region, because of the founder effects.

The scope of this chapter is to provide updated information on SCA3, also known as Machado-Joseph (MJD) disease, the most frequent SCA subtype worldwide and other rare spinocerebellar ataxias including dentatorubral-pallidoluysian atrophy (DRPLA), the X-linked fragile X tremor/ataxia syndrome (FXTAS) and the nonprogressive episodic forms of inherited ataxias (EAs). Furthermore, the therapeutic strategies that are currently being investigated to treat the ataxia and non-ataxia symptoms in SCAs arc also dcscribcd.

SPINOCEREBELLAR ATAXIA TYPE 3 (SCA3) [MIM #109150]

Clinical Features

SCA3, also known as Machado-Joseph disease (MJD) (from here we call it SCA3/ MJD), is the most common dominantly inherited cerebellar ataxia worldwide and is characterised by cerebellar ataxia and pyramidal signs variably associated with a dystonic-rigid extrapyramidal syndrome or peripheral amyotrophy.[3-6] The age of onset of SCA3/MJD is variable, but most commonly in the second to fifth decade. In a large cohort of affected individuals from the Azores, the mean age of onset was 37 years. The variable range of symptoms at onset largely reflects differences in the length of a CAG repeat located within the *ATXN3* gene which is the molecular causative defect in SCA3/MJD. Presenting features include gait problems, speech difficulties, clumsiness and often, visual blurring and diplopia. Progressive ataxia, hyperreflexia, nystagmus and dysarthria may occur early in the disease. Upper motor neuron signs often become prominent early on. Ambulation becomes increasingly difficult, leading to the need for assistive devices (including wheelchair) ten to 15 years following onset. Saccadic eye movements become slow and ophthalmoparesis develops, resulting initially in up-gaze restriction. Disconjugate eye movements result in diplopia. Since squint angles commonly vary over the years, diplopia should not be treated by eye surgery. Most patients find substantial relief from prism glasses that compensate for the main angle of strabismus. At the same time, a number of other "brain stem" signs develop, including temporal and facial atrophy, characteristic action-induced perioral twitches, vestibular symptoms, tongue atrophy and fasciculations, dysphagia and poor ability to cough and clear secretions. Often, a staring appearance to the eyes is observed, but neither this nor the perioral fasciculations are specific for SCA3/MJD. Other findings may include the following: (i) vocal cord paralysis, described in three of 19 persons with SCA3/MJD;[7] (ii) vestibular dysfunction;[8] (iii) autonomic problems, including bladder and thermoregulation disturbances;[9] (iv) a disabling sleep disturbance, rapid eye movement behaviour disorder,[10,11] and restless legs syndrome.[12,13] Some SCA3/ MJD individuals have impaired executive and emotional functioning that is unrelated to ataxia severity. Evidence of peripheral polyneuropathy may appear with loss of distal sensation, ankle reflexes and sometimes other reflexes as well and with some degree of muscle wasting. Severe ataxia of limbs and gait (with either hyperreflexia or areflexia) associated with muscle wasting is observed. Sitting posture is compromised, with affected individuals assuming various tilted positions. Late in the disease course, individuals are wheelchair bound and have severe dysarthria, dysphagia, facial and temporal atrophy, poor cough, often dystonic posturing and ophthalmoparesis and occasionally blepharospasm. The disease progresses relentlessly; death from pulmonary complications and cachexia occurs from six to 29 years after onset.[14]

Occasionally, family members with mutations of the same allele may exhibit other clinical features such as a dystonic-rigid syndrome, a parkinsonian syndrome, or a combined syndrome of dystonia and peripheral neuropathy. Individuals with later adult onset often have a disorder that combines ataxia, generalized areflexia and muscle wasting. Based on this phenotypic variability, SCA3/MJD has been classified into several types.[15,16] In some individuals one type can evolve into another during the course of the disease: (i) Type I disease (13% of cases) is characterised by onset at a young age and prominent spasticity, rigidity and bradykinesia, often with little ataxia; (ii) Type II disease, the most common (57%), is characterised by ataxia and upper motor neuron signs. Spastic paraplegia can be part of the phenotype; (iii) Type III disease (30%) manifests at a later age with ataxia and peripheral polyneuropathy. A fourth disease type characterised by DOPA-responsive Parkinsonism and neuropathy has also been described. Other features are not restricted to a specific subtype including ophthalmoplegia (56% of patients), double vision (79%), faciolingual fasciculation (35%), dysphagia (75%), weight loss without loss of appetite (54%), incontinence (29%) and restless legs syndrome (45%). Cognitive disturbances in SCA3 are mild and rarely develop into relevant dementia.

Brain imaging studies reveal pontocerebellar atrophy. The most commonly observed abnormality is enlargement of the fourth ventricle, moderate shrinkage of cerebellar vermis and hemispheres as well as pontine atrophy. The degree of brain atrophy detectable by MRI varies greatly, consistent with the wide clinical variability observed. Abnormal linear high intensity of the globus pallidus interna on T2 and FLAIR images has also been observed.[17] Nerve conduction velocity studies often reveal evidence for involvement of the sensory nerves as well as the motor neurons.

Neuropathology has been extensively studied in SCA3/MJD.[18-22] Neuropathologic studies typically reveal that the cerebellum typically shows atrophy, but in some individuals Purkinje cells and inferior olivary neurons are relatively spared.[14] Neuronal loss is revealed in the pons, substantia nigra, thalamus, anterior horn cells and Clarke's column in the spinal cord, vestibular nucleus, many cranial motor nuclei and other precerebellar brain stem nuclei.[19,20,23]

Genetics of SCA3/MJD

SCA3/MJD is associated with CAG repeat expansions in the *ATXN3* gene. Normal alleles contain fewer than 44 CAG repeats. Overall, 93.5% of normal alleles have fewer than 31 CAG repeats. Mutable normal alleles have yet to be convincingly associated with a phenotype, but can manifest meiotic instability resulting in a pathologic expansion in a subsequent generation. Alleles with 45 to 51 CAG repeats have reduced or incomplete penetrance and individuals with a reduced penetrance allele may or may not manifest the disorder during their lifetime. Abnormal expanded alleles with full penetrance contain 52 to 86 CAG repeats and are associated with the SCA3/MJD phenotype. The CAG repeat does not completely explain the age of onset (correlation ranging from −0,67 to −0,87) suggesting that additional genetic and nongenetic factors account for this variability. Repeats of more than 73 CAG motifs are frequently associated with a pyramidal phenotype whereas patients with less than 73 repeats more likely develop neuropathy.

DENTATORUBRAL-PALLIDOLUYSIAN ATROPHY (DRPLA) [MIM #125370]

Clinical Features

Dentatorubral-pallidoluysian atrophy (DRPLA), also known as Naito-Oyanagi disease, consists of progressive ataxia, choreoathetosis and dementia or character changes in adults and ataxia, myoclonus, epilepsy and progressive intellectual deterioration in children.[24,25] The age of onset is from one to 62 years with a mean age of onset of 30 years. The clinical presentation varies depending on the age of onset. The cardinal features in adults are ataxia, choreoathetosis and dementia. Cardinal features in children are progressive intellectual deterioration, behavioural changes, myoclonus and epilepsy. Atrophic changes in the cerebellum and brain stem, in particular the pontine tegmentum, are the typical MRI findings of DRPLA. Quantitative analyses revealed that both the age at MRI and the size of the expanded CAG repeat correlate with the atrophic changes. Diffuse high-intensity areas deep in the white matter are often observed on T2-weighted MRI in individuals with adult-onset DRPLA of long duration.[26] The major neuropathologic changes detected are relatively simple and consist of combined degeneration of the dentatorubral and pallidoluysian systems of the central nervous system.

Genetics of DRPLA

The diagnosis of DRPLA rests on positive family history, characteristic clinical findings and the detection of an expansion of a CAG trinucleotide/polyglutamine tract in the *ATN1* (*DRPLA*) gene. Normal alleles range from 6 to 35 CAG repeats. Mutable normal alleles contain 20-35 CAG repeats and are found in Caucasian populations. They are not associated with symptoms, but are unstable and can expand on transmission resulting in occurrence of symptoms in the next generation, albeit this is a very rare event. The CAG repeat length in individuals with DRPLA ranges from 48 to 93.

FRAGILE X TREMOR AND ATAXIA SYNDROME (FXTAS) (MIM #300623)

Clinical Features

FXTAS is characterised by late-onset progressive cerebellar ataxia and intention tremor.[27,28] Other neurologic findings include short-term memory loss, executive function deficits, cognitive decline, progressive dysarthria, dementia, parkinsonism, peripheral neuropathy, lower-limb proximal muscle weakness, spastic paraparesis and autonomic dysfunction. A definite diagnosis of FXTAS requires the presence of a premutation in the *FMR1* gene and white matter lesions on MRI in the middle cerebellar peduncles and/ or brain stem, the major neuroradiologic sign, with either intention tremor or gait ataxia which are the two major clinical signs. Other minor clinical criteria include parkinsonism, moderate to severe working memory deficits, or executive cognitive function deficits. Neuroradiologic signs including decreased cerebellar volume, increased ventricular volume and increased white matter hyper-density, all correlating with the premutation CGG repeat length.[29] Another study correlated the CGG repeat length with the peripheral nerve conduction velocities in motor and sensory nerves.[30] The penetrance varies and is age related, ranging from 5-10% in female carriers to 17-75% in male carriers of 50 years and over.[31] Tremor usually precedes ataxia onset. Life expectancy after onset of symptoms ranges from five to 25 years.

Genetics of FXTAS

The diagnosis relies on the detection of the CGG expansion within the *FMR1* gene ranging from 59 to approximately 200 repeats. Alleles of this size are not associated with mental retardation, but do convey increased risk for FXTAS. Women with alleles within this range are at risk of having offspring affected with fragile X syndrome. Because of the high prevalence of the *FMR1* premutation among individuals presenting late-onset ataxia (up to 3%) and the overlap with the clinical symptoms, FXTAS should be included onto any spinocerebellar ataxia genetic screening protocol. Early diagnosis of FXTAS patients benefits them and their relatives who may be thus advised for FRAXA.

EPISODIC ATAXIA TYPE 1 (EA1) (MIM #160120)

Clinical Features

The physiopathology and molecular genetics of known episodic ataxia syndromes have been reviewed elsewhere.[32,33] Episodic ataxia Type 1 (EA1) is a potassium channelopathy characterised by constant myokymia and dramatic episodes of spastic contractions of the skeletal muscles of the head, arms and legs with loss of both motor coordination and balance. During attacks some individuals may experience vertigo, blurred vision, diplopia, nausea, headache, diaphoresis, clumsiness, stiffening of the body, dysarthric speech and difficulty in breathing. Onset is in childhood or early adolescence. Other findings include delayed motor development, cognitive disability, choreoathetosis and carpal spasm.

Genetics of EA1

Diagnosis is based on clinical findings and molecular genetic testing of *KCNA1*, the only gene known to be associated with EA1. All affected individuals described so far are heterozygous for *KCNA1* mutations at amino acid residues highly conserved among the voltage-dependent K⁺ channel superfamily. The mutation detection frequency using sequence analysis is approximately 90%. A novel missense mutation (F414C) has been identified in an Italian EA1 family.[34] Mutations in the *KCNA1* gene have also been identified in families with myokymia without ataxia episodes.[35]

EPISODIC ATAXIA TYPE 2 (EA2) [MIM #108500]

Clinical Features

Episodic ataxia Type 2 (EA2), the most common form of episodic ataxia, is characterised by paroxysmal attacks of ataxia, vertigo and nausea typically lasting minutes to days in duration.[36,37] Stress, exertion, caffeine and alcohol may trigger attacks that can be variably associated with dysarthria, diplopia, tinnitus, dystonia, hemiplegia and headache. In fact, EA2 is allelic with two other conditions: familial hemiplegic migraine Type 1 (FHM1) characterised by complicated migraine with hemiplegia, interictal nystagmus and progressive ataxia and spinocerebellar ataxia Type 6 (SCA6) characterised by slowly progressive ataxia of late onset, some with episodic features.[38,39] Approximately 50% of individuals with EA2 have migraine headaches. Onset is typically in childhood or early adolescence. MRI demonstrates atrophy of the cerebellar vermis. The diagnosis of EA2 is most commonly made on clinical grounds.

Genetics of EA2

EA2 is inherited as an autosomal dominant manner and more than 30 different mutations have been identified causing the disease in the *CACNA1A* gene, encoding for a P/Q type voltage-gated Ca²⁺ channel alpha subunit, abundantly expressed in the cerebellum and the neuromuscular junction.[40] The majority are nonsense mutations resulting in a truncated protein product. However, a number of nontruncating mutations, such as intronic, causing exon skipping and abnormal splicing and exonic deletions have also been reported.[41] Estimated penetrance is 80-90%.

EPISODIC ATAXIA TYPE 3 (EA3) [MIM #606554]

Clinical Features

Episodic ataxia Type 3 (EA3) was described in a large Canadian Mennonite family with episodic vertigo, tinnitus and ataxia without baseline deficits.[42] The disease manifests in early adulthood. In some individuals, slowly progressive cerebellar ataxia occurs.

Genetics of EA3

A candidate region on chromosome 1q42 has been identified.[43]

EPISODIC ATAXIA TYPE 4 (EA4) [MIM #606552]

Clinical Features

Episodic ataxia Type 4 (EA4), also known as periodic vestibulocerebellar ataxia (PATX), was described in two kindreds from North Carolina, USA, with late onset episodic vertigo and ataxia as well as interictal nystagmus not responsive to acetazolamide.[44,45] The disorder is characterised by defective smooth pursuit, gaze-evoked nystagmus, ataxia and vertigo. The age of onset ranged from the third to the sixth decade.

Genetics of EA4

Linkage analysis has ruled out the EA1 and EA2 loci, but no chromosomal locus has not yet been identified.[46]

EPISODIC ATAXIA TYPE 5 (EA5) [MIM #601949]

Clinical Features

The phenotype of Episodic ataxia Type 5 (EA5) is characterised by recurrent episodes of vertigo and ataxia that can last for several hours. Interictal examination shows spontaneous down-beat and gaze-evoked nystagmus and mild dysarthria and truncal ataxia. Acetazolamide can prevent the attacks.

Genetics of EA5

EA5 results from a mutation in the *CACNB4* gene, which encodes an auxiliary beta-4 isoform of the regulatory beta subunit of voltage-activated Ca^{2+} channels. A c.311G>T (p.Cys104Phe) mutation has been described in a French-Canadian family.[47] EA5 is allelic with juvenile myoclonic epilepsy (JME) and the semiology of seizures in EA5 is similar to JME.

EPISODIC ATAXIA TYPE 6 (EA6) [MIM #612656]

Clinical Features

Episodic Ataxia Type 6 (EA6) was initially observed in a child with episodic ataxia, attacks of hemiplegia and migraine in the setting of fever and epilepsy.[48] The disease is characterised by attacks of ataxia precipitated by fever, subclinical seizures, slurred speech followed by headache and bouts of arm jerking with concomitant confusion and alternating hemiplegia.[49] MRI showed cerebellar atrophy and neurologic examination showed mild interictal truncal ataxia.

Genetics of EA6

EA6 results from mutations in the *SLC1A3* gene encoding the excitatory amino acid transporter 1 (EAAT1).[48,49] In cells expressing mutated proteins, glutamate uptake is reduced, suggesting that glutamate transporter dysfunction underlies the disease. The penetrance is incomplete.

EPISODIC ATAXIA TYPE 7 (EA7) [MIM# 611907]

Clinical Features

Episodic ataxia Type 7 (EA7) has been described in a four-generation family whose affected individuals showed episodic ataxia before age 20 years.[50] The disease is characterised by attacks associated with weakness, vertigo and dysarthria lasting hours to days. Attacks may also be brought about by exercise and/or excitement.

Genetics of EA7

A candidate region on chromosome 19q13, termed the EA7 locus, has been identified.[50]

AUTOSOMAL DOMINANT SPASTIC ATAXIA (ADSA) [MIM #108600]

Clinical Features

Affected individuals with autosomal dominant spastic ataxia (ADSA) initially show progressive leg spasticity of variable degree followed by ataxia in the form of involuntary head jerk, dysarthria, dysphagia and ocular movement abnormalities consisting of slow saccades, impaired vertical gaze and in some cases lid retraction.[51] The severity of the phenotype varies greatly and the age at onset appears from early childhood to early twenties, although most presented with onset of symptoms at age 10 to 20 years. Neuropathologic findings include degeneration of the corticospinal tracts and posterior columns. The life span and cognition of patients are not affected.

Genetics of ADSA

Linkage studies identified a locus on 12p13, termed SAX1.[51]

OTHER AUTOSOMAL DOMINANT SPINOCEREBELLAR ATAXIAS

Cerebellar ataxia with deafness, narcolepsy and optic atrophy was described in a Swedish family.[52] CT and MRI studies revealed supratentorial atrophy, more pronounced than infratentorial atrophy, pronounced dilatation of the third ventricle, low T2 signal intensity in the basal ganglia, loss of cerebral cortex-white matter differentiation and periventricular high-signal rims.[53] The gene has been linked to 6p21-p23.[52]

In four individuals in one family presenting ataxia, cerebellar atrophy, mental retardation and possible attention deficit/hyperactivity disorder (ADHD) described; associated the disease with a heterozygous 2-bp deletion mutation in exon 4 in *SCAN8A*, a gene encoding a sodium channel on 12q13 (MIM #600702).[54]

Genis et al[55] described a Spanish family with individuals presenting a late-onset cerebellar ataxia with thermoanalgesia and deep sensory loss. Unlike in SCA4, reflexes were preserved. MRI revealed cerebellar, medullar and spinal cord atrophy. Neurophysiological studies showed absence or marked reduction of the sensory nerve action potentials and somatosensory evoked potentials in lower and upper limbs but preservation of the soleus H reflex. The neuropathological study revealed severe loss of Purkinje cells and dentate neurons, extensive cell loss in the inferior olive and lower cranial nerve nuclei and demyelination of the posterior columns and spinocerebellar tracts. The genetic studies ruled out linkage of the disease in this SCA subtype to the SCA4 locus on chromosome 16 and the remaining previously identified SCA loci, therefore evidencing a new genetic identity for this ataxia subtype associated with thermoanalgesia as well as deep sensory loss with retained reflexes.

THERAPEUTIC STRATEGIES IN THE SPINOCEREBELLAR ATAXIAS

There are currently no known effective pharmacologic treatments to reverse or even substantially reduce motor disability caused by cerebellar degeneration in most of the SCAs or related cerebellar disorders, although some benefits on ataxic and non-ataxic symptoms have been reported in a few therapeutic clinical trials.[1,56-58] Some benefits regarding ataxic symptoms have been reported with acetazolamide and gabapentin in SCA6,[59] 5-hydroxytryptophan, clonazepam, buspirone or tansodpirone, sulfamethoxazole/trimethoprim or lamotrigine in SCA3, NMDA modulators or antagonists and deep brain stimulation in SCA2 with tremor. Amantadine, dopaminergic and anticholinergics drugs have been used to alleviate tremor, bradykinesia, or dystonia in SCA2 and SCA3.[60-62] Varenicline is currently being tested in SCA3 and FXTAS. Restless legs and periodic leg movements in sleep usually respond to dopaminergic treatment or tilidine.[12] Spasticity in SCAs are effectively treated with baclofen, tizanadine, or mimentine when combined with dopaminergic treatment. In selected cases where other treatments have failed, botulinum toxin has been successfully used to treat dystonia and spasticity in SCA3, although caution and small dosage is recommended since unusually severe and long lasting muscular atrophy occurs in some SCA3 patients with this treatment due to subclinical involvement of motor neurons in the anterior horn in the degenerative process. Intention tremor has been ameliorated with benzodiazepines, β-blockers, or chronic thalamic stimulation. Muscle cramps, which are often present at the onset of the condition in SCAs 2, 3, 7 and DRPLA, are alleviated with magnesium, chinine, or mexiletine.[63]

In spite of the lack of effectiveness in the treatment of ataxia symptoms in most SCAs, treatment in some spinocerebellar ataxias has proven successful. Coenzyme Q10 administration was shown to be effective in treating ataxia symptoms in patients with CoQ10 deficiency.[64] Furthermore, most autoimmune cerebellar ataxias, such as anti-glutamic acid decarboxylase (GAD)-antibody-positive cerebellar ataxia and gluten ataxia, have proven to be treatable.[65,66] In the remaining ataxias, physiotherapy is currently being used as an alternative effective treatment. Ataxia improves with daily autonomous training of gait and stance in combination with physiotherapy. Other neurological symptoms such as dysarthria and dysphagia warrant logopedic treatment to maintain the ability to communicate and to prevent pneumonia from aspiration. Valproate and piracetam have been used to treat myoclonus and/or dementia/cognitive decline.

A clinical trial with the aim of assessing the safety, tolerability and the effects of lithium in SCA1 has recently been completed and patients are being recruited to assess lithium carbonate therapy in SCA2 and SCA3. Albeit there are clinical benefits of lithium treatment; common side effects include muscle tremors, twitching, ataxia and hypothyroidism. Long term use of lithium has been linked to hyperparathyroidism http:// en.wikipedia.org/wiki/Lithium—cite_note-49, hypercalcaemia (bone loss), hypertension, kidney damage, nephrogenic diabetes insipidus (polyuria and polydipsia), seizures and weight gain. Although lithium or a bioactive analogue may be a promising drug that can potentially benefit ataxia patients, clinical and biological responses to a range of doses throughout an extended time period need to be carefully evaluated and monitored in any forthcoming clinical trial. Other ongoing clinical trials in SCAs include memantine in fragile X tremor and ataxia syndrome (FXTAS).

A few innovative approaches at the preclinical and in some cases at the clinical level include the use of RNA interference (RNAi) aiming to inhibit the expression of mutated polyglutamine-proteins in those SCAs, caused by expanded polyglutamine mutations. Prevention of protein misfolding and aggregation by over-expression of chaperones by pharmacological treatments and the regulation of gene expression by application of histone deacetylase inhibitors are giving promising results in preclinical trials and are currently being tested in some ataxia patients. In SCA1, intracerebellar injection of vectors, expressing short hairpin RNAs, profoundly improves motor coordination, restores cerebellar morphology and prevent the characteristic ataxin-1 inclusions in Purkinje cells in transgenic mice.[67] While these results show that RNAi therapy improves cellular and behavioural characteristics in preclinical trials, its application in patients to protect or even reverse disease phenotypes shall be delayed until proper toxicity tests are assessed.

Pointing to a different target, molecular chaperones provide a first line of defence against misfolded, aggregation-prone proteins. Many studies have analysed the effects of chaperone over-expression on inclusion body formation and toxicity of pathogenic polyQ fragments in cell culture and it is clear that over-expression of molecular chaperones might prove beneficial for the treatment of neurodegenerative diseases.[68] They prevent inappropriate interactions within and between nonnative polypeptides, enhance the efficiency of de novo protein folding and promote the refolding of proteins that have become misfolded as a result of the mutations and cellular stress.[69] Chemical and molecular chaperones might also prevent toxicity by blocking inappropriate protein interactions, by facilitating disease protein degradation or sequestration, or by blocking downstream signalling events leading to neuronal dysfunction and apoptosis. Congo Red, thioflavine S, chrysamine G and Direct Fast have proven effective in suppressing aggregation in vitro and in vivo,[70,71] albeit their specific efficacy in vivo is limited by their variable abilities to cross the blood-brain barrier and proper pharmacologic analogues may need to be developed for further clinical considerations.

Several low molecular mass chemical chaperones, such as the organic solvent dimethylsulfoxide (DMSO) and the cellular osmolytes glycerol, trimethylamine n-oxide and trehalose, appear effective in preventing cell death triggered by mutant ataxin 3 by increasing its stability in their native conformation.[72] Trehalose was identified in an in vitro screen for inhibitors of polyQ aggregation and its administration reduces brain atrophy, improves motor dysfunction and extends the lifespan of mice, mimicking the polyglutamine disorder Huntington's disease.[73] In vitro experiments suggest that the beneficial effects of trehalose result from its ability to bind and stabilize polyglutamine-containing proteins. More recently, a new generation of small chemical compounds, that directly target polyQ aggregation without significant cytotoxicity, have been identified in high-throughput screens using cell-free assays or by targeting cellular pathways.[74,75] These compounds inhibit polyQ aggregation in cultured cells and intact neurons and can rescue polyQ-mediated neurodegeneration in vivo.

By a different mechanism, a small molecule that acts as a co-inducer of the heat shock response by prolonging the activity of heat-shock transcription factor HSF1, arimoclomol, significantly improves behavioural phenotypes, prevents neuronal loss, extends survival rates and delays disease progression in a mouse model of neurodegeneration.[76] Similarly, activation of heat-shock responses with geldanamycin inhibits aggregation and prevents cell death.[77] This suggests that pharmacological activation of the heat shock response may be an effective therapeutic approach to treat neurodegenerative diseases. However, excessive up-regulation of chaperones might lead to undesirable side effects, such as alterations in cell cycle regulation and cancer.[78] Therefore, a delicate balance of chaperones will likely be required for a beneficial neuroprotective effect. For instance, chemical or molecular chaperones, used in combination with a pharmacological agent that up-regulates the synthesis of molecular chaperones, might be a valid therapeutic approach for treating spinocerebellar ataxias caused by polyglutamine expansions. Aggregate formation has also been successfully targeted with inhibitors of transglutaminase, such as cystamine, which reduces apoptotic cell death and alleviates disease symptoms by the expanded polyglutamine.[79,80]

Compounds targeting mitochondrial function such as coenzyme Q10,[81] creatine[82] and taurousodeoxycholic acid (TUDCA);[83] or autophagy, such as the mTor inhibitor, rapamycin and various analogous,[84] have proven effective at reducing cellular toxicity in animal models and are currently being tested in clinical trials in a few ataxia subtypes.[85] Caspase activation, which usually precedes neuronal cell death, have been targeted by inhibiting their expression, recruitment and consequent activation onto "apoptosome-like structures" or by enzymatic inhibitors all of which include minocycline, cystamine, CrmA, FADD DN and zVAD-fmk, respectively.[86] In general, the inhibitors of the different caspases have been shown to decrease microglia activation, prevent disease progression, delay onset of symptoms, enhance inclusion clearance and extend survival rates in several mouse and cell models of neurodegeneration.[87-89]

Other agents which promote the clearance of mutant proteins in the CNS or which are Ca^{2+} signalling blockers and stabilizers, such as specific inhibitors of the NR2B-subunit of N-methyl-D-aspartate glutamate receptors, blockers/antagonists of metabotropic glutamate receptor mGluR5 and inositol 1,4,5-trisphosphate receptor InsP3R1 such as remacemide; intracellular Ca^{2+} stabilizers such as dantrolene; dopamine stabilizers such as mermaid-ACR-16; dopamine depleters and agents inducing anti-excitotoxic effects such as riluzole; or agents which alleviate cognitive components such as horizon-dimebon; they all appear to be at least partially beneficial for the treatment of some neurological symptoms in spinocerebellar ataxias.[90-92] A recent clinical trial with riluzole showed a reduction of the ICARS score in patients with a wide range of cerebellar disorders.[93] Neuroprotective drugs like olesoxime have proven to increase microtubule dynamics, re-establish neuritic outgrowth, improve myelination and prevent apoptotic factor release and oxidative stress.[94] Inhibition of potassium channels with 3,4-diaminopyridine has proven efficient in normalising motor behaviours in young SCA1 mice and in restoring normal purkinje cell volume and dendrite spine density and the molecular layer thickness in older SCA1 mice. Aminopyridines, such as fampridine and diaminopyridine, increase PC excitability and are also efficient for treating down-beat nystagmus.[95-97]

The role that some ataxin proteins play in transcription and, more importantly, the effects mediated by some of their cotranscriptional regulators in the suppression of cytotoxicity are being used as targets to modulate the pathological effects of mutant ataxins, opening the path for new therapeutic strategies for treating some of the SCAs. Recent progress in histone deacetylase (HDAC) research has made possible the development of inhibitors of specific HDAC family proteins and these compounds could prove effective candidates for the treatment of spinocerebellar ataxias.[98,99] Neuroprotective and neurorestoration strategies, addressing specific bioenergetic defects, might hold particular promise in the treatment of spinocerebellar conditions. Drugs, such as rasagiline, have been shown efficient in protecting neuronal cells against apoptosis through induction of the pro-survival Bcl-2 protein and neurotrophic factors.[100] Recent alterations of the insulin growth factor (IGF-1) pathway have been reported to be implicated in both SCA1 and SCA7,[101] suggesting that in vivo neuroprotection exerted by IGF-1 through the PP2-regulated PI3K/Akt signalling pathway, could potentially be used to halt cerebellar neurodegeneration.[102,103]

Gene therapy and stem cell and grafting approaches are being considered for treating spinocerebellar neurodegenerations.[104] Delivery of proteins or compounds by viral vectors represents one such gene therapeutic approach. Neural cell replacement therapies are based on the idea that neurological function lost during neurodegeneration could be improved by introducing new cells that can form appropriate connections and replace the function of lost neurons. This strategy, although potentially effective, is still in early experimental stages. Since neurogenesis does occur in the adult nervous system, another approach is based on the stimulation of endogenous stem cells in the brain or spinal cord to generate new neurons. Studies to understand the molecular determinants and cues to stimulate endogenous stem cells are underway.[105] Although promising, we are only starting to learn the potential and challenges of these emerging therapies, especially their efficacy in treating human neurodegeneration.

Treatments for Episodic Ataxias

Several different drugs are reported to improve symptoms in EA1 and EA2, but so far there have been no controlled studies documenting or comparing efficacy of these different drugs. Carbamazepine, valproic acid and acetazolamide have been effective for EA1[35,106] and acetazolamide (ACTZ),[107] flunarizine,[108] 4-aminopyridine[95,109] and chlorzoxazone (CHZ)[110] have been effective in EA2 cases. The response to acetazolamide is often dramatic in EA2.[36,107] Acetazolamide, a carbonic-anhydrase (CA) inhibitor, may reduce the frequency and severity of the attacks in some but not all affected individuals with episodic ataxias. ACTZ should not be prescribed to individuals with liver, renal, or adrenal insufficiency. Chronic treatment with ACTZ may result in side effects including paresthesias, rash and formation of renal calculi.

Antiepileptic drugs (AEDs) such as carbamazepine may significantly reduce the frequency of the attacks in responsive individuals; however, the response is heterogeneous as some individuals are particularly resistant to drugs.[35] Antiepileptic treatment with diphenylhydantoin results in reasonable control of seizures in some individuals. In particular, phenytoin treatment may improve muscle stiffness and motor performance.[111] Nevertheless, phenytoin should be used with caution in young individuals, as it may cause permanent cerebellar dysfunction and atrophy.[112] Anticonvulsant drugs such as sulthiame may reduce the attack rates. During this treatment, abortive attacks were still noticed lasting a few seconds and troublesome side effects were paresthesias and intermittent carpal spasm.

The potassium channel blocker 4-aminopyridine was found to be effective in stopping EA2 attacks in patients.[96,109] Furthermore, 3,4-diaminopyridine was demonstrated in a placebo-controlled study to improve down-beat nystagmus, which is often observed in patients with EA2.[113]

CONCLUSION

The spinocerebellar ataxias are devastating neurological diseases for which, currently, there are no effective and selective pharmacological treatments available that reverse or even substantially reduce motor disability caused by the cerebellar neurodegeneration. Thus, physical therapy is currently the sole form of intervention that can improve walking ataxia in affected individuals. Indeed, a recent study where patients with variable forms of cerebellar degenerative disease were subjected to "intensive coordinative training" showed improvement in the ataxia and balance clinical scales, indicating that rehabilitation may be of real benefit to ataxic individuals.[114] Similarly, in a specific rehabilitation program including foot sensory stimulation and balance and gait training, 24 ataxic patients, with clinically defined sensory ataxia, improved their balance with better results in dynamic conditions.[115] These studies are of particular interest because they showed how individuals with cerebellar damage can learn to improve their movements, recover the control of their balance and proprioceptive contributions enabling them to achieve personally meaningful goals in everyday life after proper training. Until effective and selective pharmacological treatment is available for ataxia patients, which should be forthcoming in the near future, physical and sensory rehabilitation are meanwhile revealing effective approaches for improving the patient's quality of life. Taken together, all the data up-to date highlights that treatment for ataxia patients is no longer an utopia, but it is possible in a foreseeable future.

ACKNOWLEDGEMENTS

Dr. Ivelisse Sanchez's helpful comments and suggestions are kindly acknowledged. Dr. Antoni Matilla's scientific research is funded by the Spanish Ministry of Science and Innovation (BFU2008-00527/BMC), the Carlos III Health Institute (CP08/00027), the LatinAmerican Science and Technology Development Programme (CYTED) (210RT0390), the European Commission (EUROSCA project, LHSM-CT-2004-503304), and the Fundació La Marató de TV3 (Televisió de Catalunya). Antoni Matilla is indebted to the Spanish Ataxia Association (FEDAES) and the ataxia patients for their continuous support and motivation. Antoni Matilla is a Miguel Servet Investigator in Neurosciences of the Spanish National Health System.

REFERENCES

1. Matilla-Dueñas A, Goold R, Giunti P. Molecular pathogenesis of spinocerebellar ataxias. Brain 2006; 129:1357-1370.
2. Matilla-Dueñas A, Sanchez I, Corral-Juan M et al. Cellular and Molecular Pathways Triggering Neurodegeneration in the Spinocerebellar Ataxias. Cerebellum 2010; 9(2):148-166.
3. Coutinho P, Sequeiros J. Clinical, genetic and pathological aspects of Machado-Joseph disease. J Genet Hum 1981; 29(3):203-209.
4. Matilla T, McCall A, Subramony SH et al. Molecular and clinical correlations in spinocerebellar ataxia type 3 and Machado-Joseph disease. Ann Neurol 1995; 38(1):68-72.
5. Paulson HL. Dominantly inherited ataxias: lessons learned from Machado-Joseph disease/spinocerebellar ataxia type 3. Semin Neurol 2007; 27(2):133-142.
6. Riess O, Rub U, Pastore A et al. SCA3: neurological features, pathogenesis and animal models. Cerebellum 2008; 7(2):125-137.
7. Isozaki E, Naito R, Kanda T et al. Different mechanism of vocal cord paralysis between spinocerebellar ataxia (SCA 1 and SCA 3) and multiple system atrophy. J Neurol Sci 2002; 197(1-2):37-43.
8. Yoshizawa T, Nakamagoe K, Ueno T et al. Early vestibular dysfunction in Machado-Joseph disease detected by caloric test. J Neurol Sci 2004; 221(1-2):109-111.
9. Yeh TH, Lu CS, Chou YH et al. Autonomic dysfunction in Machado-Joseph disease. Arch Neurol 2005; 62(4):630-636.
10. Friedman JH. Presumed rapid eye movement behavior disorder in Machado-Joseph disease (spinocerebellar ataxia type 3). Mov Disord 2002; 17(6):1350-1353.
11. Friedman JH, Fernandez HH, Sudarsky LR. REM behavior disorder and excessive daytime somnolence in Machado-Joseph disease (SCA-3). Mov Disord 2003; 18(12):1520-1522.
12. Schols L, Haan J, Riess O et al. Sleep disturbance in spinocerebellar ataxias: is the SCA3 mutation a cause of restless legs syndrome? Neurology 1998; 51(6):1603-1607.
13. van Alfen N, Sinke RJ, Zwarts MJ et al. Intermediate CAG repeat lengths (53,54) for MJD/SCA3 are associated with an abnormal phenotype. Ann Neurol 2001; 49(6):805-807.
14. Sequeiros J, Coutinho P. Epidemiology and clinical aspects of Machado-Joseph disease. Adv Neurol 1993; 61:139-153.
15. Rosenberg RN. Machado-Joseph disease: an autosomal dominant motor system degeneration. Mov Disord 1992; 7(3):193-203.
16. Coutinho P, Andrade C. Autosomal dominant system degeneration in Portuguese families of the Azores Islands: a new genetic disorder involving cerebellar, pyramidal, extrapyramidal and spinal cord motor functions. Neurology 1978; 28:703-709.
17. Yamada S, Nishimiya J, Nakajima T et al. Linear high intensity area along the medial margin of the internal segment of the globus pallidus in Machado-Joseph disease patients. J Neurol Neurosurg Psychiatry 2005; 76(4):573-575.
18. Rub U, De Vos RA, Schultz C et al. Spinocerebellar ataxia type 3 (Machado-Joseph disease): severe destruction of the lateral reticular nucleus. Brain 2002; 125:2115-2124.
19. Rub U, Del Turco D, Del Tredici K et al. Thalamic involvement in a spinocerebellar ataxia type 2 (SCA2) and a spinocerebellar ataxia type 3 (SCA3) patient and its clinical relevance. Brain 2003; 126:2257-2272.
20. Rub U, Gierga K, Brunt ER et al. Spinocerebellar ataxias types 2 and 3: degeneration of the precerebellar nuclei isolates the three phylogenetically defined regions of the cerebellum. J Neural Transm 2005; 112(11):1523-1545.
21. Rub U, de Vos RA, Brunt ER et al. Spinocerebellar ataxia type 3 (SCA3): thalamic neurodegeneration occurs independently from thalamic ataxin-3 immunopositive neuronal intranuclear inclusions. Brain Pathol 2006; 16(3):218-227.
22. Rub U, Seidel K, Ozerden I et al. Consistent affection of the central somatosensory system in spinocerebellar ataxia type 2 and type 3 and its significance for clinical symptoms and rehabilitative therapy. Brain Res Rev 2007; 53(2):235-249.
23. Rub U, Brunt ER, Gierga K et al. The nucleus raphe interpositus in spinocerebellar ataxia type 3 (Machado-Joseph disease). J Chem Neuroanat 2003; 25(2):115-127.
24. Tsuji S. Dentatorubral-pallidoluysian atrophy: clinical aspects and molecular genetics. Adv Neurol 2002; 89:231-239.
25. Tsuji S. Dentatorubral-pallidoluysian atrophy In: Pagon R, Bird T, Dolan C et al, eds. GeneReviews [Internet]. Seattle (WA): University of Washington, Seattle, 2010.
26. Koide R, Onodera O, Ikeuchi T et al. Atrophy of the cerebellum and brainstem in dentatorubral pallidoluysian atrophy. Influence of CAG repeat size on MRI findings. Neurology 1997; 49(6):1605-1612.
27. Jacquemont S, Hagerman RJ, Leehey M et al. Fragile X premutation tremor/ataxia syndrome: molecular, clinical and neuroimaging correlates. Am J Hum Genet 2003; 72(4):869-878.

28. Jacquemont S, Hagerman RJ, Hagerman PJ et al. Fragile-X syndrome and fragile X-associated tremor/ataxia syndrome: two faces of FMR1. Lancet Neurol 2007; 6(1):45-55.
29. Cohen S, Masyn K, Adams J et al. Molecular and imaging correlates of the fragile X-associated tremor/ataxia syndrome. Neurology 2006; 67(8):1426-1431.
30. Soontarapornchai K, Maselli R, Fenton-Farrell G et al. Abnormal nerve conduction features in fragile X premutation carriers. Arch Neurol 2008; 65(4):495-498.
31. Leehey MA, Berry-Kravis E, Min SJ et al. Progression of tremor and ataxia in male carriers of the FMR1 premutation. Mov Disord 2007; 22(2):203-206.
32. Jen JC, Graves TD, Hess EJ et al. Primary episodic ataxias: diagnosis, pathogenesis and treatment. Brain 2007; 130:2484-2493.
33. Jen JC. Hereditary episodic ataxias. Ann N Y Acad Sci 2008; 1142:250-253.
34. Imbrici P, Gualandi F, D'Adamo MC et al. A novel KCNA1 mutation identified in an Italian family affected by episodic ataxia type 1. Neuroscience 2008; 157(3):577-587.
35. Eunson LH, Rea R, Zuberi SM et al. Clinical, genetic and expression studies of mutations in the potassium channel gene KCNA1 reveal new phenotypic variability. Ann Neurol 2000; 48(4):647-656.
36. Jen J, Kim GW, Baloh RW. Clinical spectrum of episodic ataxia type 2. Neurology 2004; 62(1):17-22.
37. Bertholon P, Chabrier S, Riant F et al. Episodic ataxia type 2: unusual aspects in clinical and genetic presentation. Special emphasis in childhood. J Neurol Neurosurg Psychiatry 2009; 80(11):1289-1292.
38. Jodice C, Mantuano E, Veneziano L et al. Episodic ataxia type 2 (EA2) and spinocerebellar ataxia type 6 (SCA6) due to CAG repeat expansion in the CACNA1A gene on chromosome 19p. Hum Mol Genet 1997; 6(11):1973-1978.
39. Jen JC, Yue Q, Karrim J et al. Spinocerebellar ataxia type 6 with positional vertigo and acetazolamide responsive episodic ataxia. J Neurol Neurosurg Psychiatry 1998; 65(4):565-568.
40. Ophoff RA, Terwindt GM, Vergouwe MN et al. Familial hemiplegic migraine and episodic ataxia type-2 are caused by mutations in the Ca^{2+} channel gene CACNL1A4. Cell 1996; 87(3):543-552.
41. Riant F, Lescoat C, Vahedi K et al. Identification of CACNA1A large deletions in four patients with episodic ataxia. Neurogenetics 2010; 11(1):101-106.
42. Steckley JL, Ebers GC, Cader MZ et al. An autosomal dominant disorder with episodic ataxia, vertigo and tinnitus. Neurology 2001; 57(8):1499-1502.
43. Cader MZ, Steckley JL, Dyment DA et al. A genome-wide screen and linkage mapping for a large pedigree with episodic ataxia. Neurology 2005; 65(1):156-158.
44. Farmer TW, Mustian VM. Vestibulocerebellar ataxia. A newly defined hereditary syndrome with periodic manifestations. Arch Neurol 1963; 8:471-480.
45. Small KW, Pollock SC, Vance JM et al. Ocular motility in North Carolina autosomal dominant ataxia. J Neuroophthalmol 1996; 16(2):91-95.
46. Damji KF, Allingham RR, Pollock SC et al. Periodic vestibulocerebellar ataxia, an autosomal dominant ataxia with defective smooth pursuit, is genetically distinct from other autosomal dominant ataxias. Arch Neurol 1996; 53(4):338-344.
47. Escayg A, De Waard M, Lee DD et al. Coding and noncoding variation of the human calcium-channel beta4- subunit gene CACNB4 in patients with idiopathic generalized epilepsy and episodic ataxia. Am J Hum Genet 2000; 66(5):1531-1539.
48. Jen JC, Wan J, Palos TP et al. Mutation in the glutamate transporter EAAT1 causes episodic ataxia, hemiplegia and seizures. Neurology 2005; 65(4):529-534.
49. de Vries B, Mamsa H, Stam AH et al. Episodic ataxia associated with EAAT1 mutation C186S affecting glutamate reuptake. Arch Neurol 2009; 66(1):97-101.
50. Kerber KA, Jen JC, Lee H et al. A new episodic ataxia syndrome with linkage to chromosome 19q13. Arch Neurol 2007; 64(5):749-752.
51. Meijer IA, Hand CK, Grewal KK et al. A locus for autosomal dominant hereditary spastic ataxia, sax1, maps to chromosome 12p13. Am J Hum Genet 2002; 70(3):763-769.
52. Melberg A, Hetta J, Dahl N et al. Autosomal dominant cerebellar ataxia deafness and narcolepsy. J Neurol Sci 1995; 134(1-2):119-129.
53. Melberg A, Dahl N, Hetta J et al. Neuroimaging study in autosomal dominant cerebellar ataxia, deafness and narcolepsy. Neurology 1999; 53(9):2190-2192.
54. Trudeau MM, Dalton JC, Day JW et al. Heterozygosity for a protein truncation mutation of sodium channel SCN8A in a patient with cerebellar atrophy, ataxia and mental retardation. J Med Genet 2006; 43(6):527-530.
55. Genis D, Ferrer I, Sole JV et al. A kindred with cerebellar ataxia and thermoanalgesia. J Neurol Neurosurg Psychiatry 2009; 80(5):518-523.
56. Ogawa M. Pharmacological treatments of cerebellar ataxia. Cerebellum 2004; 3(2):107-111.
57. Manto M, Marmolino D. Cerebellar ataxias. Curr Opin Neurol 2009; 22(4):419-429.
58. Trujillo-Martin MM, Serrano-Aguilar P, Monton-Alvarez F et al. Effectiveness and safety of treatments for degenerative ataxias: a systematic review. Mov Disord 2009; 24(8):1111-1124.

59. Nakamura K, Yoshida K, Miyazaki D et al. Spinocerebellar ataxia type 6 (SCA6): clinical pilot trial with gabapentin. J Neurol Sci 2009; 278(1-2):107-111.
60. Woods BT, Schaumburg HH. Nigro-spino-dentatal degeneration with nuclear ophthalmoplegia. A unique and partially treatable clinicopathological entity. J Neurol Sci 1972; 17(2):149-166.
61. Tuite PJ, Rogaeva EA, St George-Hyslop PH et al. Dopa-responsive parkinsonism phenotype of Machado-Joseph disease: confirmation of 14q CAG expansion. Ann Neurol 1995; 38(4):684-687.
62. Buhmann C, Bussopulos A, Oechsner M. Dopaminergic response in Parkinsonian phenotype of Machado-Joseph disease. Mov Disord 2003; 18(2):219-221.
63. Kanai K, Kuwabara S, Arai K et al. Muscle cramp in Machado-Joseph disease: altered motor axonal excitability properties and mexiletine treatment. Brain 2003; 126:965-973.
64. Pineda M, Montero R, Aracil A et al. Coenzyme Q(10)-responsive ataxia: 2-Year-treatment follow-up. Mov Disord 2010; 25(9):1262-8.
65. Lock RJ, Tengah DP, Williams AJ et al. Cerebellar ataxia, peripheral neuropathy, "gluten sensitivity" and anti-neuronal autoantibodies. Clin Lab 2006; 52(11-12):589-592.
66. Nanri K, Okita M, Takeguchi M et al. Intravenous immunoglobulin therapy for autoantibody-positive cerebellar ataxia. Intern Med 2009; 48(10):783-790.
67. Xia H, Mao Q, Eliason SL et al. RNAi suppresses polyglutamine-induced neurodegeneration in a model of spinocerebellar ataxia. Nat Med 2004; 10(8):816-820.
68. Muchowski PJ, Wacker JL. Modulation of neurodegeneration by molecular chaperones. Nat Rev Neurosci 2005; 6(1):11-22.
69. Chan HY, Warrick JM, Gray-Board GL et al. Mechanisms of chaperone suppression of polyglutamine disease: selectivity, synergy and modulation of protein solubility in Drosophila. Hum Mol Genet 2000; 9(19):2811-2820.
70. Heiser V, Scherzinger E, Boeddrich A et al. Inhibition of huntingtin fibrillogenesis by specific antibodies and small molecules: implications for Huntington's disease therapy. Proc Natl Acad Sci USA 2000; 97(12):6739-6744.
71. Sanchez I, Mahlke C, Yuan J. Pivotal role of oligomerization in expanded polyglutamine neurodegenerative disorders. Nature 2003; 421(6921):373-379.
72. Yoshida H, Yoshizawa T, Shibasaki F et al. Chemical chaperones reduce aggregate formation and cell death caused by the truncated Machado-Joseph disease gene product with an expanded polyglutamine stretch. Neurobiol Dis 2002; 10(2):88-99.
73. Tanaka M, Machida Y, Niu S et al. Trehalose alleviates polyglutamine-mediated pathology in a mouse model of Huntington disease. Nat Med 2004; 10(2):148-154.
74. Heiser V, Engemann S, Brocker W et al. Identification of benzothiazoles as potential polyglutamine aggregation inhibitors of Huntington's disease by using an automated filter retardation assay. Proc Natl Acad Sci USA 2002; 99:16400-16406.
75. Zhang X, Smith DL, Meriin AB et al. A potent small molecule inhibits polyglutamine aggregation in Huntington's disease neurons and suppresses neurodegeneration in vivo. Proc Natl Acad Sci USA 2005; 102(3):892-897.
76. Kieran D, Kalmar B, Dick JR et al. Treatment with arimoclomol, a coinducer of heat shock proteins, delays disease progression in ALS mice. Nat Med 2004; 10(4):402-405.
77. Rimoldi M, Servadio A, Zimarino V. Analysis of heat shock transcription factor for suppression of polyglutamine toxicity. Brain Res Bull 2001; 56(3-4):353-362.
78. Mosser DD, Morimoto RI. Molecular chaperones and the stress of oncogenesis. Oncogene 2004; 23(16):2907-2918.
79. Dedeoglu A, Kubilus JK, Jeitner TM et al. Therapeutic effects of cystamine in a murine model of Huntington's disease. J Neurosci 2002; 22(20):8942-8950.
80. Karpuj MV, Becher MW, Springer JE et al. Prolonged survival and decreased abnormal movements in transgenic model of Huntington disease, with administration of the transglutaminase inhibitor cystamine. Nat Med 2002; 8(2):143-149.
81. Shults CW. Coenzyme Q10 in neurodegenerative diseases. Curr Med Chem 2003; 10(19):1917-1921.
82. Ryu H, Rosas HD, Hersch SM et al. The therapeutic role of creatine in Huntington's disease. Pharmacol Ther 2005; 108(2):193-207.
83. Keene CD, Rodrigues CM, Eich T et al. Tauroursodeoxycholic acid, a bile acid, is neuroprotective in a transgenic animal model of Huntington's disease. Proc Natl Acad Sci USA 2002; 99(16):10671-10676.
84. Ravikumar B, Vacher C, Berger Z et al. Inhibition of mTOR induces autophagy and reduces toxicity of polyglutamine expansions in fly and mouse models of Huntington disease. Nat Genet 2004; 36(6):585-595.
85. Menzies FM, Rubinsztein DC. Broadening the therapeutic scope for rapamycin treatment. Autophagy 2010; 6(2):286-287.
86. Sanchez I, Xu CJ, Juo P et al. Caspase-8 is required for cell death induced by expanded polyglutamine repeats. Neuron 1999; 22(3):623-633.

87. Ona VO, Li M, Vonsattel JP et al. Inhibition of caspase-1 slows disease progression in a mouse model of Huntington's disease. Nature 1999; 399(6733):263-267.
88. Chen M, Ona VO, Li M et al. Minocycline inhibits caspase-1 and caspase-3 expression and delays mortality in a transgenic mouse model of Huntington disease. Nat Med 2000; 6(7):797-801.
89. Lesort M, Lee M, Tucholski J et al. Cystamine inhibits caspase activity. Implications for the treatment of polyglutamine disorders. J Biol Chem 2003; 278(6):3825-3830.
90. Gauthier S. Dimebon improves cognitive function in people with mild to moderate Alzheimer's disease. Evid Based Ment Health 2009; 12(1):21.
91. Liu J, Tang TS, Tu H et al. Deranged calcium signaling and neurodegeneration in spinocerebellar ataxia type 2. J Neurosci 2009; 29(29):9148-9162.
92. Mestre T, Ferreira J, Coelho MM et al. Therapeutic interventions for disease progression in Huntington's disease. Cochrane Database Syst Rev 2009; 3:CD006455.
93. Ristori G, Romano S, Visconti A et al. Riluzole in cerebellar ataxia: a randomized, double-blind, placebo-controlled pilot trial. Neurology 2010; 74(10):839-845.
94. Bordet T, Buisson B, Michaud M et al. Identification and characterization of cholest-4-en-3-one, oxime (TRO19622), a novel drug candidate for amyotrophic lateral sclerosis. J Pharmacol Exp Ther 2007; 322(2):709-720.
95. Strupp M, Kalla R, Glasauer S et al. Aminopyridines for the treatment of cerebellar and ocular motor disorders. Prog Brain Res 2008; 171:535-541.
96. Alvina K, Khodakhah K. The therapeutic mode of action of 4-aminopyridine in cerebellar ataxia. J Neurosci 2010; 30(21):7258-7268.
97. Tsunemi T, Ishikawa K, Tsukui K et al. The effect of 3,4-diaminopyridine on the patients with hereditary pure cerebellar ataxia. J Neurol Sci 2010; 292(1-2):81-84.
98. Dokmanovic M, Marks PA. Prospects: Histone deacetylase inhibitors. J Cell Biochem 2005; 96(2):293-304.
99. Thomas EA, Coppola G, Desplats PA et al. The HDAC inhibitor 4b ameliorates the disease phenotype and transcriptional abnormalities in Huntington's disease transgenic mice. Proc Natl Acad Sci USA 2008; 105(40):15564-15569.
100. Naoi M, Maruyama W, Yi H et al. Mitochondria in neurodegenerative disorders: regulation of the redox state and death signaling leading to neuronal death and survival. J Neural Transm 2009.
101. Gatchel JR, Watase K, Thaller C et al. The insulin-like growth factor pathway is altered in spinocerebellar ataxia type 1 and type 7. Proc Natl Acad Sci USA 2008; 105(4):1291-1296.
102. Fernandez AM, Carro EM, Lopez-Lopez C et al. Insulin-like growth factor I treatment for cerebellar ataxia: Addressing a common pathway in the pathological cascade? Brain Res Rev 2005; 50(1):134-141.
103. Leinninger GM, Feldman EL. Insulin-like growth factors in the treatment of neurological disease. Endocr Dev 2005; 9:135-159.
104. Chintawar S, Hourez R, Ravella A et al. Grafting neural precursor cells promotes functional recovery in an SCA1 mouse model. J Neurosci 2009; 29(42):13126-13135.
105. Gage FH. Neurogenesis in the adult brain. J Neurosci 2002; 22(3):612-613.
106. Klein A, Boltshauser E, Jen J et al. Episodic ataxia type 1 with distal weakness: a novel manifestation of a potassium channelopathy. Neuropediatrics 2004; 35(2):147-149.
107. Griggs RC, Moxley RT, 3rd, Lafrance RA et al. Hereditary paroxysmal ataxia: response to acetazolamide. Neurology 1978; 28(12):1259-1264.
108. Boel M, Casaer P. Familial periodic ataxia responsive to flunarizine. Neuropediatrics 1988; 19(4):218-220.
109. Strupp M, Kalla R, Dichgans M et al. Treatment of episodic ataxia type 2 with the potassium channel blocker 4-aminopyridine. Neurology 2004; 62(9):1623-1625.
110. Alvina K, Khodakhah K. KCa channels as therapeutic targets in episodic ataxia type-2. J Neurosci 2010; 30(21):7249-7257.
111. Kinali M, Jungbluth H, Eunson LH et al. Expanding the phenotype of potassium channelopathy: severe neuromyotonia and skeletal deformities without prominent Episodic Ataxia. Neuromuscul Disord 2004; 14(10):689-693.
112. De Marcos FA, Ghizoni E, Kobayashi E et al. Cerebellar volume and long-term use of phenytoin. Seizure 2003; 12(5):312-315.
113. Strupp M, Schuler O, Krafczyk S et al. Treatment of down-beat nystagmus with 3,4-diaminopyridine: a placebo-controlled study. Neurology 2003; 61(2):165-170.
114. Ilg W, Synofzik M, Brotz D et al. Intensive coordinative training improves motor performance in degenerative cerebellar disease. Neurology 2009; 73(22):1823-1830.
115. Missaoui B, Thoumie P. How far do patients with sensory ataxia benefit from so-called "proprioceptive rehabilitation"? Neurophysiol Clin 2009; 39(4-5):229-233.

NEURODEGENERATIONS INDUCED BY ORGANOPHOSPHOROUS COMPOUNDS

Alan J. Hargreaves

School of Science and Technology, Nottingham Trent University, Nottingham, UK
Email: alan.hargreaves@ntu.ac.uk

Abstract: Organophosphorous compounds (OPs) are widely used in agriculture, industry and the home. Though best known for their acute effects when used as pesticides, which target acetylcholinesterase (AChE) activity in neuromuscular junctions and the central nervous system, not all OPs are potent inhibitors of this enzyme. The widespread use of OPs has heightened concern regarding their toxicity in man, with numerous reports linking OPs to various forms of delayed neuropathy encompassing a range of neurodegenerative, psychological and neurobehavioral effects. There is mounting evidence to suggest that sub-acute levels of OPs have the ability to interact directly with a range of target proteins in addition to AChE (i.e., noncholinergic targets), causing major disruption of membrane and protein turnover, protein phosphorylation, mitochondrial dysfunction, oxidative stress and cytoskeletal re-organisation, although the mechanisms involved are not fully understood. However, major advances have been made in the study of one OP binding protein neuropathy target esterase (NTE) in terms of its true physiological role. Additionally, there is increasing evidence for the ability of OPs to cause disruption in a number of metabolic and cell signalling pathways that affect neuronal cell proliferation, differentiation and survival and to interact direct with non-esterase proteins such as tubulin. The aim of this chapter is to review our current understanding of delayed neurotoxicity, to discuss how these molecular events may relate to each other and to suggest possible future directions in mechanistic studies of OP toxicity.

INTRODUCTION

Organophosphorous compounds (OPs) have been widely used in agriculture (e.g., as pesticides) and industry (e.g., as flame retardants and lubricants) over the past half century.[1-3]

Neurodegenerative Diseases, edited by Shamim I. Ahmad.
©2012 Landes Bioscience and Springer Science+Business Media.

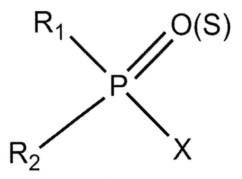

Figure 1. Generalised structure of commonly used organophosphorous compounds. Many of the commercially used OPs comprise a pentavalent phosphorous atom linked via a double bond to an oxygen atom (in organophosphates) or sulphur (organophosphorothioates). Groups R1 and R2 are typically ethoxy or methoxy groups although other substituents are possible. The group labelled 'X' is an ester linked aliphatic, homocyclic or heterocyclic arrangement and in some compounds (e.g., trichlorfon) contains halogen groups. This is the most easily hydrolysed group and, as such, becomes the 'leaving group' that is displaced when an OP binds covalently to the active site of AChE.

Figure 1 shows the general structure of commonly used OPs. A typical structure comprises a pentavalent phosphorous atom linked by a double bond to either oxygen (organophosphate) or a sulphur (organophosphorothioate) atom, with various possible combinations of aryl and alkyl substituent groups ester-linked to phosphorous via the other 3 bonds.

Organophosphorous compounds are the most widely used group of pesticides, exerting their acute toxicity via inhibition of acetylcholinesterase (AChE) in target organisms, although OPs used as industrial lubricants, flame retardants, etc., are not all classed as strong AChE inhibitors.[1-4] As exposure to many OPs can result in various forms of neurotoxicity, their widespread use has fuelled concern over their potential toxicity in humans, leading to the imposition of bans or restrictions on the use of some OPs in a number of western countries over recent years.

One example is diazinon (Fig. 2), which is used in many countries as an agricultural or domestic pesticide and is currently the only licensed sheep dip pesticide in the UK.[5] Although diazinon was considered to be only moderately toxic based on LD_{50} tests in the rat,[6] concern has been raised regarding the hazards posed by accidental exposure, particularly in pesticide handlers and workers.[7] As a result of these concerns, diazinon is now restricted or banned altogether in domestic pesticides in the USA, as an agricultural insecticide in the European Union 9 (EC 2007) and for sheep dipping in Australia.[8-12]

A well studied form of delayed neurotoxicity is OP-induced delayed neuropathy (OPIDN), the clinical symptoms of which occur 1-3 weeks following exposure to OPs such as tricresyl phosphate.[13-15] A major outbreak in the last century, referred to as 'Ginger Jake poisoning', occurred in the USA during the 1930s when alcohol was prohibited. This was found to have been caused by an alcohol-containing 'remedy' called Ginger Jake, which had been adulterated with tricresyl phosphate (TCP), causing partial paralysis in thousands of individuals.[16,17] Some OPs used as pesticides have also been linked to the induction of OPIDN, although in the case of chlorpyrifos and leptophos this required exposure to very high acutely toxic doses and diazinon has been deemed as unlikely to pose a significant risk of inducing OPIDN.[2,3,18-20] Pesticides deemed to pose a significant risk of inducing OPIDN have either been banned or restricted in use.[1-3] Although food

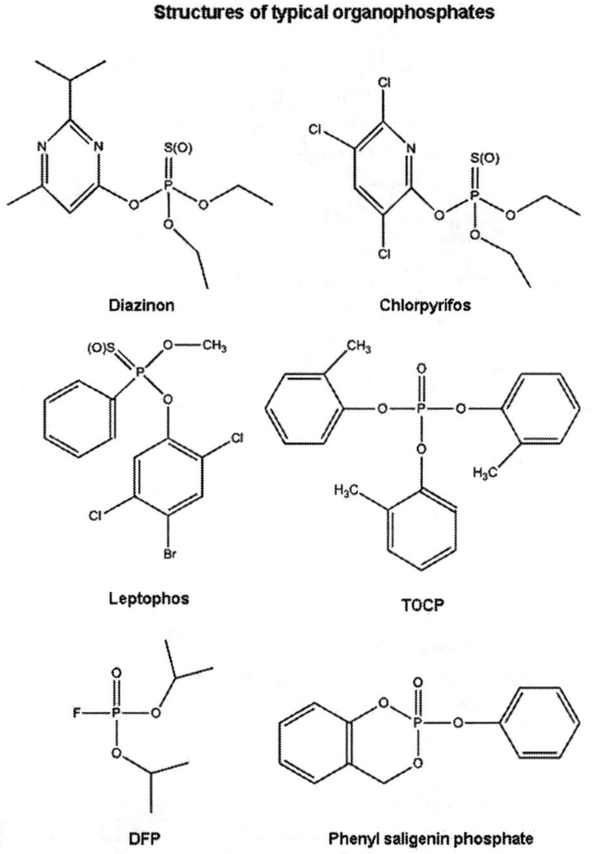

Structures of typical organophosphates

Figure 2. Structures of typical OPs. Shown are the chemical structures of some typical OPs discussed in this chapter. In order to become potent inhibitors of AChE, the organophosphorothioates (e.g., diazinon, chlorpyrifos and leptophos) require bioactivation to their 'oxon' derivatives (e.g., chlorpyrifosoxon) by specific microsomal cytochrome P450s, as a result of which the sulphur atom is replaced by an oxygen atom.

adulteration with OPs such as TCP and cases of accidentally induced OPIDN by pesticides are now thankfully very rare, OP pesticides are still in widespread use.

Furthermore, TCP is widely used in the aviation industry due to its flame retardant and lubricant properties, with the ability to function as a lubricant at very high temperatures. Though not considered a strong AChE inhibitor, this OP is a potent inhibitor of neuropathy target esterase (NTE) which, like AChE, is a member of the serine hydrolase family of enzymes. This observation, together with the fact that strong inhibitors of AChE are often weak inhibitors of NTE is consistent with the idea that individual OPs can interact with a variety of serine hydrolases with different levels of efficacy.[21-23]

TCP is, in fact, the compound of choice for studying OPIDN in animal models.[13-15] It is used in the aviation industry as a fuel and hydraulic fluid additive but is also used as a plasticiser, lubricant and flame retardant.[24-26] Isomers of TCP (such as tri-ortho-cresyl phosphate or TOCP) along with other components of aviation hydraulic fluids have been detected in air cabins and cockpits on commercial and military aircraft.[24-27] As a result,

OP poisoning has been tentatively implicated in the phenomenon of air cabin sickness.[26] Although direct involvement of OPs in air cabin sickness is not yet proven, the symptoms of air cabin sickness are similar to those caused by OP exposure.[25] Furthermore, concern has been raised about the validity of the methods of recording cockpit air quality and performing laboratory experiments on hydraulic fuel OP toxicity.[26]

Commercial preparations of TCP contain a mixture of *ortho-*, *para-* and *meta-*isomers. In response to the scientific evidence that the *ortho-* isomer of TOCP was the most potent inducer of OPIDN, it was stipulated that the level of this isomer in aviation fluids should be reduced to <1% of total TCP and the total TCP concentration to less than 3% of the total volume.[13,14,28] While the level of TCP in most cases falls within this range, it is possible that toxicity of the TOCP component may be under estimated due to the presence of more highly toxic mono- and di-*ortho*cresyl variants.[25,26] Furthermore, although recorded levels of TOCP in cabin air are lower than the recommended maximum under *normal* flying conditions, they are often within an order of magnitude of these limits.[26,29] Moreover, long term repeated exposure may be a problem for frequent fliers, predisposed individuals and/or when serious leaks enter the air cabin environment.

In addition, exposure to other TCP isomers in hydraulic fluids and their potential to interact with the *ortho* isomer in exerting a toxic effect may have been overlooked.[25,26] Indeed, little is known about the toxicity of this compound in combination with other isomers of TCP (TPCP and TMCP) and other OPs (e.g., triphenyl phosphate) present in some aviation fluids.

Thus, despite restrictions on the use of a range of OPs, there remains a significant public health concern since exposure to several of those still in use have been linked with various forms of delayed neurotoxicity.[3,13,14,30-34]

MECHANISMS OF ACUTE AND DELAYED NEUROTOXICITY OF OPs

The main neurotoxic effects reported from OP exposure in humans are:

- Acute toxicity, which occurs within hours of exposure.
- Organophosphate induced delayed neuropathy (OPIDN), which can occur up to several weeks following exposure.
- Chronic neurotoxicity including neurobehavioral and neurological deficits.

In addition, a number of OPs have been proposed to be developmental neurotoxicants but time constraints limit this chapter mainly to a discussion of effects on the mature nervous system. The first two phenomena listed above are the best understood and, together with chronic neurotoxicity, will form the main focus of the rest of this chapter. Although it is known that OPs can be used as nerve agents[2,3] in chemical warfare the review will focus on selected OPs used for beneficial purposes.

Acute Toxicity of OPs

Figure 1 shows the general structure of commonly used OPs. A typical structure comprises a pentavalent phosphorous atom linked by a double bond to either oxygen (organophosphate) or a sulphur (organophosphorothioate) atom, with various possible combinations of aryl and alkyl substituent groups attached to phosphorous via the other 3 bonds.

Table 1. Effects of receptor over-stimulation following acute exposure to OPs

Receptor System Affected	Clinical Signs
Muscarinic receptors	Bradycardia, bronchoconstriction, bronchorrhoea, hypotension, increased gastrointestinal motility, abdominal cramps, miosis and hypersalivation
Nicotinic receptors	Hypertension, tachycardia, fibrillation, fasciculation, striated muscle necrosis
Both central muscarinic and nicotinic receptors	Tremor, loss of movement co-ordination, seizures, central depression of respiration, coma, death

Organophosphorothioates such as chlorpyrifos and diazinon (Fig. 2) are commonly used as insecticides, capitalising on their ability to potently inhibit acetylcholinesterase (AChE) in target organisms. However, in order to become potent AChE inhibitors, such OPs require bioactivation, which involves the replacement of sulphur by oxygen resulting in the formation of an oxon derivative (Figs. 1 and 2). Over exposure to OPs in humans can result in the accumulation of acetylcholine (ACh) at neuromuscular junctions, causing neuromuscular block and respiratory failure in extreme cases.[3,32] Although acute exposure is relatively rare in humans, it can be followed by the development of various forms of delayed neuropathology (discussed later).

The main consequence of acute exposure to OP pesticides and nerve gases is the inhibition of AChE. This can lead to cholinergic crisis which is associated with hyperstimulation of muscarinic and nicotinic ACh receptors and related clinical effects (Table 1). If death is the result, this is usually caused by respiratory failure and/or cardiac arrest. Figure 3 shows a schematic representation of the mechanism of interaction of OPs with AChE. The OP binds irreversibly with the hydroxyl group in the side chain of the serine residue at the active site of AChE. Levels of nerve AChE

ESTERASE INHIBITION BY OPs

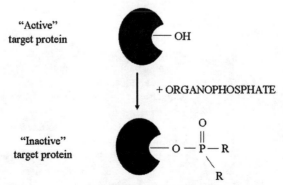

"Active" target protein   — OH

+ ORGANOPHOSPHATE

"Inactive" target protein

If target is NTE and "aging" occurs, OPIDN is induced, which involves increased negative charge at the active site

Figure 3. Mechanism of esterase inhibition by OPs. OPs bind covalently to an active site serine with the loss of the group labelled 'X' in Figure 1.

inhibition greater than approximately 70% will lead to the accumulation of ACh in synaptic clefts at neuromuscular junctions, causing neuromuscular block and respiratory failure in severe cases.

Current strategies for the treatment of acute OP toxicity include early interventions using the anticholinergic agent atropine (to block muscarinic AChE receptor activation), cholinergic re-activators (e.g., oximes) and anticonvulsant drugs such as benzodiazepines.[36,37] As organophosphorothioates can be degraded by serum paraoxonases such as PON1, individuals lacking or carrying reduced levels this enzyme are more susceptible to the delayed toxic effects of this group of agents.[2,3] Thus, another potential treatment is the injection of recombinant paraoxonase PON1, which is inactive or deficient in some individuals.[38] However, these treatments do not prevent the development of delayed neurotoxicity, consistent with the possibility that delayed effects may be triggered by the interaction of OPs with molecular targets other than AChE. Progress towards recovery following acute exposure of humans to OPs can be followed by measuring AChE, butyryl cholinesterase and neuropathy target esterase (NTE) activities in blood cells.[18,39,40]

OP-Induced Delayed Neuropathies (OPIDN)

OPIDN is a neurodegenerative condition that particularly affects nerves with long fibre tracts in both the central and peripheral nervous systems. Some but not all OPs have been found to induce OPIDN, which is characterised by delayed onset of an extended period of ataxia and upper motor neurone spasticity arising from single or repeated exposure to OPs.[1-3,21-23,41]

After exposure to OPs, apart from recognised cholinergic effects, no obvious OPIDN-related clinical changes are observed for a period of at least a few days up to several weeks. This is followed by the progressive development of symptoms outlined in Table 2, though not all of these are seen or reach the same level of severity in all patients due to a variety of reasons such as the chemical involved, the level, route and frequency of exposure, inter-individual differences in susceptibly and whether co-exposure to other toxins is involved.[15,41] After disease development there is normally

Table 2. Clinical effects observed in OPIDN

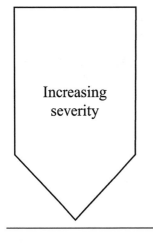

| Increasing severity | Early symptoms include cramp, burning and/or stinging sensation in the calves and possibly the ankles or feet. This may be followed by numbness and paraesthesia in the legs and feet. Weakness of the limbs develops and may later progress to the hands and arms. When walking, the toes of both feet may drag on the floor leading to a neuropathic 'steppage gait'. Symptoms may progress to abnormal balance and deterioration of certain limb reflexes. In severe cases a flaccid paralysis will eventually develop and subjects may suffer bladder and bowel problems. Any permanent CNS damage will become evident during recovery, in the form of hyper reflexia, spastic gait and increased motor tone. |

a period of stabilisation during which the acquired level of paralysis persists; provided that significant damage to the CNS has not occurred, this is followed by a gradual recovery starting in the hands and arms. However, as the CNS is incapable of nerve regeneration, if there has been significant CNS damage, neurological deficits may persist for many years and in the most severe cases may be irreversible.[15,41] Damage to the CNS would be masked in the early stages of OPIDN due to the predominantly peripheral neuropathic symptoms but would become apparent during recovery of peripheral nerve function as hyper reflexia, abnormal motor tone and the development of a spastic gait, in which the legs are kept close together and move rigidly with toes dragging along the floor.[41]

The risk of developing OPIDN following acute or repeated exposure to OPs can be monitored by measuring NTE activity in the blood cell of pesticide workers and handlers, or from individuals exposed to pesticides deliberately or accidentally.[39] A similar approach in studies of NTE in animal models of OPIDN revealed the hen to be the most sensitive species.[42] The hen has therefore become the preferred model for screening purposes and its use has led to the imposition of restrictions on the use of a number of OPs. Studies on the hen confirmed that TOCP was a potent inducer of OPIDN and that OP pesticides such as chlorpyrifos, leptophos and trichlorfon were also able to induce the condition but only following exposure to much higher often acutely toxic doses.[18,42,44]

Chronic Neurobehavioral, Cognitive and Developmental Neurotoxicity of OPs

A number of epidemiological studies have suggested that exposure to of adults to acute or repeated sub acute levels of OPs can result in long term neurological and neurobehavioral effects, affecting the CNS to a greater degree than the PNS. For example, a study on pesticide workers occupationally exposed to acute organophosphate pesticide poisoning were found to show significant impairment in a range of WHO-approved tests of behaviour and cognitive function.[45] A recent study on the effects of low level exposure to diazinon in both retired and working sheep dippers showed that sub-acute exposure was associated with increased risk of neurobehavioral dysfunction.[46] Sub acute levels of chlorpyrifos were also found to affect cognitive skills of rats exposed during weaning, raising concern about the risk of developmental effects of this pesticide.[47] This concern was further strengthened by epidemiological studies on children applying pesticides to crops showing significant impairment in tests of neurobehavioral and cognitive skills, the latter becoming worse at longer exposure times.[48]

As these types of lesion differ from both acute toxicity and OPIDN in that they can last for many years during which central effects predominate, they have been collectively referred to as OP-induced chronic neurotoxicity.[3,15] However, it should be noted that the term encompasses a wide range of neurological and neurobehavioral deficits resulting from the neurodegenerative effects of single acutely toxic or repeated sub-acute doses of OP. Many of these conditions may reflect the interaction of OPs with distinct molecular targets and/or the extent of cell death in specific populations of neurons in the CNS and the level of maturity of the nervous system at the point of exposure.

Typical symptoms and pathological lesions presented by affected individuals are shown in Table 3. These effects have been observed to varying degrees in the aforementioned studies on human and animal models of OP exposure. It is interesting to note that many of these symptoms are also exhibited by individuals suffering

Table 3. Clinical signs and pathological lesion observed in chromic OP-induced neurotoxicity

Physical	Headache, drowsiness, dizziness, fatigue, generalised weakness and tremors
Cognitive	Deficits in concentration, memory and cognitive ability
Behavioural	Anxiety, apathy, confusion, restlessness, labile emotion, lethargy, anorexia, insomnia, depression, irritability

from air cabin sickness, consistent with the possibility that the latter is due at least in part to OP exposure. As indicated earlier, the neuropathic OP of greatest concern has been the ortho isomer of tricresyl phosphate due its potent ability to induce OPIDN.[13,14] Although this isomer is now kept to a minimum level in aviation oils,[28] there are several other OPs in aviation fluids.[25,26] It is therefore clear that the ability of all aviation fluid OPs to individually or collectively induce chronic neurotoxicity needs to be investigated.

MOLECULAR TARGETS OF OPs

The fact that exposure to sub-acute levels of OPs can result in neurotoxicity supports the view that the a range of delayed neuropathic effects are due at least in part to the interaction of OPs with molecular targets other than AChE, such as other serine hydrolases.[49] In this section of the chapter we will discuss evidence to support the view that OPs have noncholinergic targets, linking these conditions to OP exposure and outline some potential future directions for research.

NEUROPATHY TARGET ESTERASE (NTE)

The primary target of OPs that induce OPIDN is considered to be NTE.[21-23,50,51] OPIDN inducers are far more potent inhibitors of NTE than they are of AChE,[52] which has formed the basis of OP screening methods using the hen model. Returning to Figure 3, if the esterase is NTE and aging occurs after binding to the active site serine residue the inhibition becomes irreversible and OPIDN will occur. Only those OPs able to induce aging of NTE, which involves increased negative charge at the active site,[21-23,50] can induce OPIDN. The fact that NTE can be inhibited significantly by non-aging OPs with no apparent effect on neural function implies that the esterase activity per se may not be essential for neural function.

Major advances in recent years led to the cloning of *NTE* gene and sequencing of protein and the use of recombinant NTE in a number of biochemical studies.[50,51,53] Such studies have shown that NTE is a membrane associated protein located mainly in the SER capable of forming an ion channel when reconstituted in artificial phospholipid vesicles.[54] Of particular interest is the fact that only neuropathic OPs (i.e., those that induce aging of NTE) were able to block ion conductance in vitro suggesting that this potential role of NTE may be impaired in OPIDN.[54] The catalytic domain resides in the C terminal region and the amino terminal region contains regulatory domains that, coupled with membrane association, are required for optimal esterase activity.[55,56] In addition to its well established OP-sensitive esterase activity,[21-23] NTE has also been found to act as a lipid hydrolase acting on phosphatidylcholine as a substrate,[56,57] suggesting that it may play a role in the regulation of phospholipid turnover in cell membranes.

The *NTE* gene sequence was found to be similar to that of Swiss Cheese (SWS) protein in *Drosophila melanogaster*; when the *SWS* gene was mutated it resulted in the development of a central neuropathy.[58] Deletion of the same gene in the mouse resulted in a similar neuropathy to that observed in Drosophila.[59] This neurodegenerative condition involved glial hyper wrapping and neuronal apoptotic cell death in large areas of the brain, indicating an essential role for NTE and similar proteins in neuronal glial interactions important in nervous system function.[51,58] Further studies on Drosophila suggested that the neuropathy was accompanied by significant inhibition of the lipid hydrolase activity exhibited by NTE,[60] suggesting that disruption of phospholipid homeostasis was a major factor in the development of the neuropathy. It remains to be established whether the same occurs in OPIDN which, as discussed earlier, affects primarily the peripheral nervous system.

Interestingly, it was found that NTE and SWS were more closely related than NTE was to the serine hydrolase family, containing a common highly conserved domain present throughout phylogeny.[51,61] As indicated earlier, doubt has been cast as to the importance of the classical esterase activity of NTE in adults as non-NTE aging OPs can inhibit NTE significantly without inducing OPIDN.[21-23] Glynn proposed that inhibition and aging of NTE might have an effect on another NTE related function such as lipid hydrolysis or ion channel activity NTE or by causing a change in function of NTE by affecting its interaction with other macromolecules,[51] which would seem to be borne out by the SWS studies outlined above. The importance of NTE per se in nerve function was further emphasized by the discovery of *NTE* mutants with disrupted esterase activity associated with motor neurone disease,[62,63] consistent with mounting evidence that NTE plays a vital role in the maintenance of neurons in adult mammals.[64,65]

A number of cellular studies have been performed in differentiating cell lines to determine the role of NTE in neuronal cell differentiation. Exposure of differentiating mouse N2a neuroblastoma cells to the OPIDN inducing OP phenyl saligenin phosphate (PSP: An analogue of the neuropathic metabolite of TOCP), resulted in reduced outgrowth of axon-like processes in association with complete inhibition of NTE.[66] A weak inducer of OPIDN chlorpyrifos was also found to inhibit neurite outgrowth and NTE in the same cellular system.[67] The observation that knock down of *NTE* in differentiating SH-SY5Y cells had no effect on the extent of neurite outgrowth but that moderately increased expression could increase the rate of outgrowth suggested that the enzyme may play a subtle regulatory role but also that other targets may be involved in the ability of OPs to inhibit neurite outgrowth.[68,69] Furthermore the increased expression of NTE caused mitotic arrest in a kidney cell line but had no effect on proliferation in human SH-SY5Y neuroblastoma cells suggesting that it may be the ability to regulate cell proliferation in certain cell types.[70] Further work is warranted in coculture systems to reflect the neuronal-glial interactions that occur in vivo more closely. Nevertheless, one of the implications of the results from manipulated NTE expression studies is that other targets may be involved in the neuropathic effects of NTE, some of which are discussed in the following section.

OTHER POTENTIAL MOLECULAR TARGETS OF ORGANOPHOSPHATES

Cytoskeletal Proteins

In early ex vivo biochemical studies of cytoskeletal enriched extracts from hens induced to exhibit OPIDN, it was found that TOCP exposure was associated with hyperphosphorylation of cytoskeletal proteins.[13,71] Histopathological studies in TOCP treated hens confirmed the presence of abnormal aggregations of highly phosphorylated neurofilament heavy chain (NFH) prior to the onset for clinical signs of OPIDN.[72,73] Inhibition of neurite outgrowth in differentiating N2a cells by phenyl saligerim phosphate (PSP) was also found to be associated with a transient increase in neurofilament phosphorylation again suggesting that disruption of cytoskeletal protein phosphorylation is a key event in OP-induced neurotoxicity.[66] However, the relationship (if any) between these changes and NTE inhibition remains to be established.

One explanation for the altered phosphorylation status of neurofilaments could be the activation of calmodulin kinase as suggested in one study.[71] However, PSP exposure in differentiating N2a cells was also found to cause increased activity of the mitogen activated protein (MEP) kinase MAP kinase ERK1/2 for which NFH is a substrate,[74] suggesting that this agent was able to affect cytoskeletal integrity by disruption of signalling pathways associated with neuronal cell differentiation and survival.[66,75] However, not all neurite inhibitory OPs appear to induce NFH hyperphosphorylation as this effect was not observed in chlorpyrifos- or leptophos-treated differentiating N2a cells.[67,76] Further work would help to determine whether increased ERK activation is an effect limited to strong inducers of OPIDN. Further evidence that OPs can disturb the phosphorylation status and/or organisation of cytoskeletal proteins was obtained in a cellular study of the effects of diazinon on neurite outgrowth.[77] Neurite inhibition was associated with increased levels and phosphorylation of the actin binding protein cofilin which regulates the dynamic properties of the microfilament network and decreased staining of neurites with anti actin antibodies suggesting a reduction in the level of actin polymer in OP treated cells. Microfilament disruption was also observed in a study on human neuroblastoma cells though distinct OPs were found to affect microfilament levels differently, again suggesting that not all OPs act in exactly the same way.[78]

Proteolytic Enzymes

One consequence of Wallerian degeneration is the influx of extracellular Ca^{2+} leading to the activation of calcium dependent enzymes such as the protease calpain. The finding that symptoms of OPIDN could be blocked by Ca^{2+} channel blockers,[79] and that calpain is activated at an early stage in TOCP treated hens,[80] supports the view that calpain-mediated proteolysis of cytoskeletal proteins such as neurofilaments is an early event in OPIDN and may explain the consistent pattern of reduced levels of neurofilament heavy chain observed in cellular and animal models exposed to a range of OPs.[66,67,80,81] However one study observed reduced calpain activity concomitant with cytoskeletal protein degradation in hens treated with the OPIDN inducer diisopropylphosphorofluoridate (DFP) suggesting that other proteases may be activated by some OPs to induce axonal degeneration.[82] One should not lose site of the fact that one of the major pathways of proteolytic processing known to be essential for neuronal form and function, the ubiquitin dependent pathway, includes a number of serine hydrolase activities that could be targetted by OPs.[83] The fact that this system is disrupted in a number of neurodegenerative disorders involving abnormal protein aggregation,[83] and the presence of abnormal NFH aggregation in cellular and animal model of OP exposure,[66,72,73,81] suggests that proteasomes might be a potential target of OPs. Further work to determine the effect of OPs on the mechanisms of protein turnover in axons would therefore be of value.

Mitochondria

Several studies have suggested that OP exposure can result in structural changes and impairment in the activity of a range of key mitochondrial enzyme activities such as succinate dehydrogenase, NADH dehydrogenase and cytochrome oxidase.[84] Other studies have suggested that OPs can disrupt energy metabolism, mitochondrial membrane organisation and respiratory activity, leading to apoptotic cell death in cultured neurons.[85] Some OPs also showed the ability to cause major disruption to the mitochondrial transmembrane potential, a good indicator of mitochondrial well being.[86] Major disruption of mitochondrial energy metabolism could lead to the reduction in cellular ATP levels that might be the trigger for many OP related effects in the rest of the cell including apoptosis and oxidative stress.[87] OPs also induced major changes in mitochondrial ultrastructure prior to cell death in lumbar spinal neurons, again suggesting that major disruption of mitochondria can occur at an early stage of OPIDN.[88] Taken together these and other studies indicate that mitochondria are targetted at least by some OPs. Further analysis of the mitochondrial proteome might help to further establish the nature of OP effects on mitochondrial function and how this relates to the various types of OP related toxicity.

OTHER POTENTIAL 'NON-ESTERASE' OP BINDING PROTEINS

Despite the well held view that OPs bind selectively to a motif in the active site of AChE, NTE and/or other serine hydrolases, there is increasing evidence to suggest that other proteins may interact directly with some OPs. It has been suggested that tyrosine and lysine residues on such proteins may provide alternative OP binding sites and that this may be relevant to the numerous proteins (e.g., of the neuronal cytoskeleton) that exhibit altered phosphorylation status.[89] A mass spectrometry approach was used to detect binding of chlorpyrifosoxon, dichlorvos, diisopropylfluorophosphate (DFP) and sarin to human serum albumin, suggesting that this approach could form the basis of a prognostic test of the response to exposure.[90] A similar outcome was found for human transferrin.[91]

Chlorpyrifosoxon disrupts microtubules in vivo and was shown to have a direct effect on microtubule proteins by preventing their ability to polymerise in vitro. Further analysis by mass spectrometry demonstrated the covalent interaction of chlorpyrifos with tyrosine residues in tubulin, the core protein of microtubules.[94,95] This may represent one of the ways in which OPs differ in the specific lesions they produce and could indicate a role for such interactions in the chronic or developmental effects of this particular OP. Further proteomic analysis of OP treated cells and organisms will no doubt reveal other potential targets of OPs and help to establish the way in which different OPs affect neural form and function.

CONCLUSION

Organophosphorus esters are capable of inducing acute toxicity by inhibition of AChE and cause various forms of delayed toxicity by interacting with a variety of noncholinergic targets which have not yet been fully characterised. Exactly what molecular changes underlie these various forms of neurotoxicity is not fully understood but a proposed scheme for the mechanism(s) of toxicity is presented in Figure 4, bearing in mind that the exact

Non-cholinergic pathways of OP toxicity

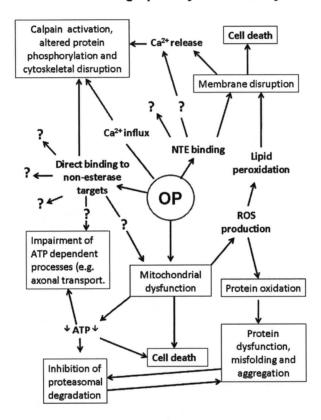

Figure 4. Noncholinergic pathways of OP toxicity. Shown is a schematic representation of the possible inter relationships between the different primary targets and pathways affected by sub-acute exposure to OPs. Binding to NTE may affect both its putative channel role in the endoplasmic reticulum and its lipid hydrolase activity, causing disruption of membranes with potential Ca^{2+} release, membrane cleavage and/or cell death, in addition to as yet unidentified roles. OPs may also cause ATP depletion, oxidative stress and/or induce apoptosis via mitochondrial impairment. If lipid peroxidation is significant this could contribute to the membrane disruption mentioned earlier while protein oxidation may result in altered protein function and/or turnover by causing mis-folding, which could in turn affect the ability of the ubiquitin-dependent proteasomal pathway of protein degradation, which may also be impaired by ATP depletion. The latter would also start to affect metabolic pathways and a range of ATP dependent processes such as axonal transport, which is essential for nerve maintenance and regeneration. Binding of OPs to non-esterase targets may induce cytoskeletal disruption (e.g., in the case of chlorpyrifosoxon binding to tubulin and a range of other effects yet to be determined. The exact mechanism will always depend on the specific OP(s), the level and duration of exposure.

effects will depend OP structure, targets affected, route, level and duration of exposure and inter individual variation. The neurodegenerative condition OPIDN is induced up to several weeks after exposure and is preceded by a number of molecular lesions including inhibition (and aging) of NTE and disruption of the organisation and/or phosphorylation status of cytoskeletal proteins important in the growth and maintenance of axons. NTE is now known to have a range of important esterase and non-esterase functions and may play a regulatory role in axon maintenance. Recent developments suggest that a number of other proteins such as transferrin, albumin and tubulin are capable of binding certain OPs in tyrosine residues, breaking the myth that an active site serine was the only motif available. Further work on NTE functions and the characterisation of novel OP targets and their role in toxicity will help to establish the molecular basis of OP-induced neurodegeneration more fully.

REFERENCES

1. Chambers JE. Organophosphate compounds an overview. In: Chambers JE, Levi P, eds. Organophosphates: Chemistry, fate and effects. San Diego: Academic Press, 1992:3-17.
2. Krieger RI, ed. Handbook of pesticide toxicology: Principles, 2nd edition. San Diego: Academic Press, 2001.
3. Gupta RC, ed. Toxicology of organophosphate and carbamate compounds. San Diego: Elsevier, 2006.
4. Aldridge WN. Tricresyl phosphates and cholinesterase. Biochem J 1954; 56:185-189.
5. Olsen KR, ed. Poisoning and drug overdose. Stamford, Connecticut: Appleton and Lange, 1998.
6. Olsen KR, ed. Poisoning and drug overdose. Stamford Connecticut; Appleton and Lange, 1998.
7. Committee on Toxicology (COT) of Chemicals in food, consumer products and he environment organophosphates. Department of Health, 2007.
8. Environmental Protection Agency. Annual Report, 2000. Office of Pesticide Programs. http://www.epa.gov/oppfead1/annual/2000/2000annual.pdf
9. Environmental Protection Agency. Reregistration Eligibility Decision for Diazinon, 2006. http://www.sps.gov/pesticides/reregistration/REDS/diazinon_red.pdf
10. European Commission (EC) decision of 6 June 2007 concerning the non inclusion of diazinon in Annex I to the Council Directive 91/414/EEC and the withdrawal of authorisations for plant protection agents containing that substance. Official Journal of the European Union 2007; 140:9-10.
11. Australian Pesticides and Veterinary Medicines Authority (APVMA). APVMA suspends the use of diazinon for sheep dipping and jetting. Media Release, 2007.
12. Department of Primary Industries (DPI) Sheep body lice: Control and eradication. Agriculture Notes of the DPI State of Victoria, 2007; AG1110.
13. AbouDonia MB, Lapadula DM. Mechanisms of organophosphorusester induced delayed neurotoxicity: Type I and II. Ann Rev Pharmacol Toxicol 1990; 30:405-440.
14. Lotti M. The pathogenesis of organophosphate polyneuropathy. Crit Rev Toxicol 1992; 21:465-487.
15. AbouDonia MB. Organophosphorus ester induced chronic neurotoxicity. Arch Env Health 2003; 58:484-497.
16. Bishop EL, Stewart HC. Incidence of partial paralysis. Am J Public Health 1930; 20:1307-1312.
17. Zeligs MA. Upper motor neuron sequelae in "Jake" paralysis: a clinical follow up study. J Nerv Ment Dis 1938; 87:464-470.
18. Richardson RJ, Moore TB, Kayyali US et al. Chlorpyrifos: Assessment of potential or delayed neurotoxicity by repeated dosing in adult hens with monitoring of brain acetylcholinesterase, brain and lymphocyte neurotoxic esterase and plasma butyrylcholinesterase activities. Toxicol Sci 1993; 21:89-96.
19. Chow E, Seiber JN, Wilson BW. Isophenos and an in vitro assay for delayed neuropathic potential. Toxicol Appl Pharmacol 1986; 83:178-183.
20. de Blaquiere GE, Waters L, Blain PG et al. Electrophysiological and biochemical effects of single and multiple doses of the organophosphate diazinon in the mouse. Toxicol Appl Pharmacol 2000; 166:81-91.
21. Johnson MK. The primary biochemical lesion leading to the delayed neurotoxic effects of some organophosphorus esters. J Neurochem 1974; 23:785-789.
22. Johnson MK. The target for initiation of delayed neurotoxicity by organophosphorus esters: biochemical studies and toxicological applications. In: Hodgson E, Bend JR, Philpot RM, Eds. Reviews in Biochemical Toxicology. New York: Elsevier, 1982; 4:141-212.
23. Johnson MK. Organophosphates and delayed neuropathy- is NTE alive and well? Toxicol Appl Pharmacol 1990; 102:385-389.

24. Committee on Air Quality in Passenger Cabins of Commercial Aircraft, Board on Environmental Studies and Toxicology, National Research Council, National Academy of Sciences. The Airliner Cabin Environment and Health of Passengers and Crew. Washington DC; National Academic Press, 2002.

25. Winder C, ed. Contaminated Air Protection: Proceedings of the Air Safety and Cabin Air Quality International Aero Industry Conference. British Airline Pilots Association (BALPA) and the University of New South Wales, Australia, 2005.

26. Winder C. Air monitoring studies for air cabin contamination. Curr Top Toxicol 2006; 3:33-48.

27. Civil Aviation Authority. Cabin air quality, CAA paper, 2004:04.

28. Harris MO, McLure P, Chessin RL et al. Toxicological profile for hydraulic fluids. ATDSR, USA; 1997.

29. Hanhela PJ, Kibby J, DeNola G et al. Organophosphate and amine contamination of cockpit air in the Hawk, F-111 and Hercules C-130 aircraft. DSTO Defence Science and Technology Organisation, Victoria, Australia; 2005.

30. Nutley BP, Crocker J. Biological monitoring of workers occupationally exposed to organophosphorous pesticides. Pesticide Science 1993; 39:315-322.

31. Pilkington A, Buchanan D, Jamal GA et al. An epidemiological study of the relations between exposure to organophosphate pesticides and indices of chronic peripheral neuropathy and neuropsychological abnormalities in sheep farmers and dippers. Occup Environ Med 2001; 58:702-710.

32. Costa LG. Current issues in organophosphate toxicology. Clin Chim Acta 2006; 366:1-13.

33. Karalliedde L, Baker D, Marrs TC. Organophosphate-induced intermediate syndrome: aetiology and relationships with myopathy. Toxicol Rev 2006; 25:1-14.

34. Slotkin TA, Ryde TI, Levin ED et al. Developmental neurotoxicity of low dose diazinon exposure of neonatal rats: Effects on serotonin systems in adolescence and adulthood. Brain Res Bull 2008; 75:640-647.

35. Slotkin T, Seidler FJ. Developmental neurotoxicants target differentiation into the serotonin phenotype: Chlorpyrifos, diazinon, dieldrin and divalent nickel. Toxicol Appl Pharmacol 2008; 233:211-219.

36. Thiermann H, Szinicz L, Eyer F et al. Modern strategies in therapy of organophosphate poisoning. Toxicol Lett 1999; 107:233-239.

37. Eddleston M, Buckley NA, Eyer P et al. Management of acute organophosphorus pesticide poisoning. The Lancet 2008; 371:597-607.

38. Akhmedova SN, Yakimovsky AK, Schwartz EI. Paraoxonase 1 Met-Leu polymorphism is associated with Parkinson's disease. J Neurol Sci 2001; 184:179-182.

39. Lotti M, Moretto R, Zoppellari R et al. Inhibition of neuropathy target esterase predicts the development of OPIDN. Arch Toxicol 1986; 59:176-179.

40. Bissbort SH, Vermaak WJH, Elias J et al. Novel test and its automation for the determination of erythrocyte acetylcholinesterase and its application to organophosphate poisoning. Clin Chim Acta 2001; 303:139-145.

41. Jokanovic M, Kosanovic M, Brkic D et al. Organophosphate induced delayed polyneuropathy in man: an overview. Clin Neurol Neurosurg 2010; in press.doi:10.1016/j.clineuro.2010.08.015

42. Lotti M, Johnson MK. Neurotoxicity of organophosphorus pesticides: predictions can be based on in vitro studies with hen and human enzymes. Arch Toxicol 1978; 41:215-221.

43. El Sebae AH, Soliman SA, AboElamayem M et al. Neurotoxicity of organophosphorus insecticides leptophos and EPN. J Env Sci Health 1977; B12:269-288.

44. Abou-Donia MB, Graham DG. Delayed neurotoxicity of sub chronic oral administration of leptophos to hens: recovery during four months after exposure. J Toxicol Env Health 1979; 5:1133-1147.

45. Rosenstock L, Keifer M, Daniell WE et al. Chronic central nervous system effects of acute organophosphate pesticide intoxication. The Lancet 1991; 338:223-227.

46. Mackenzie Ross SJ, Brewin CR, Curran HV et al. Neuropsychological and psychiatric functioning in sheep farmers exposed to low levels of organophosphate pesticides. Neuro Toxicol Teratol 2010; 32:452-459.

47. Jett DA, Navoa RV, Beckles RA et al. Cognitive function and cholinergic neurochemistry in weanling rats exposed to chlorpyrifos. Toxicol Appl Pharmacol 2001; 174:89-98.

48. Abdel Rasoul GM, Abou Salem ME, Mechael AA et al. Effects of occupational pesticide exposure on children applying pesticides. Neuro Toxicology 2008; 29:833-888.

49. Casida JE, Quistad GB. Organophosphate toxicology: Safety aspects of nonacetylcholinesterase secondary targets. Toxicology 2004; 17:983-998.

50. Glynn P. Neuropathy target esterase. Biochem J 1999; 344:625-631.

51. Glynn P. Neural development and neuropathy target esterase: two faces of neuropathy target esterase. Prog Neurobiol 2000; 61:61-74.

52. Kropp TJ, Richardson RJ. Relative inhibitory potencies of chlorpyrifosoxon, chlorpyrifos methyl oxon and mipafox for acetylcholinesterase versus neuropathy target esterase. J Toxicol Environ Health A 2003; 278:8820-8825.

53. Glynn P. Molecular cloning of neuropathy target esterase. Chem Biol Interact 1999; 119-120:513-517.

54. Forshaw PJ, Atkins J, Ray DE et al. The catalytic domain of human neuropathy target esterase mediates an organophosphate sensitive ion conductance across liposome membranes. J Neurochem 2001:400-406.

55. Atkins J, Luthjens LH, Hom ML et al. Monomers of the catalytic domain of human neuropathy target esterase are active in the presence of phospholipid. Biochem J 2002; 361:119-123.
56. Li Y, Dinsdale D, Glynn P. Protein domains, catalytic activity and subcellular distribution of neuropathy target esterase in mammalian cells. J Biol Chem 2003; 278:8820-8825.
57. Glynn P. Neuropathy target esterase and phospholipid deacylation. Biochim Biophys Acta 2005; 1736:87-93.
58. Kretzschmar D, Hasan G, Sharma S et al. The Swiss cheese mutant causes glial hyper wrapping and brain degeneration in Drosophila. J Neurosci 1997; 17:7425-7432.
59. Akassoglou K, Malester B, Xu J et al. Brain-specific degeneration of neuropathy target esterase/swiss cheese results in neurodegeneration. Proc Natl Acad Sci USA 2004; 101:5075-5780.
60. Mühlig-Versen M, Bettencourt de Cruz A, Tschäpe JA et al. Loss of swiss cheese/neuropathy target esterase activity causes disruption of phosphatidylcholine homeostasis and neuronal and glial cell death in Drosophila. J Neurosci 2005; 25:2865-2873.
61. Lush MJ, Li Y, Read DJ et al. Neuropathy target esterase and a homologous Drosophila neurodegeneration mutant protein contain a domain conserved from bacteria to man. Biochem J 1998; 332:1-4.
62. Rainier S, Bui M, Mark E et al. Neuropathy target esterase gene mutations cause motor neuron disease. Am J Human Genet 2008; 82:780-785.
63. Hein ND, Stuckley JA, Rainier SR et al. Constructs of human neuropathy target esterase catalytic domain containing mutations related to motor neuron disease have altered enzymatic properties. Toxicol Lett 2010; 196:67-73.
64. Read DJ, Li Y, Chao MV et al. Neuropathy target esterase is required for adult vertebrate axon maintenance. J Neurosci 2009; 29:11594-11600.
65. Chang PA, Wu YJ. Neuropathy target esterase: an essential enzyme for neural development and axonal maintenance. Int J Biochem Cell Biol 2010; 42:573-575.
66. Hargreaves AJ, Fowler, MJ, Sachana, M et al. Inhibition of neurite outgrowth in differentiating mouse N2a neuroblastoma cells by phenyl saligenin phosphate: Effects on neurofilament heavy chain phosphorylation, MAP kinase (ERK 1/2) activation and neuropathy target esterase activity. Biochemical Pharmacology 2006; 71:1240-1247.
67. Sachana M, Flaskos J, Alexaki E et al. The toxicity of chlorpyrifos towards differentiating mouse N2a neuroblastoma cells. Toxicol In Vitro 2001; 15:369-372.
68. Chang PA, Wu YJ, Chen R et al. Inhibition of neuropathy target esterase expressing by antisense RNA does not affect neural differentiation in human neuroblastoma (SK-N-SH) cell line. Mol Cell Biochem 2005; 272:47-54.
69. Chang PA, Chen R, Wu YJ. Reduction of neuropathy target esterase does not affect neuronal differentiation but moderate expression induces neuronal differentiation in human neuroblastoma (SK-N-SH) cell line. Mol Brain Res 2005; 141:30-38.
70. Chang PA, Liu CY, Chen R et al. Effect of over-expression of neuropathy target esterase on mammalian cell proliferation. Cell Prolif 2006; 39:429-440.
71. Suwita E, Lapadula DM, Abou-Donia MB. Calcium and calmodulin enhanced in vitro phosphorylation of hen brain cold stable microtubules and spinal cord neurofilament proteins following a single oral dose of tri-o-cresyl phosphate. Proc Natl Acad Sci USA 1986; 83:6174-6178.
72. Jensen KF, Lapadula DM, Knoth Anderson J et al. Anomalous phosphorylated neurofilament aggregations in central and peripheral axons of hens treated with tri-ortho-cresyl phosphate (TOCP). J Neurosci Res 1992; 33:455-460.
73. Jortner BS, Perkins SK, Ehrich M. Immunohistochemical study of phosphorylated neurofilaments during the evolution of organophosphorus ester-induced delayed neuropathy. Neuro Toxicology 1999; 20:971-976.
74. Veeranna, Amin ND, Ahn NG et al. Mitogen-activated protein kinases (ERK 1/2) phosphorylate Lys-Ser-Pro (KSP) repeats in neurofilament proteins NFH and NFM. J Neurosci 1998; 18:4008-4021.
75. Perron JC, Bixby JL. Distinct neurite outgrowth signalling pathways converge on ERK activation. Mol Cell Neurosci 1999; 13:362-378.
76. Sachana M, Flaskos J, Alexaki E et al. Inhibition of neurite outgrowth in N2a cells by leptophos and carbaryl: effects on neurofilament heavy chain, GAP-43, HSP-70. Toxicol In Vitro 2003; 17:115-120.
77. Harris W, Sachana M, Flaskos J et al. Proteomic analysis of differentiating neuroblastoma cells treated with sub-lethal neurite inhibitory concentrations of diazinon: Identification of novel biomarkers of effect. Toxicol Appl Pharmacol 2009; 240:159-165.
78. Carlson K, Ehrich M. Organophosphorus compounds alter F-actin content in SH-SY5Y human neuroblastoma cells. Neuro Toxicology 2001; 22:819-827.
79. El-Fawal HA, Ehrich MF. Calpain activity in organophosphorus-induced delayed neuropathy (OPIDN): Effects of a phenylalkylamine calcium channel blocker. Ann NY Acad Sci 1993; 679:325-329.
80. Song F, Yan Y, Zhao X et al. Neurofilament degradation as an early molecular event in tri-ortho-cresyl phosphate (TOCP) induced delayed neuropathy. Toxicology 2009; 258:94-100.

81. Flaskos J, Harris W, Sachana M et al. The effects of diazinon and cypermethrin on the differentiation of neuronal and glial cell lines. Toxicol Appl Pharmacol 2007; 219:172-180.

82. Gupta RP, Abou-Donia MB. Diisopropylphosphorofluoridate (DFP) treatment alters calcium activated protease activity and cytoskeletal proteins of the sciatic nerve. Brain Res 1995; 677:162-166.

83. Layfield R, Lowe J, Bedford L. The ubiquitin-proteasome system and neurodegenerative disorders. Essays Biochem 2005; 41:157-171.

84. Masoud A, Kiran R, Sandhir R. Impaired mitochondrial functions in organophosphorus induced delayed neuropathy in rats. Cell Mol Neurobiol 2009; 29:1245-1255.

85. Kaur P, Radotra B, Minz RZ et al. Impaired mitochondrial metabolism and neuronal apoptotic cell death after chronic dichlorvos (OP) exposure. Neuro Toxicology 2007; 28:1208-1219.

86. Carlson K, Ehrich M. Organophosphorus compound induced modification of SH-SY5Y human neuroblastoma mitochondrial transmembrane potential. Toxicol Appl Pharmacol 1999; 160:33-42.

87. Milatovic D, Gupta RC, Aschner M. Anticholinesterase activity and oxidative stress. Scientific World J 2006; 6:295-310.

88. Mou DL, Wang YP, Song JF et al. Triorthocresyl phosphate-induced neuronal losses in lumbar spinal cord of hens—an immunohistochemistry and ultrastructure study. Int J Neurosci 2006; 116: 1303-1316.

89. Lockridge O, Schopfer LM. Review of tyrosine and lysine as new motifs for organophosphate binding to proteins that have no active site serine. Chem Biol Interact 2010; 187:344-388.

90. Li B, Schopfer LM, Hinrichs SH et al. Matrix-assisted laser desorption/ionization time of flight mass spectrometry assay for organophosphorus toxicants bound to human albumin at Tyr411. Anal Biochem 2007; 361:263-272.

91. Li B, Schopfer LM, Grigoryan H et al. Tyrosines of human and mouse transferrin covalently labelled by organophosphorus agents: a new motif for binding to proteins that have no active site serine. Toxicol Sci 2009; 107:144-155.

92. Grigoryan H, Lockridge O. Nanoimages show disruption of tubulin polymerisation by chlorpyrifosoxon: implications for neurotoxicity. Toxicol Appl Pharmacol 2009; 240:143-148.

93. Jiang W, Duysen EG, Hansen H et al. Mice treated with chlorpyrifos or chlorpyrifosoxon have organophosphorylated tubulin in the brain and disrupted microtubule structures, suggesting a role for tubulin in neurotoxicity associated with exposure to organophosphorus agents. Toxicol Sci 2010; 115: 183-193.

94. Grigoryan H, Schopfer LM, Thompson CM et al. Mass spectrometry identifies covalent binding of soman, sarin, chlorpyrifosoxon, diisopropylfluorophosphate and FP-biotin to tyrosines on tubulin: a potential mechanism of long term toxicity by organophosphorus agents. Chem Biol Interact 2008; 175: 180-186.

95. Grigoryan H, Schopfer LM, Peeples ES et al. Mass spectrometry identifies multiple organophosphorylated sites on tubulin. Toxicol Appl Pharmacol 2009; 240:149-158.

CHAPTER 16

MITOCHONDRIAL IMPORTANCE IN ALZHEIMER'S, HUNTINGTON'S AND PARKINSON'S DISEASES

Sónia C. Correia,[1,2] Renato X. Santos,[1,2] George Perry,[3,4]
Xiongwei Zhu,*,[3] Paula I. Moreira[1,5] and Mark A. Smith[3]

[1]Center for Neuroscience and Cell Biology of Coimbra, University of Coimbra, Coimbra, Portugal; [2]Faculty of Sciences and Technology, Department of Life Sciences, University of Coimbra, Coimbra, Portugal; [3]Department of Pathology, Case Western Reserve University, Cleveland, Ohio, USA; [4]UTSA Neurosciences Institute and Department of Biology, University of Texas at San Antonio, San Antonio, Texas, USA; [5]Faculty of Medicine, Institute of Physiology, University of Coimbra, Coimbra, Portugal
Corresponding Author: Xiongwei Zhu—Email: xiongwei.zhu@case.edu

EDITOR'S NOTE

Sadly, Dr. Mark A. Smith passed away during the production of this book. Please read the full "in memoriam" located on page xxxi in the front of the book. His innovative thinking and contributions to the scientific community will be greatly missed.

Abstract: Mitochondria have been long known as "gatekeepers of life and death". Indeed, these dynamic organelles are the master coordinators of energy metabolism, being responsible for the generation of the majority of cellular ATP. Notably, mitochondria are also one of the primary producers of intracellular reactive oxygen species which are the main inducer of oxidative damage. Neurons, as metabolically active cells with high energy demands, are predominantly dependent on mitochondrial function, as reflected by the observation that mitochondrial defects are key features of chronic neurodegenerative diseases. Indeed, morphologic, biochemical and molecular genetic studies posit that mitochondria constitute a convergence point for neurodegeneration. Moreover, recent findings convey that neurons are particularly reliant on the dynamic properties of mitochondria, further emphasizing the critical role of mitochondria in neuronal functions. This chapter highlights how mitochondrial pathobiology might contribute to neurodegeneration in Alzheimer's, Parkinson's and Huntington's diseases.

Neurodegenerative Diseases, edited by Shamim I. Ahmad.
©2012 Landes Bioscience and Springer Science+Business Media.

INTRODUCTION

The prevalence of neurodegenerative diseases is rising dramatically due to the increase in life expectancy and demographic changes in the population, representing one of the major health problems. The etiology of most neurodegenerative disorders is complex and multifactorial, involving genetic predisposition, environmental and endogenous factors.[1-3] Nevertheless, mitochondria have emerged as a pivotal "convergence point" for neurodegeneration.[4,5]

Mitochondria are ubiquitous and dynamic organelles involved in many crucial cellular processes in eukaryotic organisms and are considered "gatekeepers of life and death". These organelles have as major functions, the production of over 90% of cellular ATP through the tricarboxylic acid cycle (TCA) cycle and oxidative phosphorylation, regulation of intracellular calcium (Ca^{2+}) and redox signaling and the arbitration of apoptosis.[6-8] Hence, mitochondria possess a notorious significance for neuronal function and survival, since neurons are cells with extremely high energy demands, mitochondrial oxidative phosphorylation being essential for neurons to meet their high energy requirements. That said, neurons are very vulnerable to bioenergetic crisis if there is dysfunction of mitochondrial machinery.[9,10] Dysfunctional mitochondrial energy metabolism culminates in ATP production and Ca^{2+} buffering impairment and exacerbated generation of reactive oxygen species (ROS), including hydrogen peroxide (H_2O_2), hydroxyl radical ($\cdot OH$) and superoxide anion (O_2^{-}).[7] ROS, in turn, cause cell membrane damage via lipid peroxidation and accelerates the high mutation rate of mitochondrial DNA (mtDNA).[11] Additionally, accumulation of mtDNA mutations enhances oxidative damage, induces energy crisis and exacerbates ROS generation, in a vicious cycle.[11] Additionally, the brain is especially prone to oxidative stress-induced damage as a consequence of its high levels of polyunsaturated fatty acids, high oxygen consumption, high content in transition metals and poor antioxidant defenses.[12]

Perturbations in dynamic properties of mitochondria, which include fission, fusion, motility and turnover, can lead to distinctive defects in neurons and are recognized as playing a critical role in neurodegeneration.[13] As a matter of fact, mitochondrial dynamics orchestrate a variety of vital functions required for accurate neuronal function, including maintenance of mitochondrial DNA,[14,15] involvement in apoptosis,[16] formation and function of synapses and dendritic spines and proper distribution of mitochondria.[17-21]

Since mitochondria play a critical role in the regulation of both cell survival and death, mitochondrial dysfunction has been posited to take a center stage in age related-neurodegenerative diseases. Herein, we summarize the current knowledge pertaining to the involvement of mitochondrial malfunction in the onset and progression of neurodegenerative diseases, namely Alzheimer's disease (AD), Parkinson's disease (PD) and Huntington's disease (HD). The insights from in vitro, in vivo and human studies could help to unveil the pathogenic mechanisms underlying mitochondrial dysfunction and to develop new and more effective therapeutic strategies to prevent and/or treat neurodegenerative diseases.

MITOCHONDRIAL DYSFUNCTION IN THE LIMELIGHT OF NEURODEGENERATIVE DISEASES

Alzheimer's Disease

AD is the most common form of dementia among people age 65 and older, affecting more than 35 million people worldwide.[22] Clinically, AD is characterized by a progressive cognitive deterioration, together with impairments in behavior, language and visuospatial skills, culminating in the premature death of the subject.[22] Neuropathologically, AD has as main hallmarks selective neuronal and synaptic loss, deposition of extracellular senile plaques mainly composed of amyloid-β (Aβ) peptide and the presence of intracellular neurofibrillary tangles containing hyperphosphorylated tau protein.[9,23-25]

Since the etiology of AD is complex and multifactorial, several hypotheses have been proposed over the last decades to answer one of the most intriguing questions of the actuality: What is the "culprit" of AD development? With the purpose of explaining many of the biochemical, genetic and pathological features of sporadic AD, Swerdlow and Khan presented the "mitochondrial cascade hypothesis".[26] According to this hypothesis: (1) inheritance determines mitochondrial baseline function and durability; (2) mitochondrial durability influences how mitochondria change with age; and (3) when mitochondrial alterations reach a threshold, AD histopathology and symptoms ensue.[27] Thus, mitochondrial-dependent pathogenic mechanisms are drawing increasing attention for their significant involvement on AD etiopathogenesis.

Energy Hypometabolism, Oxidative Stress and Mitochondrial Dysfunction

It is conceivable that mitochondrial abnormalities that occur in AD result from the complex nature and genesis of oxidative damage in the disease. Indeed, AD patients present reduced metabolic activity, which is believed to be a consequence of oxidative damage to vital mitochondrial components.[28-31] Positron emission tomography imaging studies revealed impaired brain glucose in AD patients, which precedes neuropsychological impairment and atrophy.[32-34] Cerebral glucose utilization is reduced by 45% and cerebral blood flow (CBF) by approximately 20%, in the early stages of AD. However, in the later stages of the disease, metabolic and physiological abnormalities aggravate, resulting in 55-65% reductions in CBF.[35] This decrease in the cerebral glucose metabolism is correlated with the altered expression and decreased activity of mitochondrial energy related proteins, including pyruvate dehydrogenase (PDH), isocitrate dehydrogenase and α-ketoglutarate dehydrogenase, also documented in postmortem AD brain and fibroblasts from AD patients.[36,37] Furthermore, Bubber and collaborators[37] found that all changes in TCA cycle activities (specifically that of PDH complex) correlated with the clinical state, suggesting a coordinated mitochondrial alteration. Moreover, these enzymes are known to be highly susceptible to oxidative modification and are altered by exposure to a range of pro-oxidants.[38]

Accumulating data from in vitro, in vivo and human studies argue that mitochondrial dysfunction and bioenergetics failure are early events implicated in AD pathogenesis. Indeed, impairment in the respiratory chain complexes I, III and IV activities was found in platelets and lymphocytes from AD patients and postmortem AD brain tissue.[39-42] In vitro studies performed in pheochromocytoma cells (PC12) also demonstrated that exposure to Aβ$_{1-40}$ and Aβ$_{25-35}$ potentiated mitochondrial dysfunction characterized by the inhibition of complexes I, III and IV of the mitochondrial respiratory chain.[43] More recently, Fattoretti and collaborators,[44] in order to establish a link between AD and mitochondrial dysfunction, investigated succinic dehydrogenase (SDH) (mitochondrial respiratory complex II) activity in mitochondria of hippocampal CA1 pyramidal neurons obtained from 3xTg-AD mice. The authors observed a decreased density (number of mitochondria/μm^3 of cytoplasm) of SDH-positive mitochondria in 3xTg-AD mice. Data from our laboratory also revealed that AD fibroblasts present high levels of oxidative stress and apoptotic markers when compared with young and age-matched controls.[45] Moreover, AD-type changes could be generated in control fibroblasts using N-methylprotoporphyrin to inhibit cytochrome c oxidase (COX) assembly, which indicates that the observed oxidative damage was associated with mitochondrial dysfunction.[45] Additionally, the effects promoted by the N-methylprotoporphyrine were reversed or attenuated by lipoic acid and *N*-acetyl cysteine.[45] Accordingly, de la Monte and Wands[46] examined postmortem brain tissue from AD patients with different degrees of severity and found that the severity of AD was related to impairments in mitochondrial gene expression, namely in complex IV, increased levels of p53 and molecular indexes of oxidative stress, including NOS and NADPH-oxidase. Thus, mitochondrial malfunction exacerbates oxidative stress and oxidative damage marked by high levels of lipid, protein and nucleic acid oxidation is increased in vulnerable neurons in AD.[47-51] Overall, these findings suggest that mitochondria are important in oxidative damage that occurs in AD and that antioxidant therapies may be promising.

Mitochondrial DNA Mutations

mtDNA mutations have also been implicated in mitochondrial dysfunction in the pathogenesis of AD. For instance, 20 point mutations in the mitochondrial-encoded cytochrome c oxidase subunits I, II and III genes have been detected in AD patients.[52] Qiu and collaborators[53] also identified two missense mutations in the mtDNA of COX in a patient with AD. Further, a high aggregate burden of somatic mtDNA mutations was observed in postmortem brain tissue from AD patients.[54,55]

Mitochondria, Amyloid-β Protein Precursor and Aβ Peptide

Mitochondria were found to be the target both for amyloid-β protein precursor (AβPP) that accumulates in the mitochondrial import channels and for Aβ that interacts with several proteins inside mitochondria and leads to mitochondrial dysfunction. For instance, Aβ was found to impair cellular respiration, energy production and mitochondrial electron chain complexes activity in human neuroblastoma cells.[56] Moreover, cultured neurons isolated from transgenic mice that overexpress a mutant form of AβPP and Aβ-binding alcohol dehydrogenase (ABAD) (Tg mAβPP/ABAD) display spontaneous generation of H_2O_2 and O_2^{-}, decreased ATP, release of cytochrome c and induction of caspase 3-like activity followed by DNA fragmentation and loss of cell viability.[57] A prominent role for mitochondrial O_2^{-} in mediating the effects of Aβ on neuronal function was reported by Massaad and collaborators.[58] In fact, it was previously reported that Aβ enters into mitochondria, compromising their integrity through the inactivation of the manganese superoxide dismutase 2 (SOD-2) and, consequently by increasing mitochondrial superoxide anion levels.[59] Moreover, some studies demonstrate that genetic reduction of SOD-2 in AD model mice can intensify AD symptoms and lead to increased plaque deposition.[60-62] Conversely, the overexpression of SOD-2 reduces hippocampal O_2^{-} and prevents memory deficits in Tg2576 mouse model of AD.[58] Recently, strong evidence for a direct link between free radicals of specific mitochondrial origin and AD-associated vascular and neuronal pathology has been reported.[63] Since SOD-2 is the main O_2^{-} scavenger in mitochondria, these authors showed that its overexpression culminates in the reduction of mitochondrial superoxide and amelioration of CBF deficits and axonal transport deficits typically exhibited by Tg2576 mice.[63] The reduction of mitochondrial superoxide also resulted in a concomitant reduction of phosphorylation of endothelial nitric oxide synthase at serine 1177, as well as phosphorylation of tau at serine 262.[63] Thus, one conclusion from this study is that mitochondrial superoxide is a key player in AD-related vascular and neuronal dysfunction, working as a downstream effector of Aβ, possibly affecting Aβ processing.[63]

Furthermore, generation of ROS is associated with dysfunction at the level of COX[57] (Fig. 1). Similarly, Crouch and colleagues[64] also found that Aβ$_{1-42}$ can disrupt mitochondrial COX activity in a sequence- and conformation-dependent manner. In an in vitro study, designed to explore the effect of the AβPP Swedish double mutation (K670M/N671L) on oxidative stress-induced cell death mechanisms in PC12 cells, increased activity of caspase 3 due to an enhanced activation of both intrinsic and extrinsic apoptotic pathways, including activation of JNK pathway was observed.[65] Moreover, apoptosis was attenuated by SP600125, a JNK inhibitor, through protection of mitochondrial dysfunction and reduction of caspase 9 activity.[65] These findings corroborate the hypothesis that the massive neurodegeneration that develops at an early age in familial AD patients could be a result of an increased vulnerability of neurons through the activation of different apoptotic pathways as a consequence of elevated levels of oxidative stress.

In addition, mitochondrial dysfunction was also linked to the accumulation of full-length and carboxy-terminally truncated AβPP across mitochondrial import channels in brain tissue from AD patients.[66] The authors observed that this accumulation of AβPP inhibited the entrance of nuclear-encoded COX subunits IV and Vb proteins, which was associated with decreased cytochrome c oxidase activity and increased H$_2$O$_2$ levels.[66] Similarly, Anandatheerthavarada et al reported an accumulation of full-length AβPP in the mitochondrial compartment in a transmembrane-arrested form that impaired mitochondrial functionality and energy metabolism.[67] Also, a progressive accumulation of Aβ monomers and oligomers was detected within the mitochondria of both transgenic mice overexpressing mutant AβPP and postmortem brains from AD patients.[64,66,68,69] More recently, Pavlov and coworkers[70] demonstrated that AβPP is a substrate for the mitochondrial γ-secretase and that AβPP intracellular domain (AICD) is produced inside mitochondria, providing a mechanistic view of the mitochondria-associated AβPP metabolism where AICD, P3 peptide and potentially Aβ are produced locally and may contribute to mitochondrial dysfunction in AD. Additionally, Iijima-Ando et al[71] reported that mislocalization of mitochondria underlies the pathogenic effects of Aβ$_{1-42}$ in a transgenic Drosophila model. Indeed, the authors found that in this Aβ$_{1-42}$ model, brain mitochondria were reduced in axons and dendrites and accumulated in the soma without severe mitochondrial damage or neurodegeneration.[71] Notably, perturbations in mitochondrial transport in neurons were sufficient to disrupt protein kinase A (PKA) signaling and induce late-onset behavioral deficits, suggesting a mechanism whereby mitochondrial mislocalization contributes to Aβ$_{1-42}$-induced neuronal dysfunction.[71] A direct link between Aβ-induced toxicity and mitochondrial dysfunction in AD pathology has been suggested by the interaction between mitochondrial Aβ and ABAD.[72,73] Moreover, this interaction was found to induce mitochondrial failure via changes in mitochondrial membrane permeability and a reduction in the activities of enzymes involved in mitochondrial respiration.[73] More recently, Hansson-Petersen et al[74] showed that Aβ peptide is imported into mitochondria via the translocase of the outer membrane import machinery and localized to mitochondrial cristae. Thus, it has been proposed that Aβ species transport to mitochondria cause mitochondrial dysfunction and oxidative damage and consequently damage neurons both structurally and functionally.[64,66,68,69,74] Previous studies from our laboratory also reported an increased susceptibility to mitochondrial permeability transition pore (mPTP) induction promoted by Aβ peptides.[75,76] In accordance, it provided a plausible mechanism underlying Aβ-induced mitochondrial dysfunction, in which Aβ interacts with cyclophilin D, a critical molecule involved in mPTP formation and cell death.[77] Du et al[77] further showed that the interaction of cyclophilin D with mitochondrial Aβ potentiates mitochondrial, neuronal and synaptic stress. Conversely, cyclophilin D ablation protects neurons from Aβ-induced mPTP formation and the resultant mitochondrial and cellular stresses. Along those same lines, cyclophilin D deficiency substantially improves learning and memory and synaptic function in an AD mouse model and alleviates Aβ-mediated reduction of long-term potentiation.[77] Another study reported that the presequence protease (PreP) is responsible for the degradation of the accumulated Aβ in mitochondria, further supporting the association of Aβ with mitochondria and mitochondrial dysfunction in AD.[78]

However, another key role of mitochondria in AD pathogenesis and the close interrelationship of this organelle and the two main pathological features of the disease were recently highlighted. Rhein et al[79] demonstrated that Aβ and tau synergistically impair mitochondrial function and energy homeostasis in 3xTg-AD mice. Accordingly, a previous study demonstrated that transgenic mice overexpressing the P301L mutant human tau protein present alterations of metabolism-related proteins including mitochondrial respiratory chain complexes, antioxidant enzymes and synaptic proteins that are associated with increased oxidative stress.[80] Moreover, mitochondria prepared from these transgenic mice displayed increased vulnerability toward Aβ insult, which reinforce a possible synergistic action of tau and Aβ pathology on the mitochondria.[80] The authors also suggest that tau pathology involves a mitochondrial and oxidative stress disorder possibly distinct from that caused by Aβ.[80]

Omi/HtrA2, a mitochondrial serine protease with chaperone activity, has also been suggested to participate in AD-associated mitochondrial dysfunction. The first evidence for the involvement of Omi/Htr2 in AD was provided by Gray et al[81] that indentified Omi/HtrA2 as a presenilin-1 (PS1)-interacting factor in a yeast two-hybrid screen. Consistently, a following study demonstrated that the C-terminus of PS1 peptide interacts with Omi/HtrA2 and stimulates Omi/HtrA2 protease activity.[82] Additionally, it was observed that Aβ also interacts with Omi/HtrA2, which results in delayed aggregation of the $Aβ_{1-42}$ peptide, indicating that besides its protease activity, Omi/HtrA2 also performs a chaperone function role in the metabolism of intracellular Aβ in AD.[83,84] Mitochondrial AβPP was shown to be a direct cleavage target of Omi/HtrA2,[85] proposing that the regulation of Omi/HtrA2 protease activity may be a therapeutic target in AD by preventing mitochondrial dysfunction caused by AβPP accumulation. More recently, it was also reported that Omi/HtrA2 interacts with PS in active γ-secretase complexes located to mitochondria.[86] Moreover, the authors found reduced AICD production in mitochondria isolated from Omi/HtrA2 knockout mouse embryonic fibroblasts, indicating a significant role of Omi/HtrA2 on γ-secretase activity.[86] Overall, these findings suggest the interactions between mitochondrial Omi/HtrA2 and Aβ, PS, or AβPP are possible links to Omi/HtrA2 in AD. These findings may contribute to a better understanding of the biochemical pathways underlying mitochondrial dysfunction in AD and may help the development of novel mitochondrial-targeted therapeutic strategies.

Mitochondrial Dynamics in Alzheimer's Disease

Ultrastructural alterations in mitochondrial morphology such as reduced size and broken internal membrane cristae were also documented in brains from AD patients.[30,87]

One reasonable explanation for these observations could be the increased mitochondrial autophagy found in AD.[88,89] Another consequence of Aβ on mitochondria is the induction of dynamic changes, including mitochondrial fission/fusion perturbations. Wang and collaborators[90] reported abnormal mitochondrial fission and fusion in fibroblasts from sporadic AD patients, marked by lower levels of dynamin-related protein 1 (Drp1), a key regulator of mitochondrial fission. The authors also observed that AD fibroblasts display elongated mitochondria which form collapsed perinuclear networks.[90,91] Accordingly, AβPP overexpression in M17 neuroblastoma cells resulted in predominantly fragmented mitochondria, decreased Drp1 and optic atrophy protein 1 (OPA1) levels and a defect in neuronal differentiation.[92] Moreover, reduced expression levels of Drp1, OPA1, mitofusin (Mfn)1 and 2 and increased mitochondria fission protein 1 (Fis1) levels were found in hippocampal tissues from AD patients compared with age-matched controls.[93] These results suggest that AD is characterized by mitochondrial fission/fusion imbalance and consequently mitochondrial fragmentation and abnormal distribution, which potentiates mitochondrial and neuronal dysfunction in this neurodegenerative disease.

Synaptic Defects in Mitochondria in Alzheimer's Disease

Synaptic defects and disruption of axonal transport have also been documented in AD pathobiology.[94-97] Indeed, a previous study reported that a brief exposure of cultured hippocampal neurons to soluble Aβ molecules resulted in rapid and severe impairment of mitochondrial transport, independent of cell death and other drastic alterations of cellular structures.[98] Similarly, it was reported that soluble oligomers of Aβ are responsible for an abnormal axonal transport of mitochondria in primary hippocampal neurons, most likely contributing to an abnormal mitochondrial distribution.[99] More recently, it was proposed that mitochondrial localization to dendritic spines may be important for the trafficking of the alpha-amino-3-hydroxy-5-methyl-4-isoxazolepropionic acid receptors (AMPARs), major ionotropic glutamate receptors involved in excitatory synaptic transmission, and Aβ disruption of mitochondrial trafficking could contribute to AMPAR removal and trafficking defects leading to synaptic inhibition.[100]

Parkinson's Disease

PD, the most frequent movement disorder, affects approximately 2% of individuals over 65 years of age and is clinically characterized by three phenotypic aspects: resting tremor, bradykinesia and rigidity. PD is caused by a progressive and massive loss of the dopaminergic neurons within the substantia nigra pars compacta (SNc) and the consequent depletion of the neurotransmitter dopamine (DA) in the striatum, which is required for normal motor function. One of the pathological hallmarks of PD and related synucleinopathies is the presence of intracellular inclusions called Lewy bodies, which are constituted of aggregates of the presynaptic soluble protein called α-synuclein.[101-105] The majority of PD cases are sporadic with unknown cause; however, mutations in several genes have been linked to familial form of PD.[104] Nonetheless, mitochondrial dysfunction is emerging as a key mechanism underlying the pathogenesis of both sporadic and familial forms of PD.[106,107]

In early 1990s, it was reported for the first time that there is reduced activity of the mitochondrial respiratory complex I (NADH-quinone oxidoreductase) in the SNc of PD patients.[108] In accordance, subsequent studies also reported an impairment of mitochondrial complex I activity in the substantia nigra,[109] platelets,[110-113] lymphocytes,[114,115] and skeletal muscle tissue[116,117] from PD patients. More recently, in highly purified mitochondria, there is a PD-specific complex I deficit in the frontal cortex.[118] Meanwhile, SNc appears to be more susceptible to complex I activity impairment than other brain regions, possibly due to the exacerbated ROS generated within dopaminergic neurons as a result of DA metabolism and iron content.[119] Consistently, cybrids containing mtDNA from PD patients present a significant impairment in complex I activity associated with increased oxidative stress levels,[120] suggesting mtDNA encoded defects in PD. Moreover, Lewy bodies within these cybrids also react positively with cytochrome c antibodies, suggesting a mitochondrial origin.[121] The use of specific complex I inhibitors, such as 1-methyl-4-phenyl-1,2,3,6-tetrahydrodropyridine (MPTP), rotenone and 6-hydroxydopamine (6-OHDA) which causes degeneration of the nigral dopaminergic neurons and PD symptoms in in vivo models, further emphasizes the involvement of mitochondria in the etiology of PD.[122-124] Furthermore, a proteomic analysis of mitochondria-enriched fractions from postmortem PD SNc revealed differential expression of multiple mitochondrial proteins in PD brain as compared to control, including complex I subunits.[125] Moreover, in vitro incubation of isolated rat brain mitochondria with recombinant human α-synuclein was shown to potentiate a dose-dependent loss of mitochondrial transmembrane potential ($\Delta\Psi$m) and phosphorylation capacity[126] (Fig. 1). However, α-synuclein did not affect the activities of respiratory chain complexes, suggesting that the former may impair mitochondrial bioenergetics by direct effect on mitochondrial membranes.[126] Finally, mortalin, a mitochondrial stress protein, is substantially decreased in PD brains and cellular models of PD,[125] being shown that the manipulation of mortalin levels in dopaminergic neurons resulted in significant alteration in sensitivity to PD phenotypes via pathways involving mitochondrial and proteasomal function as well as oxidative stress.[125]

Evidence from the literature also posits a role for mutations in genes encoding both mitochondrially targeted proteins and proteins involved in mitochondrial function and/ or oxidative stress responses in PD.[127] Indeed, mitochondrial DNA haplotype analysis revealed that certain haplogroups reduced the risk for PD, which indicated that mtDNA may contribute to PD etiology.[128] Moreover, Swerdlow and collaborators[129] reported maternally inherited mutations in mtDNA in one family with PD. Using a novel single-molecule PCR approach to quantify the total burden of mtDNA molecules with deletions, it was also shown that a high proportion of individual pigmented neurons in the aged human SNc contain very high levels of mtDNA deletions.[130] The fraction of mtDNA deletions is significantly higher in COX-deficient neurons than in COX-positive neurons, suggesting that mtDNA deletions may be directly responsible for impaired cellular respiration.[130] More recently, Ekstrand and collaborators[131] created conditional knockout "MitoPark" mice, which have a disrupted mitochondrial transcription factor A (Tfam) gene in dopaminergic neurons. These knockout mice have reduced mtDNA expression and respiratory chain deficiency in midbrain dopaminergic neurons, which lead to a Parkinsonism phenotype with adult onset and characterized by slowly progressive impairment of motor function accompanied by the formation of intraneuronal inclusions and dopamine nerve cell death.[131]

Familial forms of PD are associated with mutations in leucine-rich repeat kinase 2 (LRRK2), α-synuclein, parkin, DJ1 and PTEN-induced putative kinase 1 (PINK1), these proteins being associated with the mitochondrial outer membrane and involved in ROS production or defense[132] (Fig. 1). HtrA2 is another protein that is mutated

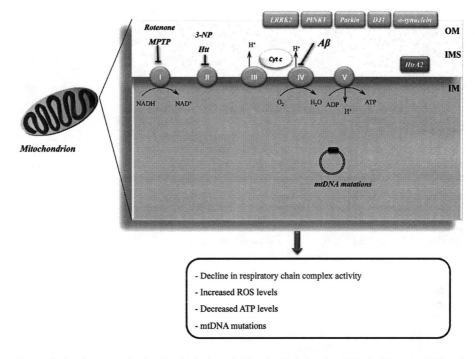

Figure 1. Involvement of mitochondrial abnormalities in Alzheimer's (AD), Parkinson's (PD) and Huntington's (HD) diseases. Impairment of the activity of respiratory chain complex IV by amyloid-β peptide (Aβ), leading to the exacerbation of reactive oxygen species (ROS) generation and ATP levels depletion, is one prominent feature of AD. Mitochondrial DNA (mtDNA) mutations also play a role in the pathogenesis of AD. Concerning PD, there is an extensively documented impairment of mitochondrial complex I activity. Indeed, the use of pharmacological inhibitors of complex I, including rotenone and 1-methyl-4-phenyl-1,2,3,6-tetrahydrodropyridine (MPTP), causes degeneration of the nigral dopaminergic neurons and PD symptoms in in vivo models. In addition, the familial forms of PD are associated with mutations in leucine-rich repeat kinase 2 (LRRK2), α-synuclein, parkin, DJ1 and PTEN-induced putative kinase 1 (PINK1), these proteins being associated with the mitochondrial outer membrane (OM) and involved in ROS production or defense. Also, HtrA2, which is localized in the intermembrane space (IMS) of mitochondria is mutated in familial PD. In HD, mutant huntingtin (Htt) induces a decline of mitochondrial respiration, particularly affecting mitochondrial respiratory complex II activity and ATP synthesis. The use of 3-nitropropionic acid (3-NP), a selective inhibitor of succinate dehydrogenase and complex II, was shown to recapitulate HD-like symptoms in several vertebrate models. Furthermore, mutant Htt is a causative factor of mtDNA damage in HD, suggesting that mtDNA damage is an early biomarker for HD-associated neurodegeneration. ADP- adenosine diphosphate; Cyt c- cytochrome c; IM- inner membrane; NAD+- oxidized nicotinamide adenine dinucleotide; NADH- reduced nicotinamide adenine dinucleotide; H+- proton.

in familial PD and localizes in the intermembrane space of mitochondria.[132] In vitro cell culture studies showed that mutant PINK1 or PINK1 knock-down induce mitochondrial respiration, ATP synthesis and proteasome function impairment and increased α-synuclein aggregation.[133] Additionally, it was reported that HtrA2 loss results in transcriptional upregulation of nuclear genes characteristic of the integrated stress response, including the transcription factor C/EBP homologous protein (CHOP), selectively in the brain.[134] HtrA2 loss also induces accumulation of unfolded proteins in the mitochondria, defective mitochondrial respiration and enhanced ROS production, which contribute to the induction of CHOP expression and neuronal cell death.[134] Previous studies also showed that the overexpression of α-synuclein in cell culture and in transgenic mice impairs mitochondrial function and increases the susceptibility to mPTP induction.[135,136] In contrast, α-synuclein-null mice are resistant to respiratory chain inhibitors implicating an involvement of mitochondria in α-synuclein-mediated toxicity.[137,138] Recently, compelling evidence demonstrated that both PINK1 and Parkin mediate the degradation of damaged mitochondria via selective autophagy (mitophagy),[139-142] the voltage-dependent anion channel 1 emerging as a target of PINK1/Parkin-mediated mitophagy.[143] Indeed, it was suggested that Parkin, together with PINK1, modulates mitochondrial trafficking, especially to the perinuclear region, a subcellular area associated with autophagy. In this way, mutations in either Parkin or PINK1 could culminate in altering mitochondrial turnover which, in turn, may cause the accumulation of defective mitochondria and, ultimately, neurodegeneration. In fact, Geisler and collaborators[144] demonstrated that PINK1 mutations compromise the selective degradation of depolarized mitochondria, mainly due to the decreased physical binding activity of PD-linked PINK1 mutations to Parkin. Thus, PINK1 mutations abrogate autophagy of impaired mitochondria upstream of Parkin. In addition to compromised PINK1 kinase activity, reduced binding of PINK1 to Parkin leads to failure in Parkin mitochondrial translocation, resulting in the accumulation of damaged mitochondria, contributing to the pathogenesis of PD.[144]

Moreover, the PINK1/Parkin pathway also regulates the mitochondrial integrity and morphology via the fission/fusion machinery. Genetic studies in *Drosophila* also demonstrated that the PINK1/Parkin pathway promotes mitochondrial fission and that the loss of mitochondrial and tissue integrity in PINK1 and parkin mutants derives from reduced mitochondrial fission.[145] Accordingly, the PINK1/parkin pathway promotes mitochondrial fission and/or inhibits fusion by negatively regulating Mfn and OPA1 function and/or positively regulating drp1 in *Drosophila*.[146] Moreover, Lutz and collaborators[147] demonstrated that Parkin- or PINK1-deficient SH-SY5Y cells showed a significant increase in the percentage of cells with truncated or fragmented mitochondria along with a decrease in cellular ATP production. The mitochondrial phenotype could morphologically and functionally be prevented by the enhanced expression of Mfn2, OPA1, or dominant negative Drp1, suggesting that a decrease in mitochondrial fusion or an increase in fission is associated with a loss of parkin or PINK1 function.[147] Notably, genetic manipulations and treatment with the small molecule mitochondrial division inhibitor (mdivi-1), which inhibits Drp1, both structural and functional mitochondrial defects induced by mutant PINK1 were attenuated, highlighting a potential therapeutic strategy for PD.[148]

Huntington's Disease

HD is an autosomal dominant neurological disorder caused by an abnormal polyglutamine (polyQ) expansion within a single gene, huntingtin (Htt), leading to the progressive loss of striatal and cortical neurons and consequent decline of cognitive and motor functions.[149] Several lines of evidence indicate that the expression of mutant Htt is associated with mitochondrial dysfunction, in both HD patients and mouse transgenic HD models[150] (Fig. 1). A pronounced decrease in glucose metabolism and a corresponding increase in lactate were documented in affected brain regions of

HD patients, which suggest a bioenergetic defect.[151] In addition, impaired activity of mitochondrial respiratory complexes II, III and IV was found in postmortem brain of HD patients.[152] Similarly, striatal cells from mtHtt mice exhibit impairment of mitochondrial respiration and ATP synthesis.[153] Conversely, it was observed that the expression of complex II subunits in striatal neurons expressing mutant Htt exon 1 restores complex II respiratory activity and protects against cell death.[151] Panov and collaborators[154] also observed that mitochondria isolated from lymphocytes of HD patients have lowered buffering capacity and their $\Delta\Psi$m depolarizes earlier at lower Ca^{2+} concentrations, proposing that mitochondrial Ca^{2+} abnormalities occur early in HD pathogenesis and may be a direct effect of mutant Htt on the organelle. To further emphasize the role of mitochondrial respiratory chain inhibition in HD pathogenesis, it has been shown that the use of 3-nitropropionic acid (3-NP), a selective inhibitor of SDH and complex II, recapitulates the loss of medium spiny neurons in the substantia nigra and HD-like symptoms in several vertebrate models.[155,156] Additionally, humans exposed to 3-NP also exhibit similar motor dysfunction to that found in HD patients.[156-158] Evidence from the literature also demonstrated that HD patients had higher frequencies of mtDNA deletions in lymphocytes in comparison to the controls, which suggest that CAG repeats instability and mutant Htt are causative factors in mtDNA damage.[159] More recently, Acevedo-Torres and coworkers[160] suggested that mtDNA damage is an early biomarker for HD-associated neurodegeneration, supporting the hypothesis that mtDNA lesions might contribute to HD pathogenesis.

Ultrastructral changes in mitochondria were also reported in HD, raising the possibility for interplay between HD and mitochondrial dynamics.[161] In fact, in rat cortical neurons it was demonstrated that 3-NP exposure leads to fragmentation and condensation of mitochondria, which can be prevented by antioxidant treatment.[162] Accordingly, Wang and collaborators[163] found that mitochondria in HeLa cells over-expressing a mutant Htt with a 74 glutamine repeat (Htt74Q) show fragmentation of mitochondria, reduced mitochondrial fusion, reduced ATP and increased cell death. On the other hand, expression of either dominant-negative Drp1 or Mfn2 restores ATP levels and attenuates cell death.[163] Additionally, mutant Htt was shown to promote a mitochondrial morphologic alteration from an elongated to a round phenotype, which correlates with a blockage in mitochondrial movement.[164,165]

Overall, mitochondrial impairment plays a key role in HD pathogenesis, such that expression of mtHtt culminates in abnormal mitochondrial ultrastructure, impaired Ca^{2+} buffering, bioenergetic defects and mtDNA deletions and damage.

CONCLUSION

Altogether, we highlighted here the clear role of mitochondrial abnormalities, including disturbances in mitochondrial machinery, dynamics and turnover in the onset and/or progression of neurodegenerative diseases, including AD, PD and HD. Mitochondrial disturbances provide a common target for combating the various abnormalities caused by the specific protein substrates of the genetic mutations, the resulting energy imbalance and the increased ROS. That said, it will be important to dissect all the key mitochondrial-dependent pathogenic mechanisms underlying neurodegeneration, which can be useful to develop new therapeutic interventions to prevent and/or mitigate age-related neurodegenerative diseases.

REFERENCES

1. Correia SC, Moreira PI. Hypoxia-inducible factor 1: a new hope to counteract neurodegeneration? J Neurochem 2010; 112:1-12.
2. Przedborski S, Vila M, Jackson-Lewis V. Neurodegeneration: what is it and where are we? J Clin Invest 2003; 111:3-10.
3. Migliore L, Coppede F. Environmental-induced oxidative stress in neurodegenerative disorders and aging. Mutat Res 2009; 674:73-84.
4. Lin MT, Beal MF. Mitochondrial dysfunction and oxidative stress in neurodegenerative diseases. Nature 2006; 443:787-795.
5. Moreira PI, Zhu X, Wang X et al. Mitochondria: a therapeutic target in neurodegeneration. Biochim Biophys Acta 2010; 1802:212-220.
6. Green DR, Kroemer G. The pathophysiology of mitochondrial cell death. Science 2004; 305:626-629.
7. Beal MF. Mitochondria take center stage in aging and neurodegeneration. Ann Neurol 2005; 58:495-505.
8. Mattson MP, Gleichmann M, Cheng A. Mitochondria in neuroplasticity and neurological disorders. Neuron 2008; 60:748-766.
9. Moreira PI, Duarte AI, Santos MS et al. An integrative view of the role of oxidative stress, mitochondria and insulin in Alzheimer's disease. J Alzheimers Dis 2009; 16:741-761.
10. Murphy AN, Fiskum G, Beal MF. Mitochondria in neurodegeneration: bioenergetic function in cell life and death. J Cereb Blood Flow Metab 1999; 19:231-245.
11. Petrozzi L, Ricci G, Giglioli NJ et al. Mitochondria and neurodegeneration. Biosci Rep 2007; 27:87-104.
12. Nunomura A, Honda K, Takeda A et al. Oxidative damage to RNA in neurodegenerative diseases. J Biomed Biotechnol 2006:82323.
13. Chen H, Chan DC. Mitochondrial dynamics—fusion, fission, movement and mitophagy—in neurodegenerative diseases. Hum Mol Genet 2009; 18:R169-176.
14. Parone PA, Da Cruz S, Tondera D et al. Preventing mitochondrial fission impairs mitochondrial function and leads to loss of mitochondrial DNA. PLoS One 2008; 3:e3257.
15. Westermann B. Merging mitochondria matters: cellular role and molecular machinery of mitochondrial fusion. EMBO Rep 2002; 3:527-531.
16. Suen DF, Norris KL, Youle RJ. Mitochondrial dynamics and apoptosis. Genes Dev 2008; 22:1577-1590.
17. Li Z, Okamoto K, Hayashi Y et al. The importance of dendritic mitochondria in the morphogenesis and plasticity of spines and synapses. Cell 2004; 119:873-887.
18. Chen H, McCaffery JM, Chan DC. Mitochondrial fusion protects against neurodegeneration in the cerebellum. Cell 2007; 130:548-562.
19. Liu QA, Shio H. Mitochondrial morphogenesis, dendrite development and synapse formation in cerebellum require both Bcl-w and the glutamate receptor delta2. PLoS Genet 2008; 4:e1000097.
20. Stowers RS, Megeath LJ, Gorska-Andrzejak J et al. Axonal transport of mitochondria to synapses depends on milton, a novel Drosophila protein. Neuron 2002; 36:1063-1077.
21. Verstreken P, Ly CV, Venken KJ et al. Synaptic mitochondria are critical for mobilization of reserve pool vesicles at Drosophila neuromuscular junctions. Neuron 2005; 47:365-378.
22. Querfurth HW, LaFerla FM. Alzheimer's disease. N Engl J Med 2010; 362:329-344.
23. Selkoe DJ. Alzheimer's disease results from the cerebral accumulation and cytotoxicity of amyloid beta-protein. J Alzheimers Dis 2001; 3:75-80.
24. Moreira PI, Honda K, Zhu X et al. Brain and brawn: parallels in oxidative strength. Neurology 2006; 66:S97-101.
25. Moreira PI, Santos MS, Oliveira CR. Alzheimer's disease: a lesson from mitochondrial dysfunction. Antioxid Redox Signal 2007; 9:1621-1630.
26. Swerdlow RH, Khan SM. A "mitochondrial cascade hypothesis" for sporadic Alzheimer's disease. Med Hypotheses 2004; 63:8-20.
27. Swerdlow RH, Khan SM. The Alzheimer's disease mitochondrial cascade hypothesis: an update. Exp Neurol 2009; 218:308-315.
28. Aksenov MY, Tucker HM, Nair P et al. The expression of key oxidative stress-handling genes in different brain regions in Alzheimer's disease. J Mol Neurosci 1998; 11:151-164.
29. Aliev G, Smith MA, Obrenovich ME et al. Role of vascular hypoperfusion-induced oxidative stress and mitochondria failure in the pathogenesis of Azheimer disease. Neurotox Res 2003; 5:491-504.
30. Hirai K, Aliev G, Nunomura A et al. Mitochondrial abnormalities in Alzheimer's disease. J Neurosci 2001; 21:3017-3023.
31. Anderson GL, Limacher M, Assaf AR et al. Effects of conjugated equine estrogen in postmenopausal women with hysterectomy: the Women's Health Initiative randomized controlled trial. JAMA 2004; 291:1701-1712.

32. Azari NP, Pettigrew KD, Schapiro MB et al. Early detection of Alzheimer's disease: a statistical approach using positron emission tomographic data. J Cereb Blood Flow Metab 1993; 13:438-447.
33. Small GW, Komo S, La Rue A et al. Early detection of Alzheimer's disease by combining apolipoprotein E and neuroimaging. Ann N Y Acad Sci 1996; 802:70-78.
34. Silverman DH, Small GW, Chang CY et al. Positron emission tomography in evaluation of dementia: Regional brain metabolism and long-term outcome. JAMA 2001; 286:2120-2127.
35. Hoyer S, Nitsch R. Cerebral excess release of neurotransmitter amino acids subsequent to reduced cerebral glucose metabolism in early-onset dementia of Alzheimer type. J Neural Transm 1989; 75:227-232.
36. Huang HM, Ou HC, Xu H et al. Inhibition of alpha-ketoglutarate dehydrogenase complex promotes cytochrome c release from mitochondria, caspase-3 activation and necrotic cell death. J Neurosci Res 2003; 74:309-317.
37. Bubber P, Haroutunian V, Fisch G et al. Mitochondrial abnormalities in Alzheimer brain: mechanistic implications. Ann Neurol 2005; 57:695-703.
38. Tretter L, Adam-Vizi V. Inhibition of Krebs cycle enzymes by hydrogen peroxide: A key role of [alpha]-ketoglutarate dehydrogenase in limiting NADH production under oxidative stress. J Neurosci 2000; 20:8972-8979.
39. Kish SJ, Bergeron C, Rajput A et al. Brain cytochrome oxidase in Alzheimer's disease. J Neurochem 1992; 59:776-779.
40. Parker WD, Jr., Mahr NJ, Filley CM et al. Reduced platelet cytochrome c oxidase activity in Alzheimer's disease. Neurology 1994; 44:1086-1090.
41. Bosetti F, Brizzi F, Barogi S et al. Cytochrome c oxidase and mitochondrial F1F0-ATPase (ATP synthase) activities in platelets and brain from patients with Alzheimer's disease. Neurobiol Aging 2002; 23:371-376.
42. Valla J, Schneider L, Niedzielko T et al. Impaired platelet mitochondrial activity in Alzheimer's disease and mild cognitive impairment. Mitochondrion 2006; 6:323-330.
43. Pereira C, Santos MS, Oliveira C. Mitochondrial function impairment induced by amyloid beta-peptide on PC12 cells. Neuroreport 1998; 9:1749-1755.
44. Fattoretti P, Balietti M, Casoli T et al. Decreased numeric density of succinic dehydrogenase-positive mitochondria in CA1 pyramidal neurons of 3xTg-AD mice. Rejuvenation Res 2010; 13:144-147.
45. Moreira PI, Harris PL, Zhu X et al. Lipoic acid and N-acetyl cysteine decrease mitochondrial-related oxidative stress in Alzheimer disease patient fibroblasts. J Alzheimers Dis 2007; 12:195-206.
46. de la Monte SM, Wands JR. Molecular indices of oxidative stress and mitochondrial dysfunction occur early and often progress with severity of Alzheimer's disease. J Alzheimers Dis 2006; 9:167-181.
47. Castellani RJ, Harris PL, Sayre LM et al. Active glycation in neurofibrillary pathology of Alzheimer disease: N(epsilon)-(carboxymethyl) lysine and hexitol-lysine. Free Radic Biol Med 2001; 31:175-180.
48. Nunomura A, Perry G, Pappolla MA et al. RNA oxidation is a prominent feature of vulnerable neurons in Alzheimer's disease. J Neurosci 1999; 19:1959-1964.
49. Nunomura A, Perry G, Aliev G et al. Oxidative damage is the earliest event in Alzheimer disease. J Neuropathol Exp Neurol 2001; 60:759-767.
50. Smith MA, Richey Harris PL, Sayre LM et al. Widespread peroxynitrite-mediated damage in Alzheimer's disease. J Neurosci 1997; 17:2653-2657.
51. Straface E, Matarrese P, Gambardella L et al. Oxidative imbalance and cathepsin D changes as peripheral blood biomarkers of Alzheimer disease: a pilot study. FEBS Lett 2005; 579:2759-2766.
52. Hamblet NS, Ragland B, Ali M et al. Mutations in mitochondrial-encoded cytochrome c oxidase subunits I, II and III genes detected in Alzheimer's disease using single-strand conformation polymorphism. Electrophoresis 2006; 27:398-408.
53. Qiu X, Chen Y, Zhou M. Two point mutations in mitochondrial DNA of cytochrome c oxidase coexist with normal mtDNA in a patient with Alzheimer's disease. Brain Res 2001; 893:261-263.
54. Lin MT, Simon DK, Ahn CH et al. High aggregate burden of somatic mtDNA point mutations in aging and Alzheimer's disease brain. Hum Mol Genet 2002; 11:133-145.
55. Coskun PE, Beal MF, Wallace DC. Alzheimer's brains harbor somatic mtDNA control-region mutations that suppress mitochondrial transcription and replication. Proc Natl Acad Sci USA 2004; 101:10726-10731.
56. Rhein V, Baysang G, Rao S et al. Amyloid-beta leads to impaired cellular respiration, energy production and mitochondrial electron chain complex activities in human neuroblastoma cells. Cell Mol Neurobiol 2009; 29:1063-1071.
57. Takuma K, Yao J, Huang J et al. ABAD enhances Abeta-induced cell stress via mitochondrial dysfunction. FASEB J 2005; 19:597-598.
58. Massaad CA, Washington TM, Pautler RG et al. Overexpression of SOD-2 reduces hippocampal superoxide and prevents memory deficits in a mouse model of Alzheimer's disease. Proc Natl Acad Sci USA 2009; 106:13576-13581.

59. Anantharaman M, Tangpong J, Keller JN et al. Beta-amyloid mediated nitration of manganese superoxide dismutase: implication for oxidative stress in a APPNLH/NLH X PS-1P264L/P264L double knock-in mouse model of Alzheimer's disease. Am J Pathol 2006; 168:1608-1618.

60. Melov S, Adlard PA, Morten K et al. Mitochondrial oxidative stress causes hyperphosphorylation of tau. PLoS ONE 2007; 2:e536.

61. Esposito L, Raber J, Kekonius L et al. Reduction in mitochondrial superoxide dismutase modulates Alzheimer's disease-like pathology and accelerates the onset of behavioral changes in human amyloid precursor protein transgenic mice. J Neurosci 2006; 26:5167-5179.

62. Li F, Calingasan NY, Yu F et al. Increased plaque burden in brains of APP mutant MnSOD heterozygous knockout mice. J Neurochem 2004; 89:1308-1312.

63. Massaad CA, Amin SK, Hu L et al. Mitochondrial superoxide contributes to blood flow and axonal transport deficits in the Tg2576 mouse model of Alzheimer's disease. PLoS One 2010; 5:e10561.

64. Crouch PJ, Blake R, Duce JA et al. Copper-dependent inhibition of human cytochrome c oxidase by a dimeric conformer of amyloid-beta1-42. J Neurosci 2005; 25:672-679.

65. Marques CA, Keil U, Bonert A et al. Neurotoxic mechanisms caused by the Alzheimer's disease-linked Swedish amyloid precursor protein mutation: oxidative stress, caspases and the JNK pathway. J Biol Chem 2003; 278:28294-28302.

66. Devi L, Prabhu BM, Galati DF et al. Accumulation of amyloid precursor protein in the mitochondrial import channels of human Alzheimer's disease brain is associated with mitochondrial dysfunction. J Neurosci 2006; 26:9057-9068.

67. Anandatheerthavarada HK, Biswas G, Robin MA et al. Mitochondrial targeting and a novel transmembrane arrest of Alzheimer's amyloid precursor protein impairs mitochondrial function in neuronal cells. J Cell Biol 2003; 161:41-54.

68. Caspersen C, Wang N, Yao J et al. Mitochondrial Abeta: a potential focal point for neuronal metabolic dysfunction in Alzheimer's disease. FASEB J 2005; 19:2040-2041.

69. Manczak M, Anekonda TS, Henson E et al. Mitochondria are a direct site of A beta accumulation in Alzheimer's disease neurons: implications for free radical generation and oxidative damage in disease progression. Hum Mol Genet 2006; 15:1437-1449.

70. Pavlov PF, Wiehager B, Sakai J et al. Mitochondrial {gamma}-secretase participates in the metabolism of mitochondria-associated amyloid precursor protein. FASEB J 2010;

71. Iijima-Ando K, Hearn SA, Shenton C et al. Mitochondrial mislocalization underlies Abeta42-induced neuronal dysfunction in a Drosophila model of Alzheimer's disease. PLoS One 2009; 4:e8310.

72. Yan SD, Stern DM. Mitochondrial dysfunction and Alzheimer's disease: role of amyloid-beta peptide alcohol dehydrogenase (ABAD). Int J Exp Pathol 2005; 86:161-171.

73. Lustbader JW, Cirilli M, Lin C et al. ABAD directly links Abeta to mitochondrial toxicity in Alzheimer's disease. Science 2004; 304:448-452.

74. Hansson Petersen CA, Alikhani N, Behbahani H et al. The amyloid beta-peptide is imported into mitochondria via the TOM import machinery and localized to mitochondrial cristae. Proc Natl Acad Sci USA 2008; 105:13145-13150.

75. Moreira PI, Santos MS, Moreno A et al. Amyloid beta-peptide promotes permeability transition pore in brain mitochondria. Biosci Rep 2001; 21:789-800.

76. Moreira PI, Santos MS, Moreno A et al. Effect of amyloid beta-peptide on permeability transition pore: a comparative study. J Neurosci Res 2002; 69:257-267.

77. Du H, Guo L, Fang F et al. Cyclophilin D deficiency attenuates mitochondrial and neuronal perturbation and ameliorates learning and memory in Alzheimer's disease. Nat Med 2008; 14:1097-1105.

78. Falkevall A, Alikhani N, Bhushan S et al. Degradation of the amyloid beta-protein by the novel mitochondrial peptidasome, PreP. J Biol Chem 2006; 281:29096-29104.

79. Rhein V, Song X, Wiesner A et al. Amyloid-beta and tau synergistically impair the oxidative phosphorylation system in triple transgenic Alzheimer's disease mice. Proc Natl Acad Sci USA 2009; 106:20057-20062.

80. David DC, Hauptmann S, Scherping I et al. Proteomic and functional analyses reveal a mitochondrial dysfunction in P301L tau transgenic mice. J Biol Chem 2005; 280:23802-23814.

81. Gray CW, Ward RV, Karran E et al. Characterization of human HtrA2, a novel serine protease involved in the mammalian cellular stress response. Eur J Biochem 2000; 267:5699-5710.

82. Gupta S, Singh R, Datta P et al. The C-terminal tail of presenilin regulates Omi/HtrA2 protease activity. J Biol Chem 2004; 279:45844-45854.

83. Park HJ, Seong YM, Choi JY et al. Alzheimer's disease-associated amyloid beta interacts with the human serine protease HtrA2/Omi. Neurosci Lett 2004; 357:63-67.

84. Kooistra J, Milojevic J, Melacini G et al. A new function of human HtrA2 as an amyloid-beta oligomerization inhibitor. J Alzheimers Dis 2009; 17:281-294.

85. Park HJ, Kim SS, Seong YM et al. Beta-amyloid precursor protein is a direct cleavage target of HtrA2 serine protease. Implications for the physiological function of HtrA2 in the mitochondria. J Biol Chem 2006; 281:34277-34287.
86. Behbahani H, Pavlov PF, Wiehager B et al. Association of Omi/HtrA2 with gamma-secretase in mitochondria. Neurochem Int 2010; 57:668-675.
87. Baloyannis SJ. Mitochondrial alterations in Alzheimer's disease. J Alzheimers Dis 2006; 9:119-126.
88. Moreira PI, Siedlak SL, Wang X et al. Increased autophagic degradation of mitochondria in Alzheimer disease. Autophagy 2007; 3:614-615.
89. Moreira PI, Siedlak SL, Wang X et al. Autophagocytosis of mitochondria is prominent in Alzheimer disease. J Neuropathol Exp Neurol 2007; 66:525-532.
90. Wang X, Su B, Fujioka H et al. Dynamin-like protein 1 reduction underlies mitochondrial morphology and distribution abnormalities in fibroblasts from sporadic Alzheimer's disease patients. Am J Pathol 2008; 173:470-482.
91. Wang X, Su B, Zheng L et al. The role of abnormal mitochondrial dynamics in the pathogenesis of Alzheimer's disease. J Neurochem 2009; 109 Suppl 1:153-159.
92. Wang X, Su B, Siedlak SL et al. Amyloid-beta overproduction causes abnormal mitochondrial dynamics via differential modulation of mitochondrial fission/fusion proteins. Proc Natl Acad Sci USA 2008; 105:19318-19323.
93. Wang X, Su B, Lee HG et al. Impaired balance of mitochondrial fission and fusion in Alzheimer's disease. J Neurosci 2009; 29:9090-9103.
94. Lassmann H, Fischer P, Jellinger K. Synaptic pathology of Alzheimer's disease. Ann N Y Acad Sci 1993; 695:59-64.
95. Blennow K, Bogdanovic N, Alafuzoff I et al. Synaptic pathology in Alzheimer's disease: relation to severity of dementia, but not to senile plaques, neurofibrillary tangles, or the ApoE4 allele. J Neural Transm 1996; 103:603-618.
96. Stokin GB, Lillo C, Falzone TL et al. Axonopathy and transport deficits early in the pathogenesis of Alzheimer's disease. Science 2005; 307:1282-1288.
97. Shankar GM, Li S, Mehta TH et al. Amyloid-beta protein dimers isolated directly from Alzheimer's brains impair synaptic plasticity and memory. Nat Med 2008; 14:837-842.
98. Rui Y, Tiwari P, Xie Z et al. Acute impairment of mitochondrial trafficking by beta-amyloid peptides in hippocampal neurons. J Neurosci 2006; 26:10480-10487.
99. Wang X, Perry G, Smith MA et al. Amyloid-beta-derived diffusible ligands cause impaired axonal transport of mitochondria in neurons. Neurodegener Dis 2010; 7:56-59.
100. Rui Y, Gu J, Yu K et al. Inhibition of AMPA receptor trafficking at hippocampal synapses by beta-amyloid oligomers: the mitochondrial contribution. Mol Brain 2010; 3:10.
101. de Rijk MC, Rocca WA, Anderson DW et al. A population perspective on diagnostic criteria for Parkinson's disease. Neurology 1997; 48:1277-1281.
102. de Lau LM, Giesbergen PC, de Rijk MC et al. Incidence of parkinsonism and Parkinson disease in a general population: the Rotterdam Study. Neurology 2004; 63:1240-1244.
103. Cardoso SM, Moreira PI, Agostinho P et al. Neurodegenerative pathways in Parkinson's disease: therapeutic strategies. Curr Drug Targets CNS Neurol Disord 2005; 4:405-419.
104. Bueler H. Impaired mitochondrial dynamics and function in the pathogenesis of Parkinson's disease. Exp Neurol 2009; 218:235-246.
105. Hardy J, Cai H, Cookson MR et al. Genetics of Parkinson's disease and parkinsonism. Ann Neurol 2006; 60:389-398.
106. Banerjee R, Starkov AA, Beal MF et al. Mitochondrial dysfunction in the limelight of Parkinson's disease pathogenesis. Biochim Biophys Acta 2009; 1792:651-663.
107. Schapira AH. Mitochondria in the aetiology and pathogenesis of Parkinson's disease. Lancet Neurol 2008; 7:97-109.
108. Schapira AH, Cooper JM, Dexter D et al. Mitochondrial complex I deficiency in Parkinson's disease. J Neurochem 1990; 54:823-827.
109. Mann VM, Cooper JM, Daniel SE et al. Complex I, iron and ferritin in Parkinson's disease substantia nigra. Ann Neurol 1994; 36:876-881.
110. Parker WD, Jr., Boyson SJ, Parks JK. Abnormalities of the electron transport chain in idiopathic Parkinson's disease. Ann Neurol 1989; 26:719-723.
111. Krige D, Carroll MT, Cooper JM et al. Platelet mitochondrial function in Parkinson's disease. The Royal Kings and Queens Parkinson Disease Research Group. Ann Neurol 1992; 32:782-788.
112. Haas RH, Nasirian F, Nakano K et al. Low platelet mitochondrial complex I and complex II/III activity in early untreated Parkinson's disease. Ann Neurol 1995; 37:714-722.

113. Blandini F, Nappi G, Greenamyre JT. Quantitative study of mitochondrial complex I in platelets of parkinsonian patients. Mov Disord 1998; 13:11-15.
114. Barroso N, Campos Y, Huertas R et al. Respiratory chain enzyme activities in lymphocytes from untreated patients with Parkinson disease. Clin Chem 1993; 39:667-669.
115. Yoshino H, Nakagawa-Hattori Y, Kondo T et al. Mitochondrial complex I and II activities of lymphocytes and platelets in Parkinson's disease. J Neural Transm Park Dis Dement Sect 1992; 4:27-34.
116. Taylor DJ, Krige D, Barnes PR et al. A 31P magnetic resonance spectroscopy study of mitochondrial function in skeletal muscle of patients with Parkinson's disease. J Neurol Sci 1994; 125:77-81.
117. Penn AM, Roberts T, Hodder J et al. Generalized mitochondrial dysfunction in Parkinson's disease detected by magnetic resonance spectroscopy of muscle. Neurology 1995; 45:2097-2099.
118. Parker WD, Jr., Parks JK, Swerdlow RH. Complex I deficiency in Parkinson's disease frontal cortex. Brain Res 2008; 1189:215-218.
119. Chinta SJ, Andersen JK. Redox imbalance in Parkinson's disease. Biochim Biophys Acta 2008; 1780:1362-1367.
120. Veech GA, Dennis J, Keeney PM et al. Disrupted mitochondrial electron transport function increases expression of anti-apoptotic bcl-2 and bcl-X(L) proteins in SH-SY5Y neuroblastoma and in Parkinson disease cybrid cells through oxidative stress. J Neurosci Res 2000; 61:693-700.
121. Trimmer PA, Borland MK, Keeney PM et al. Parkinson's disease transgenic mitochondrial cybrids generate Lewy inclusion bodies. J Neurochem 2004; 88:800-812.
122. Betarbet R, Sherer TB, MacKenzie G et al. Chronic systemic pesticide exposure reproduces features of Parkinson's disease. Nat Neurosci 2000; 3:1301-1306.
123. Gash DM, Rutland K, Hudson NL et al. Trichloroethylene: Parkinsonism and complex 1 mitochondrial neurotoxicity. Ann Neurol 2008; 63:184-192.
124. Sherer TB, Richardson JR, Testa CM et al. Mechanism of toxicity of pesticides acting at complex I: relevance to environmental etiologies of Parkinson's disease. J Neurochem 2007; 100:1469-1479.
125. Jin J, Hulette C, Wang Y et al. Proteomic identification of a stress protein, mortalin/mthsp70/GRP75: relevance to Parkinson disease. Mol Cell Proteomics 2006; 5:1193-1204.
126. Banerjee K, Sinha M, Pham Cle L et al. Alpha-synuclein induced membrane depolarization and loss of phosphorylation capacity of isolated rat brain mitochondria: implications in Parkinson's disease. FEBS Lett 2010; 584:1571-1576.
127. Thomas B, Beal MF. Parkinson's disease. Hum Mol Genet 2007; 16 Spec No. 2:R183-194.
128. Pyle A, Foltynie T, Tiangyou W et al. Mitochondrial DNA haplogroup cluster UKJT reduces the risk of PD. Ann Neurol 2005; 57:564-567.
129. Swerdlow RH, Parks JK, Davis JN, 2nd et al. Matrilineal inheritance of complex I dysfunction in a multigenerational Parkinson's disease family. Ann Neurol 1998; 44:873-881.
130. Kraytsberg Y, Kudryavtseva E, McKee AC et al. Mitochondrial DNA deletions are abundant and cause functional impairment in aged human substantia nigra neurons. Nat Genet 2006; 38:518-520.
131. Ekstrand MI, Terzioglu M, Galter D et al. Progressive parkinsonism in mice with respiratory-chain-deficient dopamine neurons. Proc Natl Acad Sci USA 2007; 104:1325-1330.
132. Knott AB, Perkins G, Schwarzenbacher R et al. Mitochondrial fragmentation in neurodegeneration. Nat Rev Neurosci 2008; 9:505-518.
133. Liu W, Vives-Bauza C, Acin-Perez R et al. PINK1 defect causes mitochondrial dysfunction, proteasomal deficit and alpha-synuclein aggregation in cell culture models of Parkinson's disease. PLoS ONE 2009; 4:e4597.
134. Moisoi N, Klupsch K, Fedele V et al. Mitochondrial dysfunction triggered by loss of HtrA2 results in the activation of a brain-specific transcriptional stress response. Cell Death Differ 2009; 16:449-464.
135. Song DD, Shults CW, Sisk A et al. Enhanced substantia nigra mitochondrial pathology in human alpha-synuclein transgenic mice after treatment with MPTP. Exp Neurol 2004; 186:158-172.
136. Hsu LJ, Sagara Y, Arroyo A et al alpha-synuclein promotes mitochondrial deficit and oxidative stress. Am J Pathol 2000; 157:401-410.
137. Klivenyi P, Siwek D, Gardian G et al. Mice lacking alpha-synuclein are resistant to mitochondrial toxins. Neurobiol Dis 2006; 21:541-548.
138. Dauer W, Kholodilov N, Vila M et al. Resistance of alpha -synuclein null mice to the parkinsonian neurotoxin MPTP. Proc Natl Acad Sci USA 2002; 99:14524-14529.
139. Vives-Bauza C, Zhou C, Huang Y et al. PINK1-dependent recruitment of Parkin to mitochondria in mitophagy. Proc Natl Acad Sci USA 2010; 107:378-383.
140. Kawajiri S, Saiki S, Sato S et al. PINK1 is recruited to mitochondria with parkin and associates with LC3 in mitophagy. FEBS Lett 2010; 584:1073-1079.
141. Matsuda N, Sato S, Shiba K et al. PINK1 stabilized by mitochondrial depolarization recruits Parkin to damaged mitochondria and activates latent Parkin for mitophagy. J Cell Biol 2010; 189:211-221.

142. Narendra DP, Jin SM, Tanaka A et al. PINK1 is selectively stabilized on impaired mitochondria to activate Parkin. PLoS Biol 2010; 8:e1000298.
143. Geisler S, Holmstrom KM, Skujat D et al. PINK1/Parkin-mediated mitophagy is dependent on VDAC1 and p62/SQSTM1. Nat Cell Biol 2010; 12:119-131.
144. Geisler S, Holmstrom KM, Treis A et al. The PINK1/Parkin-mediated mitophagy is compromised by PD-associated mutations. Autophagy 2010; 6:871-878.
145. Poole AC, Thomas RE, Andrews LA et al. The PINK1/Parkin pathway regulates mitochondrial morphology. Proc Natl Acad Sci USA 2008; 105:1638-1643.
146. Deng H, Dodson MW, Huang H et al. The Parkinson's disease genes pink1 and parkin promote mitochondrial fission and/or inhibit fusion in Drosophila. Proc Natl Acad Sci USA 2008; 105:14503-14508.
147. Lutz AK, Exner N, Fett ME et al. Loss of parkin or PINK1 function increases Drp1-dependent mitochondrial fragmentation. J Biol Chem 2009; 284:22938-22951.
148. Cui M, Tang X, Christian WV et al. Perturbations in mitochondrial dynamics induced by human mutant PINK1 can be rescued by the mitochondrial division inhibitor mdivi-1. J Biol Chem 2010; 285:11740-11752.
149. Bates GP. History of genetic disease: the molecular genetics of Huntington disease—a history. Nat Rev Genet 2005; 6:766-773.
150. Bossy-Wetzel E, Petrilli A, Knott AB. Mutant huntingtin and mitochondrial dysfunction. Trends Neurosci 2008; 31:609-616.
151. Jenkins BG, Andreassen OA, Dedeoglu A et al. Effects of CAG repeat length, HTT protein length and protein context on cerebral metabolism measured using magnetic resonance spectroscopy in transgenic mouse models of Huntington's disease. J Neurochem 2005; 95:553-562.
152. Gu M, Gash MT, Mann VM et al. Mitochondrial defect in Huntington's disease caudate nucleus. Ann Neurol 1996; 39:385-389.
153. Milakovic T, Johnson GV. Mitochondrial respiration and ATP production are significantly impaired in striatal cells expressing mutant huntingtin. J Biol Chem 2005; 280:30773-30782.
154. Panov AV, Gutekunst CA, Leavitt BR et al. Early mitochondrial calcium defects in Huntington's disease are a direct effect of polyglutamines. Nat Neurosci 2002; 5:731-736.
155. Brouillet E, Hantraye P. Effects of chronic MPTP and 3-nitropropionic acid in nonhuman primates. Curr Opin Neurol 1995; 8:469-473.
156. Rubinsztein DC. Lessons from animal models of Huntington's disease. Trends Genet 2002; 18:202-209.
157. Cattaneo E, Rigamonti D, Goffredo D et al. Loss of normal huntingtin function: new developments in Huntington's disease research. Trends Neurosci 2001; 24:182-188.
158. Sipione S, Cattaneo E. Modeling Huntington's disease in cells, flies and mice. Mol Neurobiol 2001; 23:21-51.
159. Banoei MM, Houshmand M, Panahi MS et al. Huntington's disease and mitochondrial DNA deletions: event or regular mechanism for mutant huntingtin protein and CAG repeats expansion?! Cell Mol Neurobiol 2007; 27:867-875.
160. Acevedo-Torres K, Berrios L, Rosario N et al. Mitochondrial DNA damage is a hallmark of chemically induced and the R6/2 transgenic model of Huntington's disease. DNA Repair (Amst) 2009; 8:126-136.
161. Squitieri F, Cannella M, Sgarbi G et al. Severe ultrastructural mitochondrial changes in lymphoblasts homozygous for Huntington disease mutation. Mech Ageing Dev 2006; 127:217-220.
162. Liot G, Bossy B, Lubitz S et al. Complex II inhibition by 3-NP causes mitochondrial fragmentation and neuronal cell death via an NMDA- and ROS-dependent pathway. Cell Death Differ 2009; 16:899-909.
163. Wang H, Lim PJ, Karbowski M et al. Effects of overexpression of huntingtin proteins on mitochondrial integrity. Hum Mol Genet 2009; 18:737-752.
164. Chang DT, Rintoul GL, Pandipati S et al. Mutant huntingtin aggregates impair mitochondrial movement and trafficking in cortical neurons. Neurobiol Dis 2006; 22:388-400.
165. Trushina E, Dyer RB, Badger JD, 2nd et al. Mutant huntingtin impairs axonal trafficking in mammalian neurons in vivo and in vitro. Mol Cell Biol 2004; 24:8195-8209.

CHAPTER 17

MULTIPLE SCLEROSIS

Elżbieta Miller

Neurorehabilitation Ward, III General Hospital of Lodz, Lodz, Poland, and Department of Chemistry and Clinical Biochemistry, University of Bydgoszcz, Bydgoszcz, Poland
Email: betty.miller@interia.pl

Abstract: Multiple sclerosis (MS) is a chronic, complex neurological disease with a variable clinical course in which several pathophysiological mechanisms such as axonal/neuronal damage, demyelination, inflammation, gliosis, remyelination and repair, oxidative injury and excitotoxicity, alteration of the immune system as well as biochemical disturbances and disruption of blood-brain barrier are involved.[1,2] Exacerbations of MS symptoms reflect inflammatory episodes, while the neurodegenerative aspects of gliosis and axonal loss result in the progression of disability. The precise aetiology of MS is not yet known, although epidemiological data indicate that it arises from a complex interactions between genetic susceptibility and environmental factors.[3] In this chapter the brain structures and processes involved in immunopathogenesis of MS are presented. Additionaly, clinical phenotypes and biomarkers of MS are showed.

INTRODUCTION

The worldwide prevalence of multiple sclerosis (MS) is estimated at between 1.1 and 2.5 million cases. MS is the most common neurological disease in young adults (between 20-40 years) characterized by recurrent relapses and/or progression within the central nervous system (CNS).[4] MS is a complex disease, with a variable clinical course in which several pathophysiological mechanisms such as axonal/neuronal damage, demyelination, inflammation, gliosis, remyelination and repair, oxidative stress and excitotoxicity, alteration of the immune system and disruption of blood-brain barrier are involved.[2,5] The autoimmune diseases arise from a complex interactions between genetic susceptibility and environmental factors. The geographic distribution of MS is uneven. A greater frequency is observed between 40 and 60 degrees north and south latitude.[6,7]

Geographically MS describes three frequency zones. High frequency areas have MS prevalence of 120:100,000 population, these include: Western and Northern Europe, Canada, Russia, Israel, Parts of Northern US, New Zealand and South-East Australia. Medium frequency 50:100,000 and these include southern US, most of Australia, South Africa, the southern Mediterranean basin, Russia into Siberia, the Ukraine and parts of Latin America.[5,8] Zones with a very low disease frequency of 5:100,000 population are: Asia, Sub-Saharan Africa and South America. Migrants from high to lower risk areas retain the MS risk of their birth place only if they are at least age 15 at migration.[5] Based on epidemiological considerations it is the relatively high risk of MS in the industrialised nations due to hygiene-related changes in the sequence of common infections resulting in the emergence of patterns of immune reactivity that either cause or fail to protect against the development of MS.[9-11]

In the MS pathogenesis genetic factors play very important roles. The most consistent finding in case-control genetic MS studies was the association with the major histocompatibility complex (MHC) (also called human leukocyte antigen-HLA) class II on chromosome 6p21; DR15, DQ6, Dw2 haplotype. Two genes, *HLA-DRB1* and *IL7R* (*CD127*), have been unambiguously associated with the disease susceptibility following their identification as candidates by function.[12-14]

SYMPTOMS AND THE CLINICAL PHENOTYPES OF MS

MS is variable in onset and progression. At onset of the disease the most common symptoms are impaired vision due to optic neuritis (inflammation of the optic nerve) and deficits in sensation (or over-sensation as burning or prickling).[15] The mature form of MS shows other symptoms including paresis and paralysis, ataxia, fatigue, spasticy and incontinence.[16] Cognitive impairment (difficulties with memory, concentration and other mental skills) also occurs frequently.[17] Diagnosis of MS is primarily made on clinical grounds based on the presence of multiple neurological deficits that cannot be explained by one localized CNS pathology. Symptoms must appear at more than one occasion (dissemination in time and space) for a diagnosis to be established. Radiological and laboratory tests, such as computerized tomography scan (CTS), magnetic resonance imaging (MRI), cerebrospinal fluid analyses for immunoglobulin and oligoclonal banding are helpful but only confirm clinical observations.[16] MS is classified into subtypes: Relapsing-remitting (RRMS), secondary progressive (SPMS) and primary progressive (PPMS) disorders.

Initially, more than 80%[2,4] of individuals with MS have a RRMS disease course with defined clinical exacerbations of neurologic symptoms, followed by complete or incomplete remission. RRMS is dominated by multifocal inflammation, oedema and the physiologic actions of cytokines.[2] After 10-20 years, or median age of about 39.1 years, about half of those with RRMS gradually accumulate irreversible neurological deficits in the absence of clinical relapses or new white matter lesions detected by MRI. This stage is known as SPMS which is no longer characterized by clinical attacks and remissions but by insidious progression of clinical symptoms (Fig. 1).[18-21] The remaining 20%, with progressive clinical deterioration from the onset of the disease, have PPMS. PPMS is characterized, in general, by less inflammation and earlier and more sustained axonal loss. The age at the beginnig of the PPMS is around 40 years or later than in RRMS.[22-25] Axonal injury is responsible for the transition from RRMS to progressive forms of the disease. A significantly rarer form is progressive relapsing MS, which initially presents

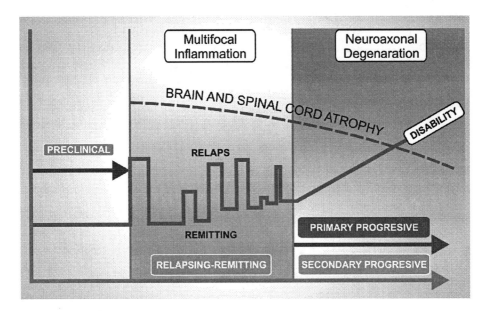

Figure 1. Pathogenical and clinical model of multiple sclerosis subtypes.

as PPMS; however, during the course of the disease these individuals develop true neurologic exacerbations.[26] Individuals with SPMS who have clinical exacerbations, followed by incomplete remission, are included in this category.[27] Axonal transection is observed in the early stage of the disease and in some cases even earlier than that (clinically isolated syndromes or CIS).[21,22] PPMS and SPMS are dominated by neuroaxonal degeneration and correlate with disability and brain and spinal cord atrophy (Fig. 1).[4,23] The most marked atrophy occurs in SPMS.[4] It is difficult to predict the clinical course of this disease.[27,28] Progression of disability seems to be increased in patients with higher number of relapses during the first and second year of the disease.[29,30]

STRUCTURES OF BRAIN INVOLVED IN IMMUNOPATHOLOGY OF MS

Blood-Brain Barier

MS is a disease of CNS where blood-brain barier (BBB) is a key structure.[31-33] The existence of BBB was first intimated by Paul Ehrlich in 1885, although the term *Bluthirnschranke* (blood-brain barrier) was first used as early as 1900 by Lewandowski (existence and cellular structure was debated well in 1960).[34,35]

The BBB is a selective physical barrier that regulates transport of blood between the circulation and CNS parenchyma. This structure is composed of capillaries surrounded by perivascular macrophages and astrocytic endfeet and in this way astrocytes can directly modulate BBB function (Fig. 2). Capillaries in CNS presents tight junctions. It allows diffusion of small gaseous molecules ($NO^{\bullet}$, O_2^- and CO_2) and smaller lipophilic compounds that easily can pass through the endothelial cell membrane across the BBB but limits the entry of large hydrophilic molecules such as proteins. Large molecules

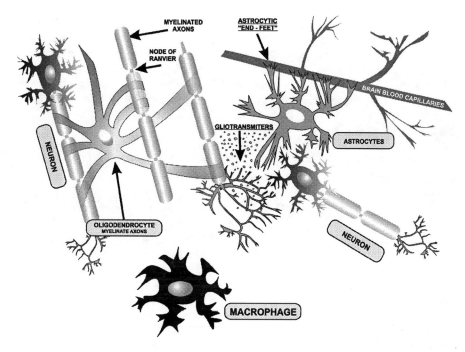

Figure 2. Brain structures involved in multiple sclerosis (MS) immunopathology.

can pass through the BBB via specific transporters located on CNS endothelial cells.[36] The wall thickness of brain capillaries is approximately 40% of that in other types of endothelial cell. It could be an adapation to the restrictive permeability of the BBB.[37]

Astrocyte

Astrocytes constitute approximately 90% of the human brain and support neural transmission, release neuromodulatory factors into the extracellular space and modulation of neurotransmission. Astrocytes play also a critical role in maintaining survival of neurons and other glia (Fig. 2). Communication among astrocytes follows from their syncytial arrangement and occurs via Ca^{2+} channels. Neuronal stimulation can initiate microvascular and endothelial cell responses via these glial elements.[38] Astrocytes release interleukine (IL)-6, tumor necrosis factor-α (TNF-α) and IL-1β and can tighten the BBB via tumor growth factor-β (TGF-β) secretion.[36]

Microglia

In healthy brain microglia consists mainly of macrophages and retract sampling and monitoring their microenvironment. Activated microglia, mainly by macrophages, secretes a variety of inflammatory and oxidative stress mediators including cytokines (TNF and IL-1β and IL-6) and chemokines (macrophage inflammatory protein MIP-1α, monocyte chemoattractant protein, MCP-1 and interferon (IFN) inducible protein IP-10) that promote the inflammatory state. The morphology of macrophages changes from ramified to amoeboid due to their phagocytic role.[39] These moderately active microglia are thought to perform beneficial functions, such as scavenging neurotoxins and reactive oxygen species (ROS), removing dying cells and cellular debris and secreting trophic factors that promote neuronal survival. Persistent activation of brain-resident microglia may increase the permeability of the BBB and promote increased infiltration of peripheral macrophages, the phenotype of which is critically determined by the CNS environment.[40]

Oligodendrocyte and Axon

Oligodendrocytes myelinate axons, increase axonal stability and induce local accumulation and phosphorylation of neurofilaments within the axon. Neuronal function is further influenced by oligodendrocyte-derived soluble factors that induce sodium channel clustering, necessary for powerfull conduction along axons and maintain this clustering even in the absence of direct axon-glial contact[41,42] (Fig. 2). Neurofilaments, the major axonal cytoskeleton proteins are consist of three components that differ in molecular size: A light chain, N-L, an intermediate chain, N-M and a heavy chain, N-H. The second major component of the axonal cytoskeleton is the microtubule, which is up 100 µm in length and consists of tubulin (α and β) subunits.[41] Actin is the major component of the microfilaments. Axons contain a large volume of membranes because of their elongated shape.[42,43] Cholesterol is the main lipid in these membranes and 24S-hydroxycholesterol is a cholesterol metabolite specific to the brain. Reduced concentrations of oxysterols are associated with brain atrophy especially with cognitive impairment.[44]

IMMUNOPATHOGENESIS OF MS

MS pathogenesis is a complex autoimmunological process, is built to a large extent on results obtained from experimental autoimmune encephalomyelitis (EAE).[45] The CNS is both immune competent and actively interactive with the peripheral immune system.[46] Potentially autoaggressive T cells exist in normal immune system. Once activated, myelin-specific T cells (activation outside the CNS) (*phase 1-activation*) can cross the BBB and migrate into the CNS (*phase 2-adhesion; phase 3-connection; phase 4-penetration*) where they become locally re-activated (*phase 5-re-activation*). After that re-activated T cells proliferate and secrete pro-inflammotory cytokines which stimulate microglia, macrophages and astrocytes and recruited B cells, ultimately result in demyelinisation, damage oligodendrocytes and axons (*phase 6-myelin injury*) with concomitant neurological deficits[47] (Fig. 3).

The most possibile hypothesis how these autoreactive myelin-specific T cells become activated are molecular mimicry processes.[48] T cells recognize short (10-20 residue) linear peptides that are derived from limited proteolysis of myelin proteins. When such peptides are presented to T cells, together with MHC on the surface of antigen-presenting cell (APC), they become activated where T cells respond to environmental antigens, that resemble self-antigens, could be a potential mechanism by which these cells get activated.[49,50] The idea of molecular mimicry is that short linear peptides, derived from viral components (a proposed etiology for MS) are sufficiently similar to myelin peptide to generate autoreactive responses.[51,52] MBP-specific CD4+ T cells are considered to be initiators of MS, but clonal expansion of CD8+ cells has been detected in MS lesions.

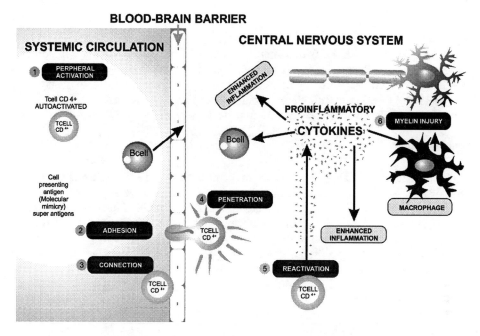

Figure 3. Pathogenesis of multiple sclerosis (MS) cascade.

CD4[+] T cells come in two varietes, TH1 and TH2. Recently discovered other types of T cells, so-called regulatory T cells (Treg.), suppress and control the naturally existing autoreactive immune cells. Another new subtype of T cells, TH17, play significant role as pathogenic effector T cells in EAE. Activated T cells express high levels of molecules such as very late antigen-4 (VLA-4) and leukocyte-function-associated antigen-1 (LFA-1) that facilitate their transmigration across the BBB.[53-55]

DISRUPTION OF BBB

Pathogenesis of migrating T cells across BBB consists of 4 phases: Phase 1-intravascular 'crawling' in which T cells do not 'roll' with the blood stream along the endothelium, rather actively crawl, often in a direction against the blood stream. Phase 2-diapedesis in which the crawling T cells stop and squeeze through the vessel wall. Phase 3-perivascular activation in which T cells crawl along the outside of the blood vessel. Phase 4-invasion of CNS parenchyma. During the course of its extravascular crawling the T cells encounter a perivascular antigen-presenting cell, e.g., a dendritic cell (competent APCs). If the antigen-specific receptor of the T cell recognizes 'its' antigen on the surface of dendritic cell, the T cells are re-activated and thereby become 'licensed' to leave the vessel area and penetrate into the surrounding tissues.[47] In the process of penetreting autoreactive cells into CNS are involved: Matrix metalloproteinases (MMPs), chemokines, cytokines, adhesive molecules and intergrins. Here an important step is activation of peripheral immune system that leads to BBB weakness by systemic production of inflammatory mediators.[56]

Metalloproteinases (MMPs)

MMPs are extracellular matrix remodeling neutral proteases which are important in normal development, angiogenesis, wound repair and a wide range of pathological processes.[57,58] MMP-9 can directly attack myelin components such as myelin-basic protein and facilitate T-cell migration across brain microvascular endothelial cells.[59] Tissue inhibitor of MMPs1 (TIMP-1) is present in patients with severe RRMS.[60,61]

Chemokines

Chemokines are small-molecular-weight chemotactic cytokines that attract leukocytes to sites of infection and inflammation. The expression of chemokines, like cytokines, is often triggered by inflammatory mediators, such as TNF, IFN-γ, microbial toxins, or trauma, although constitutively expressed chemokines, such as CXCL1 and CXCL12, are endogenous to neurons and astrocytes, respectively.[36,62] Chemokine receptors are present not only on inflammatory cells, but also on astrocytes, oligodendrocytes and neurons and are involved not only in chemoattraction but also neuronal development, modulation of cell adhesion, phagocytosis, T-cell differentiation, apoptosis and angiogenesis.[63] Analysis of MS tissue has revealed elevated levels of CXCL9, CXCL10, CCL2, CCL7 and CCL8 in reactive astrocytes within demyelinating lesions. Some receptors such as CXCR3 on lymphocytes correlate with tissue destruction, while others, such as CCR5 on macrophages and microglia, are associated with early remyelination.[63,64] CCR2 expression is elevated in T cells from patients with SPMS more than in those with RRMS and several studies have shown that both serum and cerebro-spinal fluid (CSF) levels of CCL2 are lower in RRMS patients. CCR2+ cells migrate across a BBB model with greater efficiency than CCR2 cells, suggesting that CCL2/CCR2 may be important in the pathogenesis of MS.[65]

Cytokines

Cytokines and their receptors are expressed in CNS cells and are important for development and function of the brain. There are proinflammatory cytokines IL-1β, IL-2, IL-6, IL-12, IL-18, TNF-α, IFN-γ and anti-inflammatory IL-4 and IL-10 (can be produced by macrophages, B cells).[36,66] IL-10, as an immunomodulator, inhibits Th1 proliferation. Some cytokines have pleiotropic activities such as IFN-β and even proinflammatory cytokines possess anti-inflammatory properties and vice versa. Both, infiltrating immune cells and resident glial cells whose activities are associated with inflammation, may also contribute to repair and regeneration through the secretion of neuroprotective factors, such as leukemia inhibitory factor (LIF), transforming growth factor-β (TGF-β) and ciliary neurotrophic factor (CNTF).

TNF-α is mostly known to exert inflammatory and neurotoxic effects and is also involved in normal development and function of the brain. TNF-α is highly expressed in the embryonic brain. IFN-γ is a major cytokine involved in demyelinating pathologies. Overexpression of IFN-γ in white matter of transgenic mice results in myelin ablation.[36,67] Evidence from a toxic demyelination model suggests that low levels of IFN-γ may also have surprisingly important role in protection against severe myelin loss. Although the beneficial effect of IFN-β in preventing deterioration of demyelinating diseases is well documented, little is known about its mechanism of action on myelin in the CNS.[68-70]

Adhesion Molecules

Two adhesion molecules, in particular ICAM-1 (intracellular adhesion molecule-1) and VCAM-1 (vascular cell adhesion molecule-1), play crucial roles in the extravacation and homing of T cells through the CNS parenchyma expecially in acute active plaques and ICAM-1 and VCAM-1 bind LFA-1 (CD11a/CD18) and VLA-4 (very late antigen-4) on activated lymphocytes, respectively. In MS tissue, ICAM-1 expression was observed in astrocytes at the edges of demyelinated lesions.[71,72] Leukocyte function-associated antigen-1 (LFA-1) controls the function and, in particular, the migration of immune cells.[73,74]

AXONAL INJURY

Axonal loss is significantly associated with disease progression and disability in MS.[1,4,8,21] Axonal transection can be achieved by inflammatory mediators released as part of the immune activation and lack of neurotrophic factors lead to the axons, by oligodendrocytes, in chronic demyelinated plaques.[17-20]

Oligodendrocytes produce protein Nogo which interacts with the Nogo receptor present on axons to inhibit neurite outgrowth. Since the identification of Nogo and several other molecules in myelin have been identified that suppress neurite, sprouting in the adult CNS including myelin-associated glycoprotein (MAG) and oligodendrocyte myelin glycoprotein (OMgp), both also stimulate Nogo receptors. Neurite sprouting is suppressed by Nogo and its related molecules. This process prevents formation of aberrant connections and regrowth and repair transected axons in demyelinated brain tissues. Recovery from an MS exacerbation is mainly the result of the re-establishment of transmission along the demyelinated axons by redistribution of sodium channels, permitting some degree of restoration of neuronal conduction.[75,76.]

REMYELINATION AND REPAIR

Remyelination and repair is observed in acute and also in chronic phase of MS and is connected with oligodendrocytes function. Endogenous remyelination occurs in white and grey matter in MS and fully remyelinated areas are often referred as "shadow plaques" by neuropathologists.[77,78]

Remyelination occurs in two major phases. The first phase consists of colonization of lesions by oligodendrocyte progenitor cells (OPCs). The second is the differentiation of OPCs into myelinating oligodendrocytes that connect demyelinated axons to generate functional myelin sheaths. Several intracellular and extracellular molecules such as brain-derived neurothrofic factor (BDNF), nerve growth factor (NGF) and insulin-like growth factor (IGF-1) have been identified that mediate these two phases of repair.[79]

GLIOSIS

Gliosis is the proliferation of astrocytes in the central nervous system in response to injury which results in scar formation. Astrocytes react to injury by hypertrophy and up-regulation of the glial-fibrillary acidic protein. Gliosis, along with neuronal loss, is a prominent feature of MS.[79-81]

OXIDATIVE STRESS AND EXCITITOXICITY

Reactive oxygen species (ROS) can damage lipids, proteins and nucleic acids in cells causing cell death of various cell types including the CNS.[82,83] ROS, leading to oxidative stress (OS), generated in excess primarily by macrophages, have been implicated as mediators of demyelination and axonal damage in MS.[84,85] ROS cause damage to cardinal cellular components resulting in cell death by necrosis or apoptosis.[86] CNS is particularly susceptible to ROS-induced damage due to the high oxygen demands of the brain and the concentration of antioxidants may not be as high as the demand. Brain contains antioxidants enzymes, catalase (CAT), glutathione peroxidase (GPx), glutathione reductase (GR), superoxide dismutase (SOD) and non-enzymatic such as antioxidants glutathione, vitamins A,C,D, co-enzym Q and uric acid etc.[87,88] Enzymatic and non-enzymatic antioxidants -like vitamins, micro and macro elements can regulate progress and function of different immunological cells. Extremely fast production in CNS of several ROS including: O_2^-, $HO_2^·$, H_2O_2, $^·OH$ and $NO^·$ (mainly produced by macrophages structures responsible for demyelinisation and axons disruption) takes place.[89]

Nitric Oxide

Nitric oxide (NO) is a free radical gas that at physiological concentrations is essential for many cellular processes such as neurotransmission, differentiation and signal transduction.[87] These physiological processes are regulated by NO produced by nNO synthetase (nNOS) at steady state concentrations from ~50 nM to ~500 nM. Neurons produce NO at concentration of 33 nM during normal physiological functions (low flux NO). Activated microglial and astrocytes can produce NO at steady state concentrations as high as 1 μM (high flux NO). NO, released during CNS pathology, can react with O_2^- and form reactive nitrogen species (RNS), such as peroxynitrite (ONOO⁻) which cause damage to a variety of macromolecules including proteins. ONOO⁻ is formed in a reaction that is limited only by the diffusion rates of the molecules.[90,91] Peroxynitrite has been associated with damage to neurons and is a pathogenic factor in MS. NO produced by glial cells within MS lesions also has been shown to directly reduce axonal conduction. ONOO⁻ can mediate a variety of destructive interactions including oxidation, lipid peroxidation, DNA strand breaks and nitration of amino acids, mainly tyrosine residues in proteins. Uric acid (UA) is a natural scavenger of ONOO⁻.[91] Peroxynitrite-dependent nitration of tyrosine residues, forming 3-nitrotyrosine (3NY), disrupts protein structure and function, thereby interrupting or altering cell signaling. Nitrotyrosine is found in the CNS of patients with MS and is considered a footprint for peroxynitrite mediated damage in the cell. In MS, progression and severity is tightly associated with levels of reactive nitrogen species (RNS) such as NO and peroxynitrite in the cerebrospinal fluid (CSF) and blood serum.[92]

Endogenous Antioxidants in Prevention of MS

Several enzymes, including SOD, GPx, glutatione reductase and catalase are endogenous antioxidants that possess specific free radical scavenging properties and reduce ROS levels in brain. Three types of SODs exist in brain cell. CuZn-SOD is a cytosolic enzyme that requires both copper and zinc ions as cofactors. Mn-SOD is a mitochondrial enzyme with requirements for Mn^{2+}. A copper-containing SOD is present in the extracellular space. The CuZn-SOD has been used extensively to reduce brain injury induced by ischemia. The short half-life of CuZn-SOD (6 minutes) in circulating blood and its failure to pass through the BBB make it difficult to use this enzyme in clinical therapy.[93] The glutathione levels in a phase of remission and at an exacerbation, both in RRMS as well as SPMS are low. It is accompanied by the very low activity of erythrocyte glutathione reductase, the only enzyme which provides restoration of oxidized glutathione. The compensatory activation of serum catalase was observed only in exacerbation of RRMS.[87,94] Polysaturated fatty acids, the major components of neuron membranes are higly susceptible to ROS attack and results in lipid peroxidation products. In MS lipid peroxidation is incresed.[95]

In MS low molecular weight antioxidants such as immunoregulatory vitamins C, D, E, UA, glutation and co-enzym Q also play important roles.[87]

Uric Acid

The antioxidant, UA, is a product of purine metabolizm. Serum levels of UA are relatively high, averaging around 4-5 mg/dl in women and 5-6 mg/dl in men.[96] Clinical studies show that the mean serum UA concentration is lower in the MS group and the mean serum UA level from the patients with active MS is significanthly lower than in inactive MS patients. Serum UA levels significantly increased during immunomodulatory treatment. Lower serum UA level in MS may represent a primary, by significant loss of protection against nitric oxide and peroxinitrite and the development of CNS inflammation and tissue damage since UA is an scavenger of peroxynitrite.[97,98]

Vitamin D, 1,25-dihydroxyvitamin D (1,25(OH)$_2$D)

Hypovitaminosis D is currently one of the most studied environmental risk factors for MS. This vitamin could play an immunomodulatory role in the CNS.[99] The biologically active metabolite of vitamin D, 1,25-dihydroxyvitamin D (1,25(OH)$_2$D) is able to skew the T-cell compartment into a more anti-inflammatory and regulated state, with inhibition of Th1 and Th17 cells and promotion of Th2 and Treg cells.[100,101] Studies on EAE suggest that treatment with vitamin D prevented and even cured some of the MS patients. Near the equator, where vitamin D synthesis due to sunlight is relatively higher, MS incidence is low. Furthermore, high sun exposure and a good vitamin D status in childhood and adolescence, reflected by the high serum values of 25-hydroxyvitamin D (25(OH)D), have been associated with a decreased risk for developing MS.[102,103]

BLOOD PLATELETS IN MS ETIOLOGY

In the etiology of MS blood platelets play an important role.[104,105] The platelets possess an unexpectedly large variety of receptors and release many different compounds from specific granules. Their classical role is hemostasis and thrombosis, also they participate in inflammation, immunity and tissue repair. Platelets are the first cells at sites of vascular injury, suggesting that they may be central players in neurodegenerative diseases. Like erythrocytes, they are anucleated cells but unlike erythrocytes they do possess mitochondria.[106] Platelet activation, caused by diffrent stimuli (adhesion, aggregation, secrection), is also accompanied by the release of numerous substances from specialized granules. Released platelet factor 4 (PF4; CXCL4) is a cytokine secreted in abundance from platelets upon activation. Its measurement has been taken as an index of platelet activation, also in MS. Platelets are the main source of CD40L which delivers costimulatory signals to antigen-presenting cells (APC's). Platelets can induce maturation and activation of dendritic cells (DC), probably involving else than CD40L. The role of CD40L in MS is well described in reviews[104,105] and is a target of new therapies. The discovery of Toll-like receptors (TLRs) on platelets in 2004 was another completely unexpected development. Other immune functions of platelets had been noted earlier, including generation of killer-like ROS, phagocytic activity, secretable antimicrobials (thrombocidins), interactions with leukocytes and endothelial cells by direct contact or secretory signaling. The platelets and leukocytes co-operation would facilitate disruption of endothelial junctions of BBB.[105]

The involvement of platelets in MS was first studied by Putnam in 1935,[107] who described venule thrombosis in CNS demyelination. Epidemiological studies have found principal source of CD40L (CD154). The platelet activating factor (PAF), secreted by the prevalence of immune thrombocytopenic purpura (ITP)-like thrombocytopenia in MS patients is about 25-fold higher than in the general population. The adhesion molecule, PECAM-1, may also be important in this regard. It was reported that levels of serum soluble PECAM-1 (sPECAM-1) are significantly elevated in patients with active lesions.[108] In platelets from MS the exposure of P-selectin on platelet surface has been observed; it indicates that platelets are partly activated.[106]

BIOMARKERS OF MS

It is very difficult to assign a biomarker as a surrogate for a clinical outcome in MS characterized by complex pathophysiology. An individual biomarker reflects only one of many pathogenic processes.[2-4] In diseases including MS, based on processes such as inflammation, immunopathology, oxidative stress and markers of immunological activation are signifcantly sensitive on infectious processes, menstrual cycles and may also be affected by age and sex.[3] In MS biomarkers of demyelination, axonal damage, oxidative stress, gliosis and remyelination can be extremely valuable since, based on experience with MRI markers of axonal damage,[109] they may correlate better with the development of long-term disability, may have higher prognostic value and may signifcantly enhance our understanding of the mechanism of action of employed therapies. Availiable specimens for the markers are: Urine, blood, cerebrospinal fluid (CSF) and tears.

Presently classification of biomarkers of MS, based on published studies examining pathophysiological mechanisms, has been divided into the following seven categories: Those: (i) reflecting alteration of the immune system (ii) of axonal/neuronal damage (iii) of blood-brain barrier disruption (iv) of demyelination (v) oxidative stress and excitotoxicity (vi) of gliosis (vii) of remyelination and repair (see Table 1).[2]

Biomarkers reflecting pathophysiological processes can be used for MS diagnostics and identification of disease phenotypes, prediction of disease course and onset, treatment selection and effectivity, leading to novel therapeutics.[110,111]

TREATMENT

The treatment of MS involves "acute relapse treatment" with corticosteroids and symptom management with appropriate agents and disease modification with disease-modifying drugs (DMD). DMD include β-interferon (1b,1a) and Glatatimer acetate (GA). Interferons have anti-inflammatory effect on CNS primarily by preventing autoaggressive T-cell migration through the blood-brain barrier (BBB), GA has no such limiting effect on penetration of the BBB.[22] Instead, GA appears to function via the infiltration of GA-specific T helper 2 cells which produce "bystander suppression" of the autoreactive process.[112]

The main advantage of DMD for relapsing remittance MS treatment is their established good safety profiles. Treatment of the relapsed MS with corticosteroids improves the rate of recovery mainly due to its anti-inflammmatory effect.[8,47] These drugs modify the immune response that occurs in MS through various immunomodulatory or immunosuppressive effects.[113,114]

A new generation of therapies including newly developed monoclonal antibodies such as Natalizumab and Cladribine seems to be more effective than DMD. Natalizumab is a therapeutic humanised monoclonal antibody against the adhesion molecule very late activation (VLA)-4 or integrin alpha-4beta-1. It blocks the interaction of this lymphocyte bound receptor with its endothelial ligands: Vascular cell adhesion molecule (VCAM)-1 on brain endothelium and mucosal addressin cell adhesion molecule-1 (MAdCAM-1) on vascular endothelial cells in the gut. It has been found very effective in preventing relapses of MS through its inhibition of transendothelial migration of lymphocytes across the blood-brain-barrier (BBB) and is now widely used as the presently most effective registered therapeutic agent in severe MS.[115] Cladribine, also known as 2-chlorodeoxyadenosine, is a synthetic adenosine deaminase-resistant purine nucleoside analog that preferentially depletes lymphocyte subpopulations. This sustained effect on lymphocytes is advantageous for patients with MS.[116]

Currently these therapies are rare because they have rare but serious complications such as the development of progressive multifocal leukoencephalopathy (PML) and high cost of treatment.[8] Furthermore, DMD and monoclonal antibodies can decrease the frequency of relapses and can slow down the accumulation of irreversible disability, only if employed at the early stage of onset of the disease.

Since MS is a disease with wide range of symptoms-like fatigue, paraesthesias, muscular weakness and spasticity, double vision, optic neuritis, ataxia, bladder control problems, dysphagia, dysarthria and cognitive dysfunction, the rehabilitation is very important to improve the qualiy of life.[117]

CONCLUSION

MS is a relentless lifelong neurodegenerative disease that, in the early phase, results in severe disabilities including impaired vision due to optic neuritis (inflammation of the optic nerve) and deficits in sensation (or over-sensation as burning or prickling).[15] In the mature phase of MS appears other symptoms including paresis and paralysis, ataxia, fatigue, spasticy and incontinence. Cognitive impairment (difficulties with memory, concentration and other mental skills) also occurs frequently in complex immunopathogenesis. Despite the

Table 1. Classification of biomarkers of MS

I. Biomarkers Reflecting Alteration of the Immune System	
a. Cytokines and their receptors	(IL)-1β, IL-2, IL-6, IL-12 (p40), interferon (IFN)-γ, (TNF)-α, IL-10, TGF-β, IL-4, IL-12 (p70)/IL-23
b. Chemokines and their receptors	CCR5, CXCR3, CXCL10, CCR2/CCL2, *CXCR3/CXCL10 – marker activated T cells
c. Complement-related biomarkers	C3, C4, activated neo-C9, regulators of complement activation (CD35, CD59) Activated neo-C9
d. Adhesion molecules	E-selectin, L-selectin, ICAM-1, VCAM-1, CD31, surface expression of LFA-1 and VLA-4
e. Biomarkers reflective of antigen-processing and presentation	CD40/CD40L, CD80, CD86, heat shock proteins (hsp), *CD40/CD40L (differentiate between RR and SP-MS)
f. Cell-cycle and apopto-sis-related biomarkers	Fas (CD95) and Fas-L, FLIP, Bcl-2, *TRAIL (reflective clinical response to INF β therapy in MS)
g. Antibodies	CSF IgG index, κ-light chains, oligoclonal bands, *Anti-myelinoligodendrocyte glycoprotein (anti-MOG Ab), anti-myelin basic protein (anti-MBP) (possibile marker pre-diction of definite MS after first clinical symptom (CIDS).
h. Biomarkers reflective immune-mediated neuro-protection	BDNF expresion
II. Biomarkers of Axonal/ Neuronal Damage	Cytosceletal proteins (action, tubulin and neurofilaments), tau protein, *24S-Hydroxycholesterol (marker of axonal loss-MSPP), amyloid precursor protein (APP) (marker acute amonal injury), N-acetylaspartic acid (marker of disability and axonal volume), 14-3-3 protein—predictor clinically definite MS
III. Biomarkers of Blood-Brain Barrier (BBB) Disruption	Matrix metalloproteinases (MMPs) MMP-9 and their inhibi-tors (TIMP), platelet activating factor, thrombomodulin
IV. Biomarkers of Demyelination	MBP and MBP-like material, proteolytic enzymes, endog-enous pentapeptide QYNAD, gliotoxin with Na-channel bloking properties
V. Biomarkers of Oxidative Stress and Excitotoxicity	*Nitric oxide derivatives CSF (PPMS; brain atrophy) *Isoprostanes F2-isoprostanes (markers of lipid peroxidation) *Uric acid (strong natural peroxynitrate scavenger)
VI. Biomarkers of Gliosis	Glial fibrillary acid protein (GFAP), S-100 protein,
VII. Biomarkers of Remyelination and Repair	NCAM (neural cell adhesion molecule), CNTF (Ciliary neurotrophic factor), MAP-2 + -13 (microtubuleassociated protein-2 exon 13), CPK-BB (creatine phosphatase BB), PAM (peptidylglycine α-amidating monooxygenase)

*candidate biomarker

fact a wealth of information is available describing the structure function and composition of brain and brain-associated proteins and enzymes, exact reason and mechanism for the induction of MS is still unknown.

Earlier studies on the autopsied brains showed demylination of the myeline layer appearing in the form of plaques, like the plaques form by viral infection. It was thus hypothesised that MS may be caused by some unknown class of virus, although no virus could be isolated from the disease specimens. Later the viral theory of MS was discarded and detailed molecular analysis of normal brains and brains suffered from MS pin-pointed that MS is an autoimmune disease with complex immunopathogenesis.

Further investigations are required both, to understand the root cause of MS together with the cellular immunology of neurodegeneration and also to investigate which gene(s) may be involved, mutation in which leads to MS. Such studies will help to understand not only the disease mechanisms but provide real therapeutic benefits for diseases where we can often do nothing more than palliate the irreversible loss of neurological function.

REFERENCES

1. Ramagopalan SV, Dobson R, Meier UC et al. Multiple sclerosis: risk factors, prodromes and potential causal pathways. Lanc Neurol 2010; 9:727-739.
2. Bielekova B, Martin R. Development of biomarkers in multiple sclerosis. Brain 2004; 127:1463-1478.
3. Martin R, Bielekova B, Hohlfeld R et al. Biomarkers in multiple sclerosis. Disease Markers 2006; 22:183-185.
4. Miller DH. Biomarkers and surrogate outcomes in neurodegenerative disease: les sons from multiple sclerosis. NeuroRx 2004; 1:284-294.
5. Pugliatti M, Sotgiu S, Rosati G. The worldwide prevalence of multiple sclerosis. Clin Neurol Neurosurg 2002; 104:182-191.
6. Pugliatti M, Harbo HF, Holmøy T et al. Enviromental risk factors in multiple sclerosis. Acta Neurol Scand Suppl 2008; 188:34-40.
7. Al-Omaishi J, Bashir R, Gendelman HE. The cellular immunology of multiple sclerosis. J Leukoc Biol 1999; 65:444-452.
8. Barnett MH, Parratt JDE, Pollard JD et al. MS: Is it one disease? Internat MS J 2009; 16:57-65.
9. Kurtzke JF. Multiple sclerosis In time and space—geographic clues to causa. J Neurovirol 2000; 6:134-140.
10. Ascherio A, Munger KL. Environmental risk factors for multiple sclerosis. Part II. Noninfections factors. Ann Neurol 2007; 61:504-513.
11. Munger KL, Ascherio A. Risk factors in development of multiple sclerosis. Exper Rev Clin Immunol 2007; 3:739-748.
12. Levin LI, Munger KL, O'Reilly EJ et al. Primary infection with the Epstein-Barr virus and risk of multiple sclerosis 2010; 6:824-830.
13. Oksenberg JR, Barcellos LF. The complex genetic aetiology of multiple sclerosis. J Neurovirol 2000; 6:210-214.
14. Oksenberg JR, Hauser SL. Mapping the human genome with newfound precision. Ann Neurol 2010; 67:A8-A10.
15. Baranzini SE. The genetics of autoimmune diseases: a networked perspective. Curr Opin Immunol 2009; 21:596-605.
16. Baranzini SE, Wang J, Gibson RA et al. Genome-wide association analysis of susceptibility and clinical phenotype in multiple sclerosis. Hum Mol Genet 2009; 18:767-778.
17. Al.-Omaishi J, Bashir R, Gendelman HE. The cellular immunology of multiple sclerosis. J Leuk Biol 1999; 65:444-452.
18. Arnett PA. Does cognitive reserve Apple to muli ple sclerosis? Neurology 2010; 74:1934-1935.
19. Rejdak K, Jackson S, Giovannoni G. Multiple sclerosis: a practical overview fot clinicians. Br Med Bull 2010 [Epub ahead of print].
20. Lublin FD. The incomplete nature of multiple sclerosis relapse resolution. J Neurol Sci 2007; 256:14-18.
21. Ontaneda D, Rae D, Rea AD. Management of acute exacerbations in multiple disability progression in sclerosis Ann Indian Acad Neurol 2009; 12:264-272.
22. Lerday E, Yaouanq J, Le Page E et al. Evidence for a two-stages multiple sclerosis. Brain 2010; 10:1-14.

23. Stüve O, Oksenberg J. In: Pagon RA, Bird TC, Dolan CR, Stephens K, editors. GeneReviews [Internet]. Seattle (WA): University of Washington, Seattle; 1993-2006 [updated 2010 May 11].
24. Bashir K, Whitaker JN. Clinical and laboratory features of primary progressive and secondary progressive MS. Neurology 1999; 53:765-771.
25. Liguori M, Marrosu MG, Pugliatti M et al. Age at onset in multiple sclerosis. Neurol Sci 2000; 21:825-829.
26. Tullman MJ, Oshinsky RJ, Lublin FD et al. Clinical characteristics of progressive relapsing multiple sclerosis. Mult Scler 2004; 10:451-454.
27. Lublin FD, Reingold SC. Defining the clinical course of multiple sclerosis: results of an international survey. National multiple sclerosis society (USA) advisory committee on clinical trials of new agents in multiple sclerosis. Neurology 1996; 46:907-911.
28. Martinelli V, Rodegher M, Moiola L et al. Late onset multiple sclerosis: clinical characteristics, prognostic factors and differential diagnosis. Neurol Sci 2004; 25 Suppl 4:350-355.
29. Noseworthy JH, Lucchinetti C, Rodriguez M et al. Multiple sclerosis. N Engl J Med 2000; 343:938-952.
30. Polman CH, Reingold SC, Edan G et al. Diagnostic criteria for multiple sclerosis: 2005 revisions to the "McDonald Criteria". Ann Neurol 2005; 58:840-846.
31. Wck-Guttman B, Jacobs LD. What is New in the treatment of multiple sclerosis? Drugs 2000; 59:401-410.
32. Compston A, Coles A. Multiple sclerosis. Lancet 2002; 372:1221-1231.
33. Compston A, Coles A. Multiple sclerosis. Lancet 2008; 372:1502-1517.
34. Bechmann J, Galea J, Perry VH. What is the blood-brain barier (not)? Trends Immunol 2007; 28:5-11.
35. Ribatti D, Nico B, Crivellato E. Development of the blond-brain barier: a historical point of view. Anat Rec B New Anat 2006; 289:3-8.
36. Nair A, Frederick TJ, Miller SD. Astrocytes in multiple sclerosis: A product of environment. Cell Mol Life Sci 2008; 65:2702-2720.
37. Wolburg H, Lippoldt A. Tight junctions of the blood-brain barier development, composition and regulation. Vascul Pharmacol 2002; 38:323-337.
38. Abott NJ, Ronnback L, Hansson E. Astrocyte-endothelial interactions at the blood-brain barier. Nat Rev Neurosci 2006; 7:41-53.
39. Frank-Cannon TC, Altol T, McAlpine FE. Does neuroinflammation fan the flame in neurodegenerative diseases? Mol Neurodeger 2009; 4:47.
40. Block ML, Hong JS. Microglia and inflammation-mediated neurodegeneration: multiple triggers with a common mechanism. Prog Neurobiol 2005; 76(2):77-98.
41. Bradl M, Lassman H. Oligodendrocytes: biology and pathology. Acta Neuropatol 2010; 119:37-53.
42. Bauman N, Pham-Dinh D. Biology of oligodendrocytes and myelin in the mammalian central nervous system. Physiol Rev 2001; 81:871-927.
43. Watzlawik J, Warrington AE, Rodriguez M. Importance of oligodendrocyte protection BBB. Expert Rev Neurother 2010;10(3):441-457.
44. Leoni V. Oxysterols as markers of neurological disease—a review. Scan J Cin Lab Invest 2009; 69:22-25.
45. Stadelmann C, Brück W. Interplay between mechanisms of damage and repair in multiple sclerosis. J Neurol 2008; 255:12-18.
46. Goodin DS. The casual cascade to Multiple sclerosis: a model for MS pathogenesis. PLoSOul 2009; 4:4565.
47. Hohlfeld R. Review: 'Gimme five': future challenges in multiple sclerosis. ECTRIMS Lecture 2009. Mult Scler 2010; 16:3-14.
48. Peterson LK, Fujinami RS. Inflammation, demyelination, neurodegeneration and neuroprotection in the pathogenesis of multiple sclerosis. J Neuroimmun 2007; 184:37-44.
49. Racke MK. Immunopathogenesis of multiple sclerosis. Ann Indian Neurol 2009; 12:215-220.
50. Frischer JM, Bramow S, Dal-Bianco A et al. The relation between inflamation and neurodegeneration in multiple sclerosis brains. Brain 2009; 132:1175-1189.
51. Goverman J. Autoimmune T-cell responses in the central nervous system. Nat Rev Immunol 2009; 9:393.
52. Brück W. The pathology of multiple sclerosis is the result of focal inflammatory demyelination with axonal damage. J Neurol 2005; 252 Suppl 5:3-9.
53. Christophi GP, Panos M, Hudson ChA et al. Macrophages of multiple sclerosis patients display deficient SHP-1 expression and enhanced inflammatory phenotype. Lab Invest 2009; 89:742-759.
54. Hedegaard Ch J, Krakauer M, Bendtrek LH et al. T helper cell type 1 (Th1), Th2 and TH17 responses to myelin basic protein and disease activity in multiple sclerosis. Immunology 2008; 125:161-169.
55. DeLuca GC, Wiliams K, Evangelou et al. The contribution of demyelinisation to amonal loss in multiple sclerosis. Brain 2006; 129:1507-1516.
56. Zivadinov R, Cox JL. Neuroimaging in multiple sclerosis. Int Rev Neurobiol 2007; 79:449-474.
57. Benedetti B, Rovaris M, Rocca MA et al. Evidence for stable neuroaxonal damage in the brain of patients with benign multiple sclerosis Mult Scler 2009; 15:789-794.
58. Bernal F, Elias B, Hartung HP et al. Regulation of matrix metalloproteinases and their inhibitors by interferon-β: a longitudinal study in multiple sclerosis patients. Mult Scler 2009; 15:721-727.

59. Zoppo GJ, Milner R. Integrin-matrix interactions in the cerebral microvasculature. Arterioscler Thromb Vasc Biol 2006; 26;1966-1975.

60. Arai K, Lo EH. Oligovasular signalig in white matter stroke. Biol Pharm Bull 2009; 32:1639-1644.

61. Goncalves DaSilva A, Yong VW. Matrix metalloproteinase-12 deficiency Horsens relapsing-remitting experimental autoimmune encephalomyelitis in association with cytokine and chemokine dysregulation. Am J Pathol 2009; 174:898-909.

62. Becher B, Durell BG, Noelle RJ. IL-23 produced by CNS-resident cells controls T-cell encephalitogenicity during the effector phase of experimental autoimmune encephalomyelitis. J Clin Invest 2003; 112:1186-1191.

63. Bettelli E, Oukka M, Kuchroo VK. T(H)-17 cells in the circle of immunity and autoimmunity. Nat Immunol 2007; 8:345-350.

64. Simpson JE, Newcombe J, Cuzner ML et al. Expression of the interferon-gammainducible chemokines IP-10 and Mig and their receptor, CXCR3, in multiple sclerosis lesions. Neuropathol Appl Neurobiol 2000; 26:133-142.

65. Balashov KE, Rottman JB, Weiner HL et al. CCR5(+) and CXCR3(+) T-cells are increased in multiple sclerosis and their ligands MIP-1alpha and IP-10 are expressed in demyelinating brain lesions. Proc Natl Acad Sci USA 1999; 96:6873-6878.

66. Schmitz T, Chew LJ. Cytokines and myelination in the central nervous system. ScientificWorldJournal 2008; 8:1119-1147.

67. Kort JJ, Kawamura K, Fugger L et al. Efficient presentation of myelin oligodendrocyte glycoprotein peptides but not protein by astrocytes from HLA-DR2 and HLADR4 transgenic mice. J Neuroimmunol 2006; 173:23-34.

68. Miljkovic D, Momcilovic M, Stojanovic I et al. Astrocytes stimulate interleukin-17 and interferon-gamma production in vitro. J Neurosci Res 2007; 85:3598-3606.

69. Dong Y, Benveniste EN. Immune function of astrocytes. Glia 2001; 36:180-190.

70. Hemmer B, Archelos JJ, Hartung HP. New concepts in the immunopathogenesis of multiple sclerosis. Nat Rev Neurosci 2002; 3:291-301.

71. Bullard DC, Hu X, Schoeb TR et al. Intracellular adhesion molekule—1 expresion required on multiple cell types for development of expermental autoimmune encephalomyelitis. J Immunol 2007; 178:851-857.

72. Sobel RA, Mitchell ME, Fondren G. Interacellular adhesion molecule (ICAM-1) in cellular immune reactions in the human central nervous system. Am J Pathol 1990; 136:1309-1316.

73. Allavena R, Noy S, Andrews M et al. CNS elevation of vascular and not mucosal addressin cell adhesion molecules in patients with multiple sclerosis. Am J Pathol 2010; 176:556-562.

74. Malik M, Chen YY, Kienzle MF et al. Monocyte migration and LFA-1 mediated attachment to brain microvascular endothelial is regulated by SDF-1 alpha though Lyn kinase. J Immunl 2008; 181:4632-4637.

75. Chari DM. Remyelination in multiple sclerosis. Int Rev Neurobiol 2007; 79:589-620.

76. Jakovcevski I, Filipovic R, Mo Z et al. Oligodendrocyte development and the onset of myelination in the human fetal brain. Front Neuroanat 2009; 3:5.

77. Dubois-Dalcq M, Willams A, Stadelmann Ch et al. From fish to man: understanding endogenous remyelination in central nervius system demyelinating diseases. Brain 2008; 131:1686-1700.

78. Chandran S, Hunt D, Joannides A et al. Myelin repair: the role of stem and prekursor cells in multiple sclerosis. Phil Trans R Soc 2008; 363:171-183.

79. Chang A, Smith MC, Yin X et al. Neurogenesis in the chronic lesions of multiple sclerosis. Brain 2008; 131:2366-2375.

80. Ingram G, Hakobyan S, Robertson NP et al. Complement in multiple sclerosis: its role in disease and potential as a biomarker. Clinic Exper Immunology 2008; 155:128-139.

81. Carson MJ. Microglia as leasions between the immune and central nervous systems: functional implications for multiple sclerosis. Glia 2002; 40:218-231.

82. Orhan A, Ullrich O, Infante-Durte C et al. Neuronal damage in brain inflammation. Arch Neurol 2007; 64:185-189.

83. Armstrong AC, Le TQ, Flint NC et al. Endogenous cell repair of chronic demyelination. J Neuropathol Exp Neurol 2006; 65:245-256.

84. Hooper DC, Scott GS, Zborek A et al. Uric acid, a peroxynitrite scavenger, inhibits CNS inflammation, blood-CNS barier permeability changes, and tissue damage in a mouse model of multiple sclerosis. FASEB 2004; 14:691-698.

85. Gilgun-Sherki Y, Melamed E, Offen D. The role of oxidative stress in the pathogenesis of multiple sclerosis. The need for the effective antioxidant therapy. J Neurol 2004; 251:261-268.

86. Gonsette R. Oxidative stress and excitotoxicity: a therapeutic issue in multiple sclerosis? Mult Scler 2008; 14:22-34.

87. Miller E, Mrowicka M, Żołyński K et al. Oxidative stress in multiple sclerosis. Pol Merk Lek 2009; 27:499.

88. Miller E, Mrowicka M, Malinowska K et al. Effects of whole body cryotherapy on total antioxidative status and activities of antioxidative enzymes in blood of patients with multiple sclerosis. J Med Invest 2010; 57:168-173.

89. Miller E, Rutkowski M. Contribution and part of main biochemic factors in ischemic insult. Pol Merk Lek 2006; 117:261-264.

90. Calabrese V, Cornelius C, Rizzarelli E et al. Nitric oxide in cell survival: a jams molecule. Antioxid Redox Signal 2009; 11:2717-2739.

91. Trapp BD, Sys PK. Virtual hypoxia and chronic necrosis of demyelinated axons in multiple sclerosis. Lancet Neurol 2009; 8:280-291.

92. Bishop A, Hobbs KG, Equchi A et al. Differential sensivity of oligodendrocytes and motor neurons to reactive nitrogen species: implications for multiple sclerosis. J Neurochem 2009; 109:93-104.

93. Clarcson A, Sutherland B, Appelton I. The biology and pathology of hypoxia-ischaemia-induced brain damage: an update. Arch Immunol Ther Exp 2005; 53:213-225.

94. Muniandy S, Qvist R, Ong Siok Yan G et al. The oxidative stress of hyperglycemia and the inflammatory process in endothelial cells. J Investigating Med 2009; 56:6-10.

95. Koch M, Mostert J, Arutjunyan A V et al. Plasma lipid peroxidation and progression of disability in multiple sclerosis. Eur J Neurol 2007; 14529-14533.

96. Drulovic J, Dujmovic I, Stojsavljevic N et al. Uric acid levels in serum from patients with multiple sclerosis. J Neurol 2001; 248:121-126.

97. Hooper DC, Scott GS, Zborek A et al. Uric acid, a peroxynitrite scavenger, inhibits CNS inflammation. The FASEB J 2000; 14:691-698.

98. Rentzos M, Nikolaou C, Anagnostouli M et al. Serum uric acid and multiple sclerosis. Clin Neurol Neurosurg 2006; 108:527-531.

99. Pierrot-Deseilligny C. Clinical implications of a possible role of vitamin D in multiple sclerosis. J Neurol 2009; 256:1468-1479.

100. Debouverie M. Gender as a prognostic factor and its impact on the incidence of multiple sclerosis in Lorraine, France. J Neurol Sci 2009; 286:14-17.

101. Smolders J, Damoiseaux J, Menheere P. Vitamin D as an immune modulator in multiple sclerosis. J Neuroimmunol 2008; 194:7-17.

102. Correale J, Ce'lica Ysrraelit M, Ine's Gaita'n M. Immunomodulatory effects of Vitamin D in multiple sclerosis. Brain 2009:132; 1146-1160.

103. Zittermann A. Vitamin D in preventive medicine: are we ignoring the evidence? Clinical implications of a possible role of vitamin D in multiple sclerosis. Br J Nutr 2003; 89:552-572.

104. Sheremata W, Jy W, Horstman LL et al. Evidence of platelet activation in multiple sclerosis. J Neuroinflam 2008; 5:27.

105. Horstman LL, Jy WW, Zavadinov R et al. Role of platelets in neuroinflammation: a wide-angle perspective J Neuroinflam 2010; 7:10.

106. Wachowicz B, Rywaniak JZ, Nowak P. Apoptic matkers in human blood platelets treated with peroxynitrite. Platelets 2008; 19:624-635.

107. Putnam TJ. Studies in multiple sclerosis (iv) 'encephalitis' and sclerotic plaques produced by venular obstruction. Arch Neurol Neurosurg Psychiat 1935; 33:929-940.

108. Losy J, Niezgoda A, Wender M. Increased serum levels of solube PECAM-1 in multiple sclerosis patients with brain gadolinum enhancing lesions. J Neuroimmunol 1999; 99:169-172.

109. Van Walderveen MA, Barkhof F, Pouwels PJ et al. Neuronal damage in T1-hypointense multiple sclerosis lesions demonstrated in vivo using proton magnetic resonance spectroscopy. Ann Neurol 1999; 46:79-87.

110. Martin R, Bielekova B, Hohfeld R et al. Biomarkers in multiple sclerosis. Disease Markers 2006; 22:183-185.

111. Seven A, Aslan M. Biochemical and immunological markers of multiple sclerosis. Turk J Biochem 2007; 32:112-119.

112. Markowitz CE. The current landscape and unmetneeds in multiple sclerosis. Am J Manage Care 2010; 16:211-218.

113. Vyshkina T, Kalman B. Autoantibodies and neurodegeneration in multiple sclerosis. Lab Investig 2008; 88:796-784.

114. McDonagh M, Dana T, Chan BKS et al. Drug Class Review on Disease-modifying Drugs for multiple sclerosis: Final Report. [Internet] 2007; Drug Class Reviews.

115. Abstracts: 26th Congress of the European Committee for Treatment and Research in Multiple Sclerosis (ECTRIMS) 15th Annual Conference of Rehabilitation in MS (RIMS). Mult Scler 2010; 16:S7.

116. Murphy JA, Harris JA, Cramnage AJ. Potential short-term use of oral cladribine in treatment of relapsing-remitting multiple sclerosis. Neuropsychiatr Dis Treat 2010; 6:619-25.

117. Rasova K, Feys P, Henze T et al. Emerging evidence-based physical rehabilitation for multiple sclerosis-Towards an inventory of current content across Europe. Health and Quality of Life outcomes 2010; 8:76.

CHAPTER 18

MYOTONIC DYSTROPHY TYPE 1 OR STEINERT'S DISEASE

Vincenzo Romeo

Department of Neurosciences, University of Padova, School of Medicine, Padova, Italy
Email: vincenzo.romeo@sanita.padova.it

Abstract: Myotonic Dystrophy Type 1 (DM1) is the most common worldwide autosomal dominant muscular dystrophy due to polynucleotide $[CTG]_n$ triplet expansion located on the 3′UTR of chromosome 19q13.3. A toxic gain-of-function of abnormally stored RNA in the nuclei of affected cells is assumed to be responsible for several clinical features of the disease. It plays a basic role in deregulating RNA binding protein levels and in several mRNA splicing processes of several genes, thus leading to the multisystemic features typical of DM1. In DM1, the musculoskeletal apparatus, heart, brain, eye, endocrine, respiratory and gastroenteric systems are involved with variable levels of severity. DM1 onset can be congenital, juvenile, adult or late. DM1 can be diagnosed on the grounds of clinical presentation (distal muscular atrophy and weakness, grip and percussion myotonia, ptosis, hatchet face, slurred speech, rhinolalia), EMG myotonic pattern, EKG (such as AV-blocks) or routine blood test abnormalities (such as increased CK values or hypogamma-globulinemia) and history of cataract. Its confirmation can come by DNA analysis. At present, only symptomatic therapy is possible and is addressed at correcting hormonal and glycemic balance, removing cataract, preventing respiratory failure and, above all, major cardiac disturbances. Efficacious therapies targeted at the pathogenic mechanism of DM1 are not yet available, while studies that seek to block toxic RNA intranuclear storage with specific molecules are still ongoing.

Neurodegenerative Diseases, edited by Shamim I. Ahmad.
©2012 Landes Bioscience and Springer Science+Business Media.

INTRODUCTION

Nosography of Myotonic Dystrophies

Myotonic Dystrophies (DM) represent a heterogeneous family of disturbances of muscular fibre release. Such heterogeneity is both genotypic and phenotypic. The predominant clinical aspect is the myotonic phenomenon, which is an abnormal contraction of the muscle fibre after either voluntary activation, hammer percussion (percussion myotonia), or electric stimulation (electric myotonia).

The history of DM begins in the early 1900s, when a German internist, Hans Gustav Wilhelm Steinert (1875-1911, Fig. 1), described a neuromuscular disorder characterized by dystrophic progression with myotonia at clinical examination, for the first time in 1909 (*Über das klinische und anatomische Bild des Muskel schwunds der Myotoniker*). Since then, such syndromic picture has been named 'Steinert's Disease'. Afterwards, Curschmann, Batten and Rossolimo separately described this unusual condition after

Figure 1. Hans Steinert (1875-1911). Reproduced with permission from the Leipzig University archive (H. Steinberg, A. Wagner. Hans Steinert: Zum 100. Jahrestag der Erstbeschreibung der myotonen Dystrophie. Nervenarzt 2008;79:961-970).

Steinert's original article and for this reason the disease is nowadays known as 'Steinert's Myotonic Dystrophy' or 'Steinert's disease', 'Curschmann-Batten-Steinert's syndrome', 'Myotonic Dystrophy', or 'Rossolimo-Curschmann-Batten-Steinert's syndrome', 'Myotonia Atrophica or Dystrophica'.

One major clinical characteristic is 'pleiotropism', which is the involvement of several organs and systems and an autosomal dominant inheritance, which is characterized by an earlier, more severe onset of symptoms in offspring (anticipation phenomenon).

Genetic diagnosis has been available since 1992, when Brook et al demonstrated an expanded CTG-triplet on the 3'UTR noncoding region of the 'Dystrophia Myotonica Protein Kinase' (DMPK) gene on chromosome 19q, in position 13.3 (19q13.3). This nucleotidic repeat is responsible for the disease and its clinical features.[2]

A likely phenotypic and genotypic heterogeneity of myotonic dystrophy was first suggested by Thornton et al in 1994, when he described cases characterized by a myotonic *Steinert-like* syndrome with systemic involvement and AD transmission, in which it was not possible to detect the presence of the $[CTG]_n$ expansion on chromosome 19q.[5]

Several descriptions of single or familial cases, characterized by a myotonic syndrome with multisystemic involvement and AD transmission, appeared in the early 1990s, some with mostly distal neuromuscular involvement (*Steinert-like*) and others with proximal involvement (PROximal Myotonic Myopathy 'PROMM'; Proximal Myotonic Dystrophy 'PDM').[6,7]

The finding of such pheno/genotypic heterogeneity raised the problem of how to classify DM, separating common forms (DM1, *Steinert-like*, molecularly determined) from atypical forms of DM (DM2/PROMM and PDM, not molecularly determined). Ranum et al found a link on chromosome 3q for DM2 in 1988, while Liquori et al discovered that the nucleotidic quadruplet $[CCTG]_n$, located on chromosome 3q in position 21 (3q21), in the first intron of the *zinc-finger-protein gene* (ZNF9 gene), was responsible for a large part of the *Steinert-like* or *non-Steinert-like* syndromes without a *link* on chromosome 19q and therefore not due to the nucleotidic $[CTG]_n$ expansion.[3]

The aim of the second IDMC conference (International Myotonic Dystrophy Consortium), held in 1999, was aimed to reset the previous taxonomy of myotonic dystrophies. It was established that the term 'DM2' should be adopted for all progressive multiorgan disorders linked to the DM2 locus.[8] The actual systematization of DM is presented in Table 1.

Table 1. The genotype-phenotype spectrum of myotonic dystrophies

	Genotype		Phenotype
	Chromosome	Expansion	
DM1	19q13.3	CTG	Myotonic dystrophy *(mostly distal)*
			Myotonic dystrophy *(mostly distal)*
DM2 PROMM PDM	3q21	CCTG	Proximal myotonic myopathy *(mostly proximal)*
			Proximal myotonic dystrophy *(mostly proximal)*

CLINICAL FEATURES

Steinert's Myotonic Dystrophy (DM1)

Myotonic Dystrophy Type 1 is the most frequent form of myotonic dystrophy, with an estimated minimum prevalence rate of 8-10 (9.31) affected people (up to 12, in some case series) per 100,000 population. It is characterized by autosomal-dominant (AD) inheritance with anticipation phenomenon (earlier and more severe involvement in offspring).

The disease onset in DM1 is extremely variable, but at least 4 subgroups of patients can be distinguished according to onset the: (1) Congenital Form (only maternal transmission of the expanded allele); (2) Juvenile onset; (3) Adult onset; (4) Late onset. In an asymptomatic proband with a pedigree positive for DM1, a definite diagnosis can only be made by DNA analysis, except in cases of obligate carriers.[1]

DM1 is determined by $[CTG]_n$ triplet expansion on chromosome 19q13.3, at the genomic locus of a serin-threonin kinase named DMPK[2] (Fig. 2).

DM1 patients are arbitrarily subclassified on the grounds of $[CTG]_n$ triplet expansion size:

E_1 = 50-149 CTG; E_2 = 150-1,000 CTG; E_3 = >1,000 CTG.[8]

There are at least 2 different subclassifications for E_1, E_2, E_3: The choice of adopting one rather than another, depends on the genetic-lab protocol. Another classification also includes an E_4 class, for cases with $[CTG]_n$ >1,500.

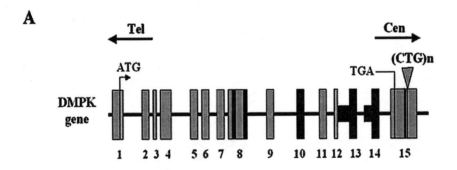

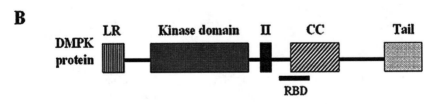

Figure 2. A) DMPK gene. B) DMPK protein. LR: leucine rich region; Kinase domain; II: substrate-specificity site; RBD: possibile rho-binding domain; CC: '*coiled-coil*'; subcellular localization domain. (Modified from: Groenen PJ et al. Constitutive and regulated modes of splicing produce six major myotonic dystrophy protein kinase (DMPK) isoforms with distinct properties. Hum Mol Genet 2000; 9(4):605-16; by permission of Oxford University Press.)

Expansions ranging from 37 to 49 $[CTG]_n$ are considered 'premutations' by some authors, since they are not sufficient to cause the clinical picture, but may be responsible for possible expanding triplets in offspring, given the peculiar instability of the $[CTG]_n$ polynucleotide.[4]

The phenotypic pleiotropic characteristics of DM1 can be summarized as follows:

1. hypotrophic muscular masses of the four limbs, with a disto>proximal distribution; weakness, with or without grip myotonia or percussion myotonia;
2. triangle-shaped or hatchet face (hypotrophy of massetere and temporal muscles);
3. blepharoptosis, mono or bilateral and/or myopathic shape of the mouth (facial weakness);
4. frontal balding;
5. slurred speech, rhinolaly;
6. respiratory failure;
7. cardiac abnormalities or arrhythmia;
8. opacity of the lens, cataract;
9. hypogonadism, diabetes and other endocrine disturbances;
10. osteoskeletal abnormalities;
11. cognitive involvement;
12. daily somnolence, hypersomnia;
13. gastroenteric disturbances.

The characteristic facial phenotype in two DM1 siblings is shown in Figure 3. Figure 4 shows the anticipation phenomenon.

An over-expanded triplet (usually >1,000 $[CTG]_n$) with early manifestation of symptoms at birth, respiratory failure and severe hypotonia (*floppy baby*) is known as 'congenital myotonic dystrophy' (DM1). Congenital myotonic dystrophy has only been documented in DM1[1,4,9,10] (Fig. 5).

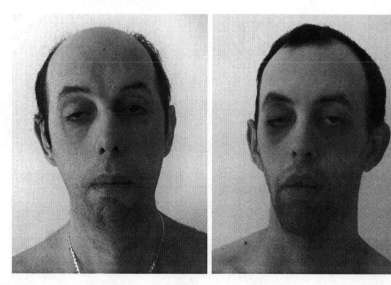

Figure 3. Two siblings affected by DM1. Both present a deficit in hypophyseal secretion of h-GH hormone, tested by GH-RH plus Arginine test.

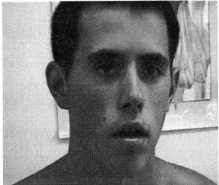

Figure 4. Woman, 45 y.o. (left), affected by myotonic dystrophy Type 1 (adult onset) and her son, 17 y.o. (right), also affected, with juvenile onset (anticipation phenomenon).

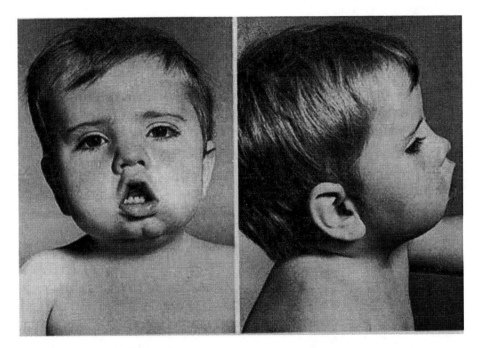

Figure 5. A case of congenital myotonic dystrophy Type 1 is shown. (Reproduced from: Peter S. Harper. Myotonic Dystrophy, 2001:figure 9; ©2001 with permission from Elsevier).

Laboratory and instrumental clinical examinations, aimed at diagnosing a patient suspected as having DM1, involve: routine blood test, electromyography, slit lamp study of the lens and muscular biopsy.

Routine blood test generally reveals hyperCKemia (rarely >1,000 UI/L), suggestive for myopathic disturbances; hypo-γ-globulinemia is not infrequently seen at serum electrophoresis.

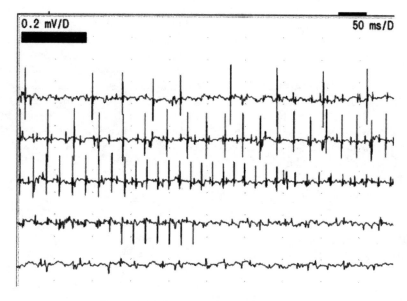

Figure 6. EMG pattern of myotonic discharge.

Electromyography shows peculiar spontaneous electrical insertional activity within the relaxing muscle, explored by electrode-needle, generally diffuse, but prominent in the small muscles of the hand and anterolateral region of the leg. Such activity is commonly referred to as 'electromyographic myotonia' and appears as multiple myotonic discharges known as '*dive bomber potentials*', which are variable for amplitude and frequency, within the single discharge (Fig. 6).

A slit lamp study documents posterior subcapsular iridescent opacities of the lens in many cases, which is very typical for DM1.

Muscle pathology findings are not specific and not pathognomonic. They are extensively described in a dedicated paragraph (see below).

In DM patients, in-depth clinical[11] and instrumental evaluation of cardiac, respiratory, neuropsychological and brain perfusion patterns are mandatory:

1. cardiologic examination (EKG; EKG dynamic-Holter; echocardiography);
2. respiratory investigation (spirometry);
3. neuroimaging (MRI, SPECT);
4. extensive neuropsychological evaluation.

Heart[1,13]

Cardiac involvement in DM1 was firstly described in 1911. Subsequently, several reports documented impaired cardiac conduction and hypotension. In addition, a wide disproportion was observed between symptomatic (16%) and asymptomatic patients with abnormalities on heart investigation. Therefore, an extensive cardiologic study is always suggested to document any anomalies of heart conduction that occur especially in the atrio-ventricular tract (generally AV conduction blocks of first degree) or branch blocks (RBB, LBB) or more complex cardiac conduction defects (at least 80% of patients). Morphofunctional abnormalities (dilatative cardiomyopathy) are rarely observed (in contrast to some other neurodegenerative or neuromuscular disorders) but, if present, they have to be seriously suspected as a potential source of major arrhythmias (atrial flutter or fibrillation, ectopic beats, sustained atrial or ventricular tachycardia). In these cases, any fibrotic change in the left or right ventricle deserves special attention, especially when associated with a dilatation. Cases of sudden death in young patients affected by DM1 have also been described (muscular performance in these patients is not a good predictor!), while the cause of death in adults and elderly is generally due to pre-existent known or unknown cardiac conduction defects.

Hence, periodic cardiologic follow-up in DM1 patients is mandatory. It should encompass standard EKG, Holter dynamic-EKG and echocardiogram. Moreover, more in-depth cardiologic studies are recommended, such as intracardiac conduction studies, in cases of progression of heart rhythm disturbances, persistence of abnormalities or critical symptomatic patients.

Heart pathology studies suggest a deterioration within the heart conduction system at any level, with prominent fatty and fibrotic changes and myocyte hypertrophy.

An unclear correlation with $[CTG]_n$ expansion has been found, although a trend of increasing risk of heart problems with increasing triplet size is likely. The typical somatic mosaicism demonstrated in several tissues of DM1 patients could reasonably explain these appearently contradictory results. Besides, familiarity for heart conduction defects apart from those due to DM1, could influence the prognosis of this latter (modifying genes theory). The frequency of coronary artery disease in DM1 is not greater than in general population expected of the same age.

Lung[1]

Respiratory problems may also occur in DM1, usually due to severe muscular atrophy, weakness, or skeletal deformity. Besides, it is still debated whether the changes observed in ventilation are mainly due to brainstem dysfunction, peripheral nerve or chemoceptor damage, or by chest and pharyngeal muscle myotonia and/or atrophy. Breathing problems become fairly frequent in the elderly, in whom they may cause severe complications and lead to a weak prognosis in about 40% of patients.

The principal mechanisms through which respiratory problems may occur are: (1) respiratory muscle atrophy/weakness/myotonia with or without diaphragm involvement (restrictive non-obstructive respiratory pattern), particularly important in infancy; (2) central factors, possibly responsible for alveolar hypoventilation, hypercapnia and hypoxaemia could also trigger hypersomnia (whose complex pathogenesis is, however, still unclear. In fact, many central factors seem to combine to determine daily somnolence, such as brainstem, hypothalamic or pituitary dysfunctions); (3) aspiration of gastric material into the bronchial tree, due to insufficient peristalsis or inefficient cardias contraction.

Brain[1,14,15]

The first report of mental disturbances was made in 1937.[27] A poor negative correlation was found between the calculated I.Q. and the $[CTG]_n$ triplet expansion, since $[CTG]_n$ is not a reliable predictor of I.Q. Avoidant personality disorder (uncommon in the general population), with obsessive-compulsive disorder, passive-aggressive and schizotypic traits, not justified by concurrent neuromuscular impairment, has also been described.[14,15]

Neuroimaging studies in DM concern both morphological (by MRI), perfusion (by SPECT) and metabolic aspects (by PET) of CNS.[14] An association between degree of cortical atrophy (MRI) and severity of intellectual impairment, which is more related to other morphological anomalies (thickening of the skull, focal lesions of the white matter, commonly seen in the temporal poles), has not been demonstrated. Perfusion studies mostly show frontal and associative temporo-parietal hypoperfusion, with major brain damage in congenital forms. The presence of white matter hyperintense lesions (WMHLs) and cortical atrophy and dilatation of ventricular spaces has been detected in DM1 patients by brain MRI. In addition, recurrence, localization, diffusion, morphology and relationship with other features of the disease appear quite controversial in several studies.[14] Cortical atrophy in DM1 is a common finding too. However, the distribution and degree of cortical atrophy do not always correlate with cognitive involvement, age at onset, disease duration, neuromuscular status and genetic condition. Only one SPECT study has demonstrated significantly low cerebral blood flow (CBF) in DM1, compared to controls.[14] The major regional-CBF defects were found in both frontal and temporo-parietal regions and more severe degrees of hypoperfusion were seen in maternally inherited DM1. Three PET studies have reported a reduction in the cortical glucose utilization rate (CMRGlu) in DM1, in a CTG-dependent manner.[14]

Brain pathology of post-mortem brains of DM1-patients shows cell loss in specific areas such as in the dorsal raphe nucleus, superior central nucleus, dorsal and ventral medullary nuclei and subtrigeminal medullary nucleus.[1] Neuronal loss in the superficial layer of the frontal, parietal and occipital cortex, as well as in the substantia nigra and locus coeruleus, have also been reported.[1] Neuronal eosinophilic inclusion bodies were described in early studies in up to 30% of the thalamic nuclei of DM1 patients.[1] Their clinical significance is still unclear. The substantia nigra and caudate nucleus may be also involved. These inclusions are made of ubiquitin and microtubule-associated proteins, which means a possible neuropathological substrate for including myotonic dystrophies among the degenerative disorders.[1] Mutant RNA accumulates as nuclear *foci* in specific brain areas where muscleblind proteins are also sequestered, leading to deregulated alternative splicing in neurons of several proteins. RNA *foci* were also detected in the subcortical white matter and corpus callosum. Neurofibrillary tangles of the Alzheimer-type have been demonstrated in DM1.[1,15] Whether the effects of a possible spliceopathy on tau transcripts alone account for the neurodegenerative aspects of patients with DM1 requires further in-depth molecular studies.

Eye[1]

Most frequent abnormality encountered in myotonic dystrophy is cataract, which can be the only manifestation of the disease, especially in elderly patients. It was first described in the early 1900s, but a familial transmission of cataract and muscle disturbances was not documented until in 1918, by Fleischer.[1] The co-existence of cataract and myotonia is suggestive but not diagnostic of DM1. On slit lamp examination early cataract appears as iridescent, multiple, dust-like opacity, usually in the posterior subcapsular layers of the lens. Moreover, the appearance of the cataract can vary in terms of symmetry, lens distribution, iridescence, density and stage of maturation. Several studies indicate that over 80% of DM1 patients, although asymptomatic, have lens abnormalities.[1] The occurrence of cataract in DM1 increases with age, as does specificity. A mature DM1-cataract cannot be distinguished from other cataracts. It should be surgically removed and replaced with a synthetic lens (about 15% require replacement). Intervention risk and outcome seem to be similar to those of non-DM1 population. The relationship between cataract and neuromuscular status is not strict, since subjects with good muscle performance can have cataract (especially in older age) and severely impaired patients may have no opacities. No clear correlation with diabetes is documented.

Various other disturbances can also occur. Blepharoptosis is susceptible of surgical treatment when the eyelid is dramatically dropping, consequently causing abnormal neck and head posture, with visual loss (personal experience of Author). In these cases blepharoplasty is needed. Lagophthalm or inability to blink is relatively frequent and can determine persistent conjunctivitis and epiphora, with or without keratitis. It has to be properly treated in order to prevent abrasion of the sclera or cornea. Myotonia of the orbicularis oculi or forehead muscles can be occasionally seen in DM1, but blepharospasm is not essential. Although some clinical evidence exists, extraocular muscles are usually not severely involved.[1] Patients quite rarely complain of problems with gaze, such as diplopia.

Glands[1,16-18]

Endocrine abnormalities are frequently found in DM1. The aim of most studies has been to document functional abnormalities in the testes, ovaries, thyroid, parathyroid, pancreas, pituitary and adrenal glands. To date, anecdotic description of single or complex involvement of these glands can be found in literature. However, the main endocrine disturbances concern insulin resistance/diabetes or gonadal functions.

The first description of diabetes mellitus in DM1 dates back to 1950. Since then, several studies have been conducted to test glucidic metabolism in DM1 patients, who seem to have a four-fold risk of developing diabetes compared to healthy controls. Globally, diabetes does not frequently occur in DM1, while the finding of glucose intolerance or insulin resistance with persistently elevated insulin plasma levels after glucose load is common, even in the case of normal blood glucose titres (over 60%). A convincing explanation has been given by the recent demonstration of atypically distributed insulin receptors on the myofibre surface, likely due to an abnormal mRNA splicing process. In DM1, the pattern of insulin resistance is more common than diabetes.

The presence of testicular atrophy has been very well documented since the first descriptions of myotonic dystrophy. Here, macro and microscopic pathologic changes are frequently associated with impaired spermatogenesis, which often leads to oligo-azoospermia or adynamia of sperms. Impotence is sometimes observed, but may be underestimated and might be due to secondary smooth muscle or gonadal dysfunction. In women, gonadal abnormalities are frequently seen and less specific. DM1 females usually complain of irregular menstruation, but the high prevalence of gynaecological symptoms in healthy controls does not permit to generalize this observation to the entire DM1 population. Parallel to primary gonadal insufficiency, follicle-stimulating hormone (FSH) is usually found to be increased in both men and women, while luteinizing hormone (LH) is mostly normal. Infertility or reduced reproductive loss is frequent. During pregnancy, some major complications have been observed significantly more often in DM1 than in the general population, such as hydramnios (15 to 30%), low foetal maturity, neonatal deaths, retained placenta or placenta praevia and the need for caesarean delivery in about 10%.

The involvement of the pituitary gland in DM1 is not well documented. Most common abnormality encountered is an increase of FSH hormone in cases of primary hypogonadism.[1] Besides, there is no consensus about the modality of h-GH secretion.

Early evidence of enlarged cranial bones (skull thickness with parasinusal hypertrophy) seemed to suggest that an excessive increase of growth hormone (GH) could be responsible, such as in acromegaly. More recent observations suggest that impaired GH secretion after proper stimulation (as GH-RH and arginin) can be seen in DM1 (about 30%), without somatic evidence of GH defect or hypopituitarism. The relationship between this abnormal GH-curve response to stimulation and insulin resistance is not clear.

Prolactin and thyroid-stimulating hormone (TSH) do not seem to significantly differ from the general population, while dehydroepiandrosterone (DHEAs) is frequently found below normal levels in up to 80% of the DM1 cases.

Gastroenteric[1]

Smooth muscle can be involved in DM1 as can skeletal muscle. Its involvement—usually of mild severity—is not necessarily consistent with striated muscle. Gastrointestinal tract (including gall bladder) and urinary pathways may be involved.

Gastrointestinal symptoms are frequently reported by patients, who usually complain of several disturbances, with variable frequency, occurrence and severity (dysphagia, vomiting, abdominal pain, constipation, diarrhoea, incontinence).

One of the most frequently reported problems is associated with pharyngeal and oesophageal functions. It is usually characterized by difficulty in swallowing, irrespective of the severity of neuromuscular involvement or disease duration/onset. The nature of this problem seems to have multiple causes. Failure or changes in peristalsis, myotonia with abnormal contraction-decontraction times, enlargement of the oesophagus, the major consequence of which may be the aspiration of material into the bronchial tree, frequently associated with severe complications or death, especially in those patients who also present respiratory restrictive insufficiency.

Malabsorption, secondary to small bowel disturbances, appears rather anecdotal and negligible. Conversely, abdominal pain is quite common, particularly in young patients. The most typical features observed are those of 'irritable bowel syndrome'. They are usually mild and generally do not need surgical intervention, and should always be taken seriously to avoid surgical or anaesthesiological complications. Cases of colonic pseudo-obstructions are described, but rare.

Abnormal behaviour in the contraction-relaxation responses of internal and external anal sphincters has also been described. Also increased values of serum bilirubin and gamma-GT have been reported, in addition to cholelithiasis (significantly higher than in the general population), to be possible signs of cholestasis, likely due to an abnormal gall bladder emptying mechanism or abnormalities in bile acid metabolism. Mini-invasive surgical procedures for cholecistectomy should be carried out when necessary.

MUSCLE PATHOLOGY[19,20]

The presence of abnormalities in muscle pathology in DM1 is typical but not pathognomonic (Fig. 7).

- Centralized and/or internalized nuclei: They can be seen even at an early stage of the disease. Usually, the greater the number of internalized nuclei, the greater the muscular involvement of the patient. In longitudinal sections, typical

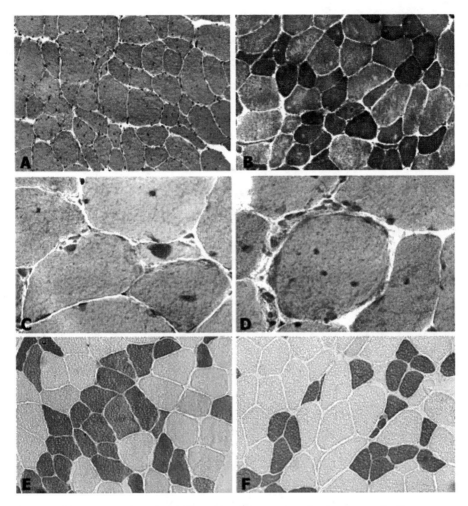

Figure 7. DM1 muscle pathological findings. Transverse section of voluntary muscle fibres, seen on optical microscopy. A) Haematoxylin-eosin (×10): internalization of nuclei, with fibre polydimensionalism. B) NADH-TR (×10): scattered central nuclei, along with Type 1 fibre hypotrophy and moth-eaten, scarcely reacting fibres. C) Gomori trichrome (×40): sarcoplasmic masses, central nuclei, atrophic fibres. D) Gomori trichrome (×40): single fibre with multiple internalized nuclei and sarcoplasmic masses, resembling a ring-fibre. E) Acid ATPase (×10): homotypic Type 1 fibre-grouping. F) Acid ATPase (×10): Type 1 fibre atrophy with Type 2 fibre hypertrophy.

chain-distributions can be observed, each of which can contain up to 20 nuclei. At present, such phenomenon may not be exclusively due to nuclear division; nuclear migration along the muscle fibre could be responsible for this pathological finding. Moreover, the presence of morphological heterogeneity of the nuclei has been reported: Some are picnotic, others appear pale and enlarged.

• Ring-fibres: Fibres with ring-shape myofibrils, first described by Heidenhain in 1918 and subsequently confirmed by Dubowitz and Brooke in 70% of biopsies, correlated with chronicization of the disease.

- Fibre-polydimensionalism: Type 1 and Type 2 fibres are clearly dishomogeneously distributed, with the former having typical small diameter. Such discrepancy evolves to marked atrophy of Type 1 fibres, whereas Type 2 fibres can occasionally hypertrophize. This association seems to be very specific for DM1, since other muscular dystrophies or other myotonic disorders do not usually have similar pictures.
- Sarcoplasmic masses can co-exist in homogeneous sarcoplasmic areas. They are frequently seen close to ring-fibres. Histochemical analyses conducted by Engel in 1962 showed that they are made of dysorganized intermyofibrillar material, where myofibrils and associated enzymes are completely absent. In 1970 Mussini et al clarified the regenerating nature of the masses, by ultrastructural microscopy.
- Several other myopathic phenomena, such as connective tissue proliferation, can be documented. In an advanced stage of the disease, angulated fibres with degeneration-regeneration aspects (signs of necrosis, basophilic fibres, phagocytosis, fibrosis, lobulated and moth-eaten fibres) can be occasionally detected. (Harper 2001, Dubowitz 2007, Mussini 1970).

PATHOGENESIS OF DM[21,22]

DM1 and DM2 are the only human genetic autosomal dominant inherited neuromuscular disorders with multi-system effect, in which the disease phenotype has been directly linked to disrupted regulation of alternative splicing, due to abnormal accumulation of toxic RNA within the affected nuclei.[22] To date, the pathogenesis of both DM has not yet been completely understood. The deposition of an abnormal transcribed and nontranslated RNA from the sequence [CTG]$_n$ at DMPK-gene *locus*, within the nuclei of affected cells that express this gene, is supposed to be responsible for the *primum movens* of the pathogenic chain. The demonstrated epiphenomenon of this process is the presence of intranuclear '*foci*' of RNA, pathologically deposited inside the nuclei of affected cells (Fig. 8).

In the beginning the description of intranuclear '*foci*' of *CTG-repeat* transcripts temporarily went unnoticed and only at a later stage received greater interest. This shows that RNA can acquire, through an aberrant deposition process, an effect of '*toxic gain-of-function*'. This would justify the mutual complexity and similarity of syndromic features of both DM1 and DM2, although the genes recognized as responsible for the two diseases are very far from each other on the genetic map and the proteins which these two genes encode for are so functionally different.

However, in the past at least two different pathogenic mechanisms had been hypothesized to explain the pathogenesis of DM1: aploinsufficiency of DMPK gene and aploinsufficiency of neighbouring genes. In the first case a transcriptional defect of DMPK-gene is enough to determine at least part of the symptoms, related to the reduction of the DMPK transcripts and, consequently, of the protein; documented decreased rates of DMPK-mRNA in the myofibres of the patients seem to support this hypothesis, as does the development of arrhythmogenic cardiopathy in the DMPK-gene *knock-out* mouse model. The second hypothesis (*neighbouring genes* like SIX5, DMWD and some others) could justify the heterogeneous clinical features of DM1.

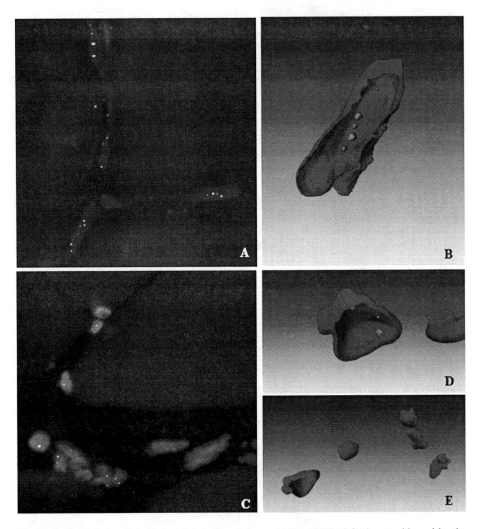

Figure 8. A) *Foci* of pathological accumulation of aberrant RNA (CTG-triplets) are evidenced by the FISH method in a case of DM1. B) A 3D-reconstruction of the detected *foci*, to better describe number, shape and intranuclear localization of the *foci*. C) *Foci* of CCTG-quadriplet repeats are shown and evidenced by FISH in a case of DM2. D,E) 3D reconstructions, in the same case.

At present, the most accredited pathogenic hypothesis for DM1 is that of a *toxic gain-of function of CUG-repeat*. Simultaneously several research studies have developed, on the role of certain endonuclear proteins targeting RNA and involved in endonuclear transfer mechanisms (*trafficking*), post-transcriptional modifications and newly-synthetized RNA catabolism. Among those, investigated are the dsRNA-BPs (double stranded *RNA binding proteins*, like PKR, TAR, RNA helicase A) and, above all, proteins of the MBNL family (muscleblind proteins), particularly MBNL1. The observation of intranuclear '*foci*' of aberrant RNA within the myonuclei and of their colocalization with aggregates of MBNL, has aroused great interest (data based on the FISH method, Fig. 8). Conversely, an increased intranuclear concentration of dsRNA-BPs and a decreased intranuclear concentration of MBNL has been documented. The hypothesis is that a 'sequestration of MBNL' by aberrant RNA, produced by the CTG triplet expansion, occurs. Therefore, the larger the CTG triplet expansion on chromosome 19q, the stronger the efficacy of the sequestration. In this way, the normal MBNL activity on the healthy RNA would be torn apart by a 'subtractive' mechanism. In this balance, dsRNA-BPs, recalled into the nucleus, would be increased, with a subsequent 'loss of stechiometric balance' of various parts and an impairment in the processing of many other neotranscribed endonuclear RNAs.

Several studies have been aiming to understand how alternative splicing dysregulation can be involved in the pathogenesis of myotonic dystrophy. To date, the results of these studies suggest that it corresponds to the clinical syndromic complexity of DM.

In healthy subjects, primary mRNA transcripts present in normal cells/myoblasts are subjected to post-transcriptional changes by alternative splicing process. This process normally produces different proteic isoforms, specific for that particular cell-line at that particular time. It is likely that in DM1-myoblasts certain primary mRNAs undergo an abnormal splicing process, resulting in aberrant secondary transcripts, unable to produce the original proteic isoform. The new proteic isoform, which can be atypical for that particular cytotype or completely original, would have little or no chance of becoming a functional protein in the cellular context where it has been produced.

This kind of anomaly has been proposed for at least 5 post-transcriptional processes: (1) chloride channel splicing; (2) insulin receptor splicing; (3) cardiac T-troponin splicing; (4) tau-protein splicing; (5) myotubularin splicing.

Each phenotype appears to correspond to specific damage, secondary to the abnormal ribonuclear processes.

In detail: (1) anomaly of the proteic isoform of the chloride channel (ClC-1) is believed to cause myotonia because of changes in the opening-closing kinetics of the voltage-dependent channel; (2) alteration of the insulin receptor is assumed to determine insulin-resistance, because the insulin receptor (IR) is unable to interact with its own ligand (insulin), for conformational reasons; (3) cardiac T-troponin (cTNT) anomaly could be implicated in the development of the arrhythmogenic cardiopathy typical of DM1; (4) abnormalities in tau-protein, associated with microtubules, appear to justify cognitive deficits; (5) myotubularin (*myotubularin-related-1*, MTMR1) deregulation mechanisms are considered to be responsible for the severe congenital phenotype and for its marked muscular atrophy and weakness.

Cataract, gonadal insufficiency and hypo-γ-globulinemia, remain unexplained.

The pathological sequence of these molecular events (deposition of aberrant RNA endonuclear '*foci*', sequestration of MBNL and consequent 'compensatory' increase in other RNA binding proteins and thereafter secondary dysregulations of RNA processing that lead to the formation of unusual, malfunctioning proteic isoforms) could be extended to the entire DM clinical spectrum of polynucleotide repeat-related diseases (Fig. 9).

Hence, although both DM1 and DM2 originate from two different genetic *loci* (chromosomes 19 and 3), they seem to determine complex and surprisingly similar syndromic features, since a 'common final pathway' can be recognized in their pathogenesis, as explained by Day et al[22] (Fig. 10).

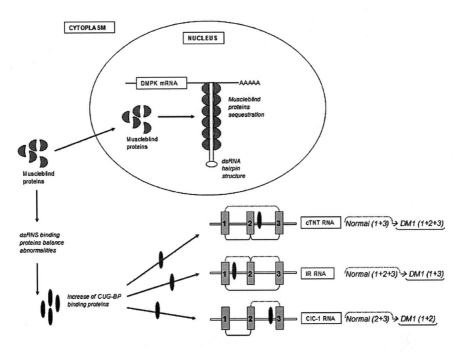

Figure 9. Pathogenic model of DM1: MBNL sequestration and increase in CUG-BP are followed by the alteration in primary mRNA transcript processing, with subsequent development of mature but pathological endonuclear mRNAs, from which several atypical proteic isoforms can derive.

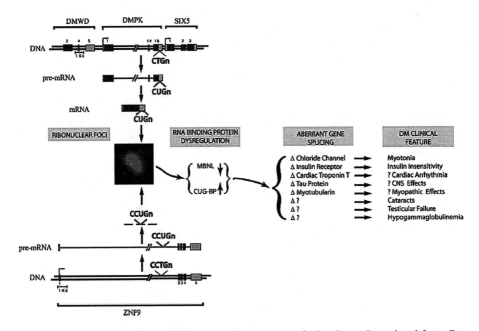

Figure 10. Pathogenesis of DM: two distinct *loci* for a common final pathway. Reproduced from: Day JW, Ranum LP. Neuromusc Disord 2005; 15:5-16;[22] ©2005 with permission from Elsevier.

MANAGEMENT, TREATMENT AND FUTURE PERSPECTIVES

At present, only symptomatic therapies are available for myotonic dystrophy. These treatments are generally safe, useful and their aim is to resolve -partially or completely-many symptoms that can occur in DM1.

Myotonia usually responds to mexiletine (200 to 600 mg/day), whose benefits have been documented in several studies.[23] A major contraindication to high doses of mexiletine is the existence of a heart conduction defect. In case of treatment with mexiletine or any other voltage-gate channels blockers, regular periodic cardiologic follow-up is highly recommended.

The presence of EKG and dynamic-EKG hallmarks of AV-blocks, that worsen over time or become complicated by additional heart disturbances or clinical symptoms, suggests more in-depth cardiologic investigations. Patients with major, potentially harmful, arrhythmic disturbances are candidates for pacemaker implantation. The implantation of an intracardiac defibrillator device must be considered in very selected cases (e.g., symptomatic patients with personal history of ventricular fibrillation).

Diabetes can be controlled by oral drugs or insulin, especially when it cannot be controlled by adequate physical activity or diet adjustment.

Gastrointestinal disturbances should always be recognized, in order to prevent aspiration of swallowed material into the bronchial tree (pharyngo-oesophageal dysfunction) and the consequences of acute colonic pseudoobstruction.[1]

Poor response to oral modafinil treatment has been documented in cases of recurrent daytime somnolence.[28]

Some peculiar conditions deserve a surgical approach. A mature cataract can be surgically removed and blepharoplasty considered in patients with severe blepharoptosis with visual loss. Surgery is also an option for tendon retractions and severe spine deformity, especially in children in association with serious respiratory impairment. In these cases, respiratory insufficiency and heart disturbances must be properly evaluated before any intervention, to avoid complications related to general anaesthesia.[12]

Hormone replacement therapy with testosterone in cases of hypogonadism has proven ineffective,[24] while small populations of DM1 patients have been treated with IGF-1, without significant side effects. Human GH replacement therapies in patients who show reduced h-GH secretion after stimulation, have been proposed, but these studies are still inconclusive.[25,26]

Unfortunately, therapies designed to arrest or invert the process of muscle atrophization are not available. Moreover, techniques capable of restoring a correct genetic condition, reducing the causative [CTG]$_n$ triplet expansion to normal values, have not been developed. However, the progressing knowledge on the pathogenic mechanisms underlying myotonic dystrophies is disclosing new perspectives on the possibility of synthesizing effective drug targeting to precisely interrupt the early pathologic cascade of events that leads from [CTG]$_n$ repeats to aberrant splicings. The principles of RNA interference have been taken into account and studies in this field are ongoing.

CONCLUSION

Myotonic dystrophy is the most frequent autosomal dominant inherited muscular dystrophy. A definite diagnosis is possible only by DNA analysis, since muscle pathology changes are typical but not pathognomonic. DM1 is due to a [CTG]$_n$ triplet expansion on chromosome 19q. This nucleotide expansion acquires a *toxic gain-of-function* within the affected myonuclei, thus leading to an abnormal RNA splicing process that involves several other genes. The severe congenital picture of myotonic dystrophy Type 1 is possible only in case of maternal transmission to the newborn. Anticipation phenomenon means earlier and severer involvement in offspring. DM1 is characterized by pleiotropism (involvement of several organs and systems) with variable severity. The patients can have major heart disturbances, endocrine involvement with insulin resistance and reduced fertility, cataract, respiratory insufficiency, cognitive impairment, gastroenteric involvement. At present, treating DM1 mostly means adopting symptomatic pharmacological or surgical strategies, or preventing the occurrence of potentially life-threatening cardiac and respiratory problems. It is reasonable to hypothesize that in the future new and safe pharmacological options—targeted to the gene defect or pathogenic mechanism—will be available to cure myotonic dystrophy.

REFERENCES

1. Harper PS. Myotonic dystrophy. Ed. W. B. Saunders, third edition.
2. Brook JD, McCurrach ME, Harley HG et al. Molecular basis of myotonic dystrophy: expansion of a trinucleotide (CTG) repeat at the 3' end of a transcript encoding a protein kinase family member. Cell 1992; 68:799-808.
3. Liquori CL, Ricker K, Moseley ML et al. Myotonic dystrophy type 2 caused by a CCTG expansion in intron 1 of ZNF9. Science 2001; 293:864-867.
4. Lavedan C, Hofmann-Radvanyi H, Shelbourne P et al. Myotonic dystrophy: size- and sex-dependent dynamics of CTG meiotic instability and somatic mosaicism. Am J Hum Genet 1993; 52:875-883.
5. Thornton CA, Griggs RC, Moxley RT 3rd. Myotonic dystrophy with no trinucleotide repeat expansion. Ann Neurol 1994; 35:269-272.
6. Udd B, Krahe R, Wallgren-Pettersson C et al. Proximal myotonic dystrophy—a family with autosomal dominant muscular dystrophy, cataracts, hearing loss and hypogonadism: heterogeneity of proximal myotonic syndromes? Neuromusc Disord 1997; 7:217-228.
7. Moxley RT 3rd, Meola G, Udd B et al. Report of the 84th ENMC workshop: PROMM (proximal myotonic myopathy) and other myotonic dystrophy-like syndromes: 2nd workshop. 13-15th October, 2000, Loosdrecht, The Netherlands. Neuromuscul Disord 2002; 12:306-317.
8. New nomenclature and DNA testing guidelines for myotonic dystrophy type 1 (DM1). The International Myotonic Dystrophy Consortium. Neurology 2000; 54:1218-1221.
9. de Die-Smulders CE, Höweler CJ, Thijs C et al. Age and causes of death in adult-onset myotonic dystrophy. Brain 1998; 121:1557-1563.
10. Reardon W, Newcombe R, Fenton I et al. The natural history of congenital myotonic dystrophy: mortality and long term clinical aspects. Arch Dis Child 1993; 68:177-181.
11. Mathieu J, Boivin H, Meunier D et al. Assessment of a disease-specific muscular impairment rating scale in myotonic dystrophy. Neurology 2001; 56:336-340.
12. Mathieu J, Allard P, Gobeil G et al. Anesthetic and surgical complications in 219 cases of myotonic dystrophy. Neurology 49:1646-1650.
13. Melacini P, Buja G, Fasoli G et al. The natural history of cardiac involvement in myotoic dystrophy: an eight-year follow-up in 17 patients. Clin Cardiol 1988; 11:231-238.
14. Romeo V, Pegoraro E, Ferrati C et al. Brain involvement in myotonic dystrophies: neuroimaging and neuropsychological comparative study in DM1 and DM2. J Neurol 2010; 257:1246-1255.
15. Meola G, Sansone V. Cerebral involvement in myotonic dystrophies. Muscle Nerve 2007; 36:294-306.
16. Mastrogiacomo I, Bonanni G, Menegazzo E et al. Clinical and hormonal aspects of male hypogonadism in myotonic dystrophy. Ital J Neurol Sci 1996; 17:59-65.
17. Barreca T, Muratorio A, Sannia A et al. Evaluation of twenty-four-hour secretory patterns of growth hormon and insulin in patients with myotonic dystrophy. J Clin Endocrinol Metab 51:1089-1092.
18. Moxley RT, Corbett AJ, Minaker KL et al. Whole body insulin resistance of myotonic dystrophy. Ann Neurology 1984; 15:157-162.
19. Dubowitz V. Muscle biopsy, 2nd edition. Philadelphia: WB Saunders, 1985.

20. Mussini I, Di Mauro S, Angelini C. Early ultrastructural and biochemical changes in muscle in dystrophia myotonica. J Neurol Sci 10:585-604.
21. Faustino NA, Cooper TA. Pre-mRNA splicing and human disease. Gen Develop 2003; 17:419-437.
22. Day JW, Ranum LP. RNA pathogenesis of the myotonic dystrophies. Neuromusc Disord 2005; 15:5-16.
23. Logigian EL, Martens WB, Moxley RT 4th et al. Mexiletine is an effective antimyotonia treatment in myotonic dystrophy type 1. Neurology 2010; 74:1441-1448.
24. Griggs RC, Pandya S, Florence JM et al. Randomized controlled trial of testosterone in myotonic dystrophy. Neurology 1989; 39:219-222.
25. Moxley RT 3rd. Potential for growth factor treatment of muscle disease. Curr Opin Neurol 1994; 7:427-434.
26. Vlachopapadopoulou E, Zachwieja JJ, Gertner JM et al. Metabolic and clinical response to recombinant human insulin-like growth factor I in myotonic dystrophy—a clinical research center study. J Clin Endocrinol Metab 1995; 80:3715-3723.
27. Maas O, Paterson AS. Mental changes in families affected by Dystrophia Myotonica. Lancet 1937; 1:21-23.
28. Annane D, Moore DH, Barnes PR et al. Psychostimulants for hypersomnia (excessive daytime sleepiness) in myotonic dystrophy. Cochrane Database Syst Rev. 2002; 4:CD003218.

CHAPTER 19

NEURODEGENERATION IN DIABETES MELLITUS

Hiroyuki Umegaki

Department of Geriatrics, Nagoya University Graduate School of Medicine, Nagoya, Aichi, Japan
Email: umegaki@med.nagoya-u.ac.jp

Abstract: Diabetes mellitus is recognized as a group of heterogeneous disorders with the common elements of hyperglycaemia and glucose intolerance due to insulin deficiency, impaired effectiveness of insulin action, or both. The prevalence of Type 2 diabetes mellitus (T2DM) increases with age and dementia also increases its incidence in later life. Recent studies have revealed that T2DM is a risk factor for cognitive dysfunction or dementia, especially those related to Alzheimer's disease (AD). Insulin resistance, which is often associated with T2DM, may induce a deficiency of insulin effects in the central nervous system (CNS). Insulin may have a neuroprotective role and may have some impact on acetylcholine (ACh) synthesis. Hyperinsulinemia, induced by insulin resistance occurring in T2DM, may be associated with insulin deficiency caused by reduced insulin transport via the blood brain barrier (BBB). Insulin has multiple important functions in the brain. Some basic research, however, suggests that insulin accelerates Alzheimer-related pathology through its effects on the amyloid beta (Aβ) metabolism and tau phosphorylation.

Asymptomatic ischemic lesions in T2DM subjects may lower the threshold for the development of dementia and this may explain the inconsistency between the basic research and clinicopathological studies.

More research to elucidate the mechanism of neurodegeneration associated with T2DM is warranted.

INTRODUCTION

Diabetes mellitus is recognized as a group of heterogeneous disorders with the common elements of hyperglycaemia and glucose intolerance due to insulin deficiency, impaired effectiveness of insulin action, or both. Diabetes mellitus is classified on the basis of

---

Neurodegenerative Diseases, edited by Shamim I. Ahmad.
©2012 Landes Bioscience and Springer Science+Business Media.

etiology and the clinical presentation of the disorder into mainly two types; Type 1 diabetes and Type 2 diabetes (T2DM). Type 1 diabetes is sometimes called insulin-dependent, immune-mediated or juvenile-onset diabetes. It is caused by destruction of the insulin-producing cells of the pancreas, typically due to an auto-immune reaction, in which they are attacked by the body's defense system. The disease can affect people of any age, but usually occurs in children or young adults. On the other hand, T2DM is characterized by insulin resistance and partial insulin deficiency, either of which may be present at the time that diabetes becomes clinically manifested. Insulin resistance is defined as an inadequate response by insulin target tissues, such as skeletal muscle, liver and adipose tissue, to the physiological effects of circulating insulin and often is accompanied by raised insulin levels. T2DM is often, but not always, associated with obesity, which itself can cause insulin resistance and lead to elevated blood glucose levels.

The prevalence of T2DM increases with age and dementia also increases its incidence in later life. Therefore, the coincidence of T2DM and dementia increases with ageing. Moreover, recent studies have indicated that older people with T2DM have a higher risk of cognitive dysfunction or dementia.[1] There is ample evidence that T2DM is related not only to vascular dementia but also to the clinical diagnosis of Alzheimer's disease (AD)-type dementia.[2]

In a large epidemiological study, The Rotterdam Study,[3] T2DM patients showed an increased risk for developing dementia. Patients treated with insulin were at an even higher relative risk, as high as 4.3-fold. In another study,[4] which examined some 800 nuns and priests longitudinally over 9 years, 15% of the cohort had or developed T2DM and showed a 65% increased risk for developing AD. A cohort of Japanese-Americans in Hawaii[5,6] showed a 1.8-fold higher risk for developing AD and a 2.3-fold higher risk for vascular dementia.

BRAIN IMAGING STUDIES IN TYPE 2 DIABETES

In T2DM patients the incidence of small vessel disease including lacunae infarcts and white matter lesions increased.[7,8] Some studies have shown that Type 2 diabetic patients compared to nondiabetic individuals show reduced volumes of the hippocampus and amygdala[9,10] and a threefold increased risk for medial temporal lobe atrophy.[11]

UNDERLYING MECHANISM OF COGNITIVE DYSFUNCTION IN T2DM

The precise mechanisms underlying T2DM-related cognitive dysfunction or the development of dementia, especially AD-type dementia, remain to be elucidated; however, several hypothetical mechanisms have been proposed.

High glucose concentration, a major pathological characteristic of diabetes, may have toxic effects on neurons in the brain through osmotic insults and oxidative stress and the maintenance of chronic high glucose also leads to the enhanced formation of advanced glycation endoproducts (AGEs), which have potentially toxic effects on neurons. AGEs are formed as the end-products of the Maillard reaction,[12] during which reducing sugars can react with the amino groups of proteins to produce cross-linked complexes and unstable compounds. AGEs have been found in the central nervous system (CNS) of diabetics. AGEs couple with free radicals and create oxidative damage, which in turn leads to neuronal injury.[13] Other than their direct toxicity, AGEs also reactivate microglia in the CNS. There is a wealth of evidence demonstrating that microglia, the resident innate immune cells in the brain, can become deleterious and damage neurons.[14] This process is implicated as an underlying mechanism in diverse neurodegenerative diseases, including AD. While microglial function is beneficial and mandatory for normal CNS functioning, microglia become toxic to neurons when they are over-activated and unregulated. In diabetes, oxidative stress also increases because of reduced antioxidant capacity.[15] It has been suggested that oxidative stress leads to neuronal injury through mitochondrial dysfunction.[16]

T2DM, especially in conjunction with obesity, is characterized by insulin resistance and/or hyperinsulinemia. The insulin molecule consists of two peptidic chains joined by two disulfide bonds. It is primarily secreted by the beta cells of the pancreas and is normally released into the circulation through the portal vein in response to a rise in blood glucose. Insulin degrading enzyme (IDE) catabolizes insulin in the liver, kidneys and muscles.[17,18] It is generally agreed that insulin located within the brain is mostly of pancreatic origin, having passed through the blood-brain barrier, although there is debate about the amount of insulin that is produced de novo within the CNS.[19] Major known actions of insulin in the brain include control of food intake (via insulin receptors located in the olfactory bulb and thalamus) and effects on cognitive functions, including memory.[20,21]

Blood glucose abnormalities and insulin resistance may also have some impact on acetylcholine (ACh) synthesis. Acetylcholine transferase (ChAT), which is an enzyme responsible for ACh synthesis, is expressed in insulin receptor-positive cortical neurons and insulin regulates the ChAT expression. Because ACh is a critical neurotransmitter in cognitive function, it may be relevant to neurocognitive disorders in diabetics.[22]

Recent basic research demonstrated that insulin signaling in the CNS prevents the pathologic binding of amyloid beta (Aβ) oligomers. Aβ oligomers are soluble molecules that attach with specificity to particular synapses, acting as pathogenic ligands.[23] The attack on synapses inhibits long-term potentiation (LTP).[24] Insulin and PPAR-gamma agonist, an insulin sensitizer, may have some protective effects on the toxic effects of Aβ oligomer.[25]

Insulin has multiple important functions in the brain, as mentioned above. These functions are disrupted in insulin-resistant states. The transport of insulin into the brain across the BBB is reduced in insulin-resistance-associated hyperinsulinemia and insulin levels in the brain are subsequently lowered.[26,27] A small pilot study demonstrated that intranasal insulin had some benefits in early AD patients.[28] With intranasal administration, insulin bypasses the periphery and the blood-brain barrier, reaching the brain and cerebrospinal fluid (CSF) within minutes via extracellular bulk flow transport along olfactory and trigeminal perivascular channels, as well as through more traditional axonal transport pathways.[29,30] Intranasal administration of insulin improved memory and attention in humans without affecting plasma glucose levels.

Some basic research, however, suggests that insulin accelerates AD-related pathology through its effects on the Aβ metabolism and tau phosphorylation.[2] Insulin reportedly raises Aβ concentrations in plasma in AD subjects and these effects may contribute to the risk of AD in T2DM. The desensitization of insulin receptors, i.e., insulin resistance, reduces the synthesis of several proteins, including insulin-degrading enzyme (IDE). IDE degrades Aβ as well as insulin and reduced amounts of IDE may result in greater amyloid deposition. Less insulin signaling may also induce increased activity of glycogen synthase kinase-3β (GSK-3β), which leads to the enhanced phosphorylation of tau protein and the formation of neurofibrillary tangles (NFTs). The results of pathological assessments in AD with or without DM are highly controversial. Several pathological studies using autopsy samples have demonstrated, however, that dementia subjects with diabetes have less AD-related neuropathology than subjects without diabetes. One study demonstrated that diabetics show significantly less AD-associated neuropathology,[31] while another failed to show any relationship between diabetes and AD-associated neuropathology.[32]

VASCULAR CONTRIBUTION TO T2DM-ASSOCIATED COGNITIVE DYSFUNCTION

AD has been thought to be a neurodegenerative disorder which can be sharply distinguished from vascular dementia. Recent studies, however, suggest that the distinction between AD and vascular dementia may not be tenable. There is now substantial and growing evidence that vascular disorders and/or impaired cerebral perfusion contribute to the development of sporadic AD. For example, cerebrovascular pathology including stroke seems to play an important role in the eventual development of the clinical symptoms of AD.[33]

On cerebral MRI, white matter hyperintensities and lacunae, both of which are frequently observed in the elderly, are generally viewed as evidence of small vessel disease in the brain (white matter lesions and lacunae). These lesions are frequently concomitant with Alzheimer-related neuropathology (senile plaques and NFTs) and contribute to cognitive impairment in AD subjects.[34] We previously reported that small vessel diseases affect cognitive function in older diabetics who have not developed either overt dementia or symptomatic stroke.[35,36] The number of asymptomatic infarcts and the extent of white matter lesions in the brain detected with magnetic resonance imaging (MRI) were found to be associated with the scores on several cognitive functional tests, especially the digit symbol substitution test, a neurocognitive test that primarily reflects declines in perceptual speed. We also reported that an inflammatory cytokine, tumor necrosis factor-α (TNF-α), which is a risk factor for atherosclerosis, is related to cognitive dysfunction in older nondemented diabetics.[37] A recent study demonstrated that T2DM subjects with a clinical diagnosis of dementia have less Alzheimer-related pathology but more ischemic lesions.[38] This finding supports the hypothesis that small vessel disease lowers the threshold for the development of dementia. That is, if subjects have the same level of cognitive dysfunction, those with a combination of two types of pathologies have fewer pathological changes in each of their pathologies than those with a single pathology that is severe enough to cause the cognitive dysfunction. Therefore, these pathological reports do not necessarily eliminate the possibility that DM accelerates the development of Alzheimer-related neuropathology in patients with a clinical diagnosis of dementia. Arvanitakis et al demonstrated in 2004 that T2DM increases the incidence of AD as determined by clinical dignosis,[39] but T2DM ameliorated perceptual speed but not global cognition. A previous study[40] and our own studies[35,36] showed that cerebral ischemic lesions are preferentially associated with a lower measure of perceptual speed. These results also suggest that small vessel disease contributes to cognitive decline in these populations.

Hypertension is often accompanied by diabetes and several longitudinal studies appear to support the notion that hypertension predisposes individuals to cognitive decline and the development of dementia.[41] Vascular alterations induced by high blood pressure may contribute to cognitive dysfunction. Hypertension is also associated with cerebrovascular disease including lacunar brain infarcts and white matter lesions, which may contribute to cognitive impairment in diabetics.

THE EFFECTS OF BLOOD GLUCOSE CONTROL

A large cohort study, the ACCORD-MIND trail, showed that HbA1c levels were cross-sectionally associated with worse performance on several cognitive functional tests that were very similar to the ones that we used in the current study.[42] Maggi et al[43] reported that higher HbA1c levels at baseline were prospectively associated with delayed verbal memory decline.[41] We also reported that higher HbA1c levels at baseline correlated with a greater decrease in scores on the DSS and Stroop tests after three years. These results suggest that diabetic disease control is important for the preservation of cognitive function in elderly diabetic patients. A recent prospective study however, reported that HbA1c levels at baseline had no effect in 5 cognitive domains.[44] Large prospective studies are warranted regarding this issue.

Another recent report suggested that a history of severe hypoglycemic episodes was associated with a greater risk of dementia.[45] Diabetic control should be balanced the merit of the treatment with the risk of hypoglycemia.

FUTURE DIRECTION

Recently, amyloid imaging technology with positron emission tomography (PET), which visualizes Aβ depositions in the human brain, has been developed and is now widely available,[44] although some limitations in resolution and specificity still exist. This technology can be used to investigate the relative contributions of ischemic and neurodegenerative changes to the increased development of dementia in T2DM subjects. Longitudinal follow-up of T2DM subjects without overt dementia using both amyloid imaging and MRI may help to elucidate these issues, especially with higher field MRI with some potential for the imaging of small vessel diseases,[46] also diffusion tensor imaging method may also provide useful data.[47] Furthermore, the CSF biomarkers total tau, hyperphosphorylated tau and the 42 amino acid form of Aβ (Aβ-42) are now established markers for AD[48] and can be used to identify AD in the early, MCI stage of the disease with high accuracy.[49] The CSF of AD and MCI patients shows decreased values of Aβ-42 and increased total tau or phosphorylated tau.[50] Following up until the development of overt dementia would make it possible to compare both the amyloid load and ischemic lesions before and after the development of dementia with these technologies. Moreover, amyloid imaging and measuring CSF biomarkers in nondemented older people with or without insulin resistance could verify the hypothesis that insulin plays a role in the processing and deposition of Aβ. These investigations are important considering the future availability of disease-modifying therapeutics such as Aβ vaccination and inhibitors for Aβ secretions.

At present, vascular risk factors may represent a therapeutic target, while neurodegenerative pathologies have not yet been amenable to treatment. Vascular risk factors including diabetes and hypertension are reportedly associated with the progression of lacunae and white matter lesions;[51] however, the beneficial effects on cognitive function of pharmaceutical interventions with antidiabetics and antihypertensives are less clear in terms of the inhibition of the progress of lacunae and white matter lesions. It remains to be investigated whether medical interventions that mitigate vascular risk factors have protective effects against the development and progress of dementia. If such protective effects do exist, the underlying mechanism of the therapeutic effects should be interesting, whether it relies on the inhibition of the development of vascular lesions or on the inhibition of the neurodegenerative process.

With the ongoing increase in the size of the older population, T2DM-associated cognitive dysfunction and dementia are becoming increasingly larger problems. A greater understanding of the relevant pathophysiology and the establishment of better therapeutic interventions are urgent needs.

CONCLUSION

T2DM is associated with cognitive dysfunction and the incidence of dementia including AD. The underlying mechanism of this association should be elucidated. This could lead to clarification of the pathogenesis of AD and to the development of a treatment or preventive method.

REFERENCES

1. Stewart R, Liolitsa D. Type 2 diabetes mellitus, cognitive impairment and dementia. Diabetic Medicine 1999; 16:93-112.
2. Li L, Hölscher C. Common pathological processes in Alzheimer disease and type 2 diabetes: a review. Brain Res Rev 2007; 56:384-402.
3. Ott A, Stolk RP, van Harskamp F et al. Diabetes mellitus and the risk of dementia: The Rotterdam Study. Neurology 1999; 58:1937-1941.
4. Arvanitakis Z, Wilson RS, Bienias JL et al. Diabetes mellitus and risk of Alzheimer's disease and decline in cognitive function. Arch Neurol 2004; 61:661-666.
5. Peila R, Rodriguez BL, Launer LJ. Honolulu-Asia Aging Study. Type 2 diabetes, APOE gene and the risk for dementia and related pathologies. Diabetes 2002; 51:1256-1262.
6. Peila R, Rodriguez BL, White LR et al. Fasting insulin and incident dementia in an elderly population of Japanese-American men. Neurology 2004; 63:228-233.
7. Vermeer SE, Koudstaal PJ, Oudkerk M et al. Prevalence and risk factors of silent brain infarcts in the population-based Rotterdam Scan Study. Stroke 2002; 33:21-25.
8. de Leeuw FE, de Groot JC, Achten E et al. Prevalence of cerebral white matter lesions in elderly people: a population based magnetic resonance imaging study. The Rotterdam Scan Study. J Neurol Neurosurg Psychiatry 2001; 70:9-14.
9. den Heijer T, Vermeer SE, van Dijk EJ et al. Type 2 diabetes and atrophy of medial temporal lobe structures on brain MRI. Diabetologia 2003; 46:1604-1610.
10. Korf ES, White LR, Scheltens P et al. Brain aging in very old men with type 2 diabetes: the Honolulu-Asia Aging Study. Diabetes Care 2006; 29:2268-2274.
11. Korf ES, van Straaten EC, de Leeuw FE et al. LADIS Study Group. Diabetes mellitus, hypertension and medial temporal lobe atrophy: the LADIS study. Diabet Med 2007; 24:166-171.
12. Yamagishi S, Ueda S, Okuda S. Food-derived advanced glycation end products (AGEs): a novel therapeutic target for various disorders. Curr Pharm Des 2007; 13:2832-2836.
13. Valente T, Gella A, Fernàndez-Busquets X et al. Immunohistochemical analysis of human brain suggests pathological synergism of Alzheimer's disease and diabetes mellitus. Neurobiol Dis 2010; 37:67-76.
14. Block ML, Zecca L, Hong JS. Microglia-mediated neurotoxicity: uncovering the molecular mechanisms. Nat Rev Neurosci 2007; 8:57-69.
15. Evans JL, Goldfine ID, Maddux BA et al. Oxidative stress and stress-activated signaling pathways: a unifying hypothesis of type 2 diabetes. Endocr Rev 2002; 23:599-622.
16. Moreira PI, Santos MS, Seiça R et al. Brain mitochondrial dysfunction as a link between Alzheimer's disease and diabetes. J Neurol Sci 2007; 257:206-214.
17. Watson GS, Craft S. The role of insulin resistance in the pathogenesis of Alzheimer's disease: implications for treatment. CNS Drugs 2003; 17:27-45.
18. Davis SN, Granner DK. Insulin, oral hypoglycemic agents and the pharmacology of the endocrine pancreas. In: Hardman JG, Gilman AG, Limbird LE, eds. Gilman and Goodman's the Pharmacological Basis of Ttherapeutics, 9th. New York: McGraw-Hill, 1996:1487-1517.

19. Woods SC, Seeley RJ, Baskin DG et al. Insulin and the blood-brain barrier (BBB). Curr Pharm Des 2003; 9:795-800.
20. Havrankova J, Roth J, Brownstein M. Insulin receptors are widely distributed in the central nervous system of the rat. Nature 1978; 272:827-829.
21. Freychet P. Insulin receptors and insulin action in the nervous system. Diab Metab Res Rev 2000; 16:390-392.
22. Rivera EJ, Goldin A, Fulmer N et al. Insulin and insulin-like growth factor expression and function deteriorate with progression of Alzheimer's disease: link to brain reductions in acetylcholine. J Alzheimers Dis 2005; 8:247-268.
23. Lacor PN, Buniel MC,Chang L et al. Synaptic targeting by Alzheimer's-related amyloid beta oligomers. J Neurosci 2004; 24:10191-10200.
24. Walsh DM, Klyubin I, Fadeeva JV et al. Naturally secreted oligomers of amyloid beta protein potently inhibit hippocampal long-term potentiation in vivo. Nature 2002; 416:535-539.
25. Sato T, Hanyu H, Hirao K et al. Efficacy of PPAR-gamma agonist pioglitazone in mild Alzheimer disease. Neurobiol Aging 2009. [Epub ahead of print]
26. Zhao WQ, Townsend M. Insulin resistance and amyloidogenesis as common molecular foundation for type 2 diabetes and Alzheimer's disease. Biochim Biophys Acta 2009; 1792:482-496.
27. Craft S, Peskind E, Schwartz MW et al. Cerebrospinal fluid and plasma insulin levels in Alzheimer's disease: relationship to severity of dementia and apolipoprotein E genotype. Neurology 1998; 50:164-168.
28. Reger MA, Watson GS, Green PS et al. Intranasal insulin improves cognition and modulates beta-amyloid in early AD. Neurology 2008; 70:440-448.
29. Thorne RG, Pronk GJ, Padmanabhan V et al. Delivery of insulin-like growth factor-I to the rat brain and spinal cord along olfactory and trigeminal pathways following intranasal administration. Neuroscience 2004; 127:481-496.
30. Benedict C, Hallschmid M, Hatke A et al. Intranasal insulin reportedly improves memory and attention in humans. Psychoneuroendocrinology 2004; 29:1326-1334.
31. Beeri MS, Silverman JM, Davis KL et al. Type 2 diabetes is negatively associated with Alzheimer's disease neuropathology. J Gerontol A Biol Sci Med Sci 2005; 60:471-475.
32. Arvanitakis Z, Schneider JA, Wilson RS et al. Diabetes is related to cerebral infarction but not to AD pathology in older persons. Neurology 2006; 7:960-1965.
33. de la Torre JC. Is Alzheimer's disease a neurodegenerative or a vascular disorder? Data, dogma and dialectics. Lancet Neurol 2004; 3:184-190.
34. Riekse RG, Leverenz JB, McCormick W et al. Effect of vascular lesions on cognition in Alzheimer's disease: a community-based study. J Am Geriatr Soc 2004; 52:1442-1448.
35. Akisaki T, Sakurai T, Takata T et al. Cognitive dysfunction associates with white matter hyperintensities and subcortical atrophy on magnetic resonance imaging of the elderly diabetes mellitus. Japanese elderly diabetes intervention trial (J-EDIT). Diabetes Metab Res Rev 2006; 22:376-384.
36. Umegaki H, Kawamura T, Mogi N et al. Glucose control levels, ischaemic brain lesions and hyperinsulinaemia were associated with cognitive dysfunction in diabetic elderly. Age Ageing 2008; 37:458-461.
37. Suzuki M, Umegaki H, Ieda S et al. Factors associated with cognitive impairment in elderly patients with diabetes mellitus. J Am Geriatr Soc 2006; 54:558-559.
38. Sonnen JA, Larson EB, Brickell K et al. Different patterns of cerebral injury in dementia with or without diabetes. Arch Neurol 2009; 66:315-322.
39. Arvanitakis Z, Wilson RS, Bienias JL et al. Diabetes mellitus and risk of Alzheimer disease and decline in cognitive function. Arch Neurol 2004; 61:661-666.
40. Schneider JA, Wilson RS, Cochran EJ et al. Relation of cerebral infarctions to dementia and cognitive function in older persons. Neurology 2003; 60:1082-1088.
41. Peters R, Beckett N, Forette F et al. Incident dementia and blood pressure lowering in the hypertension in the very elderly trial cognitive function assessment (HYVET-COG): a double-blind, placebo controlled trial. Lancet Neurol 2008; 7:683-689.
42. Cukierman-Yaffe T, Gerstein HC, Williamson JD et al. Relationship between baseline glycemic control and cognitive function in individuals with type 2 diabetes and other cardiovascular risk factors: the action to control cardiovascular risk in diabetes-memory in diabetes (ACCORD-MIND) trial; Action to Control Cardiovascular Risk in Diabetes-Memory in Diabetes (ACCORD-MIND) Investigators. Diabetes Care 2009; 32:221-226.
43. Maggi S, Limongi F, Noale M et al. LSA Study Group. Diabetes as a risk factor for cognitive decline in older patients. Dement Geriatr Cogn Disord 2009; 27:24-33.
44. Ikonomovic MD, Klunk WE, Abrahamson EE et al. Post-mortem correlates of in vivo PiB-PET amyloid imaging in a typical case of Alzheimer's disease. Brain 2008; 131:1630-1645.

45. Whitmer RA, Karter AJ, Yaffe K et al. Hypoglycemic episodes and risk of dementia in older patients with type 2 diabetes mellitus. JAMA 15; 301:1565-1572.
46. Novak V, Abduljalil AM, Novak P et al. High-resolution ultrahigh-field MRI of stroke. Magn Reson Imaging 2005; 23:539-548.
47. Kodl CT, Franc DT, Rao JP et al. Diffusion tensor imaging identifies deficits in white matter microstructure in subjects with type 1 diabetes that correlate with reduced neurocognitive function. Diabetes 2008; 57:3083-3089.
48. Zetterberg LO, Wahlund K, Blennow. Cerebrospinal fluid markers for prediction of Alzheimer's disease. Neurosci Lett 2003; 352:67-69.
49. Hansson H, Zetterberg P, Buchhave E et al. Association between CSF biomarkers and incipient Alzheimer's disease in patients with mild cognitive impairment: a follow-up study. Lancet Neurol 2006; 5:228-234.
50. Ewers M, Buerger K, Teipel SJ et al. Multicenter assessment of CSF-phosphorylated tau for the prediction of conversion of MCI. Neurology 2007; 69:2205-2212.
51. Gouw AA, van der Flier WM, Fazekas F et al. Progression of white matter hyperintensities and incidence of new lacunes over a 3-year period: the Leukoaraiosis and Disability study. Stroke 2008; 39:1414-1420.

CHAPTER 20

NEUROFIBROMATOSES

Erik J. Uhlmann and Scott R. Plotkin*

Massachusetts General Hospital, Stephen E. and Catherine Pappas Center for Neuro-Oncology, Boston, Massachusetts, USA
Corresponding Author: Scott R. Plotkin—Email: splotkin@partners.org

Abstract: The studies of familial tumor predisposition syndromes have contributed immensely to our understanding of oncogenesis. Neurofibromatosis 1, neurofibromatosis 2 and schwannomatosis are inherited autosomal dominant neurocutaneous disorders with complete penetrance. They are clinically and genetically distinct and considerable knowledge has been gathered about their pathogenesis. In this chapter, the genetics, molecular mechanism of disease, as well as clinical features, diagnosis and treatment are discussed.

INTRODUCTION

Although descriptions of patients with neurofibromatosis date back centuries, Wishart and von Recklinghausen are credited with the first description of neurofibromatosis 2 (NF2) and neurofibromatosis 1 (NF1) respectively.[1-3] Neurofibromatosis was later recognized as a heterogeneous group of inherited disorders, characterized by multiple nerve sheath tumors of the skin and spinal nerves. In reality, neurofibromatosis is a multisystem disease with both tumor and nontumor manifestations. Three specific diseases will be discussed, neurofibromatosis Type 1 (NF1), neurofibromatosis Type 2 (NF2) and schwannomatosis, as the genetic and molecular pathologic basis of these are well-established.

The genes responsible for each of these disorders were identified with a combination of genetic and molecular biology studies, allowing for detailed analysis of the gene products. Although patients with a familial tumor predisposition syndrome represent only a fraction of all patients affected by nerve sheath tumors, the significance of these disorders is enormous, as they provide crucial insights to the molecular mechanisms of

---

Neurodegenerative Diseases, edited by Shamim I. Ahmad.
©2012 Landes Bioscience and Springer Science+Business Media.

tumor formation and progression. This information is not easily obtained from studies of sporadic tumors, owing to their heterogeneity, complexity and the accumulation of a large number of random secondary genetic changes. It is likely that further studies of neurofibromatosis and schwannomatosis will continue to provide invaluable information about oncologic diseases in general.

EPIDEMIOLOGY

Neurofibromatosis Type 1 (OMIM 162200), previously known as von Recklinghausen's disease or peripheral neurofibromatosis, is one of the most common genetic conditions of the nervous system. It affects one in 3000 persons worldwide.[4] About half of patients have a family history of the disease; the other half represents founders without a known family history.

Neurofibromatosis Type 2 (OMIM 101000, 607379) is less common than NF1 with an estimated prevalence of 1 in 40,000-60,000. The birth incidence is one in 25,000-30,000.[4,5] There are no recognized geographic or ethnic differences in disease prevalence.

The annual incidence of schwannomatosis (OMIM 162091, 601607) is 1 in 2,000,000 persons.[6,7] Unlike NF1 and NF2, schwannomatosis is familial in only about 15% of patients, as most cases are thought to represent new mutations. The rate of transmission to offspring is low, likely due to a high rate of genetic mosaicism in founders.[8]

CLINICAL FEATURES

NF1 is completely penetrant; however, there is highly variable expressivity of clinical manifestations between unrelated individuals (who have different germline mutations) and between members within the same family (who share an identical germline mutation).

Patients with NF1 develop combinations of a wide spectrum of manifestations. These can be classified as tumor versus nontumor and can be categorized according to the affected organ system. Nontumor skin manifestations include café-au-lait (CAL) macules and skin fold freckling. CAL macules are homogenously hyper-pigmented lesions with regular margins and present in virtually all patients in large numbers (greater than six). They can be seen at birth and usually increase in number during the first two years of life. CAL macules are common in the general population; one or two such lesions can frequently be seen in unaffected individuals. The intensity of pigmentation is variable and tends to decrease in adulthood. Some individuals may have nearly confluent café-au-lait macules. These features can make these lesions difficult to recognize, especially in older patients.[9] Skinfold freckling is seen at sites of skin friction, such as the axilla, groin and neck and in women, under the breasts. Freckling usually develops by the age of 10 and present in most patients. Central nervous system nontumor manifestations include seizures, which occur in about 5% of patients. Cognitive impairment with learning disabilities is common. Developmental abnormalities may include aqueductal stenosis, Chiari Type 1 malformation, macrocephaly, scoliosis and glaucoma. Cardiovascular complications are also seen at an increased frequency, such as congenital heart disease, pulmonary artery stenosis, renal artery stenosis and cerebrovascular disease. The disease also affects the skeletal system with an increased frequency of short stature, pseudoarthrosis of long bones, sphenoid wing dysplasia and osteoporosis.

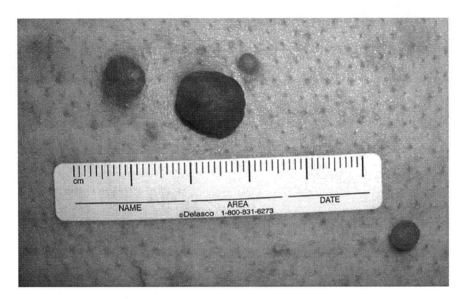

Figure 1. Photograph of cutaneous neurofibromas in an NF1 patient.

Tumor-type manifestations of NF1 include cutaneous neurofibromas, which are circumscribed masses associated with a small nerve (Fig. 1). Plexiform neurofibromas are tumors involving multiple nerve fascicles; they tend to invade local structures and can grow to large sizes. Less frequently, patients develop transient orange papules called xanthogranulomas that have been associated with the development of leukemia. The presence of multiple iris hamartomas, called Lisch nodules, is specific to NF1.[10] Optic pathway gliomas occur in about 15% of patients but are symptomatic in just 5% of patients (Fig. 2). Symptomatic tumors almost always present in childhood although progression in adulthood has been reported. Other tumor-type manifestations include astrocytomas, malignant peripheral nerve sheath tumors, pheochromocytomas, glomus tumors and gastrointestinal stromal tumors. Carcinoid tumor and breast carcinoma have been reported in association with NF1 although a direct link has not been demonstrated.

Like NF1, NF2 is completely penetrant. Adult patients commonly present in the third decade with symptoms related to vestibular schwannoma, the hallmark lesion of this condition (Fig. 3). About one-fifth of patients present under the age of 15, most often without eighth cranial nerve dysfunction but with a wide variety of complications including visual loss due to cataracts, orbital meningiomas and retinal hamartomas; seizures due to intracranial meningiomas; myelopathy due to spinal tumors; cutaneous schwannomas; foot drop; and facial mononeuropathy.[11]

Vestibular schwannoma can cause 8th cranial nerve dysfunction, resulting in progressive sensorineural hearing loss, tinnitus and imbalance. The severity of hearing loss is not proportional to tumor size.[12,13] In addition to 8th cranial nerve dysfunction, tumor compression of nearby cranial nerves and the brainstem can cause facial paralysis, facial pain, vertigo and unsteadiness. Intracranial meningiomas can exert mass effect on the cortex, resulting in seizures. Patients with NF2-associated vestibular schwannomas exhibit variable clinical and radiologic progression without correlation with specific NF2 mutation types. However, the rate of progression appears to decrease with advancing age.[14]

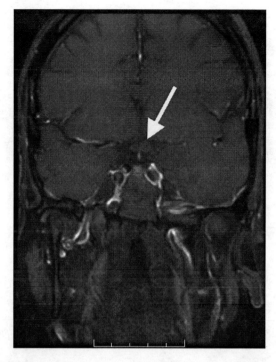

Figure 2. Coronal post-contrast MRI of the brain of an NF1 patient demonstrating optic pathway glioma affecting the optic chiasm (arrow).

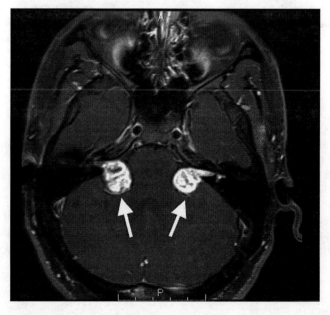

Figure 3. Axial post-contrast MRI of the brain of an NF2 patient demonstrating bilateral vestibular schwannomas (arrows).

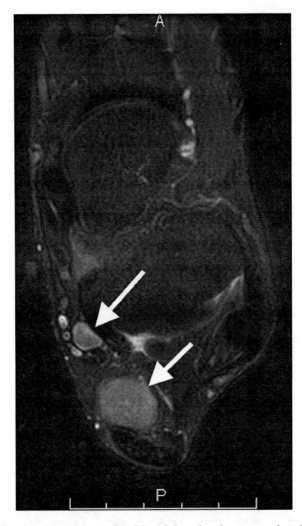

Figure 4. Axial T2-weighted fat saturated MRI of the left foot of a schwannomatosis patient demonstrating multiple peripheral schwannomas.

In contrast to NF1 and NF2, schwannomatosis demonstrates incomplete penetrance. Patients with schwannomatosis typically present with pain or with an asymptomatic mass. During the disease course, pain often remains the primary clinical problem. The disease is characterized by the development of multiple peripheral and cranial nerve schwannomas (Fig. 4) that occur in the absence of vestibular schwannomas.[15-17] Involvement of the spinal nerves including the cauda equina is commonly seen. With neuroimaging frequently performed for nonspecific symptoms, lesions are increasingly frequently found incidentally. Schwannomas can occur in peripheral nerves outside of the skull or spinal canal. Although NF2 and schwannomatosis are genetically distinct, these may be clinically overlapping conditions. For example, patients with clinical features of schwannomatosis may have mosaic NF2 on molecular testing.[11,18]

GENETIC BASIS

The *NF1* gene is on the long arm of chromosome 17 and consists of 350 kb of genomic DNA. About half of NF1 patients represent de novo mutations and have no family history.[11] The *NF2* gene is on the long arm of chromosome 22. Both the *NF1* and *NF2* genes were identified by positional cloning.[19-22] The accepted mechanism of tumorigenesis for NF1 and NF2 begins with the presence of an inactivating germline mutation to one of the alleles of these genes, resulting in heterozygosity. The second step is the occurrence of an inactivating somatic mutation in the second allele, usually seen on testing as a loss of heterozygosity. The resulting single cell thus has no functional allele and gives rise to a single tumor in clonal fashion.[23]

In schwannomatosis patients, inactivating *NF2* gene mutations are common in the tumor tissues, but not in the germline.[8,27] These findings, suggested that NF2 and schwannomatosis were distinct conditions.[8] Indeed, in some schwannomatosis patients, mutations in an unrelated gene, *SMARCB*1 has been found. *SMARCB1* is a tumor-suppressor gene associated with malignant pediatric rhabdoid tumors of the brain and kidney.[24-26] This gene is also located on the long arm of chromosome 22, close to *NF2*. The proximity of *SMARCB1* and *NF2* results in an unusual genetic complexity.

Genetic mosaicism in NF refers to the presence of inactivating NF mutations limited to certain segments of the body. Patients with mosaic NF typically have clinical phenotypes that are either segmental in presentation or attenuated in severity. Mosaic NF is thought to occur as a result of a de novo gene mutation during post-zygotic development. All of the progeny of the initial affected cell within the individual will carry the same constitutional mutation. Because the affected body region may or may not include the ovaries or testes, transmission rates are lower in patients with mosaic disease.

BIOCHEMICAL AND MOLECULAR MECHANISM OF DISEASE

The *NF1* gene product Neurofibromin is a 220-280 kD cytoplasmic protein, expressed in neurons, oligodendrocytes and Schwann cells and to a lesser degree in astrocytes, leukocytes and the adrenal medulla.[28-30] Neurofibromin functions as a negative regulator of the Ras protein, which is an important mediator of cell proliferation.[31-34] In addition, neurofibromin increases cAMP production and suppression of cAMP is sufficient to induce gliomagenesis using cells derived from NF1-deficient mice with ectopic expression of phosphodiesterase.[35,36] Neurofibromin is reported to regulate neuroglial progenitor cell self-renewal as well as astrocytic or oligodendrocytic differentiation in a brain region-specific manner.[37-40] The microRNA miR-10b appears to be an important regulator of NF1 expression and may have significance in disease severity.[41]

The *NF2* gene encodes the 69 kDa protein Merlin (moesin-ezrin-radixin-like protein).[20,22] As its name suggests, Merlin is similar to the members of the family of ERM (ezrin, radixin, moesin) proteins, which link cell membrane proteins to the actin cytoskeleton.[42,43] Merlin functions as a negative regulator of cell proliferation and motility.[44-49] The downstream pathway by which Merlin regulates proliferation and motility is an active area of research, Rac, MAPK and MLK3 have been reported as candidates.[50-53] Recent reports suggest possible links of Merlin to the Mst/Lats/Yap, MLK3 and EGFR pathways.[54-62] Merlin's tumor suppressor activity has been shown to be mediated by mTOR.[63] The E3 ubiquitin ligase CRL4/DCAF1 appears to be an important downstream target of Merlin. It has been shown that Merlin binds and inactivates CRL4/DCAF1, while tumor-derived Merlin mutations invariably disrupt this interaction. Cells lacking CRL4/DCAF1 do not show increased proliferation upon inactivation of Merlin.[64]

The *SMARCB1* gene (SWI/SNF related, Matrix associated, Actin dependent Regulator of Chromatin, subfamily B, member 1) is a tumor suppressor gene. It is known by several other names owing to its complex functions, including *SNF5*, *INI1* and *BAF47*. Its product is a 47kD nuclear protein that participates in chromatin remodeling as part of a large complex. This complex regulates the essential cellular functions such as gene expression, cell division and differentiation and plays a role in retroviral integration.[65] Expression of SMARCB1 represses transcription of cyclin D1 and causes cell cycle arrest, while increased proliferation of atypical teratoid-rhabdoid tumor cells results from inactivation of the *SMARCB1* gene and mediated by increased levels of cyclin D1.[66-68]

PATHOLOGY

Neurofibromas are tumors composed of Schwann cells, perineural cells, mast cells and fibroblasts. Cutaneous neurofibromas are unencapsulated but well-circumscribed lesions, composed of spindle cells in collagenous stroma. Plexiform neurofibromas are infiltrating lesions of the nerve. They are sparsely cellular, with myxoid stroma.

Schwannomas are neoplastic proliferations of Schwann cells. Schwannomas are well-circumscribed, encapsulated masses that are attached to the nerve but do not infiltrate it. The cells are elongated with uniform oval nuclei. The cells may show a pattern of unidirectional nuclear orientation with areas of alternating high and low cellularity, termed the Antoni A pattern. Alternatively, the tissue may exhibit lower cellularity with microcysts and myxoid changes, termed the Antoni B pattern, or a combination of both. The cells are strongly immunoreactive for S-100 and often lack Merlin expression.

DIAGNOSTIC CRITERIA

National Institutes of Health (NIH) Diagnostic Criteria for NF1

Two or more of the below criteria are required for diagnosis:

- Six or more café-au-lait macules (>5 mm in children or >15 mm in adults)
- Two or more cutaneous or subcutaneous neurofibromas or one plexiform neurofibroma
- Axillary or groin freckling
- Optic pathway glioma
- Two or more Lisch nodules
- Bony dysplasia
- First-degree relative with NF1

Manchester Diagnostic Criteria for NF2[69]

Bilateral Vestibular Schwannoma

- First-degree family relative with NF2 and unilateral VS or any two of: Meningioma, glioma, neurofibroma, schwannoma, posterior subcapsular lenticular opacities

OR

- Unilateral VS and any two of: Meningioma, glioma, neurofibroma, schwannoma, posterior subcapsular lenticular opacities

OR

- Multiple meningiomas (two or more) and unilateral VS or any two of: Glioma, neurofibroma, schwannoma, cataract

Revised Diagnostic Criteria for Schwannomatosis[70,71]

All patients with definite or possible schwannomatosis must not fulfill any of the existing sets of diagnostic criteria for NF2 and have no evidence of vestibular schwannoma on high resolution MRI, no first-degree relative with NF2 and no known constitutional NF2 mutation.

1. Definite schwannomatosis
   A. Age older than 30 years and two or more non-intradermal schwannomas, at least one with histologic confirmation
   B. One pathologically confirmed schwannoma plus a first-degree relative who meets the above criteria
2. Possible schwannomatosis
   A. Age younger than 30 years and two or more non-intradermal schwannomas, at least one with histologic confirmation
   B. Age older than 45 years and two or more non-intradermal schwannomas, at least one with histologic confirmation
   C. Radiographic evidence of a schwannoma and first degree relative meeting the criteria for definite schwannomatosis
3. Segmental schwannomatosis
   Meets criteria for either definite or possible schwannomatosis but limited to one limb or five or fewer contiguous segments of the spine.

THERAPEUTIC STRATEGY AND PROGNOSIS

Management of patients with NF1 includes active surveillance for tumor and nontumor manifestations of the disease combined with surgery reserved for tumors causing severe symptoms. Complete resection of plexiform neurofibromas is usually not possible and recurrences are common. Since the tumors are usually slow-growing, surgery can result in temporary benefit. The patients may benefit from comprehensive care delivered through a Neurofibromatosis specialty clinic owing to the wide range of possible complications, including neurologic, developmental, cognitive, oncologic, ophthalmologic, cardiovascular and orthopedic. These clinics are familiar with complications of NF and refer patients to other specialties as necessary. Surveillance for malignancies is particularly important and if found, they are typically treated with standard therapy.

Patients with an established diagnosis of NF2 should be followed by annual MRI of the brain including detailed imaging of the cerebellopontine angle (1.5 mm slice thickness). For patients with bilateral vestibular schwannomas, serial measurement of tumor volume by MRI is the best method for determining tumor growth.[72] Functional evaluation of hearing should also be performed annually and it should include standardized word recognition testing. In deafened patients, placement of an auditory brain stem implant (ABI) has been used with significant, but limited, success.[73] MRI studies of the spinal cord should be performed periodically to monitor patients with known spinal tumors. Whole body MRI is being evaluated for use in detection of larger lesions, which are thought to be more likely to become symptomatic. As with NF1, regular follow-up in a comprehensive NF2 clinic is important with referral to other specialties based on patient symptoms and examination findings.

Surgery is the standard treatment of NF2-related tumors, but the excision of all lesions is not possible or advisable. The primary goal of surgery is to preserve neurologic function and maximize quality of life. Surgery is indicated for patients with significant brainstem or spinal cord compression or with obstructive hydrocephalus. Patients with little or no neurologic dysfunction from their tumors can remain under close surveillance. Radiation therapy has been used more frequently in management of NF2-related tumors over the past few decades. Radiation therapy for vestibular schwannomas is associated with good local control but not with preservation of hearing. In addition, there have been reports of malignancies within the radiation field among NF2 patients which limits enthusiasm for this modality in younger NF2 patients. More recently, studies with bevacizumab indicate that anti-VEGF therapies may cause significant reductions in tumor volume and significant improvement in hearing function.[74] The life expectancy in NF2 patients is decreased, partly by the development of atypical meningiomas.[75]

The management of schwannomatosis focuses of improving quality of life, with the emphasis on management of chronic pain and psychiatric comorbidities such as depression and anxiety. This condition does not reduce life expectancy. Surgery is usually reserved for tumors that cause clear neurologic disability or for those with rapid growth.

Genetic counseling for patients of reproductive age should be offered and tailored specifically to the patient's preferences. Currently available reproductive techniques, such as embryo selection or chorionic villous sampling, involve ethical decisions with non-uniform acceptance and depend on testing for the specific mutation present in the patient, which may not be known. In NF1, genetic testing identifies >95% of individuals who meet clinical diagnostic criteria.[76] This information can be used to aid patients in making decisions about presymptomatic diagnosis, prenatal diagnosis and pre-implantation genetic diagnosis.[77] In NF2, genetic testing identifies about 60% of mutations in patients with no family history or if family history is not known. If at least two affected family members exist, the rate of detection increases to 90%. Testing of multiple tumor tissues may enhance the detection rate, particularly in case of mosaicism.[78]

CONCLUSION

NF1, NF2 and schwannomatosis are complex systemic conditions requiring comprehensive surveillance and treatment of complications, as necessary. Patients may benefit from visiting specialty clinics with access to a full range of clinical services. Advances in understanding of the molecular mechanism of disease, improved imaging and diagnostic techniques along with emerging targeted therapies are expected to result in better patient outcomes.

ACKNOWLEDGEMENTS

The authors would like to acknowledge Vanessa Merker for her editorial assistance in preparing this manuscript.

REFERENCES

1. Wishart JH. Case of tumours in the skull, dura mater and brain. Edinburgh Med Surg J 1822; 18:393-397.
2. von Recklinghausen F. Uber die multiplen Fibrome der Haut und ihre Beziehung zu multiplen Neuromen. Berlin, Germany: August Hirschwald; 1882.
3. Crump T. Translation of case reports in Ueber die multiplen Fibrome der Haut und ihre Beziehung zu den multiplen Neuromen by F. v. Recklinghausen. Adv Neurol 1981; 29:259-275.
4. Friedman JM, Gutmann DH, MacCollin M et al. Neurofibromatosis: Phenotype, Natural History and Pathogenesis, 3rd edition, Baltimore, Johns Hopkins Press; 1999.
5. Evans DG, Huson SM, Donnai D et al. A clinical study of type 2 neurofibromatosis. Q J Med 1992; 304:603-618.
6. Antinheimo J, Sankila R, Carpen O et al. Population-based analysis of sporadic and type 2 neurofibromatosis-associated meningiomas and schwannomas. Neurology 2000; 54:71-76.
7. Niimura M. Neurofibromatosis. Rinsho Derma 1973; 15:653-663.
8. MacCollin M, Willett C, Heinrich B et al. Familial schwannomatosis: exclusion of the NF2 locus as the germline event. Neurology 2003; 60(12):1968-1974.
9. Ferner RE. The neurofibromatoses. Pract Neurol 2010; 10(2):82-93.
10. Riccardi VM. Neurofibromatosis: past, present and future. N Engl J Med 1991; 324(18):1283-1285.
11. Evans DG, Howard E, Giblin C et al. Birth incidence and prevalence of tumour prone syndromes: estimates from a UK genetic family register service. Am J Med Genet A 2010; 152A:327-332.
12. Evans DG, Baser ME, O'Reilly B et al. Management of the patient and family with neurofibromatosis 2: a consensus conference statement. Br J Neurosurg 2005; 19:5-12.
13. Ferner RE. Neurofibromatosis 1 and neurofibromatosis 2: a twenty first century perspective. Lancet Neurol 2007; 6:340-351.
14. Mautner VF, Baser ME, Thakkar SD et al. Vestibular schwannoma growth in patients with neurofibromatosis type 2: a longitudinal study. J Neurosurg 2002; 96:223-228.
15. Huang JH, Simon SL, Nagpal S et al. Management of patients with schwannomatosis: report of six cases and review of the literature. Surg Neurol 2004; 62:353-361.
16. MacCollin M, Woodfin W, Kronn D et al. Schwannomatosis-a clinical and pathologic study. Neurology 1996; 46:1072-1079.
17. Wolkenstein P, Benchikhi H, Zeller J et al. Schwannomatosis: a clinical entity distinct from neurofibromatosis type 2. Dermatology 1997; 195:228-231.
18. Baser ME, Kuramoto L, Joe H et al. Genotype-phenotype correlations for nervous system tumors in neurofibromatosis 2: a population-based study. Am J Hum Genet 2004; 75(2):231-239.
19. Marchuk DA, Saulino AM, Tavakkol R et al. cDNA cloning of the type 1 neurofibromatosis gene: complete sequence of the NF1 gene product. Genomics 1991; 4:931-940.
20. Rouleau GA, Merel P, Lutchman M et al. Alteration in a new gene encoding a putative membrane-organizing protein causes neuro-fibromatosis type 2. Nature 1993; 363(6429):515-521.
21. Xie YG, Han FY, Peyrard M et al. Cloning of a novel, anonymous gene from a megabase-range YAC and cosmid contig in the neurofibromatosis type 2/meningioma region on human chromosome 22q12. Hum Mol Genet 1993; 2(9):1361-1368.

22. Trofatter JA, MacCollin MM, Rutter JL et al. A novel moesin-, ezrin-, radixin-like gene is a candidate for the neurofibromatosis 2 tumor suppressor. Cell 1993; 72:791-800.
23. Knudson AG. Mutation and cancer: statistical study of retinoblastoma. Proc Natl Acad Sci USA 1971; 68:820-823.
24. Biegel JA, Zhou JY, Rorke LB et al. Germ-line and acquired mutations of INI1 in atypical teratoid and rhabdoid tumors. Cancer Res 1999; 59:74-79.
25. Sevenet N, Sheridan E, Amram D et al. Constitutional mutations of the hSNF5/INI1 gene predispose to a variety of cancers. Am J Hum Genet 1999a; 65:1342-1348.
26. Fujisawa H, Takabatake Y, Fukusato T et al. Molecular analysis of the rhabdoid predisposition syndrome in a child: a novel germline hSNF5/INI1 mutation and absence of c-myc amplification. J Neurooncol 2003; 63:257-262.
27. Kaufman DL, Heinrich BS, Willett C et al. Somatic instability of the NF2 gene in schwannomatosis. Arch Neurol 2003; 60:1317-1320.
28. Cawthon RM, Weiss R, Xu GF et al. A major segment of the neurofibromatosis type 1 gene: cDNA sequence, genomic structure and point mutations. Cell 1990; 62(1):193-201.
29. Viskochil D, Buchberg AM, Xu G et al. Deletions and a translocation interrupt a cloned gene at the neurofibromatosis type 1 locus. Cell 1990; 62(1):187-192.
30. Wallace MR, Marchuk DA, Andersen LB et al. Type 1 neurofibromatosis gene: identification of a large transcript disrupted in three NF1 patients. Science 1990; 249(4965):181-186.
31. Bollag G, McCormick F. Ras regulation. NF is enough of GAP. Nature 1992; 356:663-664.
32. Kluwe L, Friedrich R, Mautner VF. Loss of NF1 allele in Schwann cells but not in fibroblasts derived from an NF1-associated neurofibroma. Genes Chromosomes Cancer 1999; 24:283-285.
33. Rutkowski JL, Wu K, Gutmann DH et al. Genetic and cellular defects contributing to benign tumor formation in neurofibromatosis type 1. Hum Mol Genet 2000; 9:1059-1066.
34. Legius E, Marchuk DA, Collins FS et al. Somatic deletion of the neurofibromatosis type 1 gene in a neurofibrosarcoma supports a tumour suppressor gene hypothesis. Nat Genet 1993; 3:122-126.
35. Brown JA, Gianino SM, Gutmann DH. Defective cAMP generation underlies the sensitivity of CNS neurons to neurofibromatosis-1 heterozygosity. J Neurosci 2010; 30(16):5579-5589.
36. Warrington NM, Gianino SM, Jackson E et al. Cyclic AMP suppression is sufficient to induce gliomagenesis in a mouse model of neurofibromatosis-1. Cancer Res 2010; 70(14):5717-5727.
37. Hegedus B, Dasgupta B, Shin JE et al. Neurofibromatosis-1 regulates neuronal and glial cell differentiation from neuroglial progenitors in vivo by both cAMP- and Ras-dependent mechanisms. Cell Stem Cell 2007; 1(4):443-457.
38. Lee DY, Yeh TH, Emnett RJ et al. Neurofibromatosis-1 regulates neuroglial progenitor proliferation and glial differentiation in a brain region-specific manner. Genes Dev 2010; 24(20):2317-2329.
39. Lee JS, Padmanabhan A, Shin J et al. Oligodendrocyte progenitor cell numbers and migration are regulated by the zebrafish orthologs of the NF1 tumor suppressor gene. Hum Mol Genet 2010; 19(23):4643-4653.
40. Staser K, Yang FC, Clapp DW. Plexiform neurofibroma genesis: questions of Nf1 gene dose and hyperactive mast cells. Curr Opin Hematol 2010; 17(4):287-293.
41. Chai G, Liu N, Ma J et al. MicroRNA-10b regulates tumorigenesis in neurofibromatosis type 1. Cancer Sci 2010.
42. Sun CX, Haipek C, Scoles DR et al. Functional analysis of the relationship between the neurofibromatosis 2 tumor suppressor and its binding partner, hepatocyte growth factor-regulated tyrosine kinase substrate. Hum Mol Genet 2002; 11(25):3167-3178.
43. Ramesh V. Merlin and the ERM proteins in Schwann cells, neurons and growth cones. Nat Rev Neurosci 2004; 5(6):462-470.
44. Lutchman M, Rouleau GA. The neurofibromatosis type 2 gene product, schwannomin, suppresses growth of NIH 3T3 cells. Cancer Res 1995; 55(11):2270-2274.
45. McClatchey AI, Saotome I, Ramesh V et al. Genes Dev 1997; 11(10):1253-1265. The Nf2 tumor suppressor gene product is essential for extraembryonic development immediately prior to gastrulation. Genes Dev 1997; 11(10):1253-1265.
46. Sherman L, Xu HM, Geist RT et al. Interdomain binding mediates tumor growth suppression by the NF2 gene product. Oncogene 1997; 15(20):2505-2509.
47. Ikeda K, Saeki Y, Gonzalez-Agosti C et al. Inhibition of NF2-negative and NF2-positive primary human meningioma cell proliferation by overexpression of merlin due to vector-mediated gene transfer. J Neurosurg 1999; 91(1):85-92.
48. Fraenzer JT, Pan H, Minimo L Jr et al. Overexpression of the NF2 gene inhibits schwannoma cell proliferation through promoting PDGFR degradation. Int J Oncol 2003; 23(6):1493-1500.
49. Xiao GH, Gallagher R, Shetler J et al. The NF2 tumor suppressor gene product, merlin, inhibits cell proliferation and cell cycle progression by repressing cyclin D1 expression. Mol Cell Biol 2005; 25(6):2384-2394.

50. Shaw RJ, Paez JG, Curto M et al. The Nf2 tumor suppressor, merlin, functions in Rac-dependent signaling. Dev Cell 2001; 1(1):63-72.
51. Kissil JL, Wilker EW, Johnson KC et al. Merlin, the product of the Nf2 tumor suppressor gene, is an inhibitor of the p21-activated kinase, Pak1. Mol Cell 2003; 12:841-849.
52. Chadee DN, Xu D, Hung G et al. Mixed-lineage kinase 3 regulates B-Raf through maintenance of the B-Raf/Raf-1 complex and inhibition by the NF2 tumor suppressor protein. Proc Natl Acad Sci USA 2006; 103(12):4463-4468.
53. Morrison H, Sperka T, Manent J et al. Merlin/neurofibromatosis type 2 suppresses growth by inhibiting the activation of Ras and Rac. Cancer Res 2007; 67(2):520-527.
54. Z Zender L, Spector MS, Xue W et al. Identification and validation of oncogenes in liver cancer using an integrative oncogenomic approach. Cell 2006; 125(7):1253-1267.
55. Dong J, Feldmann G, Huang J et al. Elucidation of a universal size-control mechanism in Drosophila and mammals. Cell 2007; 130(6):1120-1133.
56. Zhou D, Conrad C, Xia F et al. Mst1 and Mst2 maintain hepatocyte quiescence and suppress hepatocellular carcinoma development through inactivation of the Yap1 oncogene. Cancer Cell 2009; 16(5):425-438.
57. Zhang N, Bai H, David KK et al. The Merlin/NF2 tumor suppressor functions through the YAP oncoprotein to regulate tissue homeostasis in mammals. Dev Cell 2010; 19(1):27-38.
58. Lu L, Li Y, Kim SM et al. Hippo signaling is a potent in vivo growth and tumor suppressor pathway in the mammalian liver. Proc Natl Acad Sci USA 2010;107(4):1437-1442.
59. Song H, Mak KK, Topol L et al. Mammalian Mst1 and Mst2 kinases play essential roles in organ size control and tumor suppression. Proc Natl Acad Sci USA 2010; 107(4):1431-1436.
60. Zhan Y, Modi N, Stewart AM et al. Regulation of mixed lineage kinase 3 is required for Neurofibromatosis-2-mediated growth suppression in human cancer. Oncogene 2010.
61. Benhamouche S, Curto M, Saotome I et al. Nf2/Merlin controls progenitor homeostasis and tumorigenesis in the liver. Genes Dev 2010; 24(16):1718-1730.
62. Cole BK, Curto M, Chan AW et al. Localization to the cortical cytoskeleton is necessary for Nf2/merlin-dependent epidermal growth factor receptor silencing. Mol Cell Biol 2008; 28(4):1274-1284.
63. James MF, Han S, Polizzano C et al. NF2/merlin is a novel negative regulator of mTOR complex 1 and activation of mTORC1 is associated with meningioma and schwannoma growth. Mol Cell Biol 2009; 29(15):4250-4261.
64. Li W, You L, Cooper J et al. Merlin/NF2 suppresses tumorigenesis by inhibiting the E3 ubiquitin ligase CRL4 (DCAF1) in the nucleus. Cell 2010; 140(4):477-490.
65. Hulsebos TJ, Plomp AS, Wolterman RA et al. Am J Hum Genet 2007; 80(4):805-810.
66. Zhang ZK, Davies KP, Allen J et al. Cell cycle arrest and repression of cyclin D1 transcription by INI1/hSNF5. Mol Cell Biol 2002; 22(16):5975-8859.
67. Tsikitis M, Zhang Z, Edelman W et al. Genetic ablation of Cyclin D1 abrogates genesis of rhabdoid tumors resulting from Ini1 loss. Proc Natl Acad Sci USA 2005; 102(34):12129-12134.
68. Fujisawa H, Misaki K, Takabatake Y et al. Cyclin D1 is overexpressed in atypical teratoid/rhabdoid tumor with hSNF5/INI1 gene inactivation. J Neurooncol 2005; 73(2):117-124.
69. Baser ME, Friedman JM, Wallace AJ et al. Evaluation of clinical diagnostic criteria for neurofibromatosis 2. Neurology 2002; 59(11):1759-1765.
70. Baser ME, Friedman JM, Evans DG. Increasing the specificity of diagnostic criteria for schwannomatosis. Neurology 2006; 66(5):730-732.
71. MacCollin M, Chiocca EA, Evans DG et al. Diagnostic criteria for schwannomatosis. Neurology 2005; 64(11):1838-1845.
72. Harris GJ, Plotkin SR, Maccollin M et al. Three-dimensional volumetrics for tracking vestibular schwannoma growth in neurofibromatosis type II. Neurosurgery 2008; 62(6):1314-1319.
73. Colletti V. Auditory outcomes in tumor vs nontumor patients fitted with auditory brainstem implants. Adv Otorhinolaryngol 2006; 64:167-185.
74. Plotkin SR, Stemmer-Rachamimov AO, Barker FG 2nd et al. Hearing improvement after bevacizumab in patients with neurofibromatosis type 2. N Engl J Med 2009; 361(4):358-367.
75. Baser ME, Friedman JM, Aeschliman D et al. Predictors of the risk of mortality in neurofibromatosis 2. Am J Hum Genet 2002; 71(4):715-723.
76. Messiaen LM, Callens T, Mortier G et al. Exhaustive mutation analysis of the NF1 gene allows identi-fication of 95% of mutations and reveals a high frequency of unusual splicing defects. Hum Mutat 2000; 15(6):541-555.
77. Spits C, De Rycke M, Van Ranst N et al. Preimplantation genetic diagnosis for neurofibromatosis type 1. Mol Hum Reprod 2005; 11(5):381-387.
78. Moyhuddin A, Baser ME, Watson C et al. Somatic mosaicism in neurofibromatosis 2: prevalence and risk of disease transmission to offspring. J Med Genet 2003; 40(6):459-463.

CHAPTER 21

OXIDATIVE STRESS IN
DEVELOPMENTAL BRAIN DISORDERS

Masaharu Hayashi,* Rie Miyata and Naoyuki Tanuma

Department of Clinical Neuropathology, Tokyo Metropolitan Institute for Neuroscience, Tokyo, Japan
*Corresponding Author: Masaharu Hayashi—Email: hayashi-ms@igakuken.or.jp

Abstract: In order to examine the involvement of oxidative stress in developmental brain disorders, we have performed immunohistochemistry in autopsy brains and enzyme-linked immunosorbent assay (ELISA) in the cerebrospinal fluid and urines of patients. Here, we review our data on the hereditary DNA repair disorders, congenital metabolic errors and childhood-onset neurodegenerative disorders. First, in our studies on hereditary DNA repair disorders, increased oxidative DNA damage and lipid peroxidation were carried out in the degeneration of basal ganglia, intracerebral calcification and cerebellar degeneration in patients with xeroderma pigmentosum, Cockayne syndrome and ataxia-telangiectasia-like disorder, respectively. Next, congenital metabolic errors, apoptosis due to lipid peroxidation seemed to cause neuronal damage in neuronal ceroid-lipofuscinosis. Oxidative stress of DNA combined with reduced expression of antioxidant enzymes occurred in the lesion of the cerebral cortex in mucopolysaccharidoses and mitochondrial myopathy, encephalopathy, lactic acidosis and stroke-like episodes. In childhood-onset neurodegenerative disorders, increased oxidative DNA damage and lipid peroxidation may lead to motor neuron death in spinal muscular atrophy like in amyotrophic lateral sclerosis. In patients with dentatorubral-pallidoluysian atrophy, a triplet repeat disease, deposition of oxidative products of nucleosides and reduced expression of antioxidant enzymes were found in the lenticular nucleus. In contrast, the involvement of oxidative stress is not definite in patients with Lafora disease. Rett syndrome patients showed changes of oxidative stress markers and antioxidant power in urines, although the changes may be related to systemic complications.

Neurodegenerative Diseases, edited by Shamim I. Ahmad.
©2012 Landes Bioscience and Springer Science+Business Media.

INTRODUCTION

Oxygen is metabolized to generate energy in the form of ATP through a series of reductive steps at the inner membrane in the mitochondria. During these processes, reactive oxygen species (ROS) and reactive nitrogen species (RNS) are formed. Although ROS and RNS contribute to signal processing, they have also harmful effects on lipids, proteins and nucleic acids, leading to tissue damage in a process called oxidative stress.[1] Oxidative stress originates from an imbalance between the production of ROS and RNS and the antioxidant systems. ROS include superoxide anion (O_2^-), hydroxyl radicals (.OH) and hydrogen peroxide (H_2O_2), while nitric oxide (NO) and peroxynitrite ($ONOO^-$) are known as RNS. Antioxidant defenses are composed of preventive antioxidant enzymes such as catalase, superoxide dismutase (SOD), glutathione peroxidase and metal chelating proteins, in addition to radical scavenging vitamins C and E.[2] SOD converts O_2^- into H_2O_2, which is rapidly reduced by catalase and glutathione peroxidase. It may also catalyze excessive nitration of tyrosine by $ONOO^-$. The excess of ROS/RNS production over detoxification results in a shift in balance towards oxidative damage. Oxidative damage of DNA and RNA produces 8-hydroxy-2'-deoxyguanosine (8-OHdG) and 8-hydroxyguanosine (8-OHG), respectively, which are used as markers of oxidative nucleosides damage.[3] Since thymidine glycol (TG) originates from deoxythymidine in DNA but not in RNA and is not removed as easily as 8-OHdG is, TG is a stable oxidative marker specific for DNA.[4] The brain includes a large amount of lipids in cell membranes and the myelin encapsulates the neuronal fibers. Lipid peroxidation can form various aldehydes, including the early and late stage markers hexanoyl lysine adduct (HEL) and 4-hydroxynonenal (4-HNE).[1,5] Advanced glycation end products (AGE) are markers of protein glycoxidation and the generation of AGE has been described in neurological disorders in addition to aging, atherosclerosis and diabetes mellitus.[6]

Oxidative stress markers are also available for the examination of oxidative DNA damage through analysing lipid peroxidation in the urine, serum and cerebrospinal fluid (CSF), using enzyme-linked immunosorbent assay (ELISA) in children.[7-9] Potential antioxidant (PAO) is a marker of antioxidant capacity in various biologic fluids measured by colorimetry, in which Cu^{2+} is reduced by various antioxidants to Cu+. PAO enables the evaluation of not only hydrophilic antioxidants, such as vitamin C and glutathione, but also hydrophobic ones, such as vitamin E.[10] We have confirmed the involvement of oxidative stress in various developmental brain disorders and here we reviewed the data from our immunohistochemical analysis and ELISA.

HEREDITARY DNA REPAIR DISORDERS

DNA damage is implicated in pathogenesis of various neurologic disorders and neurons are targets to sustain DNA damage during oxidative stress.[11,12] Human hereditary DNA repair deficiency syndromes and ataxic disorders seem to provide a hint for linking DNA damage and DNA repair abnormalities with neurodegeneration. Xeroderma pigmentosum (XP) and Cockayne syndrome (CS) are rare, inherited neurocutaneous disorders caused by defects in nucleotide excision repair (NER) system.[13] Complementation studies by using cell hybridization have revealed the existence of eight genes in XP (groups A-G and a variant) and two in CS (A and B). NER includes global genome repair and transcription-coupled repair (TCR), which involves several

XP genes (especially *XP-A* to *XP-G*) and two CS genes (*CSA* and *CSB*). In XP, the disease starts with skin symptoms and progressive neurological manifestations, including cognitive and motor deterioration, neuronal deafness, peripheral neuropathy and brain atrophy occurs more commonly in XP-A, XP-B, XP-D and XP-G.[14] CS children develop severe growth failure with reduced subcutaneous fat, characteristic facial features (sunken eyes, sharp noses and caries teeth), mild skin symptoms and neurological disorders such as demyelinating neuropathy, ataxia, spasticity, deafness and congnitive deterioration.[15] It is likely that decreased DNA repair and persistent DNA damage can result in augmented oxidative nucleotide damage in XP and CS. Oxidative nucleotide damage and antioxidant system have been investigated in isolated skin and blood cells or their cell lines.[14] Nevertheless, protection from ultraviolet (UV) light cannot prevent development of neurodegeneration. We have neuropathologically investigated the deposition of oxidative stress markers in autopsy cases each of XP-A and CS.[16] 4-HNE and, to a lesser extent, AGE were frequently recognized in the pseudocalcified foci, neuropil free minerals and foamy spheroids in the globus pallidus in CS more predominantly than in XP-A. CS cases showed gliosis and calcification in the basal ganglia more remarkably than XP-A cases and the degree of 4-HNE deposition seemed to be in accordance with the calcification. We also found the similar deposition of 4-HNE and AGE in the calcification in the globus pallidus and/or cerebellum in autopsy case each of Fahr disease, pseudohypoparathyroidism and idiopathic intracranial calcification (Fig. 1).[17] Increased oxidative stress has been reported in vascular calcifications in bone and kidney diseases,[18,19] and lipid peroxidation and/or oxidative protein glycation may also affect the calcification subsequent to neurodegeneration in the basal ganglia and cerebellum in the developmental brain disorders including XP-A and CS. Next were examined the deposition of oxidative products in nucleotides and expression of SOD in the XP-A and CS subjects.[20] Cases of XP-A and, to a lesser extent, those of CS demonstrated nuclear deposition of 8-OHdG and TG in neurons and glial cells, in addition to cytoplasmic deposition of 8-OHG, in the globus pallidus and cerebellar cortex (Table 1). Additionally, XP-A cases exhibited reduced cytoplasmic immunoreactivity for Cu/ZnSOD in the neurons of the cerebellar cortex and the basal ganglia, although

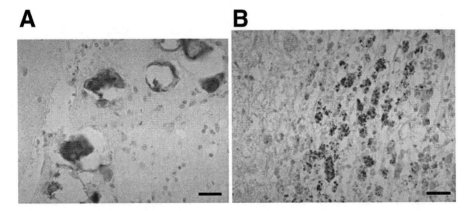

Figure 1. An autopsy case of idiopathic intracranial calcification. A) Perivascular calcification in the cerebellar cortex was immunoreactive for 4-hydroxynonenal Bar = 20 μm. B) Pseudocalcified lesion in the putamen showed granular immunoreactivity for advanced glycation end products. Bar = 20 μm.

Table 1. Summary of Immunohistochemistry for thymidine glycol in autopsy cases of xeroderma pigmentosum group A and Cockayne syndrome.

		Hippocampus		Globus pallidus		Thalamus		Dentate nucleus	
Case	Age/Sex	Neuron	Glia	Neuron	Glia	Neuron	Glia	Neuron	Glia
Xeroderma pigmentosum group A									
1	19 yrs/ Male	2+	2+	(-)	2+	1+	1+	1+	1+
2	19 yrs/ Male	1+	1+	1+	1+	(-)	1+	1+	1+
3	23 yrs/ Female	1+	1+	(-)	2+	1+	(-)	2+	(-)
4	24 yrs/ Female	(-)	(-)	(-)	(-)	(-)	1+	(-)	(-)
5	26 yrs/ Female	(-)	(-)	(-)	1+	(-)	1+	(-)	(-)
Cockayne syndrome									
1	7 yrs/ Female	1+	1+	1+	1+	(-)	(-)	(-)	(-)
2	15 yrs/ Male	(-)	(-)	(-)	(-)	(-)	(-)	(-)	(-)
3	16 yrs/ Female	(-)	(-)	(-)	2+	(-)	2+	(-)	(-)
4	18 yrs/ Male	(-)	(-)	(-)	(-)	(-)	(-)	(-)	1+
5	18 yrs/ Male	(-)	1+	(-)	1+	1+	1+	(-)	(-)

The degree of immunoreactivity for thymidine glycol was graded by the density of positively-stained nuclei of neurons or glial cells ("Glia") according to the following criteria: - = no staining visible, 1+ = a few nuclei were stained, 2+ = many nuclei were stained.

CS cases demonstrated comparatively preserved immunoreactivity for SODs, suggesting that oxidative damage to nucleotides with disturbed SOD expression can be involved in the degeneration of basal ganglia and cerebellum predominantly in XP-A. Next we started the ELISA analysis on 8-OHdG and HEL in urine samples from seven XP-A patients, one XP-D patient, five CS patients and 17 healthy controls aged 3-81 years (Fig. 2). XP-A patients aged over 20 years with long disease duration, suffering from diabetes mellitus and respiratory insufficiency, showed a remarkable increase over the mean of controls in both urinary 8-OHdG and HEL (Fig. 2). In contrast, twin CS patients aged over 20 years showing prolonged disease course demonstrated increased levels of urinary HEL but not urinary 8-OHdG. In the aforementioned autopsy study, markers of oxidative nucleoside damage and those of lipid peroxidation seemed to be deposited in XP-A and CS cases respectively and the similar tendency was speculated in the change of urinary markers from patients with XP-A and CS. In addition, we

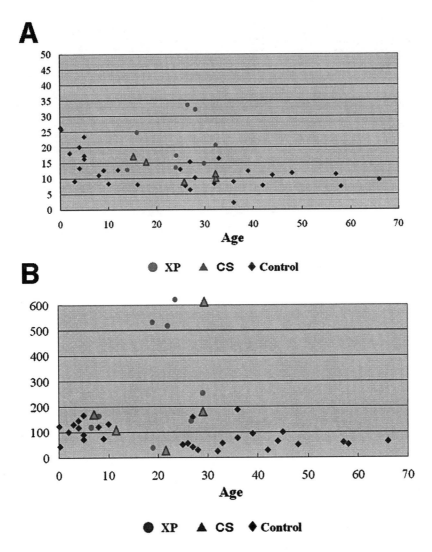

Figure 2. A) Urinary levels of 8-hydroxy-2'-deoxyguanosine (ng/mg Cre.) according to the age (years) in patients with xeroderma pigmentosum (XP, ●) and Cockyane syndrome (CS, ▲), in addition to controls (♦). B) Urinary levels of hexanoyl lysine adduct (pol/mg Cre) according to the age (years) in patients with xeroderma pigmentosum (XP, ●) and Cockyane syndrome (CS, ▲), in addition to controls (♦).

performed a preliminary analysis of the CSF levels of 8-OHdG and HEL in three and one patients of XP-A and XP-D, respectively. One XPA patient showed the increase in level of 8-OHdG in CSF over the cutoff index, whereas the levels of HEL in CSF were not elevated in four patients.

Ataxia-telangiectasia (A-T) is characterized with childhood-onset cerebellar ataxia clinically and progressive atrophy of the cerebellar cortex pathologically. Mutations in the ataxia-telangiectasia mutated (*ATM*) gene give rise to A-T and this gene encodes a protein that is a member of the phosphoinositide 3-kinase family and activation of

ATM by ionizing radiation leads to the phosphorylation of a multitude of substrates involved in recognition of double strand breaks in DNA and in cell cycle checkpoint activation.[21] *Atm*-deficient cells are hypersensitive to oxidative-stress-inducing agents specially the ionizing radiations,[22] and the cerebellum in *Atm*-deficient mice showed progressive accumulation of DNA strand breaks.[23] Antioxidants prevented Purkinje cell death in the aforementioned *Atm*-deficient mice.[24] It is possible that ATM deficiency can enhance oxidative stress and cause oxidative stress-related neuronal death. Nevertheless, oxidative stress has not been examined fully in patients with A-T.[25]

Ataxia-telangiectasia-like disorder (ATLD) is characterized by cerebellar ataxia and ATLD is one of chromosomal breakage syndrome because the patients show spontaneously occurring chromosomal aberrations and increased sensitivity to ionizing radiations.[26] ATLD is caused by mutations in the *MRE11* gene and MRE11 is one of the key components of the signaling network involved in cellular response to DNA damage.[27] We report the neuropathological findings in the first case of genetically confirmed ATLD in a pair of Japanese male siblings.[28] The siblings had the same compound heterozygous mutations of the *MRE11* gene. Brain autopsy demonstrated cerebellar atrophy in the vermis and medial part of the hemispheres, oral to the horizontal fissure. Nuclear immunoreactivity of MRE11 was absent in neurons of cerebellar cortex, cerebral cortex, basal ganglia and midbrain, whereas being widespread in normal control brains. Immunoreactivity of nuclear 8-OHdG was identified in the granule cells and Bergmann glial cells in the cerebellar cortex, both of which were functionally associated with Purkinje cells. Such 8-OHdG expression was absent in the severely affected cerebellar cortex and other brain areas. It is likely that the combination of MRE11 deficiency and oxidative DNA injury may lead to the selective cerebellar damage in patients with ATLD.

CONGENITAL METABOLIC ERRORS

Congenital metabolic errors are composed of heterogeneous diseases, such as lysosomal disorders, mitochondrial encephalomyopathy, peroxisomal disorders and disturbed metabolism of metals, manifesting both neurological and somatic abnormalities. Genes responsible for several diseases have been identified and model animals generated. Nevertheless, the pathogenesis of neurodegeneration still remains to be fully investigated; specific treatments to be developed other than bone marrow transplantation and enzyme replacement, which cannot ameliorate neurological disorders. It will be useful to exploit new therapeutics that can intervene the oxidative stress leading to neuronal damage.

Neuronal ceroid-lipofuscinosis (NCL) is a group of hereditary, lysosomal storage disorders, most of which are clinically manifested by progressive developmental retardation, visual loss, uncontrolled myoclonic epilepsy and/or cerebellar ataxia.[29] NCLs are classically classified into infantile, late-infantile, juvenile and adult forms, but several variants have recently been reported and at least 10 genetically distinct NCLs, designated CLN1 to CLN10, are presently known. We examined three autopsy cases of late-infantile NCL with progressive myoclonic epilepsy (PME), aged 8-12 years,[30] in addition to two autopsy cases of juvenile NCL suffering from the gradual progression of visual disturbances and generalized convulsion.[31] Oxidative DNA damage was observed in neurons of the cerebral cortex and, to a lesser extent, the midbrain, in both types of the disease. Protein glycation was facilitated in the Purkinje cells of the cerebellar cortex in four NCL cases, with the exception of one juvenile case. Lipid peroxidation increased in the cerebral and cerebellar cortex. Because 4-HNE can activate cell death-related caspases leading to DNA fragmentation, such coexistence of nuclei immunoreactive foe TUNEL and 4-HNE-immunoreactive cytoplasm in the frontal cortical neurons suggested the occurrence of DNA fragmentation triggered by lipid peroxidation in the late infantile form of NCL.

Mucopolysaccharidoses (MPS) are inherited neurodegenerative disorders caused by defects in specific lysosomal enzymes, resulting in the accumulation of undegraded glycosaminoglycans in lysosomes. Sanfilippo syndrome (MPS III) is an autosomal recessive disorder and comprises four subtypes (A, B, C and D), biochemically being linked to different enzymes defects. MPS III Type B (MPS IIIB) is caused by mutations in the gene encoding alpha-N-acetylglucosaminidase, a glycosidase required for the degradation of heparin sulfate.[32] The clinical features of MPS IIIB include progressive and profound neurological deterioration, with behavioral disturbances and relatively mild somatic manifestations. No effective therapy has been found yet, although oxidative stress and/or activation of microglia has been suggested to be involved in pathogenesis in model mice.[33,34]

Hunter syndrome (MPS II) is a rare, X-linked disorder caused by a deficiency of the lysosomal enzyme iduronate-2-sulfatase. In the absence of sufficient enzyme activity, glycosaminoglycans accumulate in the lysosomes of many tissues and organs and contribute to the multisystem, progressive pathologies seen in Hunter syndrome.[35] Clinically, MPS II has two subtypes: severe (Hunter A) and mild (Hunter B). Patients with the severe form of MPS II exhibit a chronic and progressive disease involving multiple organs and tissues. Patients with the mild form of MPS II have delayed onset and milder disease progression. Iduronate-2-sulfatase cannot cross the blood–brain barrier and therefore the enzyme replacement therapy is not expected to provide improvement in CNS dysfunction. Neuropathologically, patients with MPS IIIB and MPS II demonstrate neuronal swelling, dilatation of perivascular space, mild gliosis in the white matter and/ or hydrocephalus; however, pathogenesis of neurological deterioration remains elusive.[36] The involvement of oxidative damage in the brains of three cases each of MPS IIIB and MPS II and age-matched controls were examined immunochemically.[37] In cases of MPS IIIB, the density of GABAergic interneurons in the cerebral cortex immunoreactive for calbindin-D28K and parvalbumin was markedly reduced when compared with age-matched controls. It was suggested that the disturbance of GABAergic interneurons may be related to mental disturbance. The swollen neurons in the cerebral cortex demonstrated nuclear immunoreactivity for 8-OHdG and apoptotic markers. In contrast, neither lipid peroxidation nor protein glycation were observed in MPS cases. The expressions of Cu/ ZnSOD and MnSOD were reduced in two MPS II cases.

Mitochondrial myopathy, encephalopathy, lactic acidosis and stroke-like episodes (MELAS) are characterized by recurrent stroke-like episodes, epileptic seizure, short stature and deafness. Molecular genetic studies have shown that more than 80% of MELAS patients have the 3243A>G mitochondrial DNA mutation.[38] According to the cohort study in Japan, MELAS is divided into a juvenile form (onset at less than 18 years of age) and an adult form (onset at more than 18 years of age); the former form shows more severe and poor prognosis than those in the latter form.[39] The main neuropathological features of MELAS are infarct-like lesions with necrosis in the cerebral cortex and the adjacent subcortical white matter, subsequent brain atrophy and calcification in the basal ganglia. In the absence of therapeutic intervention, the infarct-like lesions spread into the neighboring region within a few weeks, irrespective of the vascular territory of cerebral

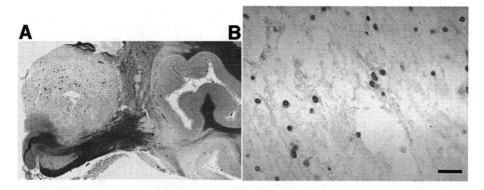

Figure 3. An autopsy case of acute necrotizing encephalopathy caused by influenza infection. A)Severe necrosis in the thalamus, Klüver-Barrera staining. B) Nuclei immunoreactive for 8-hydroxy-2'-deoxyguanosine were found in the glial cells in the necrotic lesion in the thalamus. Bar = 20 μm.

arteries. Even in the cortical and subcortical regions not affected by the infarct-like lesions, there are neuropil micro-vacuolation and increased microvasculature, whereas a few neurons remain in the infarct-like lesions (neuronal sparing). In Japan L-arginine, modulator of vascular endothelial cells, is used for the treatment of MELAS.[39] Abnormal regulation of antioxidants and mitochondrial dysfunction may be involved in oxidative neuronal loss in Huntington's disease and Friedreich ataxia.[40] Therefore, edaravone, a radical scavenger, has been tried to prevent the repetition of stroke-like episodes. The spreading of lesions seemed to occur less frequently in some patients treated with edaravone.[41] Additionally, in the autopsy brains from MELAS cases, having 3243A>G mutation, 8-OHdG was accumulated in the peri-lesional surviving neurons in the cerebral cortex, but the expressions of MnSOD and 8-oxoguanine glycosylase 1 were not up-regulated in those neurons.[41] Increased oxidative stress and insufficient defense could be the reasons of the pathogenesis of the spreading lesions in MELAS. Recently, we found the similar neuropathological characteristics with those in MELAS in the autopsy brains from cases of acute necrotizing encephalopathy caused by influenza infection (iANE).[42] Adjacent to the necrotic lesions in the thalamus and pontine tegmentum, there were neuropil micro-vacuolation, increased microvasculature and neuronal sparing; and nuclei immunoreactive for 8-OHdG was also found in the remaining neurons and glial cells (Fig. 3). In addition, two patients with iANE showed an increased level of 8-OHdG in the CSF. In pathogenesis in iANE, mitochondrial disturbance and/or oxidative stress of DNA is suggested to be involved.

CHILDHOOD-ONSET NEURODEGENERATIVE DISORDERS

Oxidative stress has been confirmed to play an important role in adult-onset neurodegenerative diseases, such as Alzheimer's disease, Parkinson's disease and amyotrophic lateral sclerosis.[43] The combination of increased oxidative damage, mitochondrial dysfunction, deposition of oxidized materials, inflammation and defects in protein clearance are most likely reasons of damaging neurons.

Spinal muscular atrophy (SMA) is a childhood-onset motor neuron disease and results from the homozygous loss of *survival motor neuron gene 1* in chromosome 5. SMA is classified into three clinical groups, depending on the age of onset and achieved motor abilities. The diseases are characterized by progressive loss of motor neurons in the spinal cord and/or brainstem, leading to symmetrical weakness and atrophy in the leg and respiratory muscles.[44] Three cases of Type I SMA (Werdnig-Hoffmann disease) and two cases of Type II SMA (intermediate type) demonstrated deposition of 4-HNE in the motor neurons of the hypoglossal nucleus and spinal anterior horn.[45,46] Furthermore, nuclei immunoreactive for 8-OHdG were observed in the motor cortex. Also lateral thalamic nucleus and cerebellar granule cells in the absence of neuronal loss and gliosis, indicate that oxidative stress may be the reason in the latent neurodegeneration other than the motor neurons in SMA.

Dentatorubral pallidoluysian atrophy (DRPLA) is a CAG-repeat disease that is classified into juvenile and early adult types showing PME and a late adult type characterized by dementia and cerebellar ataxia.[47] DRPLA patients have an expanded CAG triplet repeat (polyglutamine) on the short arm of chromosome 12 and the degree of polyglutamine expansion is involved in a variety of clinical manifestations. The pattern and distribution of neuropathological changes are region-specific and common in disease types. However intranuclear accumulation of mutant proteins, with expanded polyglutamines is recognized throughout the brain.[48] We examined accumulation of oxidative stress markers and expression of SOD in DRPLA autopsy cases, including four cases of juvenile and late adult types and two cases of early adult type.[49] Neuronal accumulation of 4-HNE was found in the hippocampus, globus pallidus and cerebellar dentate nucleus in the early and late adult types of DRPLA cases. Oxidative products of nucleosides, 8-OHdG and 8-OHG, were accumulated in the lenticular nucleus, predominantly in juvenile and early adult cases showing PME. Mitochondrial immunoreactivity of MnSOD was also reduced in the lenticular nucleus and cerebellum in cases showing PME. Expanded polyglutamine may be the reason for mitochondrial dysfunction and subsequent augmentation of oxidative stress in animal and cell models of DRPLA.[50] It is likely in the juvenile and early adult DRPLA cases the reduced MnSOD expression and increased oxidative DNA damage in the lenticular nucleus may be caused by the expanded polyglutamine, leading to the generation of PME.

Lafora disease (LD) is an autosomal recessive disorder characterized by progressive myoclonic epilepsy and presence of intracellular polyglucosan inclusions, (being called as Lafora bodies) in the brain, liver and cardiac muscles. Mutations of the *EPM2A* and *EPM2B* (*NHLRC1*) genes have been identified in LD patients.[51] Preliminary immunohistochemical analysis was performed in three autopsy cases of LD, which had a family history of LD and the abundant occurrence of Lafora bodies in the globus pallidus, cerebellar dentate nucleus and substantia nigra.[1] Nuclei immunoreactive for 8-OHdG were found in the cerebral cortex in two of three autopsy cases. In addition, two LD patients with *NHLRC1* mutations[52] displayed a mild increase in the level of 8-OHdG in urine, although there was no change in the CSF. Nevertheless, one of the two cases having *NHLRC1* mutations died of cardiac failure at the age of 36 years and immunohistochemistry for oxidative stress markers demonstrated nuclei immunoreactive for 8-OHdG in the neurons of globus pallidus and deposition of 4-HNE in the neuronal cytoplasm in the trochlear and trigeminal nuclei, irrespective of Lafora bodies (Fig. 4). In order to obtain a definite answer on the involvement of oxidative stress in LD, the comprehensive analysis in a large number of patients is necessary.

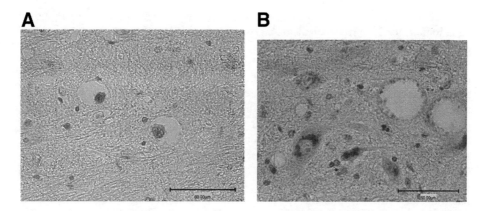

Figure 4. An autopsy case of Lafora disease with mutation of EPM2B gene. A. Nuclei immunoreactive for 8-hydroxy-2'-deoxyguanosine were found in the remaining neurons in the globus pallidus. Bar = 60 μm. B. Neurons with cytoplasm immunoreactive for 4-hydroxynonenal were scattered in the trochlear nucleus. Bar = 60 μm.

Table 2. Urinary levels of 8-hydroxy-2'-deoxyguanosine (8-OHdG), hexanoyl lysine adduct (HEL) and potential antioxidant (PAO) in patients with genetically-confirmed Rett syndrome.

		8-OHdG	HEL	PAO
Age	Mutations in *MeCP2* gene	(ng/mg Cre)	(pmol/mg Cre)	(μmol/L aop)
	Cutoff index	16.7	163.9	5187
Younger patients				
7 yrs	R168X	13.8	144.6	7563
10 yrs	R168X	24.6 Δ	153.5	6717
Older patients				
46 yrs	R306C	209.3 ↑	5929 ↑	2440 ↓
46 yrs	Frameshift (at 135)	17.7 Δ	114.7	5761
46 yrs	Frameshift (at 285)	65.3 ↑	453.4 ↑	3443 ↓
47 yrs	T158M	10.3	108.7	5896
49 yrs	R133C	12.6	227 Δ	5164

Cutoff index for each oxidative marker were the mean + 2SD value in controls aged over 6 years. The upward arrows (↑) and triangles (Δ) denote severe and mild increases of 8-OHdG and HEL, respectively, while the downward arrows (↓) mean a decrease of PAO.

Rett syndrome (RS) is a neurodevelopmental disorder mainly caused by de novo mutations in the X-chromosomal *MeCP2* gene encoding the transcriptional regulator methyl-CpG-binding protein 2 and characterized with autistic mental retardation in females.[53] Its pathogenesis remains to be investigated and no effective therapy is available to date.[54] In autopsy brains of autistic cases, oxidative stress is one of possible cause of the Purkinje cell loss in the cerebellar cortex, leading to the cognitive disturbances and several studies have shown decreased levels of antioxidants in blood cells in autistic patients.[55] A few studies on oxidative stress in RS patients, however, demonstrated increased plasma levels of lipid peroxidation markers with reduced activities of the SOD in erythrocytes,[56] and increase of intra-erythrocyte nonprotein-bound iron and protein carbonyl concentrations.[57] We performed preliminary analysis on the levels of 8-OHdG and HEL in the urine of genetically-confirmed RS patients (Table 2). There were no relationships between the phenotype and oxidative stress markers. Older RS patients, having respiratory disturbances, tend to show increased levels of 8-OHdG and/or HEL and lowered antioxidant power. RS patients are known to be associated with several systemic complications, which can alter oxidative stress markers and antioxidant abilities and the analysis in the CSF seems to be prerequisite for further investigation on oxidative stress in neurological disorders in RS.

CONCLUSION

Oxidative stress leading to modification of nucleosides, proteins and lipids may occur in hereditary DNA repair disorders, congenital metabolic errors and childhood-onset neurodegenerative disorders. It is useful for clarifying pathogenesis of neurodegeneration to examine the involvement of oxidative stress, combining immunohistochemistry in the autopsy brains and ELISA in the CSF and urine. Although the involvement of oxidative stress seems to be various, the research indicates that antioxidant therapy may play a major role in alleviating the neurodegeneration in patients with developmental brain disorders.

REFERENCES

1. Hayashi M. Oxidative stress in developmental brain disorders. Neuropathology 2009; 29(1):1-8.
2. Willcox JK, Ash SL, Catignani GL. Antioxidants and prevention of chronic disease. Crit Rev Food Sci Nutr 2004; 44(4):275-295.
3. Toyokuni S. Reactive oxygen species-induced molecular damage and its application in pathology. Pathology Int 1999; 49(2):91-102.
4. Iwai S. Synthesis and thermodynamic studies of oligonucleotides containing the two isomers of thymidine glycol. Chemistry 2001; 7(20):4343-4351.
5. Hayashi M, Tanuma N, Miyata R. The involvement of oxidative stress in epilepsy. In: Kozyrev D, Slutsky V, eds. Handbook of Free Radicals: formation, types and effects. New York: Nova Science Publishers, 2010:305-318.
6. Takeuchi M, Yamagishi S. Involvement of toxic AGEs (TAGE) in the pathogenesis of diabetic vascular complications and Alzheimer's disease. J Alzheimers Dis 2009; 16(4):845-858.
7. Shiihara T, Kato M, Ichiyama T et al. Acute encephalopathy with refractory status epilepticus: bilateral mesial temporal and claustral lesions, associated with a peripheral marker of oxidative DNA damage. J Neurol Sci 2006; 250(1-2):159-161.
8. Tanuma N, Miyata R, Hayashi M et al. Oxidative stress as a biomarker of respiratory disturbance in patients with severe motor and intellectual disabilities. Brain Dev 2008; 30(6):402-409.

9. Tanuma N, Miyata R, Kumada S et al. The axonal damage marker tau protein in the cerebrospinal fluid is increased in patients with acute encephalopathy with biphasic seizures and late reduced diffusion. Brain Dev 2010; 32(6):435-439.

10. Izuta H, Matsunaga N, Shimazawa M et al. Proliferative diabetic retinopathy and relations among antioxidant activity, oxidative stress and VEGF in vitreous body. Mol Vis 2010; 16:130-136.

11. Hayashi M, Arai N, Satoh J et al. Neurodegenerative mechanisms in subacute sclerosing panencephalitis. J Child Neurol 2002; 17(10):725-730.

12. Martin LJ. DNA damage and repair: relevance to mechanisms of neurodegeneration. J Neuropathol Exp Neurol 2008; 67(5):377-387.

13. Kraemer KH, Patronas NJ, Schiffmann R et al. Xeroderma pigmentosum, trichothiodystrophy and Cockayne syndrome: a comple genotype-phenotype relationship. Neuroscience 2007; 145(4):1388-1396.

14. Hayashi M. Role of oxidative stress in xeroderma pigmentosum. Adv Exp Med Biol 2008; 637:120-127.

15. Hayashi M, Hayakawa K, Suzuki F et al. A neuropathological study of early onset Cockayne syndrome with chromosomal anomaly 47XXX. Brain Dev 1992; 14(1):63-67.

16. Hayashi M, Itoh M, Araki S et al. Oxidative stress and disturbed glutamate transport in hereditary nucleotide repair disorders. J Neuropathol Exp Neurol 2001; 60(4):350-356.

17. Saito Y, Shibuya M, Hayashi M et al. Cerebellopontine calcification: a new entitiy of idiopathic intracranial calcification? Acta Neuropathol 2005; 110(1):77-83.

18. Mody N, Parhami F, Sarafian TA et al. Oxidative stress modulates osteoblastic differentiation of vascular and bone cells. Free Radic Biol Med 2001; 31(4):509-519.

19. Massy ZA, Maziere C, Kamel S et al. Impact of inflammation and oxidative stress on vascular calcifications in chronic kidney disease. Pediatr Nephrol 2005; 20(3):380-382.

20. Hayashi M, Araki S, Kohyama J et al. Oxidative nucleotide damage and superoxide dismutase expression in the brains of xeroderma pigmentosum group A and Cockayne syndrome. Brain Dev 2005; 27(1):34-38.

21. Ahmad S I (edt). Molecular Mechanisms of Ataxia telangiectasia, Landes Bioscience Publication, 2006.

22. Takao N, Li Y, Yamamoto K. Protective roles for ATM in cellular response to oxidative stress. FEBS Lett 2000; 472(1):133-136.

23. Stern N, Hochman A, Zemach N et al. Accumulation of DNA damage and reduced levels of nicotine adenine dinucleotide in the brains of Atm-deficient mice. J Biol Chem 2002; 277(1):602-608.

24. Chen P, Peng C, Luff J et al. Oxidative stress is responsible for deficient survival and dendritogenesis in Purkinje neurons from ataxiatelangiectasia mutated mice. J Neurosci 2003; 23(36):11453-11460.

25. Russo I, Cosentino C, Del Giudice E et al. In ataxia-teleangiectasia betamethasone response is inversely correlated to cerebellar atrophy and directly to antioxidative capacity. Eur J Neurol 2009; 16(6):755-759.

26. Taylor AM, Groom A, Byrd PJ. Ataxia-telangiectasia-like disorder (ATLD)–its clinical presentation and molecular basis. DNA Repair 2004; 3(8-9):1219-1225.

27. Iijima K, Ohara M, Seki R et al. Dancing on damaged chromatin: functions of ATM and the RAD50/MRE11/NBS1 complex in cellular responses to DNA damage. J Radiat Res 2008; 49(5):451-464.

28. Oba D, Hayashi M, Minamitani M et al. Autopsic study of cerebellar degeneration in siblings with ataxia-telangiectasia-like disorder (ATLD). Acta Neuropathol 2010; 119(4):513-520.

29. Kohlschutter A, Schulz A. Towards understanding the neuronal ceroid lipofuscinoses. Brain Dev 2009; 31(7):499-502.

30. Hachiya Y, Hayashi M, Kumada S et al. Mechanisms of neurodegeneration in neuronal ceroid-lipofuscinosis. Acta Neuropathol 2006; 111(2):168-177.

31. Anzai Y, Hayashi M, Fueki N et al. Protracted juvenile neuronal ceroid lipofuscinosis—an autopsy report and immunohistochemical analysis. Brain Dev 2006; 28(6):462-465.

32. Bunge S, Knigge A, Steglich C et al. Mucopolysaccharidosis type IIIB (Sanfilippo B): identification of 18 novel alpha-N-acetylglucosaminidase gene mutations. J Med Genet 1999; 36(1):28-31.

33. Villani GR, Gargiulo N, Faraonio R et al. Cytokines, neurotrophins and oxidative stress in brain disease from mucopolysaccharidosis IIIB. J Neurosci 2007; 85(3):612-622.

34. Ohmi K, Greenberg DS, Rajavel KS et al. Activated microglia in cortex of mouse models of mucopolysaccharidoses I and IIIB. Proc Natl Acad Sci USA 2003; 100(4):1902-1907.

35. Wraith JE, Scarpa M, Beck M et al. Mucopolysaccharidosis type II (Hunter syndrome): a clinical review and recommendations for treatment in the era of enzyme replacement therapy. Eur J Pediatr 2008; 167(3):267-277.

36. Matalon R, Kaul R, Michals K. Mucopolysaccharidosis and mucolipidosis. In: Duckett S, ed. Pediatric Neuropathology. Baltimore: Williams and Wilkins, 1995:525-544.

37. Hamano K, Hayashi M, Shioda K et al. Mechanisms of neurodegeneration in mucopolysaccharidoses II and IIIB: analysis of human brain tissue. Acta Neuropathol 2008; 115(5):547-559.

38. Iizuka T, Sakai F. Pathogenesis of stroke-like episodes in MELAS: analysis of neurovascular cellular mechanisms. Curr Neurovasc Res 2005; 2(1):29-45.

39. Koga Y, Povalko N, Nishioka J et al. MELAS and L-arginine therapy: pathophysiology of stroke-like episodes. Ann NY Acad Sci 2010; 1201:104-110.
40. Trushima E, McMurray CT. Oxidative stress and mitochondrial dysfunction in neurodegenerative diseases. Neuroscience 2007; 145(4):1233-1248.
41. Katayama Y, Maeda K, Iizuka T et al. Accumulation of oxidative stress around the stroke-like lesions of MELAS patients. Mitochondrion 2009; 9(5):306-313.
42. Mizuguchi M, Hayashi M, Nakano I et al. Concentric structure of thalamic lesions in acute necrotizing encephalopathy. Neuroradiology 2002; 44(6):489-493.
43. Halliwell B. Oxidative stress and neurodegeneration: where are we now? J Neurochem 2006; 97(6):1634-1658.
44. Wirth B, Brichta L, Hahnen E. Spinal muscular atrophy: from gene to therapy. Semin Pediatr Neurol 2006; 13(2):121-131.
45. Hayashi M, Araki S, Arai N et al. Oxidative stress and disturbed glutamate transport in spinal muscular atrophy. Brain Dev 2002; 24(8):770-775.
46. Araki S, Hayashi M, Tamagawa K et al. Neuropathological analysis in spinal muscular atrophy type II. Acta Neuropathol 2003; 106(5):441-448.
47. Hayashi M, Kumada S, Shioda K et al. Neuropathological analysis of the brainstem and cerebral cortex lesions on epileptogenesis in hereditary dentatorubral-pallidoluysian atrophy. Brain Dev 2007; 29(9):473-481.
48. Yamada M. CAG repeat disorder models and human neuropathology: similarities and differences. Acta Neuropathol 2008; 115(1):71-86.
49. Miyata R, Hayashi M, Tanuma N et al. Oxidative stress in neurodegeneration in dentatorubral-pallidoluysian atrophy. J Neurol Sci 2008; 264(1-2):133-139.
50. Puranam KL, Wu G, Strittmatter WJ et al. Polyglutamine expansion inhibits respiration by increasing reactive oxygen species in isolated mitochondria. Biochem Biophys Res Commun 2006; 341(2):607-613.
51. Ganesh S, Puri R, Singh S et al. Recent advances in the molecular basis of Lafora's progressive myoclonus epilepsy. J Hum Genet 2006; 51(1):1-8.
52. Singh S, Suzuki T, Uchiyama A et al. Mutations in the NHLRC1 gene are the common cause for Lafora disease in the Japanese population. J Hum Genet 2005; 50(7):347-352.
53. Gonzales ML, LaSalle JM. The role of MeCP2 in brain development and neurodevelopmental disorders. Curr Psychiatry Rep 2010; 12(2):127-134.
54. Armstrong DD. Neuropathology of Rett syndrome. J Child Neurol 2005; 20(9):747-753.
55. Pardo CA, Eberhart CG. The neurobiology of autism. Brain Pathol 2007; 17(4):434-447.
56. Sierra C, Vilaseca MA, Brandi N et al. Oxidative stress in Rett syndrome. Brain Dev 2001; 23(Suppl 1):S236-S239.
57. De Felice C, Ciccoli L, Leoncini S et al. Systemic oxidative stress in classic Rett syndrome. Free Radic Biol Med 2009; 47(4):440-448.

CHAPTER 22

OXIDATIVE STRESS AND MITOCHONDRIAL DYSFUNCTION IN DOWN SYNDROME

Giovanni Pagano* and Giuseppe Castello

CROM, Cancer Research Center, Mercogliano, Italy
Corresponding Author:Giovanni Pagano—Email: gbpagano@tin.it

Abstract: Down syndrome (DS) or trisomy 21 is the genetic disease with highest prevalence displaying phenotypic features that both include neurologic deficiencies and a number of clinical outcomes. DS-associated neurodegeneration recalls the clinical course of Alzheimer disease (AD), due to DS progression toward dementia and amyloid plaques reminiscent of AD clinical course. Moreover, DS represents one of the best documented cases of a human disorder aetiologically related to the redox imbalance that has long been attributed to overexpression of Cu,Zn-superoxide dismutase (SOD-1), encoded by trisomic chromosome 21. The involvement of oxidative stress has been reported both in genes located else than at chromosome 21 and in transcriptional regulation of genes located at other chromosomes. Another well documented hallmark of DS phenotype is represented by a set of immunologic defects encompassing a number of B and T-cell functions and cytokine production, together prompting a proinflammatory state. In turn, this condition can be directly interrelated with an in vivo prooxidant state.

As an essential link to oxidative stress, mitochondrial dysfunctions are observed whenever redox imbalances occur, due to the main roles of mitochondria in oxygen metabolism and this is the case for DS. Ultrastructural and biochemical abnormalities were reported in mitochondria from human DS patients and from trisomy 16 (Ts16) mice, to be reviewed in this chapter. Together, in vivo alterations of mitochondrial function are consistent with a prooxidant state as a phenotypic hallmark in DS.

INTRODUCTION

Down syndrome, resulting from complete or partial trisomy of chromosome 21 (Hsa21), displays a pleotropic phenotype including intellectual disability, various forms

Neurodegenerative Diseases, edited by Shamim I. Ahmad.
©2012 Landes Bioscience and Springer Science+Business Media.

of neurologic damage, immune disorders, propensity to lymphoblastic and myeloid leukaemias, heart defects, Type II diabetes mellitus, obesity and premature ageing.[1-4]

A number of studies have provided evidence that the DS phenotype is associated with oxidative stress, mainly due to SOD-1 overexpression that has been investigated in a number of in vitro, ex vivo and animal studies.[5-10] The ratio SOD-1 to catalase plus glutathione peroxidase (GPx) is increased, thus more hydrogen peroxide (H_2O_2) is generated by SOD-1 than catalase and glutathione peroxidase can catabolize, giving rise to oxidative stress.[12-13] Beside SOD-1 overexpression, other genes located to chromosome 21, including the amyloid precursor protein (APP), one member of the Ets family of transcription factors (ETS2) and Down syndrome critical region 1 (DSCR1), were proposed to contribute to premature neuronal loss and the development of dementia.[14] Moreover, a recent study by Conti et al demonstrated a role in DS causation for transcriptional regulation.[15]

Albeit more complex than envisioned previously, the literature maintains the evidence for a major role of nonchromosome 21 located genes in DS pathogenesis.[15]

IN VIVO PROOXIDANT STATE: A LINK WITH IMMUNODEFICIENCY IN DS PATIENTS

Direct evidence for a prooxidant state in blood cells and body fluids from DS patients was provided by a number of independent studies. A summary of the published literature in this topic is presented in Table 1. A number of oxidative stress parameters were measured in brains, blood cells and body fluids from DS patients, pointing to significant increases in oxidative DNA damage, lipid peroxidation, plasma levels of uric acid and allantoin, along with lower-than-normal levels of xanthine and hypoxanthine.[5,9,12-25] Altogether, a well-established body of evidence supports the occurrence of an in vivo multi-parameter prooxidant state in DS patients since their foetal life[14-15] and subject to age-dependent modulation.[17,18]

Another well-documented feature in DS phenotype is defects in immune functions such as abnormal IgG and IgE levels in DS patients[26,27] and defective T-cell maturation as an early integral feature of DS.[28] The investigations focussing on DS-associated immunodeficiency have been continuing in the last several decades.[29,30] In the apparent lack of updated reviews (as per the latest MedLine search, mid-October 2010), an attempt will be made to summarise the available information on DS-associated defects in immune functions.

A major feature of these defects is related to IL-1 upregulation leading to a number neurodegenerative processes that support a neuroinflammatory component in both DS and AD pathogenesis.[29,30] Other investigations highlight the relationships between immune defects and inflammation in DS pathogenesis; these are based on the following evidence: (a) pro-inflammatory cytokinaemia with excess levels of IL-1β, IL-10, IL-7, TNF-α and IFN-γ;[31,32] (b) controversial data were presented by Cetiner et al[33] who reported a significantly increased levels of antinflammatory IL-4 and IL-10 along decreased levels of proinflammatory IL-6 and TNF-α, suggesting a continuing anti-inflammatory state in DS and possibly explaining the recurrent infections seen in DS patients; (c) expression of cell surface markers in children with DS showed significantly lower counts of myeloid dendritic cells, leukocyte, lymphocytes, monocytes and granulocytes, but higher counts of proinflammatory CD14(dim)CD16(+) monocytes;[34] (d) reduced number and altered

Table 1. Reported changes in oxidative stress parameters in cells, tissues or body fluids from DS patients

Cells/Tissues/ Body Fluids	Endpoints	References
Foetal Brain	↑ SOD-1; ↔ glutathione peroxidase (GPx); ↑ MDA	5
Brain	↑ reactive carbonyls and carbonyl reductase;	9
Amniotic Fluid	↑↑ Isoprostanes	10
Erythrocytes and Neutrophils	↑ SOD-1 and GPx; ↔ catalase (CAT); ↑↑ ratio SOD-1:(GPx + CAT); ↑ MDA and lipofuscin ↑ SOD-1 and GPx; ↔ CAT; ↑ MDA	12-16
Leukocytes, Whole Blood and Plasma	Age-dependent ↑ 8-hydroxy-2′-deoxyguanosine (8-OHdG); Age-related ↑↓ GSSG:GSH; ↑ Plasma Glx levels in young patients; ↑ Plasma uric acid and ascorbic acid; ↔ Vitamin E	17-18
Plasma	↑ Uric acid and allantoin; ↓ hypoxanthine and xanthine	19
Plasma and Urine	↓ Plasma melatonin and urinary kynurenine; ↑ Urinary kynurenic acid and anthranilic acid	20
Plasma	↑ Citrulline:arginine and neopterin in demented patients; ↑ NO production	21
Serum	↑ Uric acid	22
Urine	↑ 8-OHdG and MDA	23
Urine	↑ Isoprostane 8,12-iso-iPF2alpha-VI	24
Amniotic Fluid (mRNA transcription profile)	Dysregulation of oxidative stress response genes; phospholipids, ion transport molecules, heart, muscle, structural proteins, and DNA damage repair genes	25

morphology of endothelial progenitor cells in DS, leading to the higher susceptibility to oxidative stress and to pathogen infection.[35]

　　Together, a major feature of DS-associated immunologic defects may be recognized in the overexpression of proinflammatory cytokines,[29-32,34,35] except for one report.[33] In turn, the mechanistic interplay of oxidative—and nitrosative—stress with inflammatory processes is recognized and specifically documented in some pathologies including, among others, neurodegenerative diseases, diabetes and ageing.[36-38] Thus, it can be postulated that the prooxidant state and the occurrence of excess proinflammatory cytokines in DS may be regarded as two interrelated dysfunctions leading to biochemical, immunological and clinical imbalances in DS patients. Thus an attempt of envisaging a unified mechanistic frame should lead to breakthroughs in DS clinical management.

MITOCHONDRIAL DYSFUNCTIONS IN DS

　　Early studies reported deficiencies in certain mitochondrial enzymes,[39] including monoamine oxidase, cytochrome oxidase and isocitrate dehydrogenase from blood platelets of DS patients, similar to defects in cytochrome oxidase (Complex 4) found in blood platelets and in brain tissue from patients with AD.[39]

Ultrastructural abnormalities in cultured cerebellar neurons from trisomy 16 (Ts16) mice (a model of DS) were reported,[40,41] including abnormally shaped mitochondria, diminished levels of microtubules and dense bundles of abnormal filaments. Moreover, superoxide formation was significantly increased in Ts16 neurons vs control neurons and the mitochondrial respiratory chain Complex I inhibitor, rotenone, inhibited superoxide production in diploid neurons, but the increased superoxide generation in Ts16 neurons remained. This increased production of superoxide in Ts16 neurons could be attributed to a deficient complex I of mitochondrial electron transport chain, hence to an impaired mitochondrial energy metabolism and finally neuronal cell death.[40] A selective decrease in respiration was detected with the Complex I substrates malate and glutamate but not with the Complex II substrate succinate in isolated cortex mitochondria from Ts16 mice.[41-43]

No differences in H_2O_2 production or maximal calcium uptake were detected in Ts16 mitochondria, yet a decrease in pyruvate dehydrogenase levels was detected, similar to the pattern found in Parkinson's disease.[41-43]

Astrocytes and neurons from foetal DS brain showed alterations in the processing of amyloid beta precursor protein (AbetaPP), with increased levels of AbetaPP and C99, reduced levels of secreted AbetaPP (AbetaPPs) and accumulation of insoluble Abeta.[42] These findings pointed to inhibition of mitochondrial metabolism, consistent with impaired mitochondrial function in DS astrocytes and a chronic state of increased neuronal vulnerability.[44]

Compared to age-matched control cells defective repair of oxidative damage to mitochondrial DNA (mtDNA) was reported in fibroblasts from DS patients, treated with the ROS generator menadione.[45]

Table 2. Main mitochondrial anomalies/dysfunctions reported in cells from DS patients or from trisomy 16 mice

Cells/Organisms	Endpoints	References
Platelets from DS patients	↓ Monoamine oxidase, cytochrome oxidase and isocitrate dehydrogenase	39
Trisomy 16 cerebellar neurons	↑ Levels of microtubules, abnormally shaped mitochondria and dense bundles of abnormal filaments	40
Brain of mouse trisomy 16	↑ O_2^- formation; ↓ respiration with the Complex I substrates malate and glutamate but not with the Complex II substrate succinate; ↓ the 20 kDa subunit of Complex I; ↓ pyruvate dehydrogenase levels	41-43
Astrocyte and neuronal cultures from foetal DS brain	Alterations in the processing of amyloid beta precursor protein (AbetaPP); impaired mitochondrial function in DS astrocytes	44
Fibroblasts from DS patients	Impaired repair of oxidative damage to mtDNA	45
Heart of DS fetuses	Oligonucleotide microarrays: downregulation of genes encoding mitochondrial enzymes and upregulation of genes encoding extracellular matrix proteins	46
PBMC from DS children	↑ Lucigenin-derived chemiluminescence; ↓ $\Delta\Psi(m)$	47

Conti et al[46] investigated the expression profile of chromosome 21 (Hsa21) genes using oligonucleotide microarrays in hearts of human foetuses with and without Hsa21 trisomy and found that Hsa21 gene expression was globally upregulated 1.5 fold in trisomic samples. Functional class scoring and gene set enrichment analyses of 473 genes revealed downregulation of genes encoding mitochondrial enzymes and upregulation of genes encoding extracellular matrix proteins in the foetal heart of trisomic subjects.[46]

Direct evidence for an in vivo alteration of mitochondrial function in blood cells from DS patients was reported by Roat et al,[47] who found a significant loss of mitochondrial membrane potential [$\Delta\Psi(m)$], underlying the presence of an increasing susceptibility of these organelles to damaging agents.

Table 2 summarizes the literature that establishes mitochondrial abnormalities both in cells from Ts21 patients and from Ts16 mice, which are mechanistically related to the redox abnormalities both detected in DS patients and in Ts16 mice (reviewed by Pallardó et al, ref.48).

MITOCHONDRIA-TARGETED CHEMOPREVENTION

The multi-faceted deficiencies in mitochondrial functions, detected in DS, are consistent with a set of phenotypic features related to oxidative stress as a central pathogenetic event in neurodegeneration, diabetes and propension to early ageing.[49,50]

It should be recalled that delaying the ageing process started from the early discoveries by McCay et al[51] of calorie restriction as a means for prolonging lifespan. To date, the major focus of studies into delaying ageing has been on eliciting the mechanisms for the beneficial effects of calorie restriction. Nutrient availability and energy metabolism have been found to be modulated by a family of seven NAD$^+$–dependent deacetylase enzymes termed sirtuins (SIRT1-7), three of which are associated with mitochondria[51-52] and activated by calorie restriction. Resveratrol (trans-3,5,4'-trihydroxystilbene), a polyphenol phytoalexin found mainly in grape skin and some berries, has been found as the most potent natural compound able to activate SIRT1.[52]

Resveratrol and SIRT1 exert antioxidant, anti-inflammatory and antiapoptotic effects and have been shown to protect endothelial cells against the adverse effects of cigarette smoking-induced oxidative stress. The vasoprotective effects of resveratrol probably contribute to its antiageing action in mammals and may be especially beneficial in pathophysiological conditions associated with accelerated vascular ageing[52] and by counteracting the adverse effects of a high-calorie diet in mice.[53]

The term "mitochondrial nutrient appears in a recent review[54] and refers to a group of agents that can directly or indirectly protect mitochondria from oxidative damage and improve mitochondrial function.[54-56] Direct protection includes preventing the generation of oxidants, scavenging free radicals and inhibiting oxidant reactivity, removing oxidized inactive proteins accumulation,[57] elevating cofactors of defective mitochondrial enzymes and also protecting enzymes from further oxidation. Indirect protection includes repairing oxidative damage by enhancing antioxidant defence systems either through activation of phase 2 enzymes or through increase in mitochondrial biogenesis.

Among "mitochondrial nutrients", acetyl-L-carnitine (ALC) regulates gene expression in brain and heart.[58,59] ALC provides acetyl groups for nuclear histone acetylation suggesting this as one of its mechanisms of action,[60] possibly explaining the reported role of ALC in intermediate and mitochondrial metabolism as well as its trophic and antioxidant actions.[61]

Another relevant "mitochondrial nutrient", α-lipoic acid (ALA) occurs in mitochondria and is a cofactor for pyruvate dehydrogenase and α-ketoglutarate dehydrogenase.[54,55,57,62] α-Lipoic acid features a cyclic disulfide moiety which exists in redox equilibrium with its reduced dithiol form, dihydrolipoic acid. This latter acid acts in GSH repletion and as a potent antioxidant by redox-cycling, e.g., ascorbic acid and tocopherols to their reduced forms. Increased ROS scavenging by ALA could thus counteract a number of adverse events, including mutations in mtDNA, which can exacerbate the defect in oxidative phosphorylation and loss of low molecular weight antioxidants that, together, leads to further ROS formation.[56,62,63]

Coenzyme Q_{10} (CoQ_{10}) (ubiquinone and its reduced form ubiquinol) is an electron acceptor that transfers electrons from Complexes I and II to Complex III.[64] CoQ_{10} exerts an antioxidant function in lipid and mitochondrial membranes and may also decrease electron leak from the electron transport chain via its role in bypassing defects in oxidative phosphorylation. The use of CoQ_{10} as a "mitochondrial nutrient" has been tested by recent clinical studies[65,66] of patients with DS and with other disorders.[67,68] To date, no significant clinical outcomes have been detected in the trials only utilising CoQ_{10}, except for successful mitigation of statin-induced myopathy that is induced by CoQ_{10} depletion as a side effect of statin treatment.[69] More promising results were obtained in some studies utilizing combination therapies with different "mitochondrial nutrients", providing a rationale for further long-term investigations on suitably extended patient groups.[62,70,71]

When dealing with disorders characterized by oxidative stress and mitochondrial abnormalities, a relevant pathological model is provided by a groups of diseases named mitochondrial cytopathies, displaying widespread implications for many pathological conditions.[62,72] The pre-existing clinical experience in managing patients with mitochondrial cytopathies, by means of combination nutraceuticals ("mitochondrial cocktail"), may provide useful background in planning animal experiments and in view of suitable clinical trials aimed at counteracting prooxidant state and disease progression both in DS and several age-related degenerative disorders.[72]

CONCLUSION: RESEARCH PERSPECTIVES

Multi-faceted knowledge has been generated that provide evidence for the presence of mitochondrial abnormalities in DS and in other ageing-related diseases. Changes in mitochondrial function and ultrastructure have been reported to occur in DS,[39-47] as well as in other diseases such as senile neuropathies.[73,74] These observations, however, have not been conducted using a systematic and comparative approach, thus providing limited information as to the mechanistic links of the reported mitochondrial alterations across the different disorders. Further well-targeted investigations are required.

REFERENCES

1. Hayes A, Batshaw ML. Down syndrome. Pediatr Clin North Am 1993; 40:523-535.
2. Musiani P, Valitutti S, Castellino F et al. Intrathymic deficient expansion of T-cell precursors in Down syndrome. Am J Med Genet Suppl 1990; 7:219-224.
3. Hasle H. Pattern of malignant disorders in individuals with Down's syndrome. Lancet Oncol 2001; 2:429-436.
4. Lott IT, Head E. Down syndrome and Alzheimer's disease: a link between development and aging. Ment Retard Dev Disabil Res Rev 2001; 7:172-178.

5. Brooksbank BW, Balasz R. Superoxide dismutase, glutathione peroxidase and lipoperoxidation in Down's syndrome fetal brain. Exp Brain Res 1984; 16:37-44.
6. Kedziora J, Bartosz G. Down's syndrome: A pathology involving the lack of balance of reactive oxygen species. Free Rad Biol Med 1988; 4:317-330.
7. Groner Y, Elroy-Stein O, Avraham KB et al. Cell damage by excess CuZnSOD and Down's syndrome. Biomed Pharmacother 1994; 48:231-240.
8. Busciglio J, Yankner BA. Apoptosis and increased generation of reactive oxygen species in Down's syndrome neurons in vitro. Nature 1995; 378:776-779.
9. Balcz B, Kirchner L, Cairns N et al. Increased brain protein levels of carbonyl reductase and alcohol dehydrogenase in Down syndrome and Alzheimer's disease. J Neural Transm Suppl 2001; 61:193-201.
10. Epstein CJ, Avraham KB, Lovett M et al. Transgenic mice with increased Cu/Zn-superoxide dismutase activity: Animal model of dosage effects in Down syndrome. Proc Natl Acad Sci USA 1987; 84:8044-8048.
12. Muchová J, Sustrová M, Garaiová I et al. Influence of age on activities of antioxidant enzymes and lipid peroxidation products in erythrocytes and neutrophils of Down syndrome patients. Free Radic Biol Med 2001; 31:499-508.
13. Garaiová I, Muchová J, Šustrová M et al. The relationship between antioxidant systems and some markers of oxidative stress in persons with Down syndrome. Biologia (Bratislava) 2004; 59:781-788.
14. Slonim DK, Koide K, Johnson KL et al. Functional genomic analysis of amniotic fluid cell-free mRNA suggests that oxidative stress is significant in Down syndrome fetuses. Proc Natl Acad Sci USA 2009; 106:9425-9429.
15. Conti A, Fabbrini F, D'Agostino P et al. Altered expression of mitochondrial and extracellular matrix genes in the heart of human fetuses with chromosome 21 trisomy. BMC Genomics 2007; 8:268.
14. Pastor MC, Sierra C, Dolade M et al. Antioxidant enzymes and fatty acid status in erythrocytes of Down's syndrome patients. Clin Chem 1998; 44:924-929.
15. Garcez ME, Peres W, Salvador M. Oxidative stress and hematologic and biochemical parameters in individuals with Down syndrome. Mayo Clin Proc 2005; 80:1607-1611.
16. Casado A, López-Fernández ME, Ruíz R. Lipid peroxidation in Down syndrome caused by regular trisomy 21, trisomy 21 by Robertsonian translocation and mosaic trisomy 21. Clin Chem Lab Med 2007; 45:59-62.
17. Pallardó FV, Degan P, d'Ischia M et al. Higher age-related prooxidant state in young Down syndrome patients indicates accelerated aging. Biogerontology 2006; 7:211-220.
18. Zana M, Szécsényi A, Czibula A et al. Age-dependent oxidative stress-induced DNA damage in Down's lymphocytes. Biochem Biophys Res Commun. 2006; 345:726-733.
19. Žitňanová I, Korytar P, Aruoma OI et al. Uric acid and allantoin levels in Down syndrome: antioxidant and oxidative stress mechanisms? Clin Chim Acta 2004; 341:139-146.
20. Uberos J, Romero J, Molina-Carballo A et al. Melatonin and elimination of kynurenines in children with Down's syndrome. J Pediatr Endocrinol Metab 2010; 23:277-282.
21. Coppus AM, Fekkes D, Verhoeven WM et al. Plasma levels of nitric oxide related amino acids in demented subjects with Down syndrome are related to neopterin concentrations. Amino Acids 2010; 38:923-928.
22. Nagyová A, Sustrová M, Raslová K. Serum lipid resistance to oxidation and uric acid levels in subjects with Down's syndrome. Physiol Res 2000; 49:227-231.
23. Jovanovic SV, Clements D, MacLeod K. Biomarkers of oxidative stress are significantly elevated in Down syndrome. Free Radic Biol Med 1998; 25:1044-1048.
24. Pratico D, Iuliano L, Amerio G et al. Down's syndrome is associated with increased 8,12-iso-iPF2alpha-VI levels: evidence for enhanced lipid peroxidation in vivo. Ann Neurol 2000; 48:795-798.
25. Perrone S, Longini M, Bellieni CV et al. Early oxidative stress in amniotic fluid of pregnancies with Down syndrome. Clin Biochem 2007; 40:177-180.
26. Miller ME, Mellman Cohen MM, Kohn G et al. Depressed immunoglobulin G in newborn infants with Down's syndrome. J Pediatr 1969; 75:996-1000.
27. Lopez V. Serum IgE concentration in trisomy 21. J Ment Defic Res 1974; 18:111-114.
28. Burgio GR, Lanzavecchia A, Maccario R et al. Immunodeficiency in Down's syndrome: T-lymphocyte subset imbalance in trisomic children. Clin Exp Immunol 1978; 33:298-301.
29. Mrak RE, Griffin WS. Trisomy 21 and the brain. J Neuropathol Exp Neurol. 2004; 63:679-685.
30. Griffin WS. Inflammation and neurodegenerative diseases. Am J Clin Nutr 2006; 83:470S-474S.
31. Shimada A, Hayashi Y, Ogasawara M et al. Pro-inflammatory cytokinemia is frequently found in Down syndrome patients with hematological disorders. Leuk Res 2007; 31:1199-1203.
32. Guazzarotti L, Trabattoni D, Castelletti E et al. T-lymphocyte maturation is impaired in healthy young individuals carrying trisomy 21 (Down syndrome). Am J Intellect Dev Disabil 2009; 114:100-109.
33. Cetiner S, Demirhan O, Inal TC et al. Analysis of peripheral blood T-cell subsets, natural killer cells and serum levels of cytokines in children with Down syndrome. Int J Immunogenet 2010; 37:233-237.
34. Bloemers BL, van Bleek GM, Kimpen JL et al. Distinct abnormalities in the innate immune system of children with Down syndrome. J Pediatr 2010; 156:804-809.

35. Costa V, Sommese L, Casamassimi A et al. Impairment of circulating endothelial progenitors in Down syndrome. BMC Med Genomics 2010; 3:40.
36. Roberts RA, Smith RA, Safe S et al. Toxicological and pathophysiological roles of reactive oxygen and nitrogen species. Toxicology 2010; 276:85-94.
37. Candore G, Bulati M, Caruso C et al. Inflammation, cytokines, immune response, apolipoprotein E, cholesterol and oxidative stress in Alzheimer disease: therapeutic implications. Rejuvenation Res 2010; 13:301-313.
38. Galasko D, Montine TJ. Biomarkers of oxidative damage and inflammation in Alzheimer's disease. Biomark Med 2010; 4:27-36.
39. Prince J, Jia S, Båve U et al. Mitochondrial enzyme deficiencies in Down's syndrome. J Neural Transm Park Dis Dement Sect 1994; 8:171-181.
40. Bersu ET, Ahmad FJ, Schwei MJ et al. Cytoplasmic abnormalities in cultured cerebellar neurons from the trisomy 16 mouse. Brain Res Dev Brain Res 1998; 109:115-120.
41. Schuchmann S, Heinemann U. Increased mitochondrial superoxide generation in neurons from trisomy 16 mice: a model of Down's syndrome. Free Radic Biol Med 2000; 28:235-250.
42. Capone G, Kim P, Jovanovich S et al. Evidence for increased mitochondrial superoxide production in Down syndrome. Life Sci 2002; 70:2885-2895.
43. Bambrick LL, Fiskum G. Mitochondrial dysfunction in mouse trisomy 16 brain. Brain Res 2008; 1188:9-16.
44. Busciglio J, Pelsman A, Wong C et al. Altered metabolism of the amyloid beta precursor protein is associated with mitochondrial dysfunction in Down's syndrome. Neuron 2002; 33:677-688.
45. Druzhyna N, Nair RG, LeDoux SP et al. Defective repair of oxidative damage in mitochondrial DNA in Down's syndrome. Mutat Res 1998; 409:81-89.
46. Conti A, Fabbrini F, D'Agostino P et al. Altered expression of mitochondrial and extracellular matrix genes in the heart of human fetuses with chromosome 21 trisomy. BMC Genomics 2007; 8:268.
47. Roat E, Prada N, Ferraresi R et al. Mitochondrial alterations and tendency to apoptosis in peripheral blood cells from children with Down syndrome. FEBS Lett 2007; 581:521-525.
48. Pallardó FV, Lloret A, Lebel M et al. Mitochondrial dysfunction in some oxidative stress-related genetic diseases: Ataxia-Telangiectasia, Down syndrome, Fanconi Anaemia and Werner syndrome. Biogerontology 2010; 11:401–419.
49. Harman D. Alzheimer's disease pathogenesis: role of aging. Ann NY Acad Sci 2006; 1067:454-460.
50. Ristow M. Neurodegenerative disorders associated with diabetes mellitus. J Mol Med 2004; 82:510-529.
51. McCay CM, Crowell MF, Maynard LA. The effect of retarded growth upon the length of life span and upon the ultimate body size. J Nutrition 1935; 10:63-79.
52. Csiszar A, Labinskyy N, Podlutsky A et al. Vasoprotective effects of resveratrol and SIRT1: attenuation of cigarette smoke-induced oxidative stress and proinflammatory phenotypic alterations. Am J Physiol Heart Circ Physiol 2008; 294:H2721-H2735.
53. Baur JA, Pearson KJ, Price NL et al. Resveratrol improves health and survival of mice on a high-calorie diet. Nature 2006; 444:337-342.
54. Liu J. The effects and mechanisms of mitochondrial nutrient alpha-lipoic acid on improving age-associated mitochondrial and cognitive dysfunction: an overview. Neurochem Res 2008; 33:194-203.
55. Palaniappan AR, Dai A. Mitochondrial ageing and the beneficial role of alpha-lipoic acid. Neurochem Res 2007; 32:1552-1558.
56. Plecitá-Hlavatá L, Jezek J, Jezek P. Pro-oxidant mitochondrial matrix-targeted ubiquinone MitoQ10 acts as anti-oxidant at retarded electron transport or proton pumping within Complex I. Int J Biochem Cell Biol 2009; 41:1697-1707.
57. Long J, Wang X, Gao H et al. D-galactose toxicity in mice is associated with mitochondrial dysfunction: protecting effects of mitochondrial nutrient R-alpha-lipoic acid. Biogerontology 2007; 8:373-381.
58. Gadaleta MN, Petruzzella V, Renis M et al. Reduced transcription of mitochondrial DNA in the senescent rat. Tissue dependence and effect of L-carnitine. Eur J Biochem 1990; 187:501-506.
59. Traina G, Federighi G, Brunelli M et al. Cytoprotective effect of acetyl-L-carnitine evidenced by analysis of gene expression in the rat brain. Mol Neurobiol 2009; 39:101-106.
60. Madiraju P, Pande SV, Prentki M et al. Mitochondrial acetylcarnitine provides acetyl groups for nuclear histone acetylation. Epigenetics 2009; 4:399-403.
61. Rosca MG, Lemieux H, Hoppel CL. Mitochondria in the elderly: is acetylcarnitine a rejuvenator? Adv Drug Deliv Rev 2009; 61:1332-1342.
62. Rodriguez MC, Macdonald JR, Mahoney DJ et al. Beneficial effects of creatine, CoQ10 and lipoic acid in mitochondrial disorders. Muscle Nerve 2007; 35:235–242.
63. Savitha S, Panneerselvam C. Mitigation of age-dependent oxidative damage to DNA in rat heart by carnitine and lipoic acid. Mech Ageing Dev 2007; 128:206-212.
64. Maroz A, Anderson RF, Smith RAJ et al. Reactivity of ubiquinone and ubiquinol with superoxide and the hydroperoxyl radical: implications for in vivo antioxidant activity. Free Radic Biol Med 2009; 46:105–109.

65. Miles MV, Patterson BJ, Chalfonte-Evans ML et al. Coenzyme Q10 (ubiquinol-10) supplementation improves oxidative imbalance in children with trisomy 21. Pediatr Neurol 2007; 37:398-403.
66. Tiano L, Carnevali P, Padella L et al. Effect of Coenzyme Q(10) in mitigating oxidative DNA damage in Down syndrome patients, a double blind randomized controlled trial. Neurobiol Aging 2009; doi:10.1016/j. neurobiolaging.2009.11.016.
67. Kaufmann P, Thompson JL, Levy G et al. Phase II trial of CoQ10 for ALS finds insufficient evidence to justify phase III. Ann Neurol 2009; 66:235-244.
68. Storch A, Jost WH, Vieregge P et al. Randomized, double-blind, placebo-controlled trial on symptomatic effects of coenzyme Q(10) in Parkinson disease. Arch Neurol 2007; 64:938-944.
69. Caso G, Kelly P, McNurlan MA et al. Effect of coenzyme Q10 on myopathic symptoms in patients treated with statins. Am J Cardiol 2007; 99:1409-1412.
70. Palacká P, Kucharská J, Murin J et al. Complementary therapy in diabetic patients with chronic complications: a pilot study. Bratisl Lek Listy 2010; 111:205-211.
71. Hertz N, Lister RE. Improved survival in patients with end-stage cancer treated with coenzyme Q(10) and other antioxidants: a pilot study. J Int Med Res 2009; 37:1961-1971.
72. Tarnopolsky MA. The mitochondrial cocktail: rationale for combined nutraceutical therapy in mitochondrial cytopathies. Adv Drug Deliv Rev 2008; 60:1561-1567.
73. Vila M, Ramonet D, Perier C. Mitochondrial alterations in Parkinson's disease: new clues. J Neurochem 2008; 107:317-328.
74. Wang X, Su B, Lee HG et al. Impaired balance of mitochondrial fission and fusion in Alzheimer's disease. J Neurosci 2009; 28:9090-9103.

CHAPTER 23

PICK'S DISEASE

Naoya Takeda, Yuki Kishimoto and Osamu Yokota*

Department of Neuropsychiatry, Okayama University Graduate School of Medicine, Dentistry and Pharmaceutical Sciences, Shikata-cho, Okayama, Japan
*Corresponding Author: Osamu Yokota—Email: oyokota1@yahoo.co.jp

Abstract: Picks disease is a major clinicopathological disease having circumscribed atrophy in the frontotemporal lobe. Demented patients with frontotemporal atrophy are now clinically diagnosed as frontotemporal lobar degeneration (FTLD). Other underlying pathologies in patients with FTLD include FTLD with TDP-43-positive inclusions, corticobasal degeneration, progressive supranuclear palsy, basophilic inclusion body disease, neuronal intermediate filament inclusion disease and argyrophilic grain disease.

In this chapter, recent findings regarding the distinct clinical and histopathological features of these pathological disease entities are presented including the discussion on the possibility of future antemortem diagnosis of patients with the disease.

INTRODUCTION

The term 'Pick's disease' has been historically used to refer to cases showing circumscribed atrophy in the frontal and temporal lobes, irrespective of the presence or absence of Pick bodies. However, because the detailed pathological and biochemical backgrounds of cases of 'Pick's disease with Pick bodies' and 'Pick's disease without Pick bodies' have now been clarified, it is recommended that the term 'Pick's disease' be used only for cases having tau-positive Pick bodies, irrespective of the distribution of cerebral atrophy. That is, the pathological concept and basis for the pathological diagnosis of 'Pick's disease' have drastically changed from macroscopic findings (i.e., circumscribed lobar atrophy) to molecular pathological features (e.g., the biochemical and morphological nature of abnormally accumulated proteins).

Currently, demented patients having dysfunction of the frontal and temporal lobes associated with variable degrees of cerebral atrophy in these regions are clinically

Neurodegenerative Diseases, edited by Shamim I. Ahmad.
©2012 Landes Bioscience and Springer Science+Business Media.

diagnosed as having frontotemporal lobar degeneration (FTLD). FTLD is composed of three distinct clinical subtypes: Frontotemporal dementia (FTD), progressive nonfluent aphasia (PA) and semantic dementia (SD). FTD is characterized by progressive behavioral disturbances including personality changes and uninhibited, self-centered and/or stereotypic behaviors. SD is characterized by fluent speech output and loss of concept knowledge, resulting in loss of understanding of nominal terms, impairment of recognition of faces and common objects and impairment of auditory comprehension. PA is defined as progressive, nonfluent spontaneous speech. FTD, SD and PA are clinical syndromes; therefore, each subtype can include patients having variable distinct pathological disease entities. At present, because patients having different pathologies often show similar clinical features of FTLD, it is still difficult to infer precisely the underlying pathologies in FTLD patients in life. However, several previous studies have demonstrated the possibility that the trends of clinical presentations are not always identical between pathological disease entities.

In this chapter, the term Pick's disease is used to refer only to cases having Pick bodies, while the term 'FTLD' is used to refer to pathologically heterogeneous cases with circumscribed atrophy in the frontal and temporal lobes. First, we briefly note the historical clinicopathological concept of 'Pick's disease', the clinical concept of FTLD proposed later and the present pathological classification of FTLD. Second, clinical and pathological features of major underlying pathologies in patients with the diseases are described.

PICK'S DISEASE AND CLINICOPATHOLOGICAL CONCEPT OF FTLD

In 1892, Arnold Pick reported the case of August H. who was the first case of 'Pick's disease', now called FTLD.[1] Initially this man presented with memory impairment and transcortical sensory aphasia was seen later in the course. Postmortem pathological examination demonstrated left side-predominant circumscribed atrophy in the temporo-frontal lobe. Thereafter, Pick reported eight similar cases with circumscribed atrophy in the frontal or temporal lobe. In the papers he did not note the detailed histological findings for each case, including the presence or absence of intraneuronal inclusions.

In 1911, Alois Alzheimer first reported the spherical argyrophilic intraneuronal cytoplasmic inclusions that are now called Pick bodies.[2] In 1926, Onari and Spatz proposed the term 'Pick's disease' to refer to cases having circumscribed frontotemporal atrophy, irrespective of the presence or absence of Pick bodies.[3] Indeed, among five cases examined, only two cases had Pick bodies, two cases lacked them and it was not reported whether one case had the inclusions. Their pathological definition of 'Pick's disease' was: (1) macroscopic atrophy of the temporo-frontal lobe, (2) neuronal loss in the affected cerebral cortex, which was more predominant in superficial rather than deep cortical layers, (3) none or slight arteriosclerotic change, (4) absence of inflammatory change and (5) absence of senile plaques and neurofibrillary tangles. This means their concept of 'Pick's disease' was mainly defined by macroscopic features (i.e., lobar atrophy of the frontotemporal lobe) rather than microscopic cytopathological findings (i.e., Pick bodies). Thereafter, several clinical studies were carried out based on the view that the presence of Pick bodies may not affect the clinical picture in cases having the frontotemporal lobar atrophy.[4] In 1994, about 70 years after the report by Onari and

Spatz, the clinicopathological disease entity of FTD was proposed.[5] FTD was considered to include three clinicopathological subtypes: The frontal lobe degeneration (FLD) type, Pick-type and motor neuron disease-type. In the classification system, it was noted that the FLD-type is characterized by relatively mild cerebral atrophy in the frontal and temporal lobes without Pick bodies and that Pick-type cases show severe cortical atrophy with or without Pick bodies. The motor neuron disease-type was defined by the involvement of the frontotemporal lobe and motor neurons as well. Hence, in this classification system, the presence of Pick bodies is not necessary for the diagnosis of Pick-type. This view was consistent with that of Onari and Spatz. The disease concept 'FTD' was a prototype of the clinical classification of FTLD proposed in 1998. FTLD comprises three clinical syndromes, FTD, PA and SD.[6]

HISTORY OF 'PICK'S DISEASE WITHOUT PICK BODIES' AND PRESENT PATHOLOGICAL CLASSIFICATION

Some FTLD cases show circumscribed frontotemporal atrophy but lack Pick bodies; historically it is called 'Pick disease without Pick bodies' or 'atypical Pick's disease'. However, along with the accumulation of biochemical and immunohistochemical findings, many other pathological entities such as, corticobasal degeneration (CBD),[7] progressive supranuclear palsy (PSP),[8] argyrophilic grain disease (AGD),[9] basophilic inclusion body disease (BIBD),[10] neuronal intermediate neurofilament inclusion body disease (NIFID),[11,12] and FTLD-TDP[13,14] have been extracted from pathologically heterogeneous cases based on distinct histopathological features such as neuronal or glial inclusions. A histopathological hallmark of CBD is astrocytic plaques,[7] and that of PSP is tufted astrocytes.

Cases of generalized variant of Pick's disease (now called BIBD) have intraneuronal basophilic inclusion bodies.[10] Small argyrophilic granules, so-called 'argyrophilic grains', that preferably occur in the limbic system are the pathological hallmark of AGD.[9] FTLD cases having ubiquitin-positive inclusions were called FTLD-U.[15,16] However, later, the inclusions in most FTLD-U cases were reported to be immunopositive for TDP-43,[13,14] and those cases were renamed FTLD-TDP.[17] The neurofilament-positive, tau-negative, TDP-43-negative inclusion is a pathological hallmark of neuronal intermediate neurofilament inclusion body disease (NIFID).[12,18] Recently, it has been reported that fused in sarcoma (FUS) protein is a major component of the basophilic inclusions in BIBD cases[19] and neurofilament-positive inclusions in NIFID cases as well.[20]

Dementia lacking distinctive histological features (DLDH) was defined by the absence of any inclusion.[21] DLDH was previously considered to be the most frequent pathology in FTLD cases. However, later it was reported that approximately 90% of DLDH cases actually have tau-negative ubiquitin-positive neuronal inclusions, which had been a pathological hallmark of FTLD-U.[22] Several epidemiological studies in Western countries also suggested that DLDH is a rare histological subtype in FTLD.[23-25]

The pathological classification of FTLD revised in 2010 is shown in Table 1. Each pathological disease entity is placed in a molecular class defined by the kind of abnormally accumulated proteins.[27] FTLD cases having tau pathology are included in one molecular class, FTLD-tau and Pick's disease is one of the major pathological subtypes in this class.

Table 1. Nomenclature for neuropathologic subtypes of FTLD[27]

Major Molecular Class	Recognized Subtypes	Associated Genes
FTLD-tau	Pick's disease Corticobasal degeneration (CBD) Progressive supranuclear palsy (PSP) Argyrophilic grain disease (AGD) Multiple system tauopathy with dementia (MSTD) Neurofibrillary tangle predominant dementia (NFT-dementia) White matter tauopathy with globular glial inclusions (WMT-GGI) Unclassifiable	*MAPT*
FTLD-TDP	Type 1-4 Unclassifiable	*GRN* *VCP* 9p *TARDBP*
FTLD-UPS	FTD linked to chromosome 3	*CHMP2B*
FTLD-FUS	Atypical FTLD with ubiquitinated inclusions (aFTLD-U) Neuronal intermediate filament inclusion disease (NIFID) Basophilic inclusion body disease (BIBD)	*FUS*
FTLD-ni		

CHMP2B, charged multivescicular body protein 2B gene; FUS, fused in sarcoma; *GRN*, progranulin gene; *MAPT*, microtubule-associated protein tau gene; ni, no inclusions; *TARDBP*, transactive response DNA binding protein gene; TDP, TDP-43; UPS, ubiquitin proteasome system; *VCP*, valosin-containing protein gene.

FTLD-TDP is another molecular class, which is composed of cases with inclusions containing the transactivation-responsive DNA-binding protein of Mr 43 kDa (TDP-43). TDP-43 is considered to be a major component of inclusions in amyotrophic lateral sclerosis (ALS).[13,14] Because the inclusions in BIBD and NIFID cases are FUS-positive, BIBD and NIFID are classified in the most novel molecular category, FTLD-FUS. A small proportion of FTLD cases have ubiquitin-positive, TDP-43-negative and FUS-positive neuronal inclusions. They are called atypical FTLD-U.[28,29] Like BIBD and NIFID, atypical FTLD-U is also classified as FTLD-FUS.[27] FTLD cases lacking any distinctive neuronal or glial inclusions immunoreactive for tau, TDP-43, neurofilament, ubiquitin, or FUS are classified as FTLD-ni. Therefore, the previously proposed DLDH is classified in this category.[27] However, as noticed by some researchers; it remains unclear whether FTLD-ni cases actually exist,[30] because most of the DLDH cases have been reclassified as various molecular classes based on biochemical and immunohistochemical findings that were clarified.

As mentioned above, most pathological disease entity has been defined by a kind of abnormally accumulated protein. However, it has also been demonstrated that the accumulation of abnormal proteins does not always occur in a disease-specific manner. For example, in TDP-43 accumulation of not only FTLD-TDP and ALS is observed but also in other degenerative diseases, including Alzheimer's disease (AD),[31,32] AD+ dementia with Lewy bodies (DLB),[32-34] DLB without AD,[35] CBD,[36] ALS and parkinsonism-dementia complex of Guam (ALS/PDC of Guam),[37,38] and AGD.[39] In addition, recently, it has been reported that in some cases of progressive supranuclear palsy (PSP), actually have TDP-43 pathology which was previously reported to completely lack it.[40] The distribution of TDP-43-positive pathological changes in these diseases, except for in ALS/PDC of Guam, tends to be limited to the limbic system (e.g., the amygdala and hippocampal dentate gyrus) and temporal cortex, suggesting that the mechanism of TDP-43 accumulation in tauopathy and may be different from that in FTLD-TDP and ALS. Some researchers have proposed that FTLD-TDP and ALS are called 'major TDP-43 disease'[41] or 'disease with TDP-43 pathology as a primary histopathological feature'.[42]

Several studies have demonstrated the potential effects of concurrent TDP-43 pathology on the clinical picture of each disease, such as AD[43] and PSP.[40] In general, it is considered that the co-existence of abnormal accumulation of another kind of protein can affect the clinical presentation. This can make it difficult to infer the underlying pathologies in each patient precisely.

FREQUENCIES OF PATHOLOGICAL SUBTYPES OF FTLD

It is well known that FTLD is the second most common clinical diagnosis after AD and DLB.[44,45] However, to infer the pathological background in FTLD patients, epidemiological data regarding the pathological disease entities of FTLD are useful. In previous reports, FTLD-U was the most frequent pathology in patients clinically diagnosed as having FTLD, with the frequency of 12-62%.[23-25,46-49] Further, it is considered that most of the FTLD-U cases reported previously were actually FTLD-TDP.[13,14] The second most common pathology may be Pick's disease (8-33%), followed by CBD (12-22%) and PSP (4-29%). The frequencies of BIBD and NIFID cases were very low (1-4%). It was reported in Western countries that the frequency of FTLD-U (most of which may be actually FTLD-TDP) is about twice as high as that of Pick's disease. However, interestingly, a nationwide survey conducted in Japan in 2000 demonstrated that the frequency of Pick's disease was similar to that of 'Pick's disease without Pick bodies'. (We confirmed that almost all of them were FTLD-TDP).[26] Why the proportions of FTLD-TDP in the data of Japan are different from those in Western countries is unclear. However, it is possible that this discrepancy can be explained by a difference in the frequencies of familial FTLD-TDP cases in Western countries and Japan. For example, in Japan, familial cases are very rare among clinically diagnosed FTLD patients[44,45] and autopsy-confirmed FTLD-TDP cases as well.[50] In contrast, 10-60% of clinical FTLD patients in Western countries were reported to have a family history.[51] Likewise, about 60% in a large series of FTLD-U cases had a family history, suggesting that the racial difference in the genetic background may not be negligible in FTLD cases. As described below, several previous data suggested that the clinical presentations may be different in FTLD-TDP cases with and without a family history.

PICK'S DISEASE AND FTLD-TDP

Pick Bodies and TDP-43-Positive Inclusions

The existence of Pick bodies is a principal pathological hallmark of Pick's disease and now only cases with Pick bodies are diagnosed as Pick's disease (Fig. 1a). Pick bodies are round or oval argyrophilic intraneuronal cytoplasmic inclusions. They appear to be less fibrillary than the neurofibrillary tangles in AD. In contrast to the neurofibrillary tangles, that are often observed in the cognitively normal elderly, no Pick body has been noted in a cognitively normal subject. Several silver stains, such as Bielschowsky and Bodian methods, visualize Pick bodies. The inclusions do not show argyrophilia on Gallyas-Braak silver-stained sections in most cases of Pick's disease.

In anti-tau antibodies immunostain Pick bodies (Fig. 1b) Pick bodies are strongly labeled by three-repeat tau-specific antibody, but none or weakly stained by four-repeat tau-specific anti-tau antibody, suggesting that the major component of Pick bodies is three-repeat tau. The inclusions are observed in the hippocampal dentate gyrus, CA1, subiculum, entorhinal cortex and frontal, temporal, cingulate and insular cortices. In the

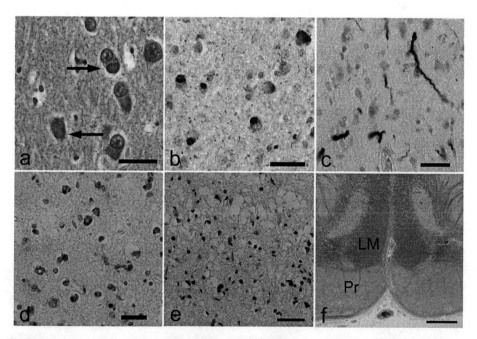

Figure 1. Pick's disease and frontotemporal lobar degeneration with TDP-43-positive inclusions a-b) Pick bodies in the hippocampus CA2-3 (arrows). c) Type 1 histology of FTLD-TDP. Relatively thick and long TDP-43-positive dystrophic neurites with few or no neuronal inclusions are seen in the frontal cortex. d) Type 2 histology of FTLD-TDP. Many TDP-43-positive intracytoplasmic inclusions with few or no dystrophic neurites are seen in the temporal cortex. e) Type 3 histology of FTLD-TDP. Both many short dystrophic neurites and intracytoplasmic inclusions are seen in the temporal cortex. f) Degeneration in the pyramidal tract at the level of the medulla oblongata. Loss of myelin is evident in the pyramidal tract, while the lemniscus medialis is spared. a) Bodian stain, b) AT-8 immunohistochemistry, c-e) phosphorylated TDP-43 immunohistochemistry, f) Klüver-Barrera stain. Pr, Pyramidal tract; LM lemniscusmedialis. Scale bars = a-d) 30 μm, e) 50 μm, f) 1 mm.

cerebral cortex, Pick bodies are often found in the superficial and deep layers. Under electron microscopy, Pick bodies are composed of mostly straight filaments with a 15-nm width and occasional twisted filaments with periodicity are also observed.[52]

In FTLD-TDP, TDP-43-positive lesions in the frontotemporal cortex are classified in to four histological subtypes based on their morphologic features. Type 1 is characterized by many long TDP-43-positive dystrophic neuritis, predominant in the superficial cortical layer in the frontotemporal cortex (Fig. 1c). TDP-43-positive neuronal cytoplasmic inclusions (NCIs) or neuronal intranuclear inclusions (NIIs) are rare or absent. Type 2 histology is characterized by many TDP-43-positive NCIs with a few or no dystrophic neurites (Fig. 1d). NIIs are rare or absent. Type 3 histology is characterized by many NCIs and short dystrophic neurites in the superficial cortical layers with a variable number of NIIs (Fig. 1e). Type 4 histology is characterized by many NIIs with few NCIs and dystrophic neurites.

Type 1 histology is frequently noted in FTLD patients clinically showing SD, while Type 2 histology is frequently observed in FTLD patients with motor neuron disease, which has also been called ALS with dementia. *PGRN* mutations are associated with Type 3 histology. Type 4 histology is associated with *VCP* mutations.

Clinicopathological Correlation in Pick's Disease and Sporadic FTLD-TDP

Pick's disease and sporadic FTLD-TDP are two major underlying pathologies in FTLD patients without a family history. Therefore, in clinical practice, clarifying the difference between the clinical features of the two diseases is useful to infer the underlying pathologies in sporadic FTLD patients.

We recently explored the differences in frequencies of clinical symptoms and evolving patterns of clinical syndromes (e.g., FTD, PA and SD) between Pick's disease and sporadic FTLD-TDP cases.[50] In the study, 20 FTLD-TDP (mean age at death or MAD, 62.6 years; mean disease duration or MDD, 9.2 years) and 19 Pick's disease cases (MAD, 66.2 years; MDD, 8.9 years) were examined. No case had a family history. There was no difference between all FTLD-TDP cases and Pick's disease cases with respect to the sex ratio, age at onset, or disease duration. In the series, most frequent early symptom that developed within one year after the onset in sporadic FTLD-TDP was naming difficulty (50%), followed by semantic memory impairment (39%), forgetfulness (28%), disinhibited behaviors (28%) and impairment of auditory comprehension (22%). Impairment of speech output (speech apraxia, phonemic paraphasias, stuttering and hesitation in utterance were included) was rare (5.6%). In contrast, the most frequent early symptoms in Pick's disease were disinhibited behavior and impairment of speech output (29%), followed by apathy (14%), stereotypic behaviors (14%) and naming difficulty (14%). Impairment of speech output in Pick's disease was five times more frequent than in sporadic FTLD-TDP in this series. Apathy was twice as frequent in FTLD-TDP cases as in the cases of Pick's disease. In contrast to sporadic FTLD-TDP, none of our Pick's disease cases showed semantic memory impairment in the early phase. In our study because the presence or absence of each symptom determined was based on the description in the medical records, the possibility that 'naming difficulty' or 'memory impairment' was actually 'impairment of semantic memory' cannot be excluded.

Most frequent clinical symptoms, throughout the course in our sporadic FTLD-TDP cases, were apathy and asymmetric motor disturbance (78% each), followed by stereotypic behaviors (61%), oral tendency and/or pica (50%) and semantic memory impairment (44%). Asymmetric motor disturbances included rigidity (44%), tremor (5.6%), upper motor neuron signs (44%), hemiparesis or hemiplegia (11%) and contracture (28%). In sporadic FTLD-TDP cases, the mean duration from the disease onset to the development of asymmetric motor disturbances was an average 6 years (4.5 years in Pick's disease) and the mean duration from the development of asymmetric motor disturbances to death was an average 2.5 years (9.5 years in Pick's disease). Of 14 FTLD-TDP cases that showed asymmetric motor disturbances, two subsequently developed unilateral spatial agnosia on the same side as the motor disturbance. It was noticed that patients could not recognize a person who approached from the side opposite the affected lobe, or disuse of a unilateral arm although lacking weakness. In the cases of Pick's disease, stereotypic behaviors (71%), apathy (64%) and oral tendency and/or pica (43%) were frequent; however, asymmetric motor disturbances and semantic memory impairment were rare in Pick's disease cases in this study (14% and 0%). Consistent with the trend in the early symptoms, impairment of speech output such as nonfluent aphasia and apraxia of speech, were frequent in Pick's disease cases (29%), but rare in sporadic FTLD-TDP cases (5.6%), throughout the course.

FTLD patients often present sequentially with several clinical syndromes of FTLD, FTD, PA and SD, during the course. The most common first symptom in our Pick's disease cases was FTD (64%), followed by PA or speech apraxia (total 21%). In contrast, the most frequent first symptom in our sporadic FTLD-TDP cases was SD (39%), followed by FTD (28%). Of sporadic FTLD-TDP cases, 50% showed SD or impairment of auditory comprehension in the early stage of the course. Of the SD cases, about 70% subsequently showed FTD as the second symptom. Asymmetric motor disturbance was observed as the second symptom in 78% of all sporadic FTLD-TDP cases and 14% of those with asymmetric motor disturbance showed unilateral spatial agnosia in the later stage of the course.

Histopathological examinations in our series of Pick's disease and sporadic FTLD-TDP cases demonstrated several differences regarding the topographical distribution of neuronal loss between sporadic FTLD-TDP and Pick's disease cases.[50] Neuronal loss in the frontal cortex tended to be more severe in Pick's disease cases, in which behavioral symptoms, nonfluent aphasia and speech apraxia (these impairments of speech output may be due to involvement of the posterior portion of the inferior frontal gyrus to the inferior portion of the primary motor cortex) are frequently noted. In sporadic FTLD-TDP cases, the temporal cortex tended to be more severely degenerated, corresponding to that of sporadic FTLD-TDP cases, which frequently showed semantic memory impairment and auditory comprehension deficit. These are due to the involvement of the temporal tip and inferior portion of the superior temporal gyrus, respectively. Of sporadic FTLD-TDP cases, 75% showed temporal-predominant neuronal loss and no case had frontal-predominant degeneration. In 25% of FTLD-TDP cases, the frontal and temporal cortices were degenerated almost equally. In contrast, of our Pick's disease cases, 44% showed temporal-predominant neuronal loss, 25% showed frontal-predominant neuronal loss and the frontal and temporal cortices were degenerated almost equally in 31%. These findings suggest that the degeneration in the cerebral cortex in sporadic FTLD-TDP tended to be temporal-predominant, while that in Pick's disease is frontal-predominant. In addition, interestingly, three sporadic FTLD-TDP and three Pick's disease cases in our series had evident neuronal loss in the parietal cortex and two of three sporadic FTLD-TDP cases showed unilateral spatial agnosia. The caudate nucleus, putamen, globus pallidus and substantia nigra were more severely degenerated in sporadic FTLD-TDP cases than in Pick's disease cases. The corticospinal tract at the level of the medulla oblongata was degenerated in 67% of sporadic FTLD-TDP cases (Fig. 1f), while it was degenerated in 6.7% in Pick's disease cases. These histopathological trends of severe degeneration of the basal ganglia, substantia nigra and corticospinal tract may be associated with the frequent motor disturbances in sporadic FTLD-TDP cases.

Several earlier studies demonstrated that the clinical picture in Pick's disease cases tends to be characterized by dysfunction of the frontal lobe rather than that of the temporal lobe. Hodges et al[46] reported that among 20 Pick's disease cases, the most common clinical phenotype was FTD (55%), followed by PA (30%) and SD (15%). Kertesz et al[48] also noted that of six Pick's disease cases, three cases showed FTD as the first syndrome, three cases showed PA and no case developed SD. Likewise, Shi et al[49] noted that of 14 Pick's disease cases, 10 cases showed FTD, three cases progressive apraxia and one case PA, while no case showed SD. These findings appear to be consistent with results in our series.

In most of the previous studies that examined the clinical features in FTLD-TDP, the subjects include a significant number of familial FTLD-TDP cases, especially those having *GRN* mutations. Somewhat unexpectedly, it remains unclear whether the clinical and pathological features of sporadic and familial FTLD-TDP cases are different. Odawara et al[53] examined seven sporadic FTLD-U cases and three Pick's disease cases and noted that the seven sporadic cases of FTLD-U tended to have temporal-predominant atrophy with semantic memory impairment. Davies et al[54] also noted that FTLD-TDP is the frequent underlying pathology in patients presenting clinically with SD: Amongst 24 patients with SD, 18 had FTLD-U, 3 had Pick's disease and 3 had AD. In addition, in a report by Godbolt et al[55] SD was observed only in sporadic FTLD-U cases (7 of 18 cases, 39%), but not in any of the FTLD-U cases with a family history. The results observed in our sporadic FTLD-TDP series are consistent with these findings.

It was also reported that a small proportion of FTLD-TDP cases showed some unusual clinical symptoms: CBD-like or PSP-like motor disturbance,[48,56,57] early memory impairment,[58] and DLB-like syndrome (e.g., parkinsonism, fluctuating cognition, parasomnia and hallucinations).[59] In our series, one sporadic FTLD-TDP case presented with motor disturbances (i.e., gait disturbance, rigidity and pyramidal signs in the left side extremities, dysarthria) and apathy within one year after the onset and another case showed rigidity in the right extremities two years after developing initial symptoms (forgetfulness, stereotypic speech output and irritability).

Interestingly, several studies support the possibility that the trend of topographical distribution of neuronal loss is not always similar in familial FTLD-TDP cases with *GRN* mutations and sporadic FTLD-TDP cases. Josephs et al[60] reported that FTLD-TDP cases with *GRN* mutations tended to show frontal-predominant degeneration, different from the trend in sporadic FTLD-TDP cases that we examined. The frontal-predominant degeneration in FTLD-TDP with *GRN* mutations appears to be quite consistent with the findings that FTD and PA are frequent in this subpopulation.[51] In FTLD cases with microtubule associated protein tau gene (*MAPT*) mutations, it was also reported that FTD was the most common clinical picture.[49,51,61] In Western countries, it has been reported that the most frequent clinical presentations in FTLD-TDP cases, including many familial cases, were FTD (50-80%) and PA (14-20%), while SD was very rare.[48,49,61,62] In contrast, in Japan, it has been considered that FTLD patients often show SD as the prominent clinical syndrome associated with the temporal-predominant circumscribed atrophy. Based on our finding that sporadic FTLD-TDP patients often show SD, the frequency of SD cases in FTLD patients may be influenced by the proportion of subjects having *GRN* or *MAPT* mutations in each study.

Clinical features associated with the involvement of the parietal lobe can also develop in FTLD-TDP cases with *GRN* mutations.[62] Neuronal loss in the parietal cortex was reported to be more severe in FTLD-TDP cases with *GRN* mutations than in FTLD-TDP cases without them.[60] Unilateral spatial agnosia also occurs in some FTLD-TDP cases with *GRN* mutations[51] and sporadic FTLD-TDP cases as well.[50] In addition, some Pick's disease cases can exhibit apraxia as the most prominent symptom from the early stage.[49,63,64]

CORTICOBASAL DEGENERATION (CBD) AND PROGRESSIVE SUPRANUCLEAR PALSY (PSP)

In CBD cases neuronal loss associated with gliosis is observed in the frontal and parietal cortices, substantia nigra, globus pallidus and thalamus. Numerous argyrophilic threads and pretangles are noted in these regions. The presence of glial inclusions, so-called astrocytic plaques, is also a significant pathological hallmark in CBD (Fig. 2a). The glial lesions preferably occur in the affected cerebral cortex and subcortical white matter, especially in the frontal cortex and striatum. Neuronal loss in PSP cases is usually prominent in the basal ganglia and brain stem nuclei, rather than in the cerebral cortex. Most frequently and strongly affected regions in PSP are the substantia nigra, subthalamic nucleus, globus pallidus, tegmentum of the midbrain, dentate nucleus in the cerebellum and to a lesser degree, the thalamus, striatum, red nucleus, locus coeruleus, pontine nucleus and inferior olivary nucleus. Neurofibrillary tangles are noted in these regions. Another significant pathological hallmark of PSP is the development of glial lesions, the so-called tufted astrocytes (Fig. 2b). The glial lesions are noted in the frontal cortex and striatum; thus, the distribution is different from that of neuronal loss and neurofibrillary tangles. Both astrocytic plaques and tufted astrocytes are immunolabelled with anti-tau antibodies. The

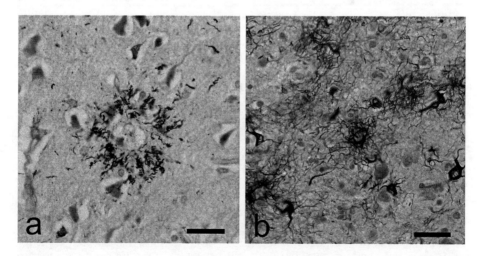

Figure 2. Corticobasal degeneration (CBD) and progressive supranuclear palsy (PSP). a) An astrocytic plaque is seen in the frontal cortex. Argyrophilic processes surround an unstained central core. b) Tufted astrocytes are seen in the frontal cortex. Argyrophilic radiating processes with a glial nucleus are noted. The processes of tufted astrocytes are finer than those of astrocytic plaques. a,b) Gallyas-Braak with hematoxylin-eosin stain. All scale bars = 30 μm.

tau-positive glial lesions in both PSP and CBD cases are strongly immunostained with a four-repeat tau-specific antibody, but not at all or only weakly stained with a three-repeat tau-specific antibody. CBD and PSP are included in the molecular class 'FTLD-tau' in the recent classification for FTLD (Table 1).[27]

The major motor symptoms of CBD are asymmetric parkinsonism, negative levodopa response, alien hand sign and limb kinetic apraxia. However, it is also important to note that these motor disturbances are not prominent in some CBD patients.[65,66] For example, it was reported that the underlying pathology in approximately 20% of patients presenting clinical FTLD were cases of CBD.[48] Josephs et al[67] reported that only 48% of pathologically-confirmed CBD cases showed CBD clinically, while the other cases showed PA (19%), PSP (14%), or FTD (10%). Kertesz et al[48] also reported that PA (38%) and FTD (31%), rather than CBD (23%) and PSP (3%), were showing more common clinical pictures in pathologically-confirmed CBD cases. PA or apraxia of speech are also noted even in the early stage of the course.[68] This trend, regarding clinical presentations in CBD, may occur because the distribution of neuronal loss tends to be more severe in the convexity of the fronto-parietal cortex, rather than the temporal cortex.[65] Thus, it may be also natural that SD and an auditory comprehension deficit are exceptional in CBD cases.[54,69] The callosum tends to be thin when associated with degeneration of the subcortical white matter in the frontoparietal lobe. The pyramidal tract is also frequently involved and the pyramidal sign was reported to be observed in 60% of pathologically-confirmed CBD cases.[70] Underlying pathologies in autopsy cases clinically diagnosed as having CBD were reported to be CBD (48%), PSP (48%) and Pick's disease (4%),[67] or CBD (54%), AD (15%), Pick's disease (7.7%) and PSP (7.7%).[71] The alien hand sign is a well-known clinical feature in CBD patients; however, it can also develop in Pick's disease,[49,63,64] BIBD,[72] NIFID,[12,72] FTLD-TDP with *GRN* mutations,[62] and AD with atypical cerebral atrophy.[71]

The clinical picture of PSP is mainly characterized by motor disturbances such as frequent falls in the early stage of the course, supranuclear vertical gaze palsy, axial rigidity and a negative levodopa response. However, reflecting the heterogeneity in the distribution pattern of pathological changes, clinical presentations in PSP cases are more variable than previously believed. Recently, a division of the clinical syndromes of PSP into the following five subtypes was proposed: (1) Richardson syndrome in which are observed lurching gait, backward falls, personality change or cognitive slowing, supranuclear gaze palsy, eyelid abnormalities such as spontaneous involuntary eyelid closure or apraxia of eyelid opening, reduced spontaneous blink rate, slurred speech, axial rigidity, absence of tremor and negative levodopa response,[8] (2) PSP-P in which can be seen normal eye movements, resting tremor, limb bradykinesia, a positive levodopa response and focal dementia, (3) PSP-PAGF with pure akinesia with gait freezing (4) PSP-CBS with CBD-like clinical picture and (5) PSP-PNFA with progressive nonfluent aphasia.[73] Richardson syndrome and PSP-PNFA tend to show cognitive impairment from the early stage of the course.[73] Although the cortical atrophy in PSP is less prominent and less asymmetrical than that in CBD, the distribution of neuronal loss and gliosis in the cerebral cortex in PSP is similar to that in CBD, usually prominent in the convexity of the fronto-parietal region. Among the pathologically-confirmed PSP cases, the most frequent clinical picture was reported to be PSP (61%), followed by CBD (20%), PA (10%) and FTD (4%).[67] Of pathologically-confirmed PSP cases 10% showed PA or apraxia of speech in the early stage of the course.[68]

BASOPHILIC INCLUSION BODY DISEASE (BIBD) AND NEURONAL INTERMEDIATE FILAMENT INCLUSION DISEASE (NIFID)

BIBD and NIFID are rare pathologies in patients with FTLD with or without motor neuron disease. Both BIBD and NIFID cases have intraneuronal cytoplasmic inclusions, which have a spherical, oval, or horseshoe-like shape. Because the inclusions in BIBD and NIFID have morphologically similar characteristics on conventional stains (e.g., hematoxylin-eosin and silver stains), the BIBD cases reported previously may actually include NIFID cases.[72] Neuronal cytoplasmic inclusions in NIFID are immuno positive for neurofilament and α-internexin antibodies, while inclusions in BIBD were negative.[16] In contrast, the neuronal inclusions in both the diseases are immunopositive for FUS (Fig. 3a,b); therefore, these diseases are now classified into the same molecular category as FTLD-FUS.[27] In BIBD, FUS-positive inclusions tend to occur mainly in the basal ganglia and motor neurons.[10] However, recent immunohistochemical studies have revealed that neuronal inclusions are frequent in the cerebral cortex, although to a lesser degree.[19] In NIFID, FUS-positive inclusions are frequently observed in the neocortex, striatum, hippocampus, substantia nigra and lower motor neurons.[20]

Clinical features in previous autopsy-confirmed BIBD cases including our three cases are as follows:[19,72,74-76] The age at onset ranged from 29 to 57 years (mean: 46.4 years) and disease duration from 5 to 12 years (mean: 7.7 years). The age at onset tended to be younger and the progression more rapid compared with those in FTLD-TDP and Pick's disease. Clinical diagnoses for previous BIBD cases were FTD, MND, MND with dementia and PSP. No case was diagnosed as having SD. Apraxia, alien hand syndrome and choreoathetosis-like involuntary movements were noted in some cases.[72] Cerebral atrophy in BIBD cases were variably distributed in the frontotemporal cortex.[72]

About 20 autopsy-confirmed cases of NIFID have been reported.[12,18,72] The age at onset was 23-67 years (mean: 42 years) and disease duration was 2.5-13 years (mean: 5.0 years). That is, like BIBD, NIFID cases may tend to show a younger age of onset and more rapid

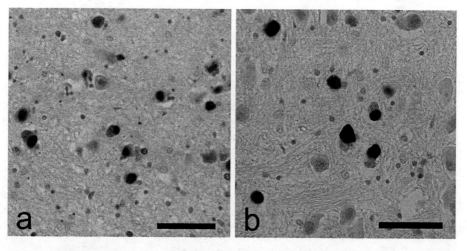

Figure 3. Basophilic inclusion body disease. a) Many intraneuronal inclusions are immunolabeled with an anti-FUS antibody in the basal nucleus of Meynert. b) FUS-positive round intraneuronal inclusions are seen in the pontine nucleus. a,b) FUS immunohistochemistry. All scale bars = 50 μm.

progression than FTLD-TDP and Pick's disease cases. The clinical diagnoses in NIFID cases were FTD, ALS with dementia, primary lateral sclerosis, CBD and progressive aphasia. In some cases, impairment of lower motor neurons and/or dysarthria preceded the development of dementia by 1-2 years.[18,72] Alien hand sign, apraxia and unilateral parkinsonism can also develop.[12,72] It may be noteworthy that previous NIFID cases whose macroscopic images are available tended to show frontal-predominant atrophy.[18,72] The basal ganglia, especially the striatum, substantia nigra and pyramidal tract were severely degenerated in most BIBD and NIFID cases. In BIBD and NIFID cases that we examined, evident atrophy of basal ganglia was observed on CT or MRI images within 1-2 years after the onset.[72]

ARGYROPHILIC GRAIN DISEASE (AGD)

AGD is histopathologically diagnosed by the presence of argyrophilic grains visualized with Gallyas-Braak silver stain (Fig. 4a). Argyrophilic grains are immunolabelled with anti-tau antibodies and are mainly composed of four-repeat tau (Fig. 4b-d).[77] Although patients having argyrophilic grains do not always develop dementia in life, it is considered that dementia is due to AGD in 5-10% of all patients with dementia.[78] Argyrophilic grains are found not only in 'pure' AGD but also other neurodegenerative diseases, such as PSP, CBD, AD, Lewy body disease and ALS.[78]

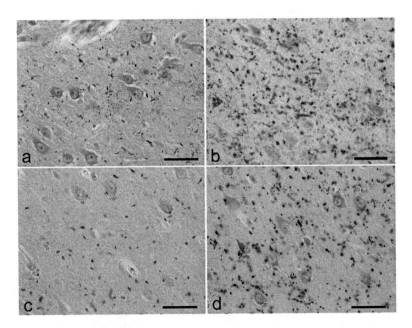

Figure 4. Argyrophilic grains in serial sections of subiculum in argyrophilic grain disease (AGD) case. a) Numerous spindle- or comma-shaped argyrophilic grains are evident. b) Argyrophilic grains are strongly immunolabeled with phosphorylation-dependent anti-tau antibody. c) Immunoreactivity with anti-three-repeat tau-specific antibody of argyrophilic grains is weak. d) In contrast, the lesions are strongly labeled with anti-four-repeat tau-specific antibody. a) Gallyas-Braak with hematoxylin-eosin stain, b) AT-8 immunohistochemistry, c) RD3 immunohistochemistry, d) RD4 immunohistochemistry. All scale bars = 50 μm.

It is assumed that argyrophilic grains develop first in the ambient gyrus and then extend to the amygdala, parahippocampal gyrus, CA1 and subiculum.[79] In some cases, argyrophilic grains can be also noted in the temporal cortex and cingulate gyrus. Enlargement of the lateral ventricles and atrophy of the parahippocampal gyrus adjacent to the amygdala are often observed, while the temporal and frontal cortices are relatively well spared. However, a small proportion of AGD cases exceptionally show severe atrophy in the temporal cortices.[80] The mean age of death is about 80 years. However, although rare, a few early-onset cases have also been reported.[81] Memory impairment, irritability and eating disturbances are common.[80,82,83] Apathy, diminished initiative, delusion, obsessive behaviors and restlessness can also develop.[82,83] The nature or frequency of motor disturbance in AGD cases remain unclear. A patient with AGD was clinically diagnosed as Parkinson's disease with dementia.[84] This patient first developed tremor at the age of 63 and in the later course, rigidity, dyskinesia and gait disturbance occurred. Pathologically, this case had AGD, but lacked any other lesions such as AD or Lewy body-related pathology.

CONCLUSION

Patients presenting clinical FTLD have variable underlying pathologies, including Pick's disease, FTLD-TDP, CBD, PSP, BIBD, NIFID and AGD and Alzheimer's disease with atypical cerebral atrophy. Although data on the clinical features in these pathological disease entities have been accumulated, yet it is difficult to identify the underlying pathology in patients with FTLD. To predict underlying pathologies precisely, further accumulation of the clinicopathological findings, as well as the exploration of disease-specific biomarkers in each pathological case, are awaited.

ACKNOWLEDGEMENTS

We would-like to thank Ms. M. Onbe (Department of Neuropsychiatry, Okayama University Graduate School of Medicine, Dentistry and Pharmaceutical Sciences) for her excellent technical assistance. This study was supported in part by a research grant from the Zikei Institute of Psychiatry.

REFERENCES

1. Pick A. Ueber die Beziehungen der senilen Hirnatrophie zur Aphasie. Prag Med Wochenschr 1892; 17: 165-167.
2. Alzheimer A. Über eigenartige Krankheitfälle des spätern Alters. Z Ges Neurol Psychiatr 1911; 4: 356-385.
3. Onari K, Spatz H. Anatomische Beiträge zur Lehre von pickschen umschriebenen Größhirnrinden Atrophie (Picksche Krankheit). Z Ges Neurol Psychiatr 1926; 101:470-511.
4. Cummings JL, Duchen LW. Klüver-Bucy syndrome in Pick disease: clinical and pathologic correlations. Neurology 1981; 31:1415-1422.
5. The Lund and Manchester Groups. Clinical and neuropathological criteria for frontotemporal dementia. J Neurol Neurosurg Psychiatry 1994; 57:416-418.
6. Neary D, Snowden JS, Gustafson L et al. Frontotemporal lobar degeneration: a consensus on clinical diagnostic criteria. Neurology 1998; 51:1546-1554.

7. Gibb WR, Luthert PJ, Marsden CD. Corticobasal degeneration. Brain 1989; 112:1171-1192.

8. Steele JC, Richardson JC, Olszewski J. Progressive supranuclear palsy. A heterogeneous degeneration involving the brain stem, basal ganglia and cerebellum with vertical gaze and pseudobulbar palsy, nuchal dystonia and dementia. Arch Neurol 1964; 10:333-359.

9. Braak H, Braak E. Argyrophilic grains: characteristic pathology of cerebral cortex in cases of adult onset dementia without Alzheimer changes. Neurosci Lett 1987; 76:124-127.

10. Munoz-Garcia D, Ludwin SK. Classic and generalized variants of Pick's disease: a clinicopathological, ultrastructural and immunocytochemical comparative study. Ann Neurol 1984; 16:467-480.

11. Cairns NJ, Perry RH, Jaros E et al. Patients with a novel neurofilamentopathy: dementia with neurofilament inclusions. Neurosci Lett 2003; 341:177-180.

12. Josephs KA, Holton JL, Rossor MN et al. Neurofilament inclusion body disease: a new proteinopathy? Brain 2003; 126:2291-2303.

13. Arai T, Hasegawa M, Akiyama H et al. TDP-43 is a component of ubiquitin-positive tau-negative inclusions in frontotemporal lobar degeneration and amyotrophic lateral sclerosis. Biochem Biophys Res Commun 2006; 351:602-611.

14. Neumann M, Sampathu DM, Kwong LK et al. Ubiquitinated TDP-43 in frontotemporal lobar degeneration and amyotrophic lateral sclerosis. Science 2006; 314:130-133.

15. Jackson M, Lennox G, Lowe J. Motor neuron disease-inclusion dementia. Neurodegeneration 1996; 5:339-350.

16. Cairns NJ, Bigio EH, Mackenzie IR et al. Neuropathologic diagnostic and nosologic criteria for frontotemporal lobar degeneration: consensus of the consortium for frontotemporal lobar degeneration. Acta Neuropathol 2007; 114:5-22.

17. Mackenzie IR, Neumann M, Bigio EH et al. Nomenclature for neuropathologic subtypes of frontotemporal lobar degeneration: consensus recommendations. Acta Neuropathol 2009; 117:15-18.

18. Cairns NJ, Grossman M, Arnold SE et al. Clinical and neuropathologic variation in neuronal intermediate filament inclusion disease. Neurology 2004; 63:1376-1384.

19. Munoz DG, Neumann M, Kusaka H et al. FUS pathology in basophilic inclusion body disease. Acta Neuropathol 2009; 118:617-627.

20. Neumann M, Roeber S, Kretzschmar HA et al. Abundant FUS-immunoreactive pathology in neuronal intermediate filament inclusion disease. Acta Neuropathol 2009; 118:605-616.

21. Knopman DS, Mastri AR, Frey WH 2nd et al. Dementia lacking distinctive histologic features: a common non-Alzheimer degenerative dementia. Neurology 1990; 40:251-256.

22. Mackenzie IR, Shi J, Shaw CL et al. Dementia lacking distinctive histology (DLDH) revisited. Acta Neuropathol 2006; 112:551-559.

23. Munoz DG, Dickson DW, Bergeron C et al. The neuropathology and biochemistry of frontotemporal dementia. Ann Neurol 2003; 54 Suppl 5:s24-s28.

24. Josephs KA, Holton JL, Rossor MN et al. Frontotemporal lobar degeneration and ubiquitin immunohistochemistry. Neuropathol Appl Neurobiol 2004; 30:369-373.

25. Forman MS, Farmer J, Johnson JK et al. Frontotemporal dementia: clinicopathological correlations. Ann Neurol 2006; 59:952-962.

26. Ikeda K. Neuropathological discrepancy between Japanese Pick's disease without Pick bodies and frontal lobe degeneration type of frontotemporal dementia proposed by Lund and Manchester Group. Neuropathology 2000; 20:76-82.

27. Mackenzie IR, Neumann M, Bigio EH et al. Nomenclature and nosology for neuropathologic subtypes of frontotemporal lobar degeneration: an update. Acta Neuropathol 2010; 119:1-4.

28. Mackenzie IR, Foti D, Woulfe J et al. Atypical frontotemporal lobar degeneration with ubiquitin-positive, TDP-43-negative neuronal inclusions. Brain 2008; 131:1282-1293.

29. Neumann M, Rademakers R, Roeber S et al. A new subtype of frontotemporal lobar degeneration with FUS pathology. Brain 2009; 132:2922-2931.

30. Frank S, Tolnay M. Frontotemporal lobar degeneration: toward the end of confusion. Acta Neuropathol 2009; 118:629.

31. Amador-Ortiz C, Lin WL, Ahmed Z et al. TDP-43 immunoreactivity in hippocampal sclerosis and Alzheimer's disease. Ann Neurol 2007; 61:435-445.

32. Arai T, Mackenzie IR, Hasegawa M et al. Phosphorylated TDP-43 in Alzheimer's disease and dementia with Lewy bodies. Acta Neuropathol 2009; 117:125-136.

33. Higashi S, Iseki E, Yamamoto R et al. Concurrence of TDP-43, tau and alpha-synuclein pathology in brains of Alzheimer's disease and dementia with Lewy bodies. Brain Res 2007; 1184:284-294.

34. Nakashima-Yasuda H, Uryu K et al. Co-morbidity of TDP-43 proteinopathy in Lewy body related diseases. Acta Neuropathol 2007; 114:221-229.

35. Yokota O, Davidson Y, Arai T et al. Effect of topographical distribution of alpha-synuclein pathology on TDP-43 accumulation in Lewy body disease. Acta Neuropathol 2010, in press.

36. Uryu K, Nakashima-Yasuda H, Forman MS et al. Concomitant TAR-DNA-binding protein 43 pathology is present in Alzheimer disease and corticobasal degeneration but not in other tauopathies. J Neuropathol Exp Neurol 2008; 67:555-564.

37. Hasegawa M, Arai T, Akiyama H et al. TDP-43 is deposited in the Guam parkinsonism-dementia complex brains. Brain 2007; 130:1386-1394.

38. Geser F, Winton MJ, Kwong LK et al. Pathological TDP-43 in parkinsonism-dementia complex and amyotrophic lateral sclerosis of Guam. Acta Neuropathol 2008; 115(1):133-145.

39. Fujishiro H, Uchikado H, Arai T et al. Accumulation of phosphorylated TDP-43 in brains of patients with argyrophilic grain disease. Acta Neuropathol 2009; 117:151-158.

40. Yokota O, Davidson Y, Bigio EH et al. Phosphorylated TDP-43 pathology and hippocampal sclerosis in progressive supranuclear palsy. Acta Neuropathol 2010; 120:55-66.

41. Geser F, Martinez-Lage M, Kwong LK et al. Amyotrophic lateral sclerosis, frontotemporal dementia and beyond: the TDP-43 diseases. J Neurol 2009; 256(8):1205-1214.

42. Chen-Plotkin AS, Lee VM, Trojanowski JQ. TAR DNA-binding protein 43 in neurodegenerative disease. Nat Rev Neurol 2010; 6(4):211-220.

43. Josephs KA, Whitwell JL, Knopman DS et al. Abnormal TDP-43 immunoreactivity in AD modifies clinicopathologic and radiologic phenotype. Neurology 2008; 70:1850-1857.

44. Ikeda M, Ishikawa T, Tanabe H. Epidemiology of frontotemporal lobar degeneration. Dement Geriatr Cogn Disord 2004; 17:265-268.

45. Yokota O, Sasaki K, Fujisawa Y et al. Frequency of early and late-onset dementias in a Japanese memory disorders clinic. Eur J Neurol 2005; 12:782-790.

46. Hodges JR, Davies RR, Xuereb JH et al. Clinicopathological correlates in frontotemporal dementia. Ann Neurol 2004; 56:399-406.

47. Lipton AM, White CL 3rd, Bigio EH. Frontotemporal lobar degeneration with motor neuron disease-type inclusions predominates in 76 cases of frontotemporal degeneration. Acta Neuropathol 2004; 108:379-385.

48. Kertesz A, McMonagle P, Blair M et al. The evolution and pathology of frontotemporal dementia. Brain 2005; 128:1996-2005.

49. Shi J, Shaw CL, Du Plessis D et al. Histopathological changes underlying frontotemporal lobar degeneration with clinicopathological correlation. Acta Neuropathol 2005; 110:501-512.

50. Yokota O, Tsuchiya K, Arai T et al. Clinicopathological characterization of Pick's disease versus frontotemporal lobar degeneration with ubiquitin/TDP-43-positive inclusions. Acta Neuropathol 2009; 117: 429-444.

51. Snowden JS, Pickering-Brown SM, Mackenzie IR et al. Progranulin gene mutations associated with frontotemporal dementia and progressive nonfluent aphasia. Brain 2006; 129:3091-3102.

52. King ME, Ghoshal N, Wall JS et al. Structural analysis of Pick's disease-derived and in vitro-assembled tau filaments. Am J Pathol 2001; 158(4):1481-1490.

53. Odawara T, Iseki E, Kanai A et al. Clinicopathological study of two subtypes of Pick's disease in Japan. Dement Geriatr Cogn Disord 2003; 15:19-25.

54. Davies RR, Hodges JR, Kril JJ et al. The pathological basis of semantic dementia. Brain 2005; 128: 1984-1995.

55. Godbolt AK, Josephs KA, Revesz T et al. Sporadic and familial dementia with ubiquitin-positive tau-negative inclusions: clinical features of one histopathological abnormality underlying frontotemporal lobar degeneration. Arch Neurol 2005; 62:1097-1101.

56. Grimes DA, Bergeron CB, Lang AE. Motor neuron disease-inclusion dementia presenting as cortical-basal ganglionic degeneration. Mov Disord 1999; 14:674-780.

57. Paviour DC, Lees AJ, Josephs KA et al. Frontotemporal lobar degeneration with ubiquitin-only-immunoreactive neuronal changes: broadening the clinical picture to include progressive supranuclear palsy. Brain 2004; 127:2441-2451.

58. Graham A, Davies R, Xuereb J et al. Pathologically proven frontotemporal dementia presenting with severe amnesia. Brain 2005; 128:597-605.

59. Claassen DO, Parisi JE, Giannini C et al. Frontotemporal dementia mimicking dementia with Lewy bodies. Cogn Behav Neurol 2008; 21(3):157-163.

60. Josephs KA, Ahmed Z, Katsuse O et al. Neuropathologic features of frontotemporal lobar degeneration with ubiquitin-positive inclusions with progranulin gene (PGRN) mutations. J Neuropathol Exp Neurol 2007; 66:142-151.

61. Pickering-Brown SM, Rollinson S, Du Plessis D et al. Frequency and clinical characteristics of progranulin mutation carriers in the Manchester frontotemporal lobar degeneration cohort: comparison with patients with MAPT and no known mutations. Brain 2008; 131:721-731.

62. Le Ber I, Camuzat A, Hannequin D et al. Phenotype variability in progranulin mutation carriers: a clinical, neuropsychological, imaging and genetic study. Brain 2008; 131:732-746.

63. Fukui T, Sugita K, Kawamura M et al. Primary progressive apraxia in Pick's disease: a clinicopathologic study. Neurology 1996; 47:467-473.
64. Kawamura M, Mochizuki S. Primary progressive apraxia. Neuropathology 1999; 19:249-258.
65. Tsuchiya K, Ikeda K, Uchihara T et al. Distribution of cerebral cortical lesions in corticobasal degeneration: a clinicopathological study of five autopsy cases in Japan. Acta Neuropathol 1997; 94:416-424.
66. Grimes DA, Lang AE, Bergeron CB. Dementia as the most common presentation of cortical-basal ganglionic degeneration. Neurology 1999; 53:1969-1974.
67. Joseph KA, Petersen RC, Knopman DS et al. Clinicopathological analysis of frontotemporal and corticobasal degenerations and PSP. Neurology 2006; 66:41-48.
68. Josephs KA, Duffy JR. Apraxia of speech and nonfluent aphasia: a new clinical marker for corticobasal degeneration and progressive supranuclear palsy. Current Opinion in Neurology 2008; 21:688-692.
69. Ikeda K, Akiyama H, Iritani S et al. Corticobasal degeneration with primary progressive aphasia and accentuated cortical lesion in superior temporal gyrus: case report and review. Acta Neuropathol 1996; 92:534-539.
70. Tsuchiya K, Murayama S, Mitani K et al. Constant and severe involvement of Betz cells in corticobasal degeneration is not consistent with pyramidal signs: a clinicopathological study of ten autopsy cases. Acta Neuropathol 2005; 109:353-366.
71. Boeve BF, Maraganore DM, Parisi JE et al. Pathologic heterogeneity in clinically diagnosed corticobasal degeneration. Neurology 1999; 53:795-800.
72. Yokota O, Tsuchiya K, Terada S et al. Basophilic inclusion body disease and neuronal intermediate filament inclusion disease: a comparative clinicopathological study. Acta Neuropathol 2008; 115:561-575.
73. Williams DR, Lees AJ. Progressive supranuclear palsy: clinicopathological concepts and diagnostic challenges. Lancet Neurol 2009; 8:270-279.
74. Kusaka H, Matsumoto S, Imai T. An adult-onset case of sporadic motor neuron disease with basophilic inclusions. Acta Neuropathol 1990; 80:660-665.
75. Kusaka H, Matsumoto S, Imai T. Adult-onset motor neuron disease with basophilic intraneuronal inclusion bodies. Clin Neuropathol 1993; 12:215-218.
76. Ishihara K, Araki S, Ihori N et al. An autopsy case of frontotemporal dementia with severe dysarthria and motor neuron disease showing numerous basophilic inclusions. Neuropathology 2006; 26: 447-454.
77. Togo T, Sahara N, Yen SH et al. Argyrophilic grain disease is a sporadic 4-repeat tauopathy. J Neuropathol Exp Neurol 2002; 61(6):547-556.
78. Togo T, Cookson N, Dickson DW. Argyrophilic grain disease: neuropathology, frequency in a dementia brain bank and lack of relationship with apolipoprotein E. Brain Pathol 2002; 12:45-52.
79. Saito Y, Ruberu NN, Sawabe M et al. Staging of argyrophilic grains: an age-associated tauopathy. J Neuropathol Exp Neurol 2004; 63:911-918.
80. Tsuchiya K, Mitani K, Arai T et al. Argyrophilic grain disease mimicking temporal Pick's disease: a clinical, radiological and pathological study of an autopsy case with a clinical course of 15 years. Acta Neuropathol 2001; 102:195-199.
81. Ishihara K, Araki S, Ihori N et al. Argyrophilic grain disease presenting with frontotemporal dementia: a neuropsychological and pathological study of an autopsied case with presenile onset. Neuropathology 2005; 25:165-170.
82. Ikeda K, Akiyama H, Arai T et al. Clinical aspects of argyrophilic grain disease. Clin Neuropathol 2000; 19:278-284.
83. Togo T, Isojima D, Akatsu H et al. Clinical features of argyrophilic grain disease: a retrospective survey of cases with neuropsychiatric symptoms. Am J Geriatr Psychiatry 2005; 13:1083-1091.
84. Uchikado H, Tsuchiya K, Tominaga I et al. Argyrophilic grain disease clinically mimicking Parkinson's disease with dementia: report of an autopsy case. Brain and Nerve 2004; 56:785-788 (in Japanese with English abstract).

CHAPTER 24

PREMATURE AGING SYNDROME

Fabio Coppedè

Department of Human and Environmental Sciences, Section of Medical Genetics, University of Pisa, Italy
Email: f.coppede@geog.unipi.it

Abstract: Hutchinson-Gilford progeria syndrome and Werner syndrome are two of the best characterized human progeroid diseases with clinical features mimicking physiological aging at an early age. Both disorders have been the focus of intense research in recent years since they might provide insights into the pathology of normal human aging. The chapter contains a detailed description of the clinical features of both disorders and then it focuses on the genetics, the resulting biochemical alterations at the protein level and the most recent findings and hypotheses concerning the molecular basis of the premature aging phenotypes. A description of available diagnostic and therapeutic approaches is included.

INTRODUCTION

Progeroid syndromes constitute a group of disorders characterized by clinical features mimicking physiological aging at an early age and have been the focus of intense research since they might provide insights into the pathology of normal aging. Hutchinson-Gilford progeria syndrome (HGPS) and Werner syndrome (WS) are two of the best characterized human progeroid diseases. HGPS is an extremely rare genetic disorder classified as a segmental progeroid syndrome as multiple organs and tissues replicate phenotypes associated with normal aging. Children with HGPS appear healthy at birth but develop distinctive clinical features during the first years of their life and die at a median age of 11-13 years. In contrast, WS represents the most studied model disease of premature aging in adulthood. Usually WS patients have a median life expectancy of approximately 47-54 years with clinical conditions starting from the second decade of life, when they develop signs and pathologies that resemble many aspects of normal human aging.

Neurodegenerative Diseases, edited by Shamim I. Ahmad.
©2012 Landes Bioscience and Springer Science+Business Media.

Another name for WS is "progeria of the adult" to distinguish it from HGPS, which is often referred to as "progeria of childhood".[1-4]

EPIDEMIOLOGY

Hutchinson-Gilford Progeria Syndrome

HGPS is an extremely rare genetic disorder with a reported prevalence of 1 in 8 million and an estimated birth prevalence of 1 in 4 million births, taking into consideration unreported or misdiagnosed cases. No differences have been observed based on ethnic background.[5]

Werner Syndrome

WS is a rare autosomal recessive disorder with a reported prevalence that varies with the level of consanguinity in populations. In the Japanese population, where a founder effect has been described, the frequency of Werner heterozygotes appears to be around 1/180 in the general population.[6] WS frequency has been roughly estimated to be 1: 100,000 in Japan and 1: 1,000,000-1: 10,000,000 outside of Japan.[7] The only exception to the latter data can be seen in the clustering of WS in Sardinia, with a total number of 18 described cases.[8]

CLINICAL FEATURES

Hutchinson-Gilford Progeria Syndrome

In 1886 Jonathan Hutchinson provided the first description of the disease, followed by a clinical report by Hastings Gilford which recognized that some of the symptoms resembled early aging and suggested naming the entity 'progeria' from the ancient Greek words 'pro' meaning early and 'geras' meaning old age.[9-11]

Children with HGPS usually appear healthy at birth but develop distinctive clinical features during the first years of their life. The first sign is often a visible vein across the nasal bridge. From 6 to 12 months or somewhat later, failure to thrive develops. HGPS children gain a mean of 0.440 kg/year reaching a mean final weight of 14.5 kg and a mean final height of 110 cm.[12] Other symptoms become apparent during the second to third year, including characteristic facies, alopecia, loss of subcutaneous fat, stiffness of joints, bone changes and abnormal tightness of the skin over the abdomen and upper thighs.[13] The mean age at diagnosis is reported to be 2.9 years.[14] With time, the skin becomes thin, dry and atrophic, with reduced turgor and sometimes with hyperkeratosis. Small, light-brown spots frequently develop on the neck and upper thorax and subsequently on the scalp and limbs.[14] Typical facial abnormalities include a small and beaked nose, micrognathia, alopecia, prominent scalp veins, prominent eyes and protruding ears that lack lobules. The facial characteristics gradually develop and both face and body change with time: the subcutaneous fat in the face disappears completely and the facial muscles decrease in size. The body shows increasing loss of subcutaneous fat and muscle bulk and the joints protrude.[14] At the bone level, patients show clavicular hypoplasia, generalized osteopenia and acro-osteolyses of distal phalanges. Motor and mental development is normal, cognitive functions are preserved and the children follow a normal psychosocial development and show a normal behaviour for their age. Dentition is delayed and crowded. HGPS individuals have a high-pitched voice, do not reproduce and their appearance becomes with time that of an older person. Most children die in their early teens from heart attacks and strokes caused by progressive atherosclerotic disease, with myocardial infarction representing the most frequent cause of death at a mean age of around 13 years. Cancer incidence is not increased and cataracts are not frequent.[13-15]

Werner Syndrome

WS represents the most studied model disease of premature aging in adulthood; the syndrome is named after Otto Werner, who described it in his doctoral thesis in 1904. Usually WS patients look normal until the second decade of life, when they develop signs and pathologies that resemble many aspects of normal human aging, including grey hair, alopecia, prematurely aged face with beaked nose, skin atrophy with scleroderma-like lesions, ischemic heart disease, osteoporosis, bilateral cataracts, Type II diabetes mellitus, lipodystrophy and hypogonadism. They also experience an increased risk of rare non-epithelial cancers, especially sarcomas and in most of the cases they die because of malignant tumours or arteriosclerosis during the fourth decade of life.[16]

The first clinical sign is a lack of the pubertal growth spurt during the teen years. A median height of 142 cm and a median body weight of 36 kg have been observed in Japanese WS subjects.[17] In their 20s and 30s, affected individuals begin to manifest alopecia, greying hair and scleroderma-like skin changes, followed by bilateral cataracts, Type II diabetes mellitus, hypogonadism, skin ulcers and osteoporosis in the 30s. The median age of diagnosis is 38 years.[17-18] The analysis of 99 WS cases revealed that bilateral cataracts were present in all of them. Skin alterations, greying and/or loss of hair and short stature were reported in more than 95% of the cases. Osteoporosis was observed in more than 90% of WS subjects and diabetes, atherosclerosis and neoplasia in 70.8, 39.5 and 43.6% of them, respectively. The median age at death had been 54.3 years.[19]

An unusual spectrum of cancers including a large number of sarcomas and very rare types of cancers in typical locations has been observed in WS subjects.[17] The specific cell type in which cancer develops could depend on the nature of the *WRN* mutation.[20] Other peculiarities include characteristic chronic ulcers around the ankles and osteoporosis that in WS patients has a more pronounced effect on long bones, particularly those of the legs, while during normal aging of the general population osteoporosis preferentially affects vertebrae.[16]

Fertility in WS patients appears to decline soon after sexual maturity. However, successful pregnancies have been reported in women and men have fathered children.[18] Controversy exists regarding the degree of brain involvement in WS (see below the section "neuropathogenesis").

AETIOLOGY: GENETIC AND BIOCHEMICAL BASIS

Hutchinson-Gilford Progeria Syndrome

The majority (approximately 90%) of classical HGPS is caused by a de novo point mutation in exon 11 of the *LMNA* gene (1824C>T, p.G608G).[22,23] *LMNA* encodes the four different A-type lamins (lamin A, AΔ10, C and C$_2$) which are intermediate filament proteins of the inner nuclear lamina. Lamin A (encoded by exons 1-12) and lamin C (encoded by exons 1-10) are the major proteins expressed in differentiated cells. Lamin AΔ10 is identical to lamin A except that it lacks exon 10 and has been detected in several cell types, including cells from human colon, placenta, tumor cells, leukocytes and fibroblasts.[24,25] Lamin C$_2$ has an alternative exon 1 compared with lamin C and is present in germ cells.[26] Lamin A proteins contain CaaX boxes at their C-terminal ends, they are synthesized as prelamin A proteins which undergo farnesylation and other posttranslational modifications to become mature proteins (Fig. 1).[27] The genetic basis of HGPS was identified in 2003 by two independent research groups. [22,23] Eriksson et al observed the G608G mutation in 18 out of 20 HGPS individuals. This mutation affects the lamin A protein and results in the activation of a cryptic splice donor site that removes 150 nucleotides from exon 11. The resulting lamin AΔ150 mRNA gives rise to a lamin A isoform containing an internal deletion of 50 amino acids, known as progerin, a protein that cannot undergo complete maturation. Particularly, the deletion eliminates the site for endoproteolitic cleavage by Zmpste24 (FACE-1); thus progerin cannot undergo further processing and retains its carboxyl-terminal farnesylcysteine methyl ester (Fig. 1).[22] The same mutation was identified by De Sandre-Giovannoli et al in two HGPS affected children.[23] One of the HGPS patients analysed by Eriksson and coworkers had a G608S mutation leading to the same cryptic splice effect.[22] The recurrent 1824C>T mutation causing HGPS is a de novo dominant point mutation, mostly originating on the paternal allele and often linked with advanced paternal age. The parents of probands are not affected and the risk to the sibs of a proband is small.[22] However, one HGPS patient was described harbouring the common G608G mutation transmitted by the mother who showed somatic and germline mosaicism without HGPS manifestations.[28]

There are several different autosomal dominant or recessive diseases caused by mutations in the *LMNA* gene, collectively called primary laminopathies and including muscular dystrophies (Emery-Dreifuss Muscular Dystrophy, Dilated Cardiomyopathy and Limb-Girdle Muscular Dystrophy), lipodysrophies (Familial Partial Lipodystrophy, Generalized Lipodystrophy Type 2 and Mandibuloacral Dysplasia Type A), neuropathies (Charcot-Marie-Tooth disease Type 2B1) and segmental progeroid syndromes (classical HGPS, atypical HGPS, atypical Werner's syndrome and Restrictive Dermopathy).[29]

Two HGPS individuals showed uniparental isodisomy (UPD) of chromosome 1 (including the *LMNA* locus), a mosaic rearrangement of chromosome 1 and a deletion involving the *LMNA* locus. They appeared to have classical HGPS.[22] A subject bearing a missense (p.T623S) mutation leading to the deletion of 35 amino acids in exon 11 of *LMNA* showed a less aggressive progeroid phenotype.[30] Slowly progressing progeria was observed also in another patient carrying the T623S mutation.[31] Restrictive Dermopathy (RD) is a perinatal disorder characterized by severe intrauterine growth delay and can be considered as an "extreme" form of premature aging.[15] RD is caused by mutations in the *ZMPSTE24* gene, encoding a protein involved in prelamin A processing (Fig. 1), leading to lamin A precursor accumulation. Two RD patients bearing dominant splicing mutations

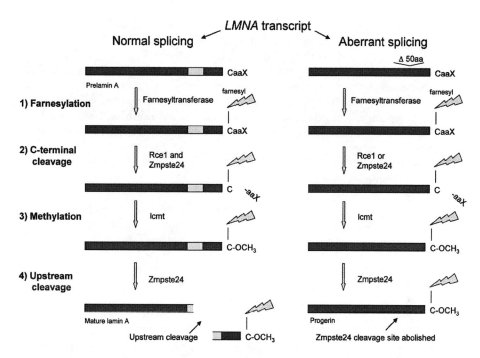

Figure 1. Schematic representation of prelamin A processing. The C terminal end of prelamin A contains a CaaX motif which is modified by farnesylation of the cysteine residue (1), followed by the cleavage of the –aaX terminal residues (2) that can be performed either by Rce1 (Ras converting enzyme 1) or by Zmpste24 (Zinc metalloprotease related to Ste24p). After the cleavage, the cysteine residue is carboxymethylated by Icmt (isoprenylcysteine carboxyl methyltransferase) (3). The last step removes the C-terminal 15 residues through cleavage by Zmpste24 (4), yielding mature lamin A. In HGPS cells bearing the *LMNA* G608G mutation the activation of the cryptic splice site results in the deletion of a 50aa region from prelamin A. The deleted region contains the second Zmpste24 cleavage site. Therefore, the use of the cryptic splice site leads to the production and the accumulation of a smaller farnesylated and carboxy-methylated mutant prelamin A protein, termed progerin, that cannot undergo complete maturation.

in the *LMNA* gene, resulting in complete or partial loss of exon 11, were described and lived longer than typical RD cases.[32-33]

Two patients with extraordinarily severe forms of progeria caused by unusual mutations in *LMNA* have been described. Both mutations (IVS11 + 1G>A and p.V607V) resulted in a strong activation of the aberrant splice site observed in typical HGPS, leading to increased progerin expression compared to typical HGPS cases.[34] Moreover, a patient with a HGPS-like phenotype, carrying homozygous null mutations in the *ZMPSTE24* gene and a heterozygous mutation in the *LMNA* gene was described. Although null *ZMPSTE24* mutations should have resulted in RD, the *LMNA* mutation resulted in a C-terminal elongation of prelamin A impairing its posttranslational modifications and resulting in a clinical picture resembling a HGPS-like phenotype.[35]

In addition to autosomal dominant forms, homozygous *LMNA* mutations resulting in autosomal recessive HGPS have been described in some families and heterozygous carrier status was detected in both of the asymptomatic parents.[36-37]

Severe progeroid syndromes can be caused by *LMNA* mutations without a prelamin A-processing defect. A two-year-old boy with HGPS due to compound heterozygous missense mutations (p.T528M and M540T) in *LMNA* has been described.[38] Both mutations affect a conserved region within the C-terminal globular domain of A-type lamins, but the nuclei of the patient showed no prelamin A accumulation.[38]

Recently, 11 patients with atypical progeroid syndrome (APS) have been described, many with novel heterozygous missense *LMNA* mutations. They had a few overlapping but some distinct clinical features as compared with HGPS patients and the pathogenesis of clinical manifestations seemed not to be related to the accumulation of mutant farnesylated prelamin A.[39] Also atypical Werner's syndrome is caused by dominant *LMNA* mutations (see below).

Werner Syndrome

The majority of WS affected individuals carry mutations in both alleles of the *WRN* gene (*RECQL2*) which encodes the WRN protein, a member of the RecQ DNA helicase family.[21] WRN is a multifunctional nuclear protein that maintains genome stability by means of DNA-dependent ATPase, 3'→5' helicase, 3'→5' exonuclease and DNA strand annealing activities.[16]

Several *WRN* gene mutations have been found in WS patients (for a review see ref. 16), these are mainly nonsense mutations, insertions and/or deletions and splice mutations resulting in the production of truncated proteins lacking the nuclear localization signal, with subsequent absence of functional WRN protein in nuclei.[7,16] WRN has several functional domains (Fig. 2) and is considered to be a 'caretaker of the genome' since it participates in distinct DNA metabolic pathways, including DNA replication, DNA recombination, telomere maintenance, apoptosis and DNA repair.[16] The majority of WS causative mutations are found across the *WRN* gene and generate stop signals or frameshift leading to truncated proteins; however, two different missense mutations (p.K125N and K135E) have been described in one patient, likely impairing protein stability.[40]

Typical WS is an autosomal recessive genetic disease. Therefore, the parents of a proband are obligate heterozygotes for a disease-causing mutation and at conception each sib of an affected individual has 25% chance of being affected and 50% chance of being an asymptomatic carrier. However, not all the clinically diagnosed WS cases are caused by *WRN* gene mutations. Approximately 15% of WS cases with no *WRN* mutations carried dominant heterozygous missense mutations in the *LMNA* gene.[41] WS caused by *LMNA* mutations was then referred as atypical Werner's syndrome on the basis of molecular criteria. Patients affected by atypical WS have early onset of aging phenotypes and an accelerated rate of disease progression than typical WS individuals; they also commonly show absence of bilateral cataracts and diabetes.[41]

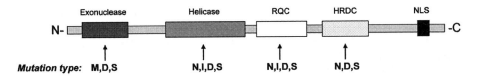

Figure 2. Diagram illustrating the functional domains of the WRN protein and the types of mutation found in each of them. Domains: Exonuclease, Helicase, RQC = RecQ C-terminal domain, HRDC = helicase and RNase D C-terminal domain. NLS = nuclear localization signal. Mutations: D = deletion, I = insertion, M = missense mutation, N = nonsense mutation, S = splice mutation.

MOLECULAR MECHANISMS

Hutchinson-Gilford Progeria Syndrome

Figure 1 shows the posttranslational processing of nuclear lamins. In HGPS individuals carrying the dominant G608G mutation a cryptic splice donor site is activated within exon 11. Sporadic use of this cryptic splice site, in splicing of *LMNA* pre-mRNA, removes a 150-bp sequence, causing a 50 amino acids deletion within the prelamin A protein. The missing region contains the second Zmpste24 cleavage site (Fig. 1). Therefore, though mature lamin A is still largely produced in HGPS individuals, the use of the cryptic splice site leads to the production and the accumulation of a smaller farnesylated and carboxy-methylated mutant prelamin A protein, termed lamin AΔ50 or progerin.[22]

We performed absolute quantification of the lamin A, C and AΔ150 transcripts during in vitro aging of primary dermal fibroblasts from HGPS patients and unaffected age-matched and parent controls.[25] Lamin C was the most highly expressed *LMNA* locus transcript in both groups. In agreement with previous studies the lamin AΔ150 transcript was present also in cells from normal individuals, but its level was >160-fold higher in samples from HGPS than in unaffected controls.[25] The level of lamin A was not affected by the high expression of lamin AΔ150 in HGPS cells. Lamin A and lamin AΔ150 were equally expressed in HGPS cells and their combined amount was about twice the amount of lamin A observed in the control group. Our quantification of lamins A, C and AΔ150 transcripts in RNA from different passages of cells showed that although the lamin A and C levels were unchanged, the lamin AΔ150 transcript levels increased in both late passage HGPS and unaffected parent control cells. This may indicate that there is a similar mechanism in progeria and normal aging, with a reduced stringency of the splicing machinery with aging.[25]

Lamins constitute the major component of the nuclear lamina: in addition to providing structure and shape to the nucleus, they are involved in chromatin organization and epigenetic regulation of chromatin, in DNA replication, in transcription, in cell proliferation and differentiation, in DNA repair and in the infection cycles of viruses.[42] Cells obtained from HGPS patients show a markedly reduced life span when grown in culture; moreover they accumulate defects in nuclear structure and architecture as they become older in culture, including lobulation of the nuclear envelope, thickening of the nuclear lamina, loss of peripheral heterochromatin and clustering of nuclear pores.[43] No single mechanism for HGPS has been identified but the farnesyl-prelamin A (progerin) is targeted to the nuclear envelope, where it interferes with the integrity of the nuclear envelope and causes misshapen cell nuclei. The retention of the farnesyl group by progerin may have numerous consequences, as farnesyl groups increase lipophilicity and are involved in membrane association and in protein interactions and is likely to be an important factor in the HGPS phenotype. Indeed, farnesyltransferase inhibition reverses the nuclear morphology in HGPS mouse and cellular models, as well as in HGPS fibroblasts.[44] Cells from normal individuals of old age have similar defects compared to those reported in cells from HGPS patients and the cryptic splice site, which is active in HGPS cells, is also used in cells and tissues from both young and old healthy individuals.[45] Moreover, the inhibition of the *LMNA* cryptic splice site in cells from old individuals reversed their nuclear abnormalities.[45] Further indication that progerin might be involved in the normal aging process comes from recent studies that show that the number of progerin protein positive cells obtained from healthy individuals increases with age, both in vitro and in vivo.[46,47]

Most studies point to an impaired ability to maintain the genome integrity, including defective repair of DNA, accumulation of DNA damage and altered gene expression. The initial response to DNA damage involves the chromatin dependent activation of complex checkpoint signaling pathways that delay the cell cycle and repair the defects. Liu et al[48] evaluated the ability to repair DNA damage following irradiation in HGPS fibroblasts and in Zmpste24-deficient mouse embryonic fibroblasts (MEFs), since lack of functional Zmpste24 also results in progeroid phenotypes in mice and humans. These authors observed that the recruitment of the repair factor p53-binding protein 1 (53BP1) and of Rad51 to sites of DNA lesion was impaired in HGPS fibroblasts and in Zmpste24$^{-/-}$ MEFs, resulting in delayed checkpoint response and defective DNA repair.[48] In addition, they observed increased sensitivity to various DNA-damaging agents in Zmpste24$^{-/-}$ mice.[48] An early response to various types of DNA damage is the phosphorylation of histone H2AX, a variant form of histone H2A, to give intranuclear y-H2AX foci. This phosphorylation is mediated by ATM (for 'ataxia telangiectasia mutated'), ATR (for 'ATM and Rad3 related') and DNA dependent DNA kinases. Phosphorylated histone H2AX represents a hallmark of DNA double-strand breaks (DSBs) and was found to be increased in cells obtained from HGPS patients and from individuals affected by restrictive dermopathy.[49] The unresolved question is what features of the DNA damage response pathways are deficient in HGPS cells. A plausible mechanism suggests that as progerin accumulates it sequester replication and repair factors, leading to stalled replication forks which collapse into DSBs. These accessible DSB termini bind Xeroderma pigmentosum group A (XPA) protein which excludes normal binding by DNA repair proteins. The bound XPA also signals activation of ATM and ATR, arresting cell cycle progression and leading to arrested growth. In addition, the effective sequestration of XPA at these DSB damage sites makes HGPS cells more sensitive to ultraviolet light and other mutagens normally repaired by the nucleotide excision repair pathway of which XPA is a necessary and specific component.[50,51]

It has also been observed that progerin causes changes in histone methylation patterns and in the transcriptional activation of normally heterochromatic regions, as well as some decondensation of the inactive X chromosome, pointing to epigenetic effects. These changes preceding the formation of misshapen cell nuclei.[52] Several genes have been shown to have altered expression levels in HGPS, particularly transcription factors, and late passage HGPS fibroblasts show a marked loss of peripheral heterochromatin.[53] Changes in histone methylation caused by the expression of mutant lamins might also account for telomere shortening reported in cells from HGPS and a well known marker of normal aging.[42,54] Recent in vitro assays indicate that the 50 amino acids deletion results in the loss of a DNA binding site within the tail of prelamin A that contributes to altered heterochromatin anchorage by progerin, offering a molecular mechanism for heterochromatin alterations related to HGPS.[55]

The average telomere length in fibroblasts from HGPS patients is shorter than in age-matched controls.[54] How mutations in lamin A lead to shortened telomere lengths is not known nor is the contribution of individual chromosome ends to the low average length understood. In contrast the telomere length in hematopoietic cells, which typically do not express lamin A, was found to be within the normal range in HGPS patient samples, suggesting that mutant lamin A decreases telomere length via a direct effect and that expression of progerin is necessary for telomere loss in HGPS.[54] It has also been observed that while progerin induces early senescence associated with increased DNA-damage signalling, telomerase extends HGPS cellular lifespan. Particularly, telomerase extends HGPS cellular lifespan by decreasing progerin-induced DNA-damage signalling and activation of p53 and Rb pathways that otherwise mediate the onset of premature senescence. Moreover, progerin-induced DNA-damage signalling is localized to telomeres and is associated with telomere aggregates and chromosomal aberrations. Telomerase amelioration of DNA-damage signalling is relatively rapid and correlates in time with the acquisition by HGPS cells of the ability to proliferate. All of these findings suggest that HGPS premature cellular senescence might result from progerin-induced telomere dysfunction.[56]

During interphase progerin anchors to the nuclear membrane and causes characteristic nuclear blebbing. However, during mitosis progerin mislocalizes into insoluble cytoplasmic aggregates and membranes and causes abnormal chromosome segregation and binucleation. These abnormal association of progerin with membranes during mitosis delays the onset and progression of cytokinesis.[46] Furthermore, it has been demonstrated that the targeting of nuclear envelope/lamina components into daughter cell nuclei in early G1 is impaired in cells expressing progerin. The mutant protein also appears to be responsible for defects in the retinoblastoma protein-mediated transition into S-phase.[57]

Another hypothesis for the involvement of progerin in accelerated aging suggests that progerin interferes with the function of human mesenchymal stem cells (hMSCs). Induction of progerin in hMSCs would change their molecular identity and differentiation potential, supporting a model in which accelerated ageing in HGPS patients and possibly also physiological ageing, is the result of adult stem cell dysfunction and progressive deterioration of tissue functions.[58] Another model links HGPS to stem cell-driven tissue regeneration.[59] According to this model, nuclear structural and mechanical properties of HGPS cells would increase apoptotic cell death to levels that exhaust tissues' ability for stem cell-driven regeneration. Tissue-specific differences in cell death or regenerative potential, or both, would result in the tissue-specific segmental aging pattern seen in HGPS.[59]

Several authors have investigated the possible contribution of structural and mechanical properties of the nuclear lamina in HGPS. Structural and mechanical properties of the lamina are altered in HGPS cells, lamins A and C become trapped at the nuclear periphery and the lamina has a significantly reduced ability to rearrange under mechanical stress.[60] An increased mechanical sensitivity in vascular cells could contribute to loss of smooth muscle cells and the development of arteriosclerosis, the leading cause of death in HGPS patients.[61] It was found that skin fibroblasts from HGPS patients developed progressively stiffer nuclei with increasing passage number. Moreover, they had decreased viability and increased apoptosis under repetitive mechanical strain, as well as attenuated wound healing and these defects preceded changes in nuclear stiffness. These data suggest that increased mechanical sensitivity in HGPS cells could provide a potential mechanism for the progressive loss of vascular smooth muscle cells in HGPS patients.[61]

Werner Syndrome

Cells isolated from WS individuals display increased chromosomal aberrations and premature senescence in culture, as well as accelerated telomere shortening.[62] In addition, several defects in DNA replication have been observed, including dysfunctions in DNA replication initiation, in the establishment of replication foci and in the resolution of stalled replication forks during replication and DNA repair.[3] WRN is a member of the RecQ DNA helicase family and possesses ATPase, $3' \rightarrow 5'$ helicase, $3' \rightarrow 5'$ exonuclease and DNA strand annealing activities. While several different mutations within the *WRN* gene are seen in WS patients, most of them result in the production of a truncated protein lacking the nuclear localization signal, with subsequent absence of functional WRN protein in nuclei.[63] WRN has several functional domains and is active in resolving a variety of DNA substrates, including forks, flaps, D-loops, bubbles, Holliday junctions and G-quadruplexes. As a consequence WRN is considered to be a 'caretaker of the genome' since it participates in distinct DNA metabolic pathways, including DNA replication, DNA recombination, telomere maintenance, apoptosis and DNA repair.[64]

Concerning DNA replication, WRN is involved in the recovery from replication fork stalling. Particularly, it functions in protecting cells from DNA DSBs formation that can occur as a result of replication fork stalling and collapse. Moreover, the protein is involved in the response to replication fork stalling induced by agents that generate DNA crosslinks. During S-phase WRN interacts with both ATM and ATR proteins that mediate the S-phase signalling checkpoints that are activated in response to DNA damage at stalled forks. WRN is also associated with some of the major players in DNA replication, particularly in lagging strand synthesis, including the polymerase Pol δ, flap endonuclease 1 (FEN1), the replication protein A (RPA) and the proliferating nuclear antigen (PCNA).[64]

The WRN protein has also a role in transcription processes and might be involved in the transcription of genes upstream in the network of aging pathways.[64]

We recently reviewed the multiple roles of WRN in DNA repair processes.[65] Indeed, WRN is involved in various repair pathways, including base excision repair (BER), homologous recombination (HR) and nonhomologous end-joining (NHEJ). BER is the major mammalian pathway for the repair of single-strand lesions caused by oxidation, alkylation and methylation and proceeds either via short-pathway which involves single nucleotide replacement, or via long-patch pathway, which involves multiple nucleotide strand displacement. Overall, WRN interacts with several BER proteins (including the polymerase Polβ, the endonuclease APE1, the glycosylase NEIL1 and others) and has a role either in short- and in long-patch BER; its deficiency results in cellular sensitivity to DNA alkylating agents, as well as in increased accumulation of oxidative DNA damage. WRN is also involved in DNA DSB repair. DNA DSBs are repaired either by HR or NHEJ pathways. WRN participates in the assembly of repair complexes at y-H2AX foci and in resolving DNA recombination intermediates in RAD51 mediated HR. WRN interacts with several HR proteins or protein complexes (including the Mre11-RAD50-NBS1 complex and the HR protein RAD52) and its deficiency in WS cells results in defective recombination resolution, limited cell division potential and chromosomal instability. There is also indication that WRN participates in NHEJ. Indeed, WRN interacts with two key players of NHEJ, the Ku70–Ku80 heterodimer and DNA-PKcs. These interactions stimulate WRN activities, facilitating efficient processing of DSBs.[65]

Another pivotal role of the WRN protein is in telomere maintenance, with WRN being involved in replication, recombination and repair processes at telomeres.[64] Telomeres are nucleoprotein complexes situated at the ends of chromosomes that prevent chromosome termini from being recognized as broken ends. WRN seems to be recruited to the replicating telomeres in response to replication stress to resolve lethal events during replication. In addition WRN participates in telomere recombination, thus protecting telomeres from aberrant HR and excessive shortening. Moreover, WRN has been also involved in the repair of oxidative DNA damage at telomeres, likely via interactions with BER proteins.[64] Indeed, cells from WS subjects show accelerate telomere shortening.[62]

NEUROPATHOGENESIS

Hutchinson-Gilford Progeria Syndrome

HGPS individuals usually die in early teens and do not show signs of neuropathogenesis. Both motor and mental development is normal and cognitive functions are preserved.[13-14]

Werner Syndrome

Neurological abnormalities are not constant findings in WS, though some cases of dementia, brain atrophy or peripheral neuropathy have been reported.[65-74] Haustein et al published the neuropathological data of a 49-year-old patient with a history of Werner syndrome: the main features were arteriosclerosis of the small intracerebral arteries, cerebral atrophy and vascular myelopathy. In addition, there was marked lipofuscin deposition and corpora amylacea, but no neurofibrillary tangles or amyloid plaques (the two cardinal hallmarks of Alzheimer's disease).[66] Subsequent analysis of two WS cases for amyloid beta peptide (Aβ) and hyperphosphorylated tau protein revealed extensive frontal and temporal lobe Aβ deposition in the oldest WS case (age 57 years) and restricted neurofibrillary pathology in the medial temporal lobe of both cases.[67] However, the analysis of another 55-year-old WS patient revealed no central nervous system pathology, such as amyloid plaques or neurofibrillary tangles.[68] Overall, WS individuals do not appear to be usually susceptible to Alzheimer's disease and the incidence of dementia of the Alzheimer type is believed to be not increased in WS subjects; moreover, a common polymorphism in the *WRN* gene was not associated with increased risk of Alzheimer's disease.[21,68,69] Individuals with WS may however have central nervous system complications of arteriosclerosis. The neurological manifestations of three WS patients were described by Anderson and Haas and included transient ischemic attacks secondary to atherosclerosis in the common carotid arteries in one patient, sensory peripheral neuropathy in the second patient and peripheral neuropathy with a possible myelopathy in the third one.[70] Peripheral neuropathy was also described by Malandrini et al[71] in a 32-year-old patient and by Umehara et al that described a WS patient with spastic paraparesis and polyneuropathy with slowing of central and peripheral nerve conduction. Sural nerve biopsy revealed a significantly higher frequency of demyelination and remyelination and a loss of myelinated fibers.[72] A 39-year-old WS patient had a hoarse voice, a diffuse muscle weakness and atrophy in the upper and lower limbs with chronic ulcers on the legs. Cranial MRI revealed several lesions in the white matter. Sural nerve biopsy revealed muscle atrophy and the loss of myelinated fibers. Thus, central and peripheral nervous systems were affected in this case.[73] A 56-year-old woman with Werner's syndrome presented dementia consisting of childish behaviour, loss of intelligence and severe amnesia. Brain analysis gave a diagnosis of progressive subcortical vascular encephalopathy of the Binswanger type, which seemed to be the cause of her dementia. After death, mild to moderate demyelinization was found in the subcortical area of the autopsied cerebrum, confirming the clinical diagnosis.[74]

DIAGNOSIS AND THERAPEUTICAL APPROACHES

Hutchinson-Gilford Progeria Syndrome

The clinical diagnosis of HGPS is based on the recognition of the clinical features and on detection of *LMNA* gene mutations. The molecular genetic testing for the G608G mutation in exon 11 is clinically available (see ref. 13 for details). Unfortunately there is actually no cure for HGPS. Therapeutic approaches, based on the use of farnesyltransferase inhibitors (FTIs), revealed that the treatment was capable of improving the nuclear abnormalities in several cellular models of the disease. Some in vivo effects have also been observed when using *Zmpste24*-deficient mice, but these studies revealed that alternative processing by geranylgeranylation of prelamin A occurs instead of farnesylation in the presence of FTIs and this alternative processing also results in the production of progerin, likely explaining the low efficiency of FTIs in ameliorating the phenotypes of progeroid mouse models. The combined use of statins and aminobisphosphonates that inhibits both farnesyltransferase and geranylgeranyltransferase improved the ageing-like phenotype in *Zmpste24*-deficient mice and has led to the approval of a clinical trial to test this combination of drugs for the treatment of progeria.[75,76] Other approaches are based on the possibility to reduce the utilization of the alternative splice donor site in exon 11 or to reduce prelamin A or progerin transcripts by means of RNA interference techniques.[77]

Werner Syndrome

The clinical diagnosis of WS is currently based on the presence of four cardinal signs (cataracts, skin changes, short stature and greying or loss of hair) which are observable in more than 95% of the cases, as well as additional signs (osteoporosis, voice change, atherosclerosis, Type II diabetes mellitus, etc). A definite diagnosis is made when all the cardinal signs (onset over ten years old) and two additional signs are present. In addition, sequencing of the *WRN* gene can be performed and the absence of normal WRN protein can be further confirmed by Western blot analysis (see ref. 16 for details). There is no cure for WS, but only treatment of symptoms exists. These include aggressive treatment of skin ulcers, control of Type II diabetes mellitus, cholesterol-lowering drugs if lipid profile is abnormal, surgical treatment of cataracts and treatment of malignancies in a standard fashion.[16,21] Smoking avoidance, regular exercise and weight control are encouraged to reduce atherosclerosis risk.[16,21] A recent paper suggests that vitamin C restores healthy aging in a mouse model of Werner syndrome, suggesting that vitamin C supplementation could be beneficial for patients with WS.[78]

CONCLUSION

HGPS and WS have been the focus of intense investigation in recent years following the discovery of the disease causative genes and the possibility to create cellular and animal models of both diseases, thus improving our understanding of the molecular mechanisms leading to either premature or physiological aging in humans. Based on recent findings on WS and HGPS, Ding and Shen suggested that human aging can be triggered by two main mechanisms, telomere shortening and DNA damage.[3] In telomere-dependent aging, telomere shortening and dysfunction may lead to DNA damage responses which induce cellular senescence. In DNA damage-initiated aging, DNA damage accumulates, along with DNA repair deficiencies, resulting in genomic instability and accelerated cellular senescence. These two mechanisms can also act cooperatively to increase the overall level of genomic instability, triggering the onset of human aging phenotypes.[3] However, as largely discussed in this chapter other possible mechanisms have been hypothesized to explain the progeroid phenotypes observed in these diseases. One of the most exciting areas of current and future research remains that of the development and testing of therapeutical approaches in cell cultures and animal models of premature aging diseases.

REFERENCES

1. Navarro CL, Cau P, Lévy N. Molecular bases of progeroid syndromes. Hum Mol Genet 2006; 15:R151-161.
2. Kudlow BA, Kennedy BK, Monnat RJ Jr. Werner and Hutchinson-Gilford progeria syndromes: mechanistic basis of human progeroid diseases. Nat Rev Mol Cell Biol 2007; 8:394-404.
3. Ding SL, Shen CY. Model of human aging: recent findings on Werner's and Hutchinson-Gilford progeria syndromes. Clin Interv Aging 2008; 3:431-444.
4. Domínguez-Gerpe L, Araújo-Vilar D. Prematurely aged children: molecular alterations leading to Hutchinson-Gilford progeria and Werner syndromes. Curr Aging Sci 2008; 1:202-212.
5. Pollex RL, Hegele RA. Hutchinson–Gilford progeria syndrome. Clin Genet 2004; 66:375-381.
6. Satoh M, Imai M, Sugimoto M et al. Prevalence of Werner's syndrome heterozygotes in Japan. Lancet 1999; 353:1766.
7. Goto M. Clinical aspects of Werner's syndrome: Its natural history and the genetics of the disease. In: Lebel M, ed. Molecular Mechanisms of Werner's Syndrome. New York: Kluver Academic Plenum Publishers, 2004:1-11.
8. Masala MV, Scapaticci S, Olivieri C et al. Epidemiology and clinical aspects of Werner's syndrome in North Sardinia: description of a cluster. Eur J Dermatol 2007; 17:213-216.
9. Hutchinson J. Congenital absence of hair and mammary glands with atrophic condition of the skin and its appendages. Trans Med Chir Soc Edinburgh 1886; 69:473-477.
10. Gilford H. On a condition of mixed premature and immature development. Med Chirurg Trans 1987; 80:17-45.
11. Gilford H. Progeria: A form of senilism. Practitioner 1904; 73:188-217.
12. Gordon LB, McCarten KM, Giobbie-Hurder A et al. Disease progression in Hutchinson-Gilford progeria syndrome: impact on growth and development. Pediatrics 2007; 120:824-833.
13. Brown WT, Gordon LB, Collins FS. Hutchinson-Gilford Progeria syndrome. In: Pagon RA, Bird TC, Dolan CR et al, eds. GeneReviews [Internet]. Seattle: University of Washington, Seattle; 1993-2003 [updated 2006].
14. Hennekam RC. Hutchinson-Gilford progeria syndrome: review of the phenotype. Am J Med Genet A 2006; 140:2603-2624.
15. Pereira S, Bourgeois P, Navarro C et al. HGPS and related premature aging disorders: from genomic identification to the first therapeutic approaches. Mech Ageing Dev 2008; 129:449-459.
16. Muftuoglu M, Oshima J, von Kobbe C et al. The clinical characteristics of Werner syndrome: molecular and biochemical diagnosis. Hum Genet 2008; 124:369-377.
17. Goto M. Hierarchical deterioration of body systems in Werner's syndrome: implications for normal ageing. Mech Ageing Dev 1997; 98:239-254.
18. Epstein CJ, Martin GM, Schultz AL et al. Werner's syndrome a review of its symptomatology, natural history, pathologic features, genetics and relationship to the natural aging process. Medicine (Baltimore) 1966; 45:177-221.
19. Huang S, Lee L, Hanson NB et al. The spectrum of WRN mutations in Werner syndrome patients. Hum Mutat 2006; 27:558-567.
20. Ishikawa Y, Sugano H, Matsumoto T et al. Unusual features of thyroid carcinomas in Japanese patients with Werner syndrome and possible genotype-phenotype relations to cell type and race. Cancer 1999; 85:1345-1352.
21. Leistritz DF, Hanson N, Martin GM et al. Werner Syndrome. In: Pagon RA, Bird TC, Dolan CR et al, eds. GeneReviews [Internet]. Seattle:University of Washington, 1993-2002 [updated 2007].
22. Eriksson M, Brown WT, Gordon LB et al. Recurrent de novo point mutations in lamin A cause Hutchinson–Gilford progeria syndrome. Nature 2003; 423:293-298.
23. De Sandre-Giovannoli A, Bernard R, Cau P et al. Lamin a truncation in Hutchinson-Gilford progeria. Science 2003; 300:2055.

24. Machiels BM, Zorenc AH, Endert JM et al. An alternative splicing product of the lamin A/C gene lacks exon 10. J Biol Chem 1996; 271:9249-9253.
25. Rodriguez S, Coppedè F, Sagelius H et al. Increased expression of the Hutchinson-Gilford progeria syndrome truncated lamin A transcript during cell aging. Eur J Hum Genet 2009; 17:928-937.
26. Furukawa K, Inagaki H, Hotta Y. Identification and cloning of an mRNA coding for a germ cell-specific A-type lamin in mice. Exp Cell Res 1994; 212:426-430.
27. Beck LA, Hosick TJ, Sinensky M. Isoprenylation is required for the processing of the lamin A precursor. J Cell Biol 1990; 110:1489-1499.
28. Wuyts W, Biervliet M, Reyniers E et al. Somatic and gonadal mosaicism in Hutchinson-Gilford progeria. Am J Med Genet A 2005; 135:66-68.
29. Jacob KN, Garg A. Laminopathies: multisystem dystrophy syndromes. Mol Genet Metab 2006; 87:289-302.
30. Fukuchi K, Katsuya T, Sugimoto K et al. LMNA mutation in a 45 year old Japanese subject with Hutchinson-Gilford progeria syndrome. J Med Genet 2004; 41:e67.
31. Shalev SA, De Sandre-Giovannoli A, Shani AA et al. An association of Hutchinson-Gilford progeria and malignancy. Am J Med Genet A 2007; 143A:1821-1826.
32. Navarro CL, De Sandre-Giovannoli A, Bernard R et al. Lamin A and ZMPSTE24 (FACE-1) defects cause nuclear disorganization and identify restrictive dermopathy as a lethal neonatal laminopathy. Hum Mol Genet 2004; 13:2493-503.
33. Navarro CL, Cadiñanos J, De Sandre-Giovannoli A et al. Loss of ZMPSTE24 (FACE-1) causes autosomal recessive restrictive dermopathy and accumulation of Lamin A precursors. Hum Mol Genet 2005; 14:1503-1513.
34. Moulson CL, Fong LG, Gardner JM et al. Increased progerin expression associated with unusual LMNA mutations causes severe progeroid syndromes. Hum Mutat 2007; 28:882-889.
35. Denecke J, Brune T, Feldhaus T et al. A homozygous ZMPSTE24 null mutation in combination with a heterozygous mutation in the LMNA gene causes Hutchinson-Gilford progeria syndrome (HGPS): insights into the pathophysiology of HGPS. Hum Mutat 2006; 27:524-531.
36. Plasilova M, Chattopadhyay C, Pal P et al. Homozygous missense mutation in the lamin A/C gene causes autosomal recessive Hutchinson-Gilford progeria syndrome. J Med Genet 2004; 41:609-614.
37. Liang L, Zhang H, Gu X. Homozygous LMNA mutation R527C in atypical Hutchinson-Gilford progeria syndrome: evidence for autosomal recessive inheritance. Acta Paediatr 2009; 98:1365-1368.
38. Verstraeten VL, Broers JL, van Steensel MA et al. Compound heterozygosity for mutations in LMNA causes a progeria syndrome without prelamin A accumulation. Hum Mol Genet 2006; 15:2509-2522.
39. Garg A, Subramanyam L, Agarwal AK et al. Atypical progeroid syndrome due to heterozygous missense LMNA mutations. J Clin Endocrinol Metab 2009; 94:4971-4983.
40. Huang S, Lee L, Hanson NB et al. The spectrum of WRN mutations in Werner syndrome patients. Hum Mutat 2006; 27:558-567.
41. Chen L, Lee L, Kudlow BA et al. LMNA mutations in atypical Werner's syndrome. Lancet 2003; 362:440-445.
42. Dechat T, Pfleghaar K, Sengupta K et al. Nuclear lamins: major factors in the structural organization and function of the nucleus and chromatin. Genes Dev 2008; 22:832-853.
43. Goldman RD, Shumaker DK, Erdos MR et al. Accumulation of mutant lamin A causes progressive changes in nuclear architecture in Hutchinson– Gilford progeria syndrome. Proc Natl Acad Sci USA 2004; 101:8963-8968.
44. Young SG, Meta M, Yang SH et al. Prelamin A farnesylation and progeroid syndromes. J Biol Chem 2006; 281:39741-39745.
45. Scaffidi P, Misteli T. Lamin A-dependent nuclear defects in human aging. Science 2006; 312:1059-1063.
46. Cao K, Capell BC, Erdos MR et al. A lamin A protein isoform overexpressed in Hutchinson-Gilford progeria syndrome interferes with mitosis in progeria and normal cells. Proc Natl Acad Sci USA 2007; 104:4949-4954.
47. McClintock D, Ratner D, Lokuge M et al. The mutant form of lamin A that causes Hutchinson– Gilford progeria is a biomarker of cellular aging in human skin. PLoS ONE 2007; 2:e1269.
48. Liu B, Wang J, Chan KM et al. Genomic instability in laminopathy-based premature aging. Nat Med 2005; 11:780-785.
49. Liu Y, Rusinol A, Sinensky M et al. DNA damage responses in progeroid syndromes arise from defective maturation of prelamin A. J Cell Sci 2006; 119:4644-4649.
50. Liu Y, Wang Y, Rusinol AE et al. Involvement of xeroderma pigmentosum group A (XPA) in progeria arising from defective maturation of prelamin A. FASEB J 2008; 22:603-611.
51. Musich PR, Zou Y. Genomic instability and DNA damage responses in progeria arising from defective maturation of prelamin A. Aging 2009; 1:28-37.
52. Shumaker DK, Dechat T, Kohlmaier A et al. Mutant nuclear lamin A leads to progressive alterations of epigenetic control in premature aging. Proc Natl Acad Sci USA 2006; 103:8703-8708.

53. Csoka AB, English SB, Simkevich CP et al. Genome-scale expression profiling of Hutchinson-Gilford progeria syndrome reveals widespread transcriptional misregulation leading to mesodermal/mesenchymal defects and accelerated atherosclerosis. Aging Cell 2004; 3:235-243.
54. Decker ML, Chavez E, Vulto I et al. Telomere length in Hutchinson-Gilford progeria syndrome. Mech Ageing Dev 2009; 130:377-383.
55. Bruston F, Delbarre E, Ostlund C et al. Loss of a DNA binding site within the tail of prelamin A contributes to altered heterochromatin anchorage by progerin. FEBS Lett 2010; 584:2999-3004.
56. Benson EK, Lee SW, Aaronson SA. Role of progerin-induced telomere dysfunction in HGPS premature cellular senescence. J Cell Sci 2010; [Epub ahead of print].
57. Dechat T, Shimi T, Adam SA et al. Alterations in mitosis and cell cycle progression caused by a mutant lamin A known to accelerate human aging. Proc Natl Acad Sci USA 2007; 104:4955-4960.
58. Scaffidi P, Misteli T. Lamin A-dependent misregulation of adult stem cells associated with accelerated ageing. Nat Cell Biol 2008; 10:452-459.
59. Halaschek-Wiener J, Brooks-Wilson A. Progeria of stem cells: stem cell exhaustion in Hutchinson-Gilford progeria syndrome. J Gerontol A Biol Sci Med Sci 2007; 62:3-8.
60. Dahl KN, Scaffidi P, Islam MF et al. Distinct structural and mechanical properties of the nuclear lamina in Hutchinson-Gilford progeria syndrome. Proc Natl Acad Sci USA 2006; 103:10271-10276.
61. Verstraeten VL, Ji JY, Cummings KS et al. Increased mechanosensitivity and nuclear stiffness in Hutchinson-Gilford progeria cells: effects of farnesyltransferase inhibitors. Aging Cell 2008; 7:383-393.
62. Melcher R, von Golitschek R, Steinlein C et al. Spectral karyotyping of Werner syndrome fibroblast cultures. Cytogenet. Cell Genet 2000; 91:180-185.
63. Yu CE, Oshima J, Wijsman EM et al. Mutations in the consensus helicase domains of the Werner syndrome gene. Werner's Syndrome Collaborative Group. Am J Hum Genet 1997; 60:330-341.
64. Rossi ML, Ghosh AK, Bohr VA. Roles of Werner syndrome protein in protection of genome integrity. DNA Repair 2010; 9:331-344.
65. Coppedè F, Migliore L. DNA repair in premature aging disorders and neurodegeneration. Curr Aging Sci 2010; 3:3-19.
66. Haustein J, Pawlas U, Cervos-Navarro J. The Werner syndrome: a case study. Clin Neuropathol 1989; 8:147-151.
67. Leverenz JB, Yu CE, Schellenberg GD. Aging-associated neuropathology in Werner syndrome. Acta Neuropathol 1998; 96:421-424.
68. Mori H, Tomiyama T, Maeda N et al. Lack of amyloid plaque formation in the central nervous system of a patient with Werner syndrome. Neuropathology 2003; 23:51-56.
69. Payão SL, de Labio RW, Gatti LL et al. Werner helicase polymorphism is not associated with Alzheimer's disease. J Alzheimers Dis 2004; 6(6):591-4; discussion 673-681.
70. Anderson NE, Haas LF. Neurological complications of Werner's syndrome. J Neurol 2003; 250: 1174-1178.
71. Malandrini A, Dotti MT, Villanova M. Neurological involvement in Werner's syndrome: clinical and biopsy study of a familial case. Eur Neurol 2000; 44:187-189.
72. Umehara F, Abe M, Nakagawa M et al. Werner's syndrome associated with spastic paraparesis and peripheral neuropathy. Neurology 1993; 43:1252-1254.
73. Just A, Canaple S, Joly H et al. Neurologic complications in a case of Werner syndrome. Rev Neurol (Paris) 1996; 152:634-636.
74. Kawamura H, Mori S, Murano S et al. Werner's syndrome associated with progressive subcortical vascular encephalopathy of the Binswanger type. Nippon Ronen Igakkai Zasshi 1999; 36:648-651.
75. Varela I, Pereira S, Ugalde AP et al. Combined treatment with statins and aminobisphosphonates extends longevity in a mouse model of human premature aging. Nat Med 2008; 14:767-772.
76. Osorio FG, Obaya AJ, López-Otín C et al. Accelerated ageing: from mechanism to therapy through animal models. Transgenic Res 2009; 18:7-15.
77. Worman HJ, Fong LG, Muchir A et al. Laminopathies and the long strange trip from basic cell biology to therapy. J Clin Invest 2009; 119:1825-1836.
78. Massip L, Garand C, Paquet ER et al. Vitamin C restores healthy aging in a mouse model for Werner syndrome. FASEB J 2010; 24:158-217.

CHAPTER 25

THE SAVANT SYNDROME AND ITS POSSIBLE RELATIONSHIP TO EPILEPSY

John R. Hughes

Department of Neurology, University of Illinois Medical Center, Chicago, Illinois, USA
Email: jhughes@uic.edu

Abstract: The goal of this chapter is to review the Savant syndrome (SS), characterized by outstanding islands of mental ability in otherwise handicapped individuals. Two forms exist: The congenital and acquired form. Among the many examples of the congenital form are the calendar calculators, who can quickly provide the day of the week for any date in the past. Other examples are the musical savants with perfect pitch and the hyperlexics, who (in one case) can read a page in 8 seconds and recall the text later at a 99% level. Other types of talents and artistic skills can be found, involving 3-D drawing, map memory, poetry, painting, sculpturing, including one savant who could recite without error the value of Pi to 22,514 places. The acquired form refers to the development of outstanding skills after some brain injury or disease, usually involving the left fronto-temporal area. This type of injury seems to inhibit the 'tyranny of the left hemisphere', allowing the right hemisphere to develop the savant skills. One other way to inhibit the left fronto-temporal area is to use transcranial magnetic stimulation in normal subjects and nearly one-half of these subjects can then perform new skills during the stimulation that they could not perform before. This type of finding indicates the potentiality in all of us for the development of savant skills under special circumstances. Explanations of the congenital SS include enhanced local connectivity as a compensation for underconnectivity of long-range fibers, but also weak central coherence, replaced by great attention to details, enhanced perceptual functioning and obsessive pre-occupation with specific interests.

INTRODUCTION

The Savant syndrome (SS) is characterized by remarkable islands of mental ability in otherwise mentally handicapped individuals.[1] Treffert and Wallace[2] have traced the history of the SS, pointing out that in 1789 Dr. Benjamin Rush, often considered the

Neurodegenerative Diseases, edited by Shamim I. Ahmad.
©2012 Landes Bioscience and Springer Science+Business Media.

Father of American Psychiatry, described the outstanding mathematical ability of one man, named Thomas Fuller. Nearly 100 years later in 1887 Dr. J. Langdon Down, well known for the syndrome with his own name and supervisor of the Earlswood Asylum in London, referred to 10 patients with the SS as "idiot savants". An even earlier reference was added by Foerstl[3] who discussed a man named Jedediah Buxton, described in February 1751 in the magazine, "Gentleman". Mr. Buxton could not write his own name but could do quick mathematical calculations, like multiplying a 39 digit number by itself.

The goal of this study is to describe the SS and also to investigate the possible relationship of this syndrome to epilepsy, especially because the SS are often found in autistic patients who often have seizures.

METHOD

The Medline section on the internet provided references to published studies on the "Savant syndrome".

DIFFERENT FORMS

Congenital SS

Treffert[1] has indicated that nearly one-half of those with the SS are individuals with autism (A) and the other one-half have some other type of developmental disability.

Calendar Calculators (CC)

One of the most common forms of the SS are the CC. These individuals can readily identify the day of the week for any given date in the past. Different authors have reported various characteristics of the groups that they have studied. Dubischar-Krivec et al[4] reported shorter reaction times and fewer errors for present or past dates than controls, but no differences from controls were found for future dates. Thioux et al[5] described Donny, a young autistic savant, "who is possibly the fastest and most accurate calendar prodigy ever described". The title of this report likely justifies the latter statement: "The day of the week when you were born in 700 msec". O'Connor and Hermelin[6] reported that calculating speeds were similar in 10 year-old savants as in adult savants. Performance could not be accounted for by practice alone and no improvement was detectable over time.

Other authors have pointed out various normal characteristics in this group. An example is that savants did not differ from controls for short—or—long term memory. However, the savants showed a recall superiority for long-term retention of calendrical material.[7] Others[8] have described in one savant poor explicit knowledge of calendar structure and also that error rate and response latency increased with temporal remoteness of dates under consideration. Mottron et al[9] also dealt with errors, reporting that testing of all the dates in a year revealed a random distribution of errors which were not stable across time. Also, that particular savant was able to answer "reversed" questions that could not be solved by a classical algorithm.

Another group[10] described a Chinese calendar savant who had exceptional proficiency in converting a date from the Gregorian calendar to the Chinese calendar. The results did not support the hypotheses of rote memorization, eidetic imagery, high speed calculation and anchoring strategy, use of calendar irregularities or monthly configurations. The authors proposed that this particular savant was familiar with 14 calendar templates and the knowledge of matching these templates to every year. For dates beyond the 20th century the calculation was accomplished by regressing the date to a corresponding year in the 20th century by adding or subtracting 28 or 700 years. This explanation from the authors may be difficult for the reader to follow and it would seem even very much more difficult to put the calculations into play to accomplish this conversion of dates.

Other explanations for the CC have not been so complex, but neither have they been very informative. Pring and Hermelin[11] concluded that savants adopt a cognitive style of weak central coherence, protecting single representations from being retained in the form of stable enduring wholes. This strategy allows for transformation, reorganization and reconstruction of the relationship between single items of information. In other words, savants are more likely to process details at the expense of global information. Casey et al[12] discussed the dysfunctional attention hypothesis to explain the skills of CC by the ability to divide, shift, direct and sustain attention. They concluded that a deficit in shifting selective attention from one stimulus to another was attributed to an inability to disengage attention as a result of a deficient orientation and an overselectivity. In other words, this latter explanation may come down to a focused attention. Finally, Hermelin and O'Connor[13] concluded that these human calculators used rule-based strategies. One last "explanation" was given by another group,[9] suggesting that the CC are "solving the problems in a non-algorithmic way", meaning that no logical set of rules are followed in arriving at any answers. It may be helpful to indicate how the savants do not perform their skills, as one step closer to understanding how they do perform these same skills.

Musical Savants

Absolute or perfect pitch is the ability to properly identify any musical note after hearing it. On a test involved with absolute pitch, investigated by Pring et al[14] the savants scored higher than nonmusicians, but not statistically better than a group of professional musicians. Also, in short-term memory tests for musical phrases the savants and musicians performed indistinguishably. The authors concluded that the skills of musical savants are separate from general intelligence, but also confirming that absolute pitch can, in fact, exist in these individuals. According to Heaton et al[15] the autistic savants performed in a highly superior way, compared to controls who had self-reported absolute pitch. Five years earlier, Heaton[16] included other data, claiming that the savant group showed enhanced pitch memory and pitch labeling and also superior chord segmentation. Young and Nettelbeck[17] reported on a savant with perfect pitch, exceptional recall and performance of structured music and also an increased speed of information processing. Although this savant had high levels of concentration and memory, he had difficulties in verbal reasoning.

Hyperlexica

The life of one of the most famous savants, Kim Peek, was dramatized in the popular movie, "Rain Man", played by actor Dustin Hoffman. Kim reads the left side of a page with his left eye and simultaneously the right side of the page with his right eye (without a corpus callosum). The time taken for these two pages for Kim is usually 8 seconds and upon testing for retention he was 99% correct of the material just read.[2] These values are in contrast to 45 seconds for reading and 45% correct on testing seen in a group of normal individuals. Goldberg[18] described the condition of hyperlexia that the skill arises without any practice, is not integrated with other areas of knowledge, is associated with a dysfunctional procedural memory system, but intact declarative memories. Finally, O'Connor and Hermelin[19] described two autistic children whose reading speeds were considerably faster than that seen in controls. To these investigators efficient grapheme-phoneme (written-sound) conversion is likely a modular component of the reading skill responsible for the hyperlexia in these savants.

Various Other Skills

One group[20] mentioned hypermnesia as one of the special skills found among those with the SS. An excellent example is Kim Peek who remembers nearly everything that he has read in 9,000 books.[2] That same group also mentioned several branches of the arts, as part of the SS, such as drawing, painting and sculpturing. Mottron and Belleville[21] described one individual who could draw objects rotated in 3-dimensional space more accurately than controls. In addition, this individual could detect a perspective incongruency between an object and a landscape at a superior level. The authors claimed that this skill was reached without the use of explicit or implicit perspective rules. Two years earlier the same authors[22] described the same patient with the same outstanding ability for 3-dimensional drawing. Their complex conclusion (difficult to understand) was that the performance was an "anomaly with hierarchical organization of the local and global parts of a figure, a local interference effect in incongruent hierarchical visual stimuli, a deficit in relating local parts to global form information in impossible figures and an absence of feature grouping in graphical recall".

O'Connor and Hermelin[23] discussed gifted artists who were outstanding in the perception of a recognition memory for shapes. Results indicated that reproduction ability depended on artistic ability, independent of intelligence. Somewhat related to the skill of drawing was the talent of a 4-year-old boy[24] who had an exceptional memory for maps and spatial locations, although the boy had a low-average intelligence and impaired motor skills. Luszki[25] mentioned a savant who was excellent on a block design test, but scored only 76 on a performance IQ test. Dowker et al[26] described a savant poet who referred more often to aspects of self-analyses, while descriptions of other people not related to the self were less frequently mentioned. This latter emphasis on self is consistent with one of the basic core features of autism. Another special talent is the mechanical ability to repair everything and anything. Brink[27] described such a mechanical genius with a right-sided paralysis and transient loss of speech from a bullet into the left temporal area. Hoffman and Reeves[28] reported on a similar savant who could not speak until 20 years of age, but had great mechanical ability. Finally, other investigators[29] have described artists who were savants, commenting that their specific ability can occur at any level of intelligence, usually between 30-75 IQ levels.

From all of these examples, it seems clear that nearly any specific skill or talent can occur in the SS.

Acquired SS

Without Seizures

Treffert[30] has written extensively about the congenital form of SS, likely present at around birth, but also he has discussed the acquired form occurring later in life. In this instance "after some brain injury or brain disease, savant skills unexpectedly emerge, sometimes at a prodigious level, when no such skills were present before injury or illness". Many examples were given by Treffert, including a 10-year-old boy knocked unconscious by a baseball, then later could do quick calendar calculations, an 8-year-old boy with the similar talent after a left hemispherectomy and a 3-year-old child after meningitis was later considered a musical genius. Also included was a 9-year-old boy who was shot with a bullet to the left temporal area of the brain, leaving him with a right-sided hemiparesis, later developing special mechanical abilities. Finally, two painters were mentioned who had significant qualitative improvements after strokes involving the left occipital lobe and thalamus.

A surprising number of acquired cases had been earlier diagnosed with fronto-temporal degeneration (FTD). These cases included a 68-year-old man without any interest in art who became an accomplished artist after FTD and also two other similar patients. In addition, another case was an example of "paradoxical functional facilitation "as a release phenomenon "where loss of some skills permits emergence of others", like artistry. One investigator, mentioned by Treffert,[30] had collected 12 cases of FTD with later emergent artistic ability. One particular patient was a 51-year-old artist whose skills surfaced for the first time after a subarachnoid hemorrhage associated with cerebral artery aneurysms. The frontal damage in this latter patient was similar to that seen in FTD. Finally, a talented art teacher at age 49 years at the beginning of a progressive FTD process, later showed an "impressive artistic growth".

With Seizures

Only very few, if any, examples can be found in the world literature of acquired SS with epilepsy. One possible example comes from Oliver Sachs[31] who described a person with "high fever, weight loss, delirium and perhaps seizures". Following this illness this individual began painting extremely accurate scenes that colleagues believed could not have been painted before the illness.

The case of Patrick Obyedkov, a 35-year-old musical savant after the onset of seizures, was a fictional character on the TV serial, "House".

The example of Daniel Tammet should be mentioned, but may not be relevant here, because his childhood seizures at 3 years of age had always been considered as not related to the SS but instead to his synesthesia. The latter condition occurred after his seizures. For only 3 years Daniel was on anti-epileptic drugs and he has been seizure free since early childhood.[32] He experienced this synesthesia in the form that every number that he experiences has its own color, shape and texture. At 4 years of age he developed rapid calendar calculating ability, in addition to an incredible memory for numbers. On American TV many viewers witnessed Daniel at 26 years of age in front of Oxford University dons reciting (without a single mistake) the value of Pi to 22,514 decimal places over a 5-hour period. Thus, this prodigious memory was not likely related to his early seizures at 3 years of age. Since nearly one-half of savants have A, which is closely associated with seizures, speculation would be the savants might have seizures, but they usually do not have such attacks.

Thus, only one case of possible seizures can be found in the world literature of either the acquired SS or the congenital type. On the other hand, many examples can be found of acquired SS without seizures, but with some damage or disturbance to the brain, usually the left frontal temporal area. Since a seizure condition involves a hyperexcitable brain, this condition is not likely the kind of circumstance to release savant skills and this reviewer is not aware of any patients with a left temporal lobectomy who have later developed the SS. However, a hypoexcitable left frontal-temporal area from a head injury, CVA, FTD, etc., seems sufficient to release the normal right hemisphere to develop the SS, escaping the "tyranny of the left hemisphere".

One way to escape this "tyranny" is to stimulate the left frontal-temporal area with repetitive magnetic transcranial pulses. Snyder et al[33] reported that 8 of 12 normal individuals improved their ability to accurately guess the number of discrete items that were placed together. In a study by Young et al[34] 5 of 17 subjects developed savant types of skills, including declarative memory, drawing, mathematical and calendar calculating during the stimulation. In another study published 3 years earlier Snyder et al[35] reported that 4 of 11 subjects displayed enhanced proofreading ability during the magnetic pulses. These findings argue that all of us under certain circumstances, like inhibiting the left frontal-temporal area may have the possibility of savant types of skills! This possibility should stimulate the readers to consider the potentiality of their own cognitive powers or artistic talents.

POSSIBLE CAUSES OF THE SS

Takahata and Kato[36] have discussed the acquired SS, exemplified by patients with FTD who then develop savant skills and have provided a few explanations to account for this phenomenon. First is the hypermnesic model that these skills develop from existing or dormant cognitive functions, like memory. Second is the paradoxical functional facilitation model, which emphasizes the role of reciprocal inhibitory interactions among cortical regions. Third is the autistic model, including the weak central coherence theory, also emphasizing the underconnectivity or disruption of long-range fibers, but with an enhanced local connectivity in given areas. The authors speculated that the enhanced local connectivity results from the specialized and facilitated cognitive processes responsible for savant skills.

The hypermnesic model with skills developing from existing functions finds some support from Miller[37] who concluded that skills exhibited by savants shared many characteristics found in normal subjects and that the skills are usually accompanied by normative levels of performance in various subtests. Also, this investigator claimed that it is unclear that savants have distinctive cognitive strengths or motivational dispositions. Pring and Hermelin[38] similarly concluded that there may be no differences between savants and normal subjects regarding the nature of mental structures underlying specific talents, which are independent of the level of general cognitive functioning. These latter 2 studies do not therefore clarify how the savants perform their outstanding skills.

The third autistic model is based, in part, on the weak central coherence theory, which some authors[39] believe would predispose individuals to develop their talents with obsessive pre-occupation. Others[40] have drawn a similar conclusion, that the excessive use of cognitive processing would be "instigated by the probable failure of central executive control mechanisms". Still others[11] concluded that a "cognitive style of weak central coherence may protect single representations from being retained in the form of stable enduring wholes". Happe[41] also concluded that a cognitive style biased towards local rather than global information processing (called weak central coherence) may be involved in the SS. Also, other authors[22] have referred to an anomaly in the hierarchical organization of the local and global parts of the skill involved. The other part of the autistic model is the presence of long-range fiber underconnectivity, emphasized by Hughes,[42] also associated with enhanced short-range or local connectivity.

Many other explanations have been suggested with various depths of insight into the fascinating phenomenon of the savant syndrome. One group[43] hypothesized that a mutation gives rise to the development of the positive aspects of the SS, but has a deleterious effect on the development of other phenotypic traits, resulting in autism. Nurmi et al[44] reported that chromosome 15q11-q13 involving the GABRB3 gene provided a genetic contribution to a subset of affected individuals with savant skills. Two years later Ma et al[45] failed to demonstrate a linkage to 15q11-q13. Casanova et al[46] had an interesting approach to the possible explanation of these skills by an analysis of the morphometry of 3 distinguished scientists who died and had an autopsy. They reported significant differences in these scientists' cortex from comparison groups in both smaller minicolumn width and mean cell spacing. These conditions would be expected to enhance neuronal processing, as would be anticipated in the SS.

Mottron et al[47] presented the concept of "enhanced perceptual functioning" to explain the SS. Included were possibilities of locally oriented visual and auditory perception, enhanced low-level discrimination and the use of a more posterior network in complex visual tasks. Also included were enhanced perception of static stimuli, decreased perception of complex movement, autonomy of low-level information processing toward higher-order operations and increased perceptual expertise. Consistent with the complexity of the latter explanations, Kelly et al[48] found no evidence that the savants used short-cut strategies.

The "dysfunctional attention hypothesis" of Casey et al[12] addressed the deficiencies of savants in shifting selective attention from one stimulus to another. This hypothesis does not clarify the mechanisms used by savants to explain their talents, except to imply that savants are locked into one stimulus. The 3 hypotheses offered by Kehrer[49] were that savants have a different perception and storing of perceived impulses, an abnormal memory process and storing function and these outstanding abilities are reinforced by the environment. However, these points do not easily explain the positive aspects of the SS. O'Connor and Hermelin[50] claimed that savants use strategies founded on "deduction and application of rules governing the material upon which their special ability operates", also generating "novel examples of such rule-based structures". Perhaps, these comments are not too helpful in understanding the SS, but another point is more helpful that these savants have an obsessional pre-occupation with a limited section of the environment. The theory of Goldberg[18] was that savants have a dysfunctional procedural memory system (related to skills or procedures) though their declarative memories (of facts) are intact. O'Connor and Hermelin one year earlier[51] than the later report[50] concluded that the savants have a superior image memory and ready access to a 'picture lexicon'. The example of savants like Daniel Tammet, who can quickly multiply a long number by itself, requires comment. It is as if the problem is presented to these savants and they 'read off' the answer as the brain 'automatically' does the calculation. While listening to a familiar opera, this reviewer often 'automatically' knew the next note, but this knowledge derives from learning the sequence of notes after many times hearing those same sequences. The parallel with the SS is that they also deal with sequences, but the major difference here is that the savants show these prodigious skills without any rehearsal of the problem just presented to them.

CHARACTERISTICS OF SAVANTS

Intelligence

The intellectual functioning was viewed as "average",[52] or "limited",[5,48] "independent"[50,51,53,54] and also "low average".[24] O'Connor[55] has added that all savants can abstract to some degree.

Memory

Memory for the specific talent was described as exceptional,[24] but as normal for general and verbal memory, independent of the measured verbal IQ.[56] Also, skills were not based on efficient rote memory.[50] Another characteristic is that skills of savants do not change with practice.[50,55]

Sensation and Perception

As previously mentioned, Mottron et al[47] proposed "enhanced perceptual functioning" in savants. Included are the possibilities of locally oriented perceptions, use of a more posterior network, enhanced perception of static stimuli, decreased perception of movements and differences between perception and general intelligence. These conditions add up to increased perceptual processing. On the other hand, Chen et al[57] reported normal auditory and visual evoked potentials in savants.

Focus

Bor et al[58] described one savant under discussion as having a "propensity to focus on local detail". Casanova et al[46] had a similar phrase, a "focused attention", likely related to an "obsessive personality profile".[40] Hou et al[59] stated this characteristic as a "focus on one topic at the expense of other interests", similar to the "pre-occupations" of O'Connor and Hermelin.[50,51,53]

ANATOMICAL RELATIONSHIPS

One of the most fascinating and significant points in this review has been emphasized by Treffert[29] who has summarized the condition of an acquired SS. He has stated that damage to the left frontal-temporal area seems required to escape the "tyranny of the left hemisphere" and thus to permit the development of a savant skill by the right hemisphere. After studying all types of the SS, Bujas-Petkovic[60] concluded that "the functioning of the right cerebral hemisphere is most probably responsible for those abilities". Treffert[1] has summarized the situation in another way, by concluding that left-brain damage occurs with right brain compensation. As previously mentioned, Snyder et al[33] was one of the investigators who reported inhibiting the left anterior temporal area by repetitive magnetic stimulation, could result in the improvement of various abilities. Hou et al[59] agreed that "the anatomic substrate for the savant syndrome may involve the loss of function also in the left temporal lobe", but added that there is also enhanced function of the posterior neocortex.

The prefrontal area has also been emphasized in some reports. Takahata and Kato[36] referred to the paradoxical functional facilitation model in which reciprocal inhibitory interactions occur among adjacent or distant cortical regions, especially that of the prefrontal cortex and the posterior regions of the brain. The same prefrontal area was mentioned by Bor et al[58] reporting on a celebrated savant (D.T.) whose fMRI showed hyperactivity in the lateral prefrontal cortex when demonstrating one of his skills by encoding digits.

INCIDENCE

Treffert[1] has summarized data on the incidence of the SS. As previously mentioned, he has stated that approximately 10% of autistic persons exhibit savant abilities, that roughly 50% of those with the SS have autism and the remaining 50% have other forms of a developmental disability. The remaining cases of the SS include the acquired group that was previously discussed. The same author indicated that only 50-100 individuals in the world have these exceptionally prodigious talents. One other report[61] judged the general incidence of the SS in Finland, concluding that there were 45 cases in that country with an incidence rate of 1.4/1000 (1 in 714) who also had mental retardation.

CONCLUSION

The goal of this report is to review the Savant syndrome (SS) and to explore its possible relationship to epilepsy, especially because nearly one-half of savants are autistic and many autistic individuals have seizures. There are 2 different forms, congenital and acquired. Among the congenital types, calendar calculators are one of the most common forms and one individual could provide the day of the week for any past event in 700 msec! The explanations for the way in which this is done have varied from the simple to the complex. Some claim that a 'weak central coherence' is associated with a mental bias toward details at the expense of global information and others have suggested that the process involves complex matching templates to every year also with regression of dates. Musical savants with absolute or perfect pitch have been identified, at times with superior chord segmentation and increased speed of information processing. Hyperlexia is another form, exemplified by the 'Rain Man', Kim Peek, who reads the left side of a page with his left eye and simultaneously the right side with the right eye in 8 seconds with a 99% correct recall. Various other skills include hypermnesia, 3-D drawing, map memory, outstanding poetry, painting and all types of other artistic talents.

The acquired form refers to the development of outstanding skills after some brain injury or disease. Examples include a number of individuals with meningitis, bullets in their brains, CVAs, etc. who later develop savant skills. Also, savants who have had fronto-temporal dementia, usually involving the left side are found and represent an example of the acquired type, releasing the right hemisphere to develop savant skills by avoiding the 'tyranny of the left hemisphere'. Only one possible case of seizures could be found in the world literature associated with the savant syndrome and a hyperexcitable cortex with a seizure disorder does not seem to be compatible with a SS. In order to escape this tyranny of the left fronto-temporal area, transcranial magnetic stimulation inhibiting this area has led to nearly one-half of normal subjects developing savant types of skills during the actual stimulation. This finding suggests the great potentiality that all of us may have in cognitive powers or artistic talents if conditions were right for these skills to emerge.

The stated causes or explanations of the SS have been varied, including a 'paradoxical functional facilitation', but also an enhanced local connectivity as a compensation to the underconnectivity of long-range fibers. This latter point is usually mentioned in conjunction with a weak central coherence with attention to detail rather than the stable whole. Other explanations include an 'enhanced perceptual functioning' or a dysfunctional attention or an obsessive pre-occupation.

Characteristics of savants include average or limited intelligence, exceptional memory of their special talent, but otherwise normal without any change in the talent with practice. Also included is a propensity to focus on local detail.

The anatomical relationships involve the development of skills in the right hemisphere after inhibiting the damaged left fronto-temporal area. Also, the prefrontal area may be special by the hyperactivity noted in that same area while performing the skills.

The incidence of the SS is that approximately 10% of autistic persons exhibit savant abilities. Roughly one-half of the SS are autistic and the rest have other forms of developmental disability. One new and exciting category is the acquired SS.

REFERENCES

1. Treffert DA. The savant syndrome and autistic disorder. CNS Spectr 1999; 4(12):57-60.
2. Treffert DA, Wallace GL. Islands of genius. Sci Am 2002; 286(6):76-85.
3. Foerstl J. Early interest in the idiot savant. Am J Psychiatry 1989; 146(4):566.
4. Dubischar-Krivec AM, Neumann N, Poustka F et al. Calendar calculating in savants with autism and healthy calendar calculators. Psychol Med 2009; 39(8):1355-1363.
5. Thioux M, Stark DE, Klaiman C et al. The day of the week when you were born in 700 ms: calendar computation in an Autistic savant. J Exp Psychol Hum Percept Perform 2006; 32(5):1155-1168.
6. O'Connor N, Hermelin B. Do young calendrical calculators improve with age? J Child Psychol Psychiatry 1992; 33(5):907-912.
7. Heavey L, Pring L, Hemelin B. A date to remember: the nature of memory in savant calendrical calculators. Psychol Med 1999; 29(1):145-160.
8. Iavarone A, Patruno M, Galeone F et al. Brief report: error pattern in an autistic savant calendar calculator. J Autism Dev Disord 2007; 37(4):775-779.
9. Mottron L, Lemmens K, Gagnon L et al. Non-algorithmic access to calendar information in a calendar calculator with autism. J Autism Dev Disord 2006; 36(2):239-247.
10. Ho ED, Tsang AK, Ho DY. An investigation of the calendar calculation ability of a Chinese calendar savant. J Autism Dev Disord 1991; 21(3):315-327.
11. Pring L, Hermelin B. Numbers and letters: exploring an autistic savant's unpracticed ability. Neurocase 2002; 8(4):330-337.
12. Casey BJ, Gordon CT, Mannheim GB et al. Dysfunctional attention in autistic savants. J Clin Exp Neuropsychol 1993; 15(6):933-946.
13. Hermelin B, O'Connor N. Idiot savant calendrical calculators: rules and regularities. Psychol Med 1986; 16(4):885-893.
14. Pring L, Woolf K, Tadic V. Melody and pitch processing in five musical savants with congenital blindness. Perception 2008; 37(2):290-307.

15. Heaton P, Davis RE, Happe FG. Research note: exceptional absolute pitch perception for spoken words in an able adult with autism. Neuropsychologia 2008; 46(7):2095-2098.
16. Heaton P. Pitch memory, labelling and disembedding in autism. J Child Psychol Psychiatry 2003; 44(4):543-551.
17. Young RL, Nettelbeck T. The abilities of a musical savant and his family. J Autism Dev Disord 1995; 25(3):231-248.
18. Goldberg TE. On hermetic reading abilities. J Autism Dev Disord 1987; 17(1):29-44.
19. O'Connor N, Hermelin B. Two autistic savant readers. J Autism Dev Disord 1994; 24(4):501-515.
20. Etchepareborda MC, Diaz-Lucero A, Pascuale MJ et al. Asperger's syndrome, little teachers: special skills. Rev Neurol 2007; 44 Suppl 2:S43-S47.
21. Mottron L, Belleville S. Perspective production in a savant autistic draughtsman. Psychol Med 1995; 25(3):639-648.
22. Mottron L, Belleville S. A study of perceptual analysis in a high-level autistic subject with exceptional graphic abilities. Brain Cogn 1993; 23(2):279-309.
23. O'Connor N, Hermelin B. Visual memory and motor programmes: their use by idiot-savant artists and controls. Br J Psychol 1987; 78:307-323.
24. Rovet J, Krekewich K, Perlman K et al. Savant characteristics in a child with developmental delay and deletion in the short arm of chromosome 20. Dev Med Child Neurol 1995; 37(7):637-644.
25. Luszki WA. An idiot savant on the WAIS? Psychol Rep 1966; 19(2):603-609.
26. Dowker A, Hermelin B, Pring L. A savant poet. Psychol Med 1996; 26(5):913-924.
27. Brink TL. Idiot savant with unusual mechanical ability: an organic explanation. Am J Psychiatry 1980; 137(2):250-251.
28. Hoffman E, Reeves R. An idiot savant with unusual mechanical ability. Am J Psychiatry 1979; 136(5): 713-714.
29. Hermelin B, O'Connor N. The idiot savant: flawed genius or clever Hans? Psychol Med 1983; 13(3): 479-481.
30. Treffert DA. The "Acquired" Savant—"Accidental" Genius. Madison WIL: Wisconsin Medical Society, 2009.
31. Sacks O. An Anthropologist on Mars, Seven Paradoxical Tales. New York: Knopf, 1995.
32. Tammet D. Born on a Blue Day: Inside the Extraordinary Mind of an Autistic Savant. New York: Free Press, 2006.
33. Snyder A, Bahramali H, Hawker T et al. Savant-like numerosity skills revealed in normal people by magnetic pulses. Perception 2006; 35(6):837-845.
34. Young RL, Ridding MC, Morrell TL. Switching skills on by turning off part of the brain. Neurocase 2004; 10(3):215-222.
35. Snyder AW, Mulcahy E, Taylor JL et al. Savant-like skills exposed in normal people by suppressing the left fronto-temporal lobe. J Integr Neurosci 2003; 2(2):149-158.
36. Takahata K, Kato M. Neural mechanism underlying autistic savant and acquired savant syndrome. Brain Nerve 2008; 60(7):861-869.
37. Miller LK. The savant syndrome: intellectual impairment and exceptional skill. Psychol Bull 1999; 125(1):31-46.
38. Pring L, Hermelin B. Bottle, tulip and wineglass: semantic and structural picture processing by savant artists. J Child Psychol Psychiatry 1993; 34(8):1365-1385.
39. Pring L. Savant talent. Dev Med Child Neurol 2005; 47(7):500-503.
40. Gonzalez-Garrido AA, Ruiz-Sandoval JL, Gomez Valezques FR et al. Hypercalculia in savant syndrome: central executive failure? Arch Med Res 2002; 33(6):586-589.
41. Happe F. Autism: cognitive deficit or cognitive style? Trends Cogn Sci 1999; 3(6):216-222.
42. Hughes JR. Autism: the first firm finding = underconnectivity? Epilepsy Behav 2007; 11(1):20-24.
43. Ploeger A, van der Maas HL, Raijmakers ME et al. Why did the savant syndrome not spread in the population? A psychiatric example of a developmental constraint. Psychiatry Res 2009; 166(1):85-90.
44. Nurmi EL, Dowd M, Tadevosyan-Leyfer O et al. Exploratory subsetting of autism families based on savant skills improves evidence of genetic linkage to 15q11-q13. J Am Acad Child Adolesc Psychiatry 2003; 42(7):856-863.
45. Ma DQ, Jaworski J, Menold MM et al. Ordered-subset analysis of savant skills in autism for 15q11-q13. Am J Med Genet B Neuropsychiatr Genet 2005; 135B(1):38-41.
46. Casanova MF, Switala AE, Trippe J et al. Comparative minicolumnar morphometry of three distinguished scientists. Autism 2007; 11(6):557-569.
47. Mottron L, Dawson M, Soulieres I et al. Enhanced perceptual functioning in autism: an update and eight principles of autistic perception. J Autism Dev Disord 2006; 36(1):27-43.
48. Kelly SJ, Macaruso P, Sokol SM. Mental calculation in an autistic savant: a case study. J Clin Exp Neuropsychol 1997; 19(2):172-184.

49. Kehrer HE. Savant capabilities of autistic persons. Acta Paedopsychiatr 1992; 55(3):151-155.
50. O'Connor N, Hermelin B. Low intelligence and special abilities. J Child Psychol Psychiatry 1988; 29(4):391-396.
51. O'Connor N, Hermelin B. Visual and graphic abilities of the idiot savant artist. Psychol Med 1987; 17(1):79-90.
52. Heaton P, Wallace GL. Annotation: the savant syndrome. J Child Psychol Psychiatry 2004; 45(5):899-911.
53. O'Connor N, Hermelin B. Idiot savant calendrical calculators: maths or memory? Psychol Med 1984; 14(4):801-806.
54. Hermelin B, Pring L. The pictorial context dependency of savant artists: a research note. Percept Mot Skills 1998; 87:995-1001.
55. O'Connor N. The 1988 Jansson memorial lecture. The performance of the 'idiot-savant': implicit and explicit. Br J Disord Commun 1989; 24(1):1-20.
56. O'Connor N, Hermelin B. The memory structure of autistic idiot-savant mnemonists. Br J Psychol 1989; 80:97-111.
57. Chen X, Zhang M, Wang J et al. Normalization of auditory evoked potential and visual evoked potential in patients with idiot savant. Chin Med J (Engl) 1999; 112(3):246-248.
58. Bor D, Bilington J, Baron-Cohen S. Savant memory for digits in a case of synaesthesia and Asperger syndrome is related to hyperactivity in the lateral prefrontal cortex. Neurocase 2007; 13(5):311-319.
59. Hous C, Miller BL, Cummings JL et al. Autistic savants. Neuropsychiatry Neuropsychol Behav Neurol 2000; 13(1):29-38.
60. Bujas-Petkovic Z. Special talents of autistic children (autistic-savant) and their mental functions. Lijec Vjesn 1994; 116(1-2):26-29.
61. Saloviita T, Ruusila L, Ruusila U. Incidence of savant syndrome in Finland. Percept Mot Skills 2000; 91(1):120-122.

CHAPTER 26

SJOGREN-LARSSON SYNDROME

Lívia Almeida Dutra,* Pedro Braga-Neto, José Luiz Pedroso
and Orlando Graziani Povoas Barsottini

Department of Neurology and Neurosurgery, Federal University of São Paulo, São Paulo, Brazil
Corresponding Author: Livia Almeida Dutra—Email: liviaadutra@hotmail.com

Abstract: Sjogren-Larsson syndrome is a rare disease characterized by the occurrence of mental retardation, spastic diplegia and ichthyosis. The involvement of brain and skin is justified by a mutation in *FALDH* gene that affects the metabolism of fatty acids and leads to abnormal accumulation of lipids. The normal formation of multilamellar membranes in the stratum corneum and myelin is impaired. The aim of this chapter is to review the classical manifestation of the disease and its differential diagnosis.

INTRODUCTION

Sjogren-Larsson syndrome (SLS) is a rare autosomal recessive disorder characterized by ichthyosis, mental retardation and spastic diplegia or quadriplegia. Other common features include short stature, kyphoscoliosis, pigmentary degeneration of retina and abnormal or thin scalp hair.[1]

SLS was first described fifty years ago by Sjögren, in a preliminary description of patients from Northern Sweden. Larsson, in the following year, established the autosomal recessive inheritance of the disease.[2-6] Three decades after its initial description, SLS was found to be an inborn error of lipid metabolism.

Neurodegenerative Diseases, edited by Shamim I. Ahmad.
©2012 Landes Bioscience and Springer Science+Business Media.

CLINICAL MANIFESTATIONS

Skin Characteristics

The hallmark of SLS is the occurrence of ichthyosis, which is generalized in distribution especially at neck, axillary folds and lower abdomen.[7] Face tends to be spared. After birth ichthyosis is mild, often erythematous, unlike other forms of ichthyosis.[8] Patients with SLS tend to be born preterm, without a collodion membrane covering their skin.[4,9,10] During the first year of life, however, the skin presents one highly characteristic appearance with brownish yellow discoloration and a markedly wrinkle hyperkeratosis (Fig. 1).[11]

Pruritus is an accompanying disabling feature as well as thin scalp hair.[1,2,11] The presence of pruritus in SLS contrasts with other ichthyotic skin disorders, which are generally non-itching. For this reason, the presence of pruritus is recognized as an important additional symptom that suggests SLS.[11]

Patients with SLS present impairment in the sweating capacity. This is a frequent problem in patients with other types of congenital ichthyosis and may lead to heat intolerance in hot climate and during exercise.[7]

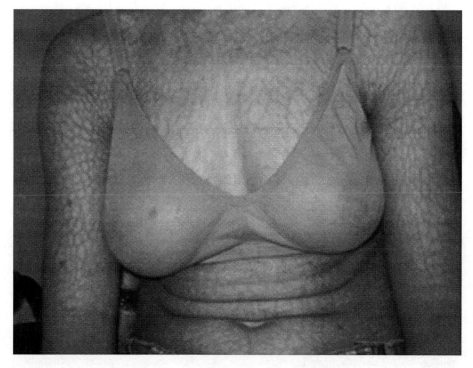

Figure 1. Ichthyosis. Reproduced from: Dutra LA et al. The Neurologist 2009; 15:332-334;[1] with permission of Lippicott Williams & Wilkins.

Neurologic Abnormalities

A slight spasticity may be seen in the first months, but clear neurologic symptoms appear after two years of life and consist of delay in reaching motor milestones due to spastic diplegia.[2,7] Motor disability with spasticity and muscular paresis were most pronounced in the legs and fairly slight in the arms.[6] Approximately one-half of the patients are non-ambulatory and most others require braces or crutches to walk.[12]

Mild, moderate or profound intellectual retardation may be seen and rare patients are normal.[12] Vocal, based on pseudobulbar dysartria, in combination with cognitive deficits, is affected in the majority of patients.[9] Seizures occur in about 40% of the patients.[2]

Ocular Manifestations

A distinctive ophthalmologic finding is the presence of retinal crystalline inclusions, so-called glistening white dots, surrounding the fovea. Not all SLS patients do not have the retinal inclusions, but their presence is a pathognomonic feature for this neurocutaneous disease. SLS has also been associated to pigmentary degeneration of retina and important photophobia.[1,2]

Radiological Findings

SLS presents demyelination of the cerebral white matter and of the corticospinal and vestibulospinal tracts. Magnetic resonance imaging (MRI) reveals abnormal high signal intensity in the periventricular white matter of the frontal lobes and at the level of the centrum semiovale and corpus callosum. Spectroscopy may reveal a lipid peak.[2] Typically, the subcortical white matter (subcortical U fibers) are spared.[1,11,13] There is evidence that myelination might be delayed in some patients. Small areas of subcortical white matter in all regions remained insufficiently myelinated even in the oldest patients in their forties.[14]

MOLECULAR INSIGHTS INTO LIPID METABOLISM

The manifestations of SLS are attributed to a deficiency of the fatty aldehyde dehydrogenase (FALDH), also known as ALDH3A2 or ALDH10, a component of the fatty alcohol oxidoreductase (FAO) complex, because of a recessive mutation in the *FALDH* gene located on the short arm of chromosome 17 (17p11.2).[1] There are more than 70 mutations described, including deletions, insertions, missense mutations, splicing defects and complex rearrangements in this gene. Missense mutations account for the largest group of mutations (38%) found in *ALDH3A2* and result in amino acid substitutions that are scattered throughout the gene.[15]

The FALDH oxidizes fatty aldehyde substrates arising from metabolism of fatty alcohols, leukotriene B4, ether glycerolipids and other potential sources such as sphingolipids. The pathogenesis of the cutaneous and neurologic symptoms is thought to result from abnormal lipid accumulation in the membranes of skin and brain.[1] Recently it has been described that the deficiency of FALDH determines impairment in lamellar body formation and secretion in the striatum corneum of the skin, which alters skin permeability, thus elucidating the occurrence of ichthyosis in SLS.[9]

The human *ALDH3A2* gene is 31 kb long and consists of 11 exons that are numbered 1-10 with an additional exon (exon 9#x2019;) situated between exons 9 and 10.[16,17] Alternative splicing of exon 9#x2019; results in the production of two transcripts, which encode protein isoforms that differ at their carboxy-terminal domains. The most abundant transcript is derived from splicing of exons 1-10 and produces a protein of 485 amino acids. A minor transcript that accounts for less than 10% of the total FALDH mRNA is produced by splicing of exon 9#x2019; between exons 9 and 10 and encodes a variant protein isoform (FALDHv) of 508 amino acids.[6]

Density gradient fractionation studies indicated that human liver FALDH is located solely in the endoplasmic reticulum. The isoform FALDH-V is localized in the peroxisome and contributes to the oxidization of pristanal, an intermediate of the alpha-oxidation pathway.[2]

Transcriptional activation of FALDH is directly regulated by peroxisomal proliferator-activated receptor α through a direct repeat-1 site, located in the FALDH promoter. In addition, FALDH is efficiently induced by linoleic acid in rat hepatoma Fao cells through transcriptional activation by peroxisomal proliferator-activated receptor α. Data suggest that the autocatalytic nature of the FALDH-N system against endoplasmic reticulum stress that is induced by polyunsaturated fatty acid; polyunsaturated fatty acid binds to peroxisomal proliferator-activated receptor α to activate the expression of FALDH-N, which then detoxifies polyunsaturated fatty acid-derived fatty aldehydes and protects cells from endoplasmic reticulum stress.[18]

DIAGNOSIS

SLS is confirmed by measuring FALDH or FAO activity in cultured fibroblasts or leukocytes, using fluorometric or gas chromatography mass spectrometry assays. Histochemical staining for FAO activity in a fresh skin biopsy is an alternative method.[1,2]

DIFFERENTIAL DIAGNOSIS

Ichthyosis might be present in other neurocutaneous syndromes. In the setting of skin and neurologic abnormalities, some differential diagnosis as IBIDS syndrome, Refsum disease, Dorfman-Chanarin syndrome, Rud syndrome and Passwell syndrome should be considered, as well carbohydrate-deficient glycoprotein syndrome Type 1, multiple sulphatase deficiency, neural lipid storage disorder and mitochondrial disorders.[1]

In the context of ichthyosis is important to verify the occurrence of gynecology and urolologic abnormalities, cardiomyopathy, polyneuropathy, dwarfism and retinitis pigmentosa because these findings might suggest other diagnosis than SLS (Table 1). One should remember the following syndromes that we briefly review.

Refsum Disease

Patients with Refsum disease present ichtyosis, progressive peripheral neuropathy, cerebellar ataxia, constriction of the visual fields and night blindness, suggesting retinitis pigmentosa. There can be present nonneurologic findings such as cardiomyopathy and

Table 1. Clinical findings in ichthyosis-related diseases

	Sjogren-Larsson	Tay	Refsum	Dorf-Chanarin	Rud	Passwell
Diplegia	+					
Deafness			+	+		
Retinitis pigmentosa			+			
Ataxia			+	+		
Peripheral neuropathy			+			
Anomalies of hair		+				
Hepatomegaly				+		
Cardiomyopathy			+	+		
Kidney tubular defects						+
Hypogonadism		+			+	+
Mental retardation	+	+	+	+	+	+
Dwarfism		+			+	+
Epilepsy	+			+	+	
Myopathy						
Cataracts		+				
Hypoplasia of subcutaneous tissue		+				

Reproduced from: Dutra LA et al. The Neurologist 2009; 15:332-334;[1] with permission of Lippicott Williams & Wilkins.

electrocardiographic abnormalities that cause sudden death and skeletal defects. It is a peroxisomal alpha-oxidation of phytanic acid or tetramethylhexadecanoic acid is defective. A gene of phytanoyl-CoA hydroxylase is localized on chromosome 10p1.[1,19,20]

Dorfman Chanarin Syndrome

In Dorfman-Chanarin syndrome, neutral lipids are deposited in the liver, spleen, muscle, skin, central nervous system and leukocytes. Also there is congenital ichthyosis. Other symptoms are myopathy, ataxia, nystagmus, sensorineural deafness, cataracts, mental retardation, cardiomyopathy and aortic insufficiency. It is inherited as an autosomal recessive trait. It shares some clinical findings with Refsum disease, such as deafness, cardiomyopathy and ataxia. It is more common in the Middle East and Mediterranean countries.[1,21]

Tay Syndrome

In this syndrome there is a congenital ichthyosis inherited as an autosomal recessive trait and characterized by an anomaly of hair growth known as trichothiodystrophy. Other features are short stature, mental retardation, delayed neuromuscular development, hypoplasia of subcutaneous fatty tissue, prematurely aged facial appearance, hypogonadism, cataracts, osteosclerosis and increased susceptibility to infections.[1,22,23]

Rud Syndrome

Patients with Rud syndrome present with ichthyosis, hypogonadism, mental retardation, epilepsy and dwarfism. Some cases suggest an autosomal recessive and

X-linked disorder. The genetic defect is suspected to involve a deletion of the sulfatase locus. The deficiency of steroid sulfatase can be demonstrated in some cases.[1,24,25]

Paswell Syndrome

In Passwell syndrome, there is congenital ichthyosis associated with neuroectodermal, gynecologic and biochemical abnormalities. In 1973, Justen Passwell et al reported a syndrome of congenital ichthyosis with cerebral atrophy, mental retardation, dwarfism, generalized aminoaciduria and infantile genitals with no secondary sexual characters, in 3 siblings of an Iraqi Jewish descent family.[10] Renal glycosuria and uricosuria were noted in some affected and unaffected members.[1,22,26,27]

TREATMENT

Historically, ichthyosis improved notably after the start of oral retinoid treatment, usually in the 1980s. For the cutaneous manifestation the use of oral acitretin and topical therapy consisting of ureia cream and lotion for hydration of the skin are widely accepted.[7] There are no treatments for the neurologic manifestation except for the use of antiepileptic drugs whenever needed and use of botulinum toxin and orthesis for spasticity.

CONCLUSION

Ichthyosis is one of the clues for diagnosis in neurology. Diagnoses other than SLS should be investigated when ichthyosis is associated with hypogonadism, ataxia, retinitis, cardiomyopathy, renal abnormalities and dwarfism.

REFERENCES

1. Dutra LA, Aquino CCH, Barsottini OGP. Sjogren-Larsson syndrome: case report and review of neurologic abnormalities and ichthyosis. The Neurologist 2009; 15:332-334.
2. Rizzo WB. Sjogren-Larsson syndrome: molecular genetics and biochemical pathogenesis of fatty aldehyde dehydrogenase deficiency. Mol Genet Metab 2007; 90:1-9.
3. Sjögren T. Oligophrenia combined with congenital ichthyosiform erythrodermia, spastic syndrome and macular-retinal degeneration. Acta Genetica 1956; 6:80-91.
4. Sjögren T, Larsson T. Oligophrenia in combination with congenital ichthyosis and spastic disorders. Acta Psychiatr Neurol Scand 1957; 32(Suppl 113):1-113.
5. Theile U. Sjögren-Larsson syndrome. Oligophrenia—ichthyosis—di-tetraplegia. Humangenetik 1974; 22:91-118.
6. Haug S, Braun-Falco M. Restoration of fatty aldehyde dehydrogenase deficiency in Sjögren-Larsson syndrome. Getie Ther 2006; 13:1021-1026.
7. Rizzo WB, Carney G, Lin Z. The molecular basis of Sjogren-Larsson syndrome: mutation analysis of the fatty aldehyde dehydrogenase gene. Am J Hum Genet 1999; 65:1547-1560.
8. Liden S, Jagell S. The Sjogren-Larsson syndrome. Int J Dermatol 1984; 23:247-253.
9. Rizzo WB, S'Aulis D, Jennings MA et al. Ichthyosis in Sjogren-Larsson syndrome reflects defective barrier function due to abnormal lamellar body structure and secretion. Arch Dermatol Res 2010; 302(6):443-451.
10. Willemsen MA, Rotteveel JJ, van Domburg PH et al. Preterm birth in Sjögren-Larsson syndrome. Neuropediatrics 1999; 30:325-327.
11. Willemsen MA, IJlst L, Steijlen PM et al. Clinical, biochemical and molecular genetic characteristics of 19 patients with the Sjogren-Larsson syndrome. Brain 2001; 124:1426-1434.

12. Jagell S, Heijbel J. Sjögren-Larsson syndrome: physical and neurological features. A survey of 35 patients. Helv Paediatr Acta 1982; 37:519-530.
13. Van Mieghem F, Van Goethem JW, Parizel PM et al. MR of brain in Sjogren-Larsson syndrome. Am J Neuroradiol 1997; 18:1561-1563.
14. Willemsen MAAP, van der Graaf M, van der Knaap MS et al. MR imaging and proton MR spectroscopic studies in Sjogren-Larsson syndrome: characterization of the leukoencephalopathy. AJNR 2004; 25:649-657.
15. Rizzo WB, Carney G. Sjögren-Larsson syndrome: diversity of mutations and polymorphisms in the fatty aldehyde dehydrogenase gene (ALDH3A2). Hum Mutat 2005; 26(1):1-10.
16. Rogers GR, Markova NG, De VL et al. Genomic organization and expression of the human fatty aldehyde dehydrogenase gene (FALDH). Genomics 1997; 39:127-135.
17. Chang C, Yoshida A. Human fatty aldehyde dehydrogenase gene (ALDH10): organization and tissue-dependent expression. Genomics 1997; 40:80-85.
18. Ashibe B, Motojima K. Fatty aldehyde dehydrogenase is up-regulated by polyunsaturated fatty acid via peroxisome proliferator-activated receptor alpha and suppresses polyunsaturated fatty acid-induced endoplasmic reticulum stress. FEBS J 2009; 276(23):6956-6970.
19. Cakirer S, Savas MR. Infantile refsum disease: serial evaluation with MRI. Pediatr Radiol 2005; 35:212-215.
20. Wanders RJ, Jansen GA, Skjeldal OH. Refsum disease, peroxisomes and phytanic acid oxidation: a review. J Neuropathol Exp Neurol 2001; 60:1021-1031.
21. Düzovali O, Ikizoglu G, Turhan AH et al. Dorfman-Chanarin syndrome: a case with hyperlipidemia. Turk J Pediatr 2006; 48:263-265.
22. Jorizzo JI, Crounse RG, Wheeler CE Jr. Lamellar ichthyosis, dwarfism, mental retardation and hair shaft abnormalities. J Am Acad Dermatol 1980; 2:309-317.
23. Happle R, Traupe H, Gröbe H et al. The Tay syndrome (congenital ichthyosis with trichothiodystrophy). Eur J Pediatr 1984; 141:147-152.
24. Marxmiller J, Trenkle I, Ashwal S. Rud syndrome revisited: ichthyosis, mental retardation, epilepsy and hypogonadism. Dev Med Child Neurol 1985; 27:335-343.
25. Andria G, Ballabio A, Parenti G et al. Steroid sulphatase deficiency is present in patients with the syndrome 'ichthyosis and male hypogonadism' and with 'Rud syndrome'. J Inherit Metab Dis 1984; 7 (Suppl 2):159-160.
26. Passwell J, Ziprkowskin L, Katzelson D et al. A syndrome characterized by congenital ichthyosis with atrophy, mental retardation, dwarfism and generalized amino aciduria. J Pediatr 1973; 82:466-471.
27. Muhammed K, Safia B. Passwell syndrome. Indian J Dermatol Venereol Leprol 2003; 69:180-181.
28. Gånemo A, Jagell S, Vahlquist A. Sjögren-Larsson syndrome: a study of clinical symptoms and dermatological treatment in 34 Swedish patients. Acta Derm Venereol 2009; 89(1):68-73.

CHAPTER 27

THE SPINOCEREBELLAR ATAXIAS:
Clinical Aspects and Molecular Genetics

Antoni Matilla-Dueñas,*,1 Marc Corral-Juan,1 Victor Volpini2
and Ivelisse Sanchez1

1Basic, Translational and Neurogenetics Research Unit, Department of Neurosciences, Health Sciences Research
Institute Germans Trias i Pujol (IGTP), Universitat Autònoma de Barcelona, Barcelona, Spain;
2Molecular Diagnosis Center of Inherited Diseases, Institut d'Investigacions Biomèdiques de Bellvitge (IDIBELL),
L'Hospitalet de Llobregat, Barcelona, Spain
*Corresponding Author: Antoni Matilla-Dueñas—Email: amatilla@igtp.cat; amatilla@neurodeg.net

Abstract: Spinocerebellar ataxias (SCAs) are a highly heterogeneous group of inherited
neurological disorders, based on clinical characterization alone with variable degrees
of cerebellar ataxia often accompanied by additional cerebellar and noncerebellar
symptoms which in most cases defy differentiation. Molecular causative deficits in
at least 31 genes underlie the clinical symptoms in the SCAs by triggering cerebellar
and, very frequently, brain stem dysfunction. The identification of the causative
molecular deficits enables the molecular diagnosis of the different SCA subtypes
and facilitates genetic counselling. Recent scientific advances are shedding light into
developing therapeutic strategies. The scope of this chapter is to provide updated
details of the spinocerebellar ataxias with particular emphasis on those aspects
aimed at facilitating the clinical and genetic diagnoses.

INTRODUCTION

Ataxia is a neurological disorder characterised by loss of control of body movements.
Patients suffering from ataxia are clumsy and unable to walk steadily, have slurred speech
and eventually lose the ability to swallow and breathe smoothly. Ataxia results from
variable degeneration of neurons in the cerebellar cortex, brain stem, spinocerebellar
tracts and their afferent/efferent connections. Such neurodegeneration can result from
multiple sclerosis, brain tumour, alcoholism, or an inherited genetic defect. There are
over 60 different types of inherited ataxias striking during childhood or adulthood.

Neurodegenerative Diseases, edited by Shamim I. Ahmad.
©2012 Landes Bioscience and Springer Science+Business Media.

Differential diagnosis of hereditary ataxia includes acquired, nongenetic causes of ataxia, such as alcoholism, vitamin deficiencies, multiple sclerosis, vascular disease, primary or metastatic tumours, or paraneoplastic diseases associated with occult carcinoma of the ovary, breast, or lung. The possibility of an acquired cause of ataxia needs to be considered in each individual with ataxia because a specific treatment may be available. The hereditary ataxias have been extensively reviewed,[1-5] and a comprehensive description can be found online.[6,7] The hereditary ataxias are usually subdivided by their mode of inheritance (i.e., autosomal dominant, autosomal recessive, X-linked and mitochondrial) and the causative gene or chromosomal locus. The term "spinocerebellar ataxias" is commonly used for those inherited ataxias presenting an autosomal dominant inheritance. Synonyms for spinocerebellar ataxias (SCAs) used prior to the identification of the molecular genetic basis of these disorders are Marie's ataxia, inherited olivopontocerebellar atrophy, cerebello-olivary atrophy, autosomal dominant cerebellar ataxias (ADCAs) or the more generic term, spinocerebellar degenerations.

In all SCAs, ataxia is the predominant clinical manifestation which patients frequently present with additional non-ataxia symptoms. Several scales have been developed to measure the severity of ataxia. Among them, the Scale for the Assessment and Rating of Ataxia (SARA) has been recently developed and validated and has proven useful to measure ataxia severity by assigning scores ranging from 0 to 40 with 0 indicating absence of ataxia and 40 the most severe degree of ataxia.[8] Non-ataxia symptoms have been conveniently assessed with the Inventory of Non-Ataxia Symptoms (INAS).[9] INAS consists of 30 items, each of which is related to one of the following 16 symptoms or syndromes: areflexia, hyperreflexia, extensor plantar response, spasticity, paresis, amyotrophy, fasciculations, myoclonus, rigidity, chorea, dystonia, resting tremor, sensory symptoms, brainstem oculomotor signs (horizontal and vertical ophthalmoparesis, slowing of saccades), urinary dysfunction and cognitive impairment.[10] These recently developed rating scales and SCA functional indexes are proving very valuable to validate neurological assessment methods and therapeutic interventions by measuring the severity of the ataxia and non-ataxia symptoms in ongoing and future clinical trials in SCA patients.

The treatment of hereditary ataxia is currently primarily supportive and is the scope of the following chapter. With very few exceptions (e.g., ataxia associated with vitamin E deficiency), there are no disease-modifying therapies. Despite the lack of disease-modifying treatments, obtaining an accurate diagnosis of the specific hereditary ataxia subtype is of great value. Benefits include determining prognosis, facilitating family counselling, improving research access and providing some psychological benefit in ending the often frustrating search for an accurate aetiology. Spinocerebellar ataxias may have certain clinical features that respond very well to symptomatic medical therapy. Parkinsonism, dystonia, spasticity, urinary urgency, sleep pathology, fatigue and depression are all common in many of the ataxia subtypes and very often respond to pharmacologic intervention as in other diseases. Much of the clinical interaction between the neurologist and ataxia patients should focus on identifying and treating these symptoms. Treatment of the core clinical feature of these diseases—ataxia—is predominantly rehabilitative. The value of good physical therapy far exceeds any potential benefit from medications that a physician might prescribe to improve balance and coordination. Speech and swallowing are often affected. In more severe cases, aspiration risk can be very significant and life-threatening. Routine monitoring of swallowing by speech therapists, often including modified barium swallowing tests, is indicated in most patients. Recently there have been very encouraging advances in clinical ataxia research. Collaborative study groups throughout the world have developed and validated ataxia rating scales and instrumented outcome measures and have begun to rigorously define the natural history of these diseases, thus laying the foundation for well-designed clinical trials. The promise of disease-modifying treatments is closer than ever.

The scope of this chapter is to provide an updated summary of the spinocerebellar ataxias with particular emphasis on those aspects aimed at facilitating the clinical and genetic diagnoses. Laboratories offering molecular genetic testing for SCAs are recommended to refer to and be acquainted with the OECD (Organization for Economic Co-operation and Development) Guidelines and recommendations for Quality Assurance of Molecular Genetic Testing, 2007, as well as with EMQN (European Molecular Quality Genetics Network) reporting and internal quality control guidelines.[11,12]

SPINOCEREBELLAR ATAXIA TYPE 1 (SCA1) [MIM #164400]

Clinical Features

SCA1 is characterised by progressive cerebellar ataxia, dysarthria and eventual deterioration of bulbar functions including atrophy of facial and masticatory muscles, perioral fasciculations and severe dysphagia leading to frequent aspiration.[13] Onset is typically in the third or fourth decade, although childhood onset has been reported.[14] Interval from onset to death varies from ten to 30 years and individuals with juvenile onset present a more rapid progression and a more severe disease and die before age 16 years. Early in the disease, affected individuals may have gait disturbance, slurred speech, difficulty with balancing, brisk deep tendon reflexes, hypermetric saccades, nystagmus and mild dysphagia.[14] Later symptoms include slowing of saccadic velocity, development of up-gaze palsy, dysmetria, dysdiadochokinesia and hypotonia. In advanced stages, muscle atrophy, decreased deep tendon reflexes, loss of proprioception, cognitive impairment, chorea, dystonia and bulbar dysfunction are seen. Executive dysfunction may also occur. Visual evoked potentials and motor evoked potentials following transcranial magnetic stimulation are abnormal in most individuals with SCA1. CT and MRI of the brain reveal atrophy of the brachia pontis and anterior lobe of the cerebellum and enlargement of the fourth ventricle. Neuropathologic studies reveal cerebellar atrophy with definite loss of Purkinje cells and dentate nucleus neurons, eosinophilic spheres or "torpedoes" in the axons of degenerating Purkinje cells and severe neuronal degeneration in the inferior olive. Additional features include mild neuronal loss in cranial nerve nuclei III, IV, IX, X and XII and demyelination of the restiform body and brachium conjunctivum, dorsal and ventral spinocerebellar tracts and to a lesser degree the posterior columns. Anticipation, an increase in the severity and earlier onset of the phenotype in progressive generations, is observed. SCA1 represents approximately 6% of individuals with autosomal dominant cerebellar ataxia, although this figure varies considerably based on geographic location and ethnic background.

Genetics of SCA1

Expansion of the CAG repeat in the *ATXN1* gene is the mutational mechanism in all families with SCA1 examined to date.[15–18] Normal alleles range 6-44 CAG repeats and the repeat configuration in alleles with 21 or more repeats is interrupted by 1-3 CAT trinucleotides, whereas disease-causing alleles show a perfectly uninterrupted CAG repeat configuration. Distinguishing normal interrupted alleles from mutable normal uninterrupted alleles in the 36-44 repeat range requires additional evaluation by Sfa NI restriction analysis. Mutable normal (intermediate) alleles range from 36-38 CAG repeats without CAT interruptions. Mutable normal alleles have not been associated with symptoms, but can expand into the abnormal range on transmission to offspring. Penetrance is considered to be greater than 95%, but is age dependent. A woman with 44 CAG repeats with CAT repeat interruptions had an affected father but was herself asymptomatic until age 66 years;[19] thus, she may have reduced penetrance. Full penetrance alleles range from 39-91 CAGs. An allele with 39 CAG repeats without the CAT repeat interruptions has the lowest number of repeats to be associated with symptoms.[20] Thus, alleles of 39-44 uninterrupted CAG repeats are considered abnormal and are likely associated with symptoms.

SPINOCEREBELLAR ATAXIA TYPE 2 (SCA2) [MIM #183090]

Clinical Features

SCA2 is characterised by slowly progressive ataxia and dysarthria associated with the ocular findings of nystagmus, slow saccadic eye movements and in some individuals, ophthalmoparesis can be seen.[21,22] Pyramidal findings are present. Tendon reflexes are brisk during the first years of life, but absent later. Mean age of onset is typically in the fourth decade with a ten- to 15-year disease duration. The disease progresses more rapidly when onset occurs before the age of 20 years. In the original study from Cuba, the earliest symptoms included gait ataxia often accompanied by leg cramps.[23] More than 50% of affected individuals developed a kinetic or postural tremor, decreased muscle tone, decreased tendon reflexes and abnormal eye movements with slowed saccades progressing to supranuclear ophthalmoplegia. Detailed analyses of the eye movement abnormalities have been reported.[24] It is difficult and often impossible to distinguish SCA2 from the other hereditary ataxias and the diagnosis of SCA2 rests upon the use of molecular genetic testing to detect an abnormal CAG trinucleotide repeat expansion in the ATXN2 gene.

Post-mortem examinations have been reported in the Holguin population of Cuba and a marked reduction in the number of cerebellar Purkinje cells was observed.[23] In silver preparations, Purkinje cell dendrites had poor arborisation and torpedo-like formation of their axons as they passed through the granular layer. Parallel fibers were scanty. Granule cells were decreased in number, whereas Golgi and basket cells were well preserved, as were neurons in the dentate and other cerebellar nuclei. In the brainstem, marked neuronal loss in the inferior olive, oculomotor nuclei and pontocerebellar nuclei was observed. Six of seven brains also had marked loss in the substantia nigra. In five spinal cords that were available for analysis, marked demyelination was present in the posterior columns and to a lesser degree in the spinocerebellar tracts. Motor neurons and neurons in Clarke's column were reduced in size and number. In the lumbar and sacral segments, anterior and posterior roots were partially demyelinated. Degeneration in the thalamus (sensory thalamic nuclei) and reticulotegmental nucleus of the pons has also been reported.[22] In addition, Orozco et al[23] noted severe gyral atrophy, most prominent in the frontotemporal lobes. The cerebral cortex was thinned, but without neuronal rarefaction. The cerebral white matter was atrophic and gliotic. Degeneration in the nigro-luyso-pallidal system mainly involved the substantia nigra. One brain showed patchy loss in parts of the third nerve nuclei. A case with white matter pathology has been described.[25] Nerve biopsy has shown moderate loss of large myelinated fibers.[26]

Genetics of SCA2

ATXN2 is the gene known to be associated with SCA2. The presence of one abnormal allele is diagnostic. Normal alleles contain 31 or fewer CAG repeats. Mutable normal alleles, previously called intermediate alleles, are not associated with clinical manifestations, but can be meiotically unstable, resulting in expansion of the allele size on transmission to offspring. Full penetrant alleles contain 32 or more CAG repeats without CAA interruption. Alleles of 32 and 33 CAG repeats are considered "late onset" (after age 50 years). The most common disease-causing alleles are 37 to 39 repeats. Extreme CAG repeat expansion (>200) has been reported.

SPINOCEREBELLAR ATAXIA TYPE 3 (SCA3) [MIM #109150]

SCA3, also known as Machado-Joseph disease, is reviewed in Chapter 14, as *Machado-Joseph Disease and Other Rare Autosomal Dominant Spinocerebellar Ataxias*, in this volume.

SPINOCEREBELLAR ATAXIA TYPE 4 (SCA4) [MIM #600223]

Clinical Features

SCA4 has been associated with a specific SCA subtype presenting ataxia with sensory axonal neuropathy.[27,28] Disease onset is typically in the fourth or fifth decade, but age at onset ranges from 19 to 59 years, with a median age at onset of 39 years. The earliest symptom is usually gait disturbance, followed by difficulty with fine motor tasks and often by dysarthria. All patients have vibratory and joint position sense loss and 95% have at least a minimal pinprick sensation loss. All patients also have absent ankle-jerk reflexes but knee-jerk reflexes are absent in 85% and complete areflexia is seen in 25%. In a SCA4 German family the mean age at onset was 38.3 years (ranging 20 to 61).[28] All patients have cerebellar ataxia with limb dysmetria, dysarthria and cerebellar atrophy, as well as sensory neuropathy with hypo- or areflexia, decreased sensation and absent sural sensory nerve action potentials. Post-mortem examination of a SCA4 German patient showed demyelination of cerebellar and brainstem fibre tracts, widespread cerebellar and brainstem neurodegeneration with marked neuronal loss in the substantia nigra and ventral tegmental area, central raphe and pontine nuclei, all auditory brainstem nuclei, in the abducens, principal trigeminal, spinal trigeminal, facial, superior vestibular, medial vestibular, interstitial vestibular, dorsal motor vagal, hypoglossal and prepositus hypoglossal nuclei, as well as in the nucleus raphe interpositus, all dorsal column nuclei and in the principal and medial subnuclei of the inferior olive.[29] Severe neuronal loss was seen in the Purkinje cell layer of the cerebellum, in the cerebellar fastigial nucleus, in the red, trochlear, lateral vestibular and lateral reticular nuclei, the reticulotegmental nucleus of the pons and the nucleus of Roller.

Genetics of SCA4

The *SCA4* gene has been mapped to chromosome 16q22.1 by linkage analysis studies.[27,28] The molecular defect remains to be identified.

SPINOCEREBELLAR ATAXIA TYPE 5 (SCA5) [MIM# 600224]

Clinical Features

SCA5 is a slowly progressive cerebellar syndrome characterized by gait and limb ataxia, dysarthria and uncoordinated eye movements and arises from dysfunction and degeneration of the cerebellum.[30] The common clinical feature is down-beat nystagmus, gait, stance, limb ataxia, dysarthria, intention and resting tremors, impaired smooth pursuit and gaze-evoked nystagmus. MRI shows atrophy of the cerebellar vermis and hemispheres. Onset usually is in the third decade (range: 14 to 50 years).

Genetics of SCA5

Different mutations in the gene encoding β-III spectrin (*SPTBN2*) were identified as the genetic cause of SCA5 in three independent families.[31] These include a 39-bp deletion in exon 12 of the SPTBN2 gene caused an in-frame 13-amino acid deletion (E532_M544del) within the third of the 17 spectrin repeats. This mutation is detectable by PCR. In a French family, SCA5 is related to a 15-bp deletion in exon 14 (1886_1900del; L629_R634delinsW) in the same spectrin repeat.[31] With the exception of an insertion of a tryptophan, this deletion did not disrupt the remainder of the open reading frame. In a German family, SCA5 spinocerebellar ataxia is related to a T-to-C transition (758T-C) in exon 7 of the *SPTBN2* gene that caused a leucine-to-proline change (L253P) in the calponin homology domain containing the actin/ARP1-binding site.[31] There is still controversy whether Abraham Lincoln's unusual gait was due to early midline cerebellar dysfunction caused by *SPTBN2* mutations or by orthopedic or myopathic features.

SPINOCEREBELLAR ATAXIA TYPE 6 (SCA6) [MIM #183086]

Clinical features

SCA6 is characterised by adult-onset, slowly progressive cerebellar ataxia, dysarthria and nystagmus.[32] The range in age of onset is from 19 to 71 years. The mean age of onset is between 43 and 52 years. Age of onset and clinical picture vary even within the same family; sibs with the same size full-penetrance allele may differ in age of onset by as much as 12 years, or exhibit, at least initially, an episodic course. Initial symptoms are gait unsteadiness, stumbling and imbalance in approximately 90% of individuals; the remainder present with dysarthria. Symptoms progress slowly and eventually all persons have gait ataxia, upper-limb incoordination, intention tremor and dysarthria. Dysphagia and choking are common. Diplopia occurs in approximately 50%

of individuals. Others experience visual disturbances related to difficulty fixating on moving objects, as well as horizontal gaze-evoked nystagmus (70%-100%) and vertical nystagmus (65%-83%), which is observed in fewer than 10% of those with other forms of SCA.[33] Other eye movement abnormalities, including periodic alternating nystagmus and rebound nystagmus, have also been described.[34] Hyperreflexia and extensor plantar responses occur in up to 40%-50% of individuals with SCA6. Basal ganglia signs, such as dystonia and blepharospasm, are noted in up to 25% of individuals. Mentation is generally preserved. Individuals with SCA6 do not have significant cognitive deficits,[35] sensory complaints, restless legs, stiffness, migraine, primary visual disturbances, or muscle atrophy. Life span is not shortened.

Neuropathologic studies in individuals with SCA6 have demonstrated either selective Purkinje cell degeneration or a combined degeneration of Purkinje cells and granule cells.[36,37] Selective floccular atrophy is associated with impaired pursuit and gaze-holding abnormalities in SCA6.[38] Two unrelated cases that presented with Parkinsonism and cerebellar ataxia were attributable to nigral loss and dopaminergic dysfunction.[39]

Genetics of SCA6

CACNA1A is the only gene known to be associated with SCA6. The diagnosis of SCA6 relies on the use of molecular genetic testing to detect an abnormal CAG trinucleotide repeat expansion in *CACNA1A*. Normal alleles contain 18 or fewer CAG repeats. Affected individuals have 20 to 33 CAG repeats. Although the age of onset of symptoms of SCA6 correlates inversely with the length of the expanded CAG repeat, the same broad range of onset has been noted for individuals with 22 CAG repeats, the most common disease-associated allele. In the few individuals with $(CAG)_{30}$ or $(CAG)_{33}$, onset has been later than in individuals with $(CAG)_{22}$ and $(CAG)_{23}$. A recent retrospective study showed even closer correlation of age of onset with the sum of the two allele sizes.[40] Several individuals who are homozygous for an abnormal expansion in the *CACNA1A* gene have been reported. Anticipation has not been detected in SCA6. Penetrance is nearly 100%, although symptoms may not appear until the seventh decade.

SPINOCEREBELLAR ATAXIA TYPE 7 (SCA7) [MIM #164500]

Clinical Features

SCA7 is unique among CAG/polyglutamine (polyQ) repeat diseases due to dramatic intergenerational instability in repeat length and an associated cone-rod dystrophy retinal degeneration phenotype.[41,42] It is characterised by progressive cerebellar ataxia, including dysarthria and dysphagia, dysmetria and dysdiadochokinesia and progressive central visual loss resulting in blindness in affected adults. The phenotype ranges from onset in infancy with an accelerated course and early death to onset in the fifth or occasionally sixth decade with slowly progressive retinal degeneration and cerebellar ataxia. In infancy or early childhood, ataxia may not be obvious but muscle wasting, weakness and hypotonia are common. Two infants with severe disease and expansions of more than 200 and 306 CAG repeats had neonatal hypotonia, developmental delay, poor feeding, dysphagia, congestive heart failure, cerebral and cerebellar atrophy and retinal disease,[43,44] and one of these child has died in infancy with 180 CAG repeats.[45]

Retinal degeneration in adults is characterised by abnormalities in colour vision and central visual acuity, often presenting in the late teens or early 20's before the onset of cerebellar findings. The retinal degeneration is a progressive, cone-rod dystrophy that results in blindness. During the earliest stages of retinal degeneration, adults may have no symptoms, but may have subtle granular changes in the macula and make errors in the tritan (blue-yellow) axis on detailed colour vision testing using the Farnsworth dichromatous (D15) test. Electroretinogram (ERG) shows a decrease in the photopic (cone) response initially, followed by a decrease in the scotopic (rod) response. As cone function decreases over time, central visual acuity decreases to the 20/200 (legally blind) range, more prominent macular changes appear and all colour discrimination is lost. Eventually, blindness is total. When initial symptoms occur at or before adolescence, blindness can occur within a few years. Individuals showing symptoms in their teens may be blind within a decade or less. In infantile onset, the cerebellar and brainstem degeneration is so rapid that retinal degeneration and related vision loss may not be evident. Progressive cerebellar ataxia in adults (i.e., dysmetria, dysdiadochokinesia and poor coordination) may precede, but usually follow, the onset of visual symptoms. While the rate of progression varies, the eventual result is severe dysarthria, dysphagia and a bedridden state with loss of motor control. Brisk tendon reflexes and spasticity become evident as the disease progresses. Ocular saccades may become markedly slowed. Cognitive decline and psychosis are not common, but have been reported.[44]

Neuronal loss and gliosis are observed in the cerebellum (especially Purkinje cells), inferior olivary and dentate nucleus and pontine nuclei and to a lesser extent in the globus pallidus, substantia nigra and red nucleus. Degeneration is evident in the posterior columns and spinocebellar tracks of the spinal cord. Degeneration of photoreceptors and bipolar and granular cells is evident in the retina, especially in the foveal and parafoveal regions.

Genetics of SCA7

ATXN7 is the only gene associated with SCA7. The mutation consists of a highly polymorphic CAG repeat. Normal alleles display 19 or fewer CAG repeats. To date, no normal allele with greater than 19 CAG repeats has been reported. Approximately 75% of normal alleles have ten CAG repeats. Mutable normal alleles span from 28 to 33 CAG repeats. Previously called "intermediate alleles," mutable normal alleles are meiotically unstable and not convincingly associated with a phenotype. Because of the instability of alleles in the mutable normal range, an asymptomatic individual with a mutable normal allele may be predisposed to having a child with an expanded allele.[46] Reduced penetrance alleles spanning 34-36 CAG repeats may be provisionally defined as alleles with reduced penetrance. When present in an individual with a reduced penetrance allele, symptoms are more likely to be later in onset and milder than average. Full penetrant alleles range from 36 to 460 CAG repeats.

SPINOCEREBELLAR ATAXIA TYPE 8 (SCA8) [MIM #60876]

Clinical Features

SCA8 is suspected in individuals with slowly progressing cerebellar ataxia with disease onset typically occurring in adulthood, scanning dysarthria characterised by a drawn-out slowness of speech, marked truncal instability, hyperactive tendon reflexes, cognitive deficits, family history of ataxia consistent with single occurrence in the family or either an autosomal recessive or autosomal dominant pattern of inheritance.[47,48] Ataxic symptoms of the lower extremities appear to be more pronounced than those of upper extremities. Onset ranges from age one to 65 years. The progression is typically over decades regardless of the age of onset. Some individuals present with nystagmus, dysmetric saccades and, rarely, ophthalmoplegia. Life span is typically not shortened. Because of the reduced penetrance, a single occurrence in a family is the most common presentation of the disease. MRI and CT Scan have consistently shown cerebellar atrophy, specifically in the cerebellar hemisphere and vermis in individuals with SCA8.

Genetics of SCA8

The expansion mutation associated with the SCA8 phenotype is located in both the 3' untranslated region of the *ATXN8OS* gene and a short polyglutamine ORF in a newly identified overlapping *ATXN8* gene. The expansion mutation associated with the SCA8 phenotype involves two overlapping genes: in the CTG direction, *ATXN8OS* (formerly known as *SCA8*) expresses transcripts containing the CUG expansion.[48] In the CAG direction, *ATXN8* expresses the mutation and encodes a nearly pure polyglutamine expansion protein. The CTG×CAG repeat is adjacent to a CTA×TAG repeat that is highly polymorphic but stable when transmitted from one generation to the next. The reduced penetrance of this mutation and the presence of the polymorphic CTA×TAG repeat makes it difficult to determine the pathogenic size range of the CTG×CAG repeat and important caveats for using it as a molecular diagnostic tool. Normal alleles display a range of 15 to 50 combined $(CTA \times TAG)_n (CTG \times CAG)_n$ repeats. It is not yet clear whether repeat sizes ranging from 50 to 70 repeats are pathogenic. Reduced penetrance is found for $(CTA \times TAG)_n (CTG \times CAG)_n$ repeats of all sizes.[49] Although the length of the repeat tract does not correlate with the age of onset, severity, symptoms, or disease progression, within the 75 to 250 repeat range, affected individuals tend to have longer repeat tracts than asymptomatic individuals. Individuals with $(CTA \times TAG)_n (CTG \times CAG)_n$ repeat sizes ranging from 71 to 80 repeats have been reported to have ataxia,[47,50] however, in some families this range appears to be less likely to be associated with ataxia than repeat sizes of over 100 repeats. Higher penetrance allele size, between 80 to 250 $(CTA \times TAG_n (CTG \times CAG)_n$ repeats are most often seen in individuals with ataxia; however, repeat sizes ranging from 71 to more than 1300 repeats have been found both in individuals who develop ataxia and in those who do not. A few expanded alleles have been identified in healthy individuals and in patients presenting Parkinsonism, multiple system atrophy and severe childhood onset.[51-56] These findings question the usefulness of the molecular diagnosis of the expanded allele in SCA8.

SPINOCEREBELLAR ATAXIA TYPE 10 (SCA10) [MIM #603516]

Clinical Features

SCA10 has been identified in Mexican, Mexican-American and South American (Brazilian and Argentinean) patients presenting slowly progressive cerebellar ataxia that usually starts as poor balance and unsteady gait, followed by upper-limb ataxia, scanning dysarthria and dysphagia.[57,58] Abnormal tracking eye movements are common. Recurrent seizures after the onset of gait ataxia have been reported in 20% to 100% of affected individuals. Some individuals have cognitive dysfunction, behavioural disturbances, mood disorders, mild pyramidal signs and peripheral neuropathy. Onset ranges from age 12 to 48 years. Progressive pan-cerebellar atrophy with preservation of the cerebrum and brain stem is observed and evidence of cortical dysfunction with or without focal epileptiform discharges on interictal electroencephalography in some affected individuals.

Genetics of SCA10

Diagnosis of SCA10 is based on clinical findings and confirmed by molecular genetic testing to detect an abnormal ATTCT pentanucleotide repeat expansion in intron 9 of the *ATXN10* gene, the only gene currently known to be associated with the disorder. Affected individuals have expanded alleles with the number of repeats up to 4,500 ATTCT pentanucleotide repeats, although intermediate alleles (280 to 850 repeats) may show reduced penetrance. Clinically available molecular genetic testing can detect approximately 100% of affected individuals.

SPINOCEREBELLAR ATAXIA TYPE 11 (SCA11) [MIM #604432]

Clinical Features

SCA11 is a relatively benign, late-onset, slowly progressive neurological disorder characterized by an uncomplicated cerebellar syndrome characterized by progressive cerebellar ataxia and abnormal eye signs including jerky pursuit, horizontal and vertical nystagmus.[59] Pyramidal features, peripheral neuropathy and dystonia are seen on occasion. Only two families from Britain have been reported to date.[60,61] In them, age of onset ranged from the early teens to the mid 20s. Life span is normal. Brain MRI shows mild to severe cerebellar atrophy. Neuropathologic examination of the brain of one affected individual showed marked cerebellar and brain stem loss with Purkinje cell degeneration and abnormal tau deposition in the brain stem and cortex.

Genetics of SCA11

SCA11 is caused by an adenosine insertion in exon 13 or a frameshift deletion of 2 bases (GA) both in the gene encoding tau tubulin kinase-2 (*TTBK2*).[61]

SPINOCEREBELLAR ATAXIA TYPE 12 (SCA12) [MIM #604326]

Clinical Features

SCA12 is characterised by onset of action tremor of the upper extremities in the fourth decade, slowly progressing to include ataxia and other cerebellar and cortical signs.[62] Given the small number of individuals known to have SCA12, it is possible that other clinical manifestations remain to be identified. Initially, the diagnosis of SCA12 was considered in individuals of Indian descent developing action tremor of the upper extremities in mid-life and later develop a wide range of findings, including mild cerebellar dysfunction, hyperreflexia and parkinsonian features.[63,64] More recently, SCA12 has been extended to individuals who are not Indian, but have a similar clinical presentation that may also include: a mild gait abnormality and more prominent parkinsonian features, psychiatric disorders and dementia in some of the oldest individuals.[65]

Neuroimaging reveals cerebral and cerebellar atrophy without evidence of focal lesions and in many SCA12 individuals the cerebral cortex is more atrophic than the cerebellar cortex.[63,64,66] Atrophy of the cerebellar vermis is more prominent than atrophy of the cerebellar hemispheres. Basal ganglia, thalamus, brain stem nuclei and other subcortical brain regions are relatively spared.

Genetics of SCA12

The *PPP2R2B* gene associates with SCA12. The pathological allele is a CAG expansion mutation located within or just upstream from exon 7. The repeat appears to fall in a functional promoter for one of the multiple *PPP2R2B* splice variants. Normal alleles display 4-32 CAG repeats. The most common repeat length in all samples studied to date is ten triplets. The expansion of 51 or more CAG triplets within the *PPP2R2B* gene is diagnostic, however, the boundary between the normal and mutant allele size remains to be determined and because the incomplete penetrance, diagnosis of borderline alleles should be taken cautiously. Two Northern Germans with ataxia had CAG repeats of 40 and 41 triplets, but it is unclear if their ataxia was related to the expansions.[67] A 28-year-old unaffected individual from India with no known family history of degenerative neurologic disorder had a CAG repeat of 45 triplets, a finding of uncertain significance.[68] An individual with typical Creutzfeld-Jacob disease (CJD) had an SCA12 allele of 49 triplets.[67] A 52-year-old woman from India with an allele of 62 CAG triplets did not show symptoms. Full penetrance alleles contain 51-78 CAG triplets. Repeats in the expanded range appear to be modestly unstable, with length variations of a few triplets among members of the same family. Of note, an Iranian woman with unipolar depression and her monozygotic twin sons with schizophrenia all had a *PPP2R2B* allele with a 53-CAG triplet repeat.[69] It is unknown if the expansion is causally related to the psychiatric disorders in this family, or if the family had undiagnosed features of SCA12.

SPINOCEREBELLAR ATAXIA TYPE 13 (SCA13) [MIM #605259]

Clinical Features

The phenotype of SCA13 overlaps with other infantile or adult-onset ataxias and varies from slowly progressive childhood-onset cerebellar gait ataxia associated with cerebellar dysarthria and often accompanied by moderate mental retardation (IQ 62-76), mild developmental delays in motor acquisition and occasional seizures to adult-onset progressive ataxia.[70] Nystagmus and pyramidal signs are also observed in some patients. Life span is not shortened and many persons live beyond the age 70 years. Brain MRIs in all affected persons have shown mild to moderately severe cerebellar, primarily midline and pontine atrophy. Atrophy of the brain stem or cerebral cortex is not observed.

Genetics of SCA13

The correct diagnosis of SCA13 can only be established by molecular genetic testing of the *KCNC3* gene. All known mutations cluster in exon 2; the p.R420H mutation associates with adult-onset progressive ataxia in one family[71] and the p.F448L mutation is associated with childhood onset ataxia and often mental retardation and seizures. The mutations appear to present complete penetrance.

SPINOCEREBELLAR ATAXIA TYPE 14 (SCA14) [MIM #605361]

Clinical Features

SCA14 should particularly be considered if the proband or an affected relative displays axial myoclonus or cognitive impairment.[72] SCA14 is characterised by slowly progressive cerebellar ataxia, dysarthria and nystagmus or saccadic intrusions. Axial myoclonus, cognitive impairment, tremor and mild or moderate sensory loss (mainly decreased vibration sense), may also be observed. Parkinsonian features including rigidity and tremor have also been described in some families. Findings seen in other ataxia disorders including dysphagia, dysphonia, dystonia, facial fasciculations with or without myokymia and mental retardation, may also occur in SCA14. Age of onset ranges from childhood to the sixth decade. Life span is not shortened. Brain MRIs, in all affected persons had shown mild to moderately severe cerebellar atrophy that is primarily midline. Atrophy of the brain stem or cerebral cortex is not observed.

Genetics of SCA14

The only gene associated with SCA14 is *PRKCG*, which encodes protein kinase C gamma type (PCKγ).[73] PCR amplification and direct sequencing of all 18 exons and flanking splice sites of *PRKCG* are the most sensitive approach for identifying mutations. As 16 of 23 (70%) different missense and deletion mutations cluster in exon 4 observed, it is reasonable to start sequencing from this exon. Because data are limited, the proportion of mutations identified is not known. SCA14 probably accounts for less than 1% of all SCAs.

SPINOCEREBELLAR ATAXIA TYPE 15 (SCA15) [MIM #606658]

Clinical Features

SCA15 is a pure ataxia characterised by very slow progression that has been identified in nine families worldwide to date. SCA15 is characterised by slowly progressive gait and limb ataxia, often in combination with ataxic dysarthria, titubation, upper limb postural tremor, mild hyperreflexia, gaze-evoked nystagmus and impaired vestibulo-ocular reflex gain.[74] Onset is between ages seven and 66 years, usually with gait ataxia but sometimes with tremor. Affected individuals remain ambulant for ten to 54 years after onset. Mild dysphagia and movement-induced oscillopsia after several decades of symptoms have been observed in members of two of the seven families known to have SCA15. The diagnosis of SCA15 should be considered in individuals with the following findings: very slowly progressive ataxia (e.g., still independently ambulant after 20-30 years of symptoms), tremor or mild hyperreflexia (typically without spasticity or extensor plantar responses), family history consistent with autosomal dominant inheritance. The phenotype may also reveal buccolingual dyskinesias, facial myokymias and pyramidal signs.[75] Neuroimaging typically reveals atrophy of the rostral and dorsal vermis of the cerebellum. The cerebellar hemispheres may appear normal or be mildly atrophic. The brain stem and cerebral hemispheres are unaffected. Nerve conduction studies are typically normal, although minor slowing of sural sensory and median motor conduction velocity have been noted.

Genetics of SCA15

Heterozygous sequence variants (14%) or exonic/whole-gene deletions (86%) within the inositol triphosphate receptor Type I (*ITPR*) gene have been found in SCA15. SCA15 may account for up to 2%-3% of all SCA subtypes.

SPINOCEREBELLAR ATAXIA TYPE 17 (SCA17) [MIM #607136]

Clinical Features

SCA17 is characterised by ataxia, dementia and involuntary movements, including chorea and dystonia (blepharospasm, torticollis, writer's cramp, foot dystonia).[76] Psychiatric symptoms, pyramidal signs and rigidity are common. The age of onset ranges from age three to 55 years. Individuals with full-penetrant alleles develop neurologic and/or psychiatric symptoms by age 50 years. Ataxia and psychiatric abnormalities are frequently the initial findings, followed by involuntary movement, Parkinsonism, dementia and pyramidal signs. MRI shows variable atrophy of the cerebrum, brain stem and cerebellum. The brain shows atrophy of the striatum (more apparent in the caudate nucleus) and cerebellum. Histologically, neuronal loss is observed in the striatum and Purkinje cell layer. Loss of cerebral cortical neurons is seen in some individuals.

Genetics of SCA17

The clinical features correlate with the length of the CAA/CAG repeat within the *TBP* gene, encoding the TATA-box-binding protein, which is the only gene associated with SCA17. The expansion of a CAA/CAG repeat is the only mutation observed.[77,78] The structure of the repeat sequence is $(CAG)_3$ $(CAA)_3$ $(CAG)_x$ CAA CAG CAA (CAG) $_y$ CAA CAG. Loss of this interruption may be a prerequisite of instability in SCA17 as in other diseases caused by repeat expansions.[79-81] Normal alleles contain 25 to 42 CAG/CAA repeats. Reduced penetrance alleles reveal 43 to 48 CAA/CAG repeats and individuals with an allele in this range may or may not develop symptoms. The significance of alleles of 43 and 44 repeats is particularly controversial because penetrance is estimated to be 50%, making genotype-phenotype correlations difficult. Full penetrant alleles include 49 or greater CAA/CAG repeats. The largest repeat size reported to date is 66.[79] A recent study has described a patient with camptocormia and mild features of SCA17 harbouring 43 CAG repeats within the *TBP* gene.[82]

SPINOCEREBELLAR ATAXIA TYPE 18 (SCA18) [MIM #607458]

Clinical Features

An Irish-American family with motor and sensory neuropathy features, gait difficulty dysmetria, hyporeflexia, muscle weakness and atrophy and decreased vibratory and proprioceptive sense displaying an autosomal dominant pattern of inheritance was associated with SCA18.[83] Several affected members had *pes cavus*. The age of onset is in the second and third decades. Muscle biopsy revealed neurogenic atrophy and brain MRI showed mild cerebellar atrophy.

Genetics of SCA18

An A-to-G transition at nucleotide 743 that resulted in an isoleucine to valine substitution at codon 172 (I172V) in the human interferon-related developmental regulator gene 1 (*IFRD1*) has been proposed as the causative deficit in SCA18.[84]

SPINOCEREBELLAR ATAXIA TYPE 19 (SCA19) [MIM #607346]

Clinical Features

SCA19 members of a Dutch family show a relatively mild ataxia syndrome with cognitive impairment, poor performance on the Wisconsin Card Sorting Test, myoclonus and a postural irregular tremor of low frequency.[85] In a SCA19 Chinese family (previously noted as SCA22), its members have gait and limb ataxia, hyporeflexia, dysphagia, dysarthria and gaze-evoked horizontal nystagmus. MRI revealed cerebellar atrophy.[86]

Genetics of SCA19

The molecular deficit in SCA19 remains to be identified yet, but it has been localized on chromosome 1p21-q21.

SPINOCEREBELLAR ATAXIA TYPE 20 (SCA20) [MIM #608687]

Clinical Features

SCA20 is characterised by a slow progressive ataxia and dysarthria. Approximately two-thirds of those affected also display palatal tremor ("myoclonus") and/or abnormal phonation clinically resembling spasmodic adductor dysphonia.[87,88] Dysarthria, which may be abrupt in onset, precedes the onset of ataxia in about two-thirds of affected individuals, sometimes by a number of years. Hypermetric horizontal saccades without nystagmus or disturbance of vestibulo-ocular reflex gain are seen in about half of affected persons. Although minor pyramidal signs including brisk knee jerks, crossed adductor spread may be seen, spasticity and extensor plantar responses are not. Cognition is normal. Clinical information is based on the findings in 16 personally examined affected members of a single Australian family of Anglo-Celtic descent.[89] A similar SCA20 phenotype has also been described in ataxia patients of six Portuguese families.[90] All affected individuals revealed spinocerebellar ataxia and spasmodic coughing episodes.

The diagnosis of SCA20 is based on clinical findings and neuroimaging. CT-scans show, within five years of disease onset, pronounced dentate calcification, typically without concomitant pallidal calcification. In addition to evidence of dentate calcification, MRI shows mild to moderate pan cerebellar atrophy and normal cerebrum and brain stem (except for increased inferior olivary T_2 signal in those with palatal tremor).

Genetics of SCA20

The SCA20 locus lies within the pericentromeric region of chromosome 11; but the gene is unknown. A 260-kb duplication of 11q12.2-11q12.3 has been identified as the probable cause of SCA20 in the index family.[91]

SPINOCEREBELLAR ATAXIA TYPE 21 (SCA21) [MIM #607454]

Clinical Features

SCA21 is a slowly progressive and mild ataxia associated with extrapyramidal signs.[92,93] Affected subjects exhibit variable symptoms of cerebellar ataxia, limb ataxia and akinesia, dysarthria, dysgraphia, hyporeflexia, postural tremor, rigidity, resting tremor, cognitive impairment and cerebellar atrophy. Eye movements are generally normal.

Genetics of SCA21

The responsible gene has been assigned to a 19 Mbases interval on chromosome 7p in one French family.[94] No evidence of significant linkage to this locus was found in 21 other families obtained from the EUROSCA consortium. The locus interval contains several candidate genes that could be responsible for the disease.

SPINOCEREBELLAR ATAXIA TYPE 23 (SCA23) [MIM #610245]

Clinical Features

Members of a Dutch family presenting with gait and limb ataxia, with variable dysarthria, slow saccades, ocular dysmetria, decreased vibratory sense below the knees, hyperreflexia and extensor plantar responses were assigned to the SCA23 subtype.[95,96] MRI findings reveal severe cerebellar atrophy. Neuropathological studies show frontotemporal atrophy, atrophy of the cerebellar vermis, pons and spinal cord. There is neuronal loss in the cerebellar vermis, dentate nuclei and inferior olives, but not in the pons. There is also thinning of the cerebellopontine tracts and demyelination of the posterior and lateral columns of the spinal cord.

Genetics of SCA23

The SCA23 locus has been mapped at 20p13-p12.3.

SPINOCEREBELLAR ATAXIA TYPE 25 (SCA25) [MIM #608703]

Clinical Features

SCA25 was described in a large French family with patients presenting an autosomal dominant form of spinocerebellar ataxia with sensory involvement, in particular areflexia of the lower limbs and peripheral sensory neuropathy.[97] Other variable features included nystagmus, decreased visual acuity, facial tics, extensor plantar responses, urinary urgency and gastrointestinal problems. Sural nerve biopsy of 1 patient showed loss of myelinated fibers and EMG showed sensory involvement. Brain imaging showed cerebellar atrophy.

Genetics of SCA25

The responsible deficit has been mapped at 2p21-p13.

SPINOCEREBELLAR ATAXIA TYPE 26 (SCA26) [MIM #609306]

Clinical Features

SCA26 is characterised by slowly progressive gait ataxia, upper limb ataxia and dysarthria.[98] Age at onset ranges from 26 to 60 years (mean, 42 years). The disorder has been characterized by pure cerebellar signs, including ataxia of the trunk and limbs, dysarthria and irregular visual pursuit movements. MRI showed atrophy of the cerebellum sparing the pons and medulla.

Genetics of SCA26

The *SCA26* gene has been mapped at 19p13.3.

SPINOCEREBELLAR ATAXIA TYPE 27 (SCA27) [MIM #609307]

Clinical Features

This SCA subtype was first described in a large 3-generation Dutch family in which some members had early-onset tremor, dyskinesia and slowly progressive cerebellar ataxia inherited in an autosomal dominant pattern.[99] Neurologic examination showed dysmetric saccades, disrupted ocular pursuit movements, gaze-evoked nystagmus, cerebellar dysarthria and a high-frequency, small-amplitude tremor in both hands in most of the patients. Some patients showed head tremor, subtle orofacial dyskinesias and *pes cavus*. Nerve conduction studies revealed mild axonal sensory neuropathy. Two patients showed cerebellar atrophy on MRI.

Genetics of SCA27

A heterozygous frameshift mutation (F145S) in the fibroblast growth factor 14 (*FGF14*) gene was identified in this family.[99] Some patients of this family also present dyskinesia, mental retardation and deficits in memory and executive functioning.[100] SCA27 has also been associated with a truncating frameshift (c.487delA) mutation in exon 4 of the *FGF14* gene in affected members of a German family with a juvenile onset of ataxia and mild mental retardation.[101] A reported case revealed an early onset cerebellar ataxia with microcephaly, severe mental retardation, tremor, dysarthria and pyramidal signs, which resulted from a chromosomal translocation disrupting the *FGF14* gene.[102] These genetic findings facilitate the molecular diagnosis of SCA27.

SPINOCEREBELLAR ATAXIA TYPE 28 (SCA28) [MIM #610246]

Clinical Features

SCA28 was described in an Italian family with a juvenile-onset SCA (19.5 years) with ataxia, dysarthria, slow and lower limb hyperreflexia and variable degrees of gaze-evoked nystagmus and dysmetric saccades, slow saccades, ophthalmoparesis and ptosis.[103] The disorder is slowly progressive and there is no evidence of sensory involvement or cognitive impairment. Brain MRI showed cerebellar atrophy.

Genetics of SCA28

In affected members of 5 unrelated families with SCA28, 5 different heterozygous mutations in the *AFG3L2* gene on 18p11 were identified.[104] These data are suggestive that the mutations may pathogenesis by both dominant-negative and loss of function mechanisms. A p.E700K missense mutations within the *AFG3L2* gene have been identified in a German family with early onset and slow progression.[105]

SPINOCEREBELLAR ATAXIA TYPE 29 (SCA29) [MIM #117360]

Clinical Features

Several families with individuals presenting an early-onset nonprogressive form of ataxia have been identified.[106-108] SCA29 patients have normal intelligence, truncal ataxia, mild limb dysmetria, up-beating nystagmus and gaze-provoked horizontal nystagmus. Affected individuals reveal localized atrophy of the cerebellar vermis on brain MRI.

Genetics of SCA29

By genome-wide analysis of a large kindred with congenital nonprogressive cerebellar ataxia, genetic linkage to an 18.9-cM region on chromosome 3p was identified.[109] Further studies have shown genetic heterogeneity for autosomal dominant nonprogressive congenital ataxia.[108]

SPINOCEREBELLAR ATAXIA TYPE 30 (SCA30) [MIM #613371]

Clinical Features

SCA30 has been identified in an Australian family of Anglo-Celtic origin in which affected individuals show a relatively pure, slowly evolving form of ataxia with mild to moderate dysarthria.[110] Four individuals had mild lower limb hyperreflexia; none had evidence of neuropathy. All patients had hypermetric saccades with normal vestibulocular reflex gain. One patient had slight gaze-evoked nystagmus. Brain MRI of 2 patients showed cerebellar atrophy with preservation of the brainstem. The mean reported age at onset was 52 years (ranging 45 to 76).

Genetics of SCA30

By genome-wide linkage analysis of this family a locus, termed SCA30, was identified on chromosome 4q34.3-q35.1.[110]

SPINOCEREBELLAR ATAXIA TYPE 31 (SCA31) [MIM #117210]

Clinical Features

SCA31, formerly known as 16q22.1-linked autosomal dominant cerebellar ataxia (16q-ADCA), has been identified in Japanese families with cerebellar ataxia without obvious evidence of extracerebellar neurological dysfunction.[111,112] The average age at onset of ataxia in these kindreds is >55 years. A substantial number of patients show progressive sensorineural hearing impairment. Neuropathologic examination showed moderate cerebellar atrophy with Purkinje cell degeneration, abnormal dendrites and somatic sprouts of Purkinje cells. Purkinje cells undergo shrinkage and are surrounded by amorphous materials composed of Purkinje-cell somato-dendritic sprouts and an increased number of presynaptic terminals.

Genetics of SCA31

In affected patients from unrelated Japanese families a heterozygous variation in the *PLEKHG4* gene (-16C-T) has been identified.[113-115] SCA31 is also caused by a complex pentanucleotide insertion ranging 2.5-3.8 kb containing a (TGGAA)n pentanucleotide repeats within the intergene region between two genes on chromosome 16q22.1 *BEAN* and *TK2*.[116] The length of the insertion is inversely correlated with the age at onset. The analysis for the presence of the insertion can be done by long-range PCR amplification or Southern blot analysis. In a nationwide survey of Japanese patients, the prevalence of all forms of spinocerebellar degenerations is 4.53 per 100,000.[117] Of these, 7.5% were estimated to have pure cerebellar ataxia, with onset after young adulthood. Cerebellar atrophy was appreciable on brain imaging. SCA31 is the third most common SCA subtype in Japan, after MJD and SCA6.[114]

TREATMENT

The treatment of hereditary ataxia is currently primarily supportive. With few exceptions (e.g., ataxia associated with vitamin E deficiency) there are no cures. Indeed for any treatment, obtaining first an accurate diagnostic of the specific hereditary ataxia subtype is of great value. Benefits include determining prognosis, facilitating family counselling, improving research access and providing some psychological benefits in ending the often frustrating search for an accurate aetiology. SCA may have certain clinical features that response very well to symptomatic medical theray. Parkinsonisms, dystonia, spasticity, urinary urgency, sleep pathology, fatigue and depression are all common in many of the ataxia subtypes and very often response to pharmalogical intervention as in other diseases. Much of the clinical interactions between the neurologists and ataxia patients should focus on identifying and treating these symptoms. Treatment of the core clinical features of these diseases, ataxia is predominantly rehabilitative. The value of good physical therapy far exceeds any potential benefits from medication that a physician might prescribe to improve balance and coordination. Speech and swallowing are often affected. In more severe cases, aspiration risk can be very significant and life threatening. Routine monitoring of swallowing by speech therapists, often including modified barium swallowing test is suggested for most patients.

Recently there have been some encouraging advances in clinical ataxia research. Collaborative study groups throughout the world have developed and validated ataxia rating scales and instrumented outcome measures and have begun to rigorously define the natural history of these diseases; thus laying the foundation for well designed clinical trials. Hence the promise for curable treatments is closer than ever.

CONCLUSION

Herein, we present a summary of the main clinical and genetic aspects of the spinocerebellar ataxias. Currently, more than 35 genes underlying autosomal dominant SCAs have been identified evidencing the high clinical and genetic heterogeneity. Although many SCA genes have now been identified and the diagnosis and genetic counselling are facilitated in these SCA subtypes, there are still several clinically distinct forms of inherited ataxias for which the specific gene has not been found and new forms of the disorder are still being clinically described. Therefore, the areas of detailed clinical and genetic diagnosis are currently very active research lines in SCAs. Testing of at-risk asymptomatic adult relatives of individuals with autosomal dominant cerebellar ataxia is possible only after the specific mutation for the disorder has been identified in the proband. Such testing should be performed in the context of formal genetic counselling. This testing is not useful in predicting the specific age of onset, severity, type of symptoms, or rate of progression in asymptomatic individuals. Testing of asymptomatic at-risk individuals with nonspecific or equivocal symptoms is predictive testing, not diagnostic testing. When testing at-risk individuals, an affected family member should be tested first to confirm that the mutation is identifiable by currently available techniques.[118] Testing of asymptomatic individuals during childhood because they may be at risk for adult-onset disorders for which there are no treatments is not appropriate. The principal arguments against testing asymptomatic individuals who are younger than 18 years of age are: i) the prevention of their mature choice of whether to know this information; ii) the possible stigmatization within the family and in other social settings; and, iii) the serious educational and career implications. In contrast, individuals who are symptomatic during childhood usually benefit from having a specific diagnosis established. Fortunately, the intensive ongoing clinical and neurogenetic research together with applied molecular approaches are sure to yield scientific advances that will be translated into developing effective treatments for the spinocerebellar ataxias and other similar neurological conditions.

ACKNOWLEDGEMENTS

We acknowledge funding to our scientific research by the Spanish Ministry of Science and Innovation (BFU2008-00527/BMC), the Carlos III Health Institute (CP08/00027), the LatinAmerican Science and Technology Development Programme (CYTED) (210RT0390), the European Commission (EUROSCA project, LHSM-CT-2004-503304), and the Fundació La Marató de TV3 (Televisió de Catalunya). We are indebted to the Spanish Ataxia Association (FEDAES) and the ataxia patients for their continuous support and motivation. Antoni Matilla is a *Miguel Servet* Investigator in Neurosciences of the Spanish National Health System.

REFERENCES

1. Zoghbi HY, Orr HT. Spinocerebellar ataxias. In: Scriver CR, Sly WS, Childs AL et al, eds. The Metabolic and Molecular Basis of Inherited Disease. New York: MaGraw Professionals, 2001:5741-5758.
2. Matilla-Dueñas A, Goold R, Giunti P. Molecular pathogenesis of spinocerebellar ataxias. Brain 2006; 129:1357-1370.
3. Soong BW, Paulson HL. Spinocerebellar ataxias: an update. Curr Opin Neurol 2007; 20(4):438-446.
4. Manto M, Marmolino D. Cerebellar ataxias. Curr Opin Neurol 2009; 22(4):419-429.
5. Matilla-Dueñas A, Sanchez I, Corral-Juan M et al. Cellular and Molecular Pathways Triggering Neurodegeneration in the Spinocerebellar Ataxias. Cerebellum 2010; 9(2):148-166.
6. McKusick V. Online Mendelian Inheritance in Man, OMIM (TM): McKusick-Nathans Institute of Genetic Medicine Johns Hopkins University (Baltimore, MD) and National Center for Biotechnology Information, National Library of Medicine (Bethesda, MD), http://www.ncbi.nlm.nih.gov/omim/2010.
7. Pagon RA, Bird TC, Dolan CR et al. Gene Reviews [Internet]. Seattle: University of Washington, Seattle, 1993-2010.
8. Schmitz-Hubsch T, du Montcel ST, Baliko L et al. Scale for the assessment and rating of ataxia: development of a new clinical scale. Neurology 2006; 66(11):1717-1720.
9. Schmitz-Hubsch T, Coudert M, Bauer P et al. Spinocerebellar ataxia types 1, 2, 3 and 6: disease severity and nonataxia symptoms. Neurology 2008; 71(13):982-989.
10. Schmitz-Hubsch T, Fimmers R, Rakowicz M et al. Responsiveness of different rating instruments in spinocerebellar ataxia patients. Neurology 2010; 74(8):678-684.
11. Sequeiros J, Martindale J, Seneca S. EMQN Best Practice Guidelines for molecular genetic testing of SCAs. Eur J Hum Genet 2010; 18(11):1173-6.
12. Sequeiros J, Seneca S, Martindale J. Consensus and controversies in best practices for molecular genetic testing of spinocerebellar ataxias. Eur J Hum Genet 2010; 18(11):1188-95.
13. Matilla-Dueñas A, Goold R, Giunti P. Clinical, genetic, molecular and pathophysiological insights into spinocerebellar ataxia type 1. Cerebellum 2008; 7(2):106-114.
14. Genis D, Matilla T, Volpini V et al. Clinical, neuropathologic and genetic studies of a large spinocerebellar ataxia type 1 (SCA1) kindred: (CAG)n expansion and early premonitory signs and symptoms. Neurology 1995; 45(1):24-30.
15. Matilla T, Volpini V, Genis D et al. Presymptomatic analysis of spinocerebellar ataxia type 1(SCA1) via the expansion of the SCA1 CAG-repeat in a large pedigree displaying anticipation and parental male bias. Hum Mol Genet 1993; 2(12):2123-2128.
16. Jodice C, Malaspina P, Persichetti F et al. Effect of trinucleotide repeat length and parental sex on phenotypic variation in spinocerebellar ataxia 1. Am J Hum Genet 1994; 54:959-965.
17. Ranum LPW, Chung M-y, Banfi S et al. Molecular and clinical correlations in spinocerebellar ataxia type 1 (SCA1): evidence for familial effects on the age of onset. Am J Hum Genet 1994; 55:244-252.
18. Orr HT, Zoghbi HY. SCA1 molecular genetics: a history of a 13 year collaboration against glutamines. Hum Mol Genet 2001; 10(20):2307-2311.
19. Goldfarb LG, Vasconcelos O, Platonov FA et al. Unstable triplet repeat and phenotypic variability of spinocerebellar ataxia type 1. Ann Neurol 1996; 39(4):500-506.
20. Zuhlke C, Dalski A, Hellenbroich Y et al. Spinocerebellar ataxia type 1 (SCA1): phenotype-genotype correlation studies in intermediate alleles. Eur J Hum Genet 2002; 10(3):204-209.
21. Pulst SM. Spinocerebellar ataxia type 2 In: Pagon R, Bird T, Dolan C et al, eds. GeneReviews [Internet]. Seattle: University of Washington, Seattle, 2006.
22. Lastres-Becker I, Rub U, Auburger G. Spinocerebellar ataxia 2 (SCA2). Cerebellum 2008; 7(2):115-124.
23. Orozco G, Estrada R, Perry TL et al. Dominantly inherited olivopontocerebellar atrophy from eastern Cuba. Clinical, neuropathological and biochemical findings. J Neurol Sci 1989; 93:37-50.
24. Velazquez-Perez L, Seifried C, Santos-Falcon N et al. Saccade velocity is controlled by polyglutamine size in spinocerebellar ataxia 2. Ann Neurol 2004; 56(3):444-447.

25. Armstrong J, Bonaventura I, Rojo A et al. Spinocerebellar ataxia type 2 (SCA2) with white matter involvement. Neurosci Lett 2005; 381(3):247-251.
26. Filla A, De Michele G, Banfi S et al. Has spinocerebellar ataxia type 2 a distinct phenotype? Genetic and clinical study of an Italian family. Neurology 1995; 45(4):793-796.
27. Flanigan K, Gardner K, Alderson K et al. Autosomal dominant spinocerebellar ataxia with sensory axonal neuropathy (SCA4): clinical description and genetic localization to chromosome 16q22.1. Am J Hum Genet 1996; 59(2):392-399.
28. Hellenbroich Y, Bubel S, Pawlack H et al. Refinement of the spinocerebellar ataxia type 4 locus in a large German family and exclusion of CAG repeat expansions in this region. J Neurol 2003; 250(6):668-671.
29. Hellenbroich Y, Gierga K, Reusche E et al. Spinocerebellar ataxia type 4 (SCA4): Initial pathoanatomical study reveals widespread cerebellar and brainstem degeneration. J Neur Trans 2006; 113(7):829-843.
30. Ranum LPW, Schut LJ, Lundgren JK et al. Spinocerebellar ataxia type 5 in a family descended from the grandparents of President Lincoln maps to chromosome 11. Nat Genet 1994; 8:280-284.
31. Ikeda Y, Dick KA, Weatherspoon MR et al. Spectrin mutations cause spinocerebellar ataxia type 5. Nat Genet 2006; 38(2):184-190.
32. Gomez CM. Spinocerebellar ataxia type 6. In: Pagon R, Bird T, Dolan C et al, eds. GeneReviews [Internet]. Seattle: University of Washington, Seattle, 2008.
33. Yabe I, Sasaki H, Takeichi N et al. Positional vertigo and macroscopic downbeat positioning nystagmus in spinocerebellar ataxia type 6 (SCA6). J Neurol 2003; 250(4):440-443.
34. Hashimoto T, Sasaki O, Yoshida K et al. Periodic alternating nystagmus and rebound nystagmus in spinocerebellar ataxia type 6. Mov Disord 2003; 18(10):1201-1204.
35. Globas C, Bosch S, Zuhlke C et al. The cerebellum and cognition. Intellectual function in spinocerebellar ataxia type 6 (SCA6). J Neurol 2003; 250(12):1482-1487.
36. Gomez CM, Thompson RM, Gammack JT et al. Spinocerebellar ataxia type 6: gaze-evoked and vertical nystagmus, Purkinje cell degeneration and variable age of onset. Ann Neurol 1997; 42(6):933-950.
37. Sasaki H, Kojima H, Yabe I et al. Neuropathological and molecular studies of spinocerebellar ataxia type 6 (SCA6). Acta Neuropathol 1998; 95(2):199-204.
38. Ying SH, Choi SI, Lee M et al. Relative atrophy of the flocculus and ocular motor dysfunction in SCA2 and SCA6. Ann N Y Acad Sci 2005; 1039:430-435.
39. Khan NL, Giunti P, Sweeney MG et al. Parkinsonism and nigrostriatal dysfunction are associated with spinocerebellar ataxia type 6 (SCA6). Mov Disord 2005; 20(9):1115-1119.
40. Takahashi H, Ishikawa K, Tsutsumi T et al. A clinical and genetic study in a large cohort of patients with spinocerebellar ataxia type 6. J Hum Genet 2004; 49(5):256-264.
41. Bird TD, Pagon RA, La Spada AR. Spinocerebellar ataxia type 7. In: Pagon R, Bird T, Dolan C et al, eds. GeneReviews [Internet]. Seattle: University of Washington, Seattle, 2007.
42. Garden GA, La Spada AR. Molecular pathogenesis and cellular pathology of spinocerebellar ataxia type 7 neurodegeneration. Cerebellum 2008; 7(2):138-149.
43. Babovic-Vuksanovic D, Snow K, Patterson MC et al. Spinocerebellar ataxia type 2 (SCA 2) in an infant with extreme CAG repeat expansion. Am J Med Genet 1998; 79(5):383-387.
44. Benton CS, de Silva R, Rutledge SL et al. Molecular and clinical studies in SCA-7 define a broad clinical spectrum and the infantile phenotype. Neurology 1998; 51(4):1081-1086.
45. Ansorge O, Giunti P, Michalik A et al. Ataxin-7 aggregation and ubiquitination in infantile SCA7 with 180 CAG repeats. Ann Neurol 2004; 56(3):448-452.
46. Mittal U, Roy S, Jain S et al. Post-zygotic de novo trinucleotide repeat expansion at spinocerebellar ataxia type 7 locus: evidence from an Indian family. J Hum Genet 2005; 50(3):155-157.
47. Ikeda S, Dalton JC, Day JW et al. Spinocerebellar ataxia type 8. In: Pagon R, Bird T, Dolan C et al, eds. GeneReviews [Internet]. Seattle: University of Washington, Seattle, 2007.
48. Ikeda Y, Daughters RS, Ranum LP. Bidirectional expression of the SCA8 expansion mutation: one mutation, two genes. Cerebellum 2008; 7(2):150-158.
49. Ikeda Y, Dalton JC, Moseley ML et al. Spinocerebellar ataxia type 8: molecular genetic comparisons and haplotype analysis of 37 families with ataxia. Am J Hum Genet 2004; 75(1):3-16.
50. Moseley ML, Schut LJ, Bird TD et al. SCA8 CTG repeat: en masse contractions in sperm and intergenerational sequence changes may play a role in reduced penetrance. Hum Mol Genet 2000; 9(14):2125-2130.
51. Worth PF, Houlden H, Giunti P et al. Large, expanded repeats in SCA8 are not confined to patients with cerebellar ataxia. Nat Genet 2000; 24(3):214-215.
52. Sulek A, Hoffman-Zacharska D, Zdzienicka E et al. SCA8 repeat expansion coexists with SCA1—not only with SCA6. Am J Hum Genet 2003; 73(4):972-974.
53. Corral J, Genis D, Banchs I et al. Giant SCA8 alleles in nine children whose mother has two moderately large ones. Ann Neurol 2005; 57(4):549-553.
54. Baba Y, Uitti RJ, Farrer MJ et al. Atypical Parkinsonism and SCA8. Parkinsonism Relat Disord 2006; 12(6):396.
55. Ohnari K, Aoki M, Uozumi T et al. Severe symptoms of 16q-ADCA coexisting with SCA8 repeat expansion. J Neurol Sci 2008; 273(1-2):15-18.

56. Munhoz RP, Teive HA, Raskin S et al. CTA/CTG expansions at the SCA 8 locus in multiple system atrophy. Clin Neurol Neurosurg 2009; 111(2):208-210.
57. Matsuura T, Yamagata T, Burgess DL et al. Large expansion of the ATTCT pentanucleotide repeat in spinocerebellar ataxia type 10. Nat Genet 2000; 26(2):191-194.
58. Lin X, Ashizawa T. Recent progress in spinocerebellar ataxia type-10 (SCA10). Cerebellum 2005; 4(1):37-42.
59. Johnson J, Wood N, Giunti P et al. Clinical and genetic analysis of spinocerebellar ataxia type 11. Cerebellum 2008; 7(2):159-164.
60. Worth PF, Giunti P, Gardner-Thorpe C et al. Autosomal dominant cerebellar ataxia type III: linkage in a large British family to a 7.6-cM region on chromosome 15q14-21.3. Am J Hum Genet 1999; 65(2):420-426.
61. Houlden H, Johnson J, Gardner-Thorpe C et al. Mutations in TTBK2, encoding a kinase implicated in tau phosphorylation, segregate with spinocerebellar ataxia type 11. Nat Genet 2007; 39(12):1434-1436.
62. Margolis RL, O'hearn E, Holmes SE et al. Spinocerebellar ataxia type 12. In: Pagon R, Bird T, Dolan C et al, eds. GeneReviews [Internet]. Seattle: University of Washington, Seattle, 2007.
63. O'Hearn E, Holmes SE, Calvert PC et al. SCA-12: Tremor with cerebellar and cortical atrophy is associated with a CAG repeat expansion. Neurology 2001; 56(3):299-303.
64. Srivastava AK, Choudhry S, Gopinath MS et al. Molecular and clinical correlation in five Indian families with spinocerebellar ataxia 12. Ann Neurol 2001; 50(6):796-800.
65. Brussino A, Graziano C, Giobbe D et al. Spinocerebellar ataxia type 12 identified in two Italian families may mimic sporadic ataxia. Mov Disord 2010; 25(9):1269-1273.
66. Fujigasaki H, Martin JJ, De Deyn PP et al. CAG repeat expansion in the TATA box-binding protein gene causes autosomal dominant cerebellar ataxia. Brain 2001; 124(10):1939-1947.
67. Hellenbroich Y, Schulz-Schaeffer W, Nitschke MF et al. Coincidence of a large SCA12 repeat allele with a case of Creutzfeld-Jacob disease. J Neurol Neurosurg Psychiatry 2004; 75(6):937-938.
68. Fujigasaki H, Verma IC, Camuzat A et al. SCA12 is a rare locus for autosomal dominant cerebellar ataxia: a study of an Indian family. Ann Neurol 2001; 49(1):117-121.
69. Holmes S, O'Hearn E, Brachmachari S et al. SCA12. In: Pulst S, ed. Genetics of Movement Disorders. San Diego: Academic Press, 2002.
70. Waters MF, Pulst SM. Sca13. Cerebellum 2008; 7(2):165-169.
71. Waters MF, Minassian NA, Stevanin G et al. Mutations in voltage-gated potassium channel KCNC3 cause degenerative and developmental central nervous system phenotypes. Nat Genet 2006; 38(4):447-451.
72. Chen D-H, Bird TD, Raskind WH. Spinocerebellar ataxia type 14. In: Pagon R, Bird T, Dolan C et al, eds. GeneReviews [Internet]. Seattle: University of Washington, Seattle, 2010.
73. Chen DH, Brkanac Z, Verlinde CL et al. Missense Mutations in the Regulatory Domain of PKCgamma: A New Mechanism for Dominant Nonepisodic Cerebellar Ataxia. Am J Hum Genet 2003; 72(4):839-849.
74. Storey E. Spinocerebellar ataxia type 15. In: Pagon R, Bird T, Dolan C et al, eds. GeneReviews [Internet]. Seattle: University of Washington, Seattle, 2009.
75. Di Gregorio E, Orsi L, Godani M et al. Two Italian families with ITPR1 gene deletion presenting a broader phenotype of SCA15. Cerebellum; 9(1):115-123.
76. Toyoshima Y, Onodera O, Yamada M et al. Spinocerebellar ataxia type 17. In: Pagon R, Bird T, Dolan C et al, eds. GeneReviews [Internet]. Seattle: University of Washington, Seattle, 2007.
77. Koide R, Kobayashi S, Shimohata T et al. A neurological disease caused by an expanded CAG trinucleotide repeat in the TATA-binding protein gene: a new polyglutamine disease? Hum Mol Genet 1999; 8(11):2047-2053.
78. Nakamura K, Jeong SY, Uchihara T et al. SCA17, a novel autosomal dominant cerebellar ataxia caused by an expanded polyglutamine in TATA-binding protein. Hum Mol Genet 2001; 10(14):1441-1448.
79. Maltecca F, Filla A, Castaldo I et al. Intergenerational instability and marked anticipation in SCA-17. Neurology 2003; 61(10):1441-1443.
80. Zuhlke CH, Spranger M, Spranger S et al. SCA17 caused by homozygous repeat expansion in TBP due to partial isodisomy 6. Eur J Hum Genet 2003; 11(8):629-632.
81. Zuhlke C, Dalski A, Schwinger E et al. Spinocerebellar ataxia type 17: report of a family with reduced penetrance of an unstable Gln49 TBP allele, haplotype analysis supporting a founder effect for unstable alleles and comparative analysis of SCA17 genotypes. BMC Med Genet 2005; 6:27.
82. Gamez J, Sierra-Marcos A, Gratacos M et al. Camptocormia associated with an expanded allele in the TATA box-binding protein gene. Mov Disord 2010; 25(9):1293-1295.
83. Brkanac Z, Bylenok L, Fernandez M et al. A new dominant spinocerebellar ataxia linked to chromosome 19q13.4-qter. Arch Neurol 2002; 59(8):1291-1295.
84. Brkanac Z, Spencer D, Shendure J et al. IFRD1 is a candidate gene for SMNA on chromosome 7q22-q23. Am J Hum Genet 2009; 84(5):692-697.
85. Schelhaas HJ, Ippel PF, Hageman G et al. Clinical and genetic analysis of a four-generation family with a distinct autosomal dominant cerebellar ataxia. J Neurol 2001; 248(2):113-120.
86. Chung MY, Lu YC, Cheng NC et al. A novel autosomal dominant spinocerebellar ataxia (SCA22) linked to chromosome 1p21-q23. Brain 2003; 126(Pt 6):1293-1299.
87. Storey E, Knight MA, Forrest SM et al. Spinocerebellar ataxia type 20. Cerebellum 2005; 4(1):55-57.

88. Storey E. Spinocerebellar ataxia type 20. In: Pagon R, Bird T, Dolan C et al, eds. GeneReviews [Internet]. Seattle: University of Washington, Seattle, 2009.
89. Knight MA, Gardner RJ, Bahlo M et al. Dominantly inherited ataxia and dysphonia with dentate calcification: spinocerebellar ataxia type 20. Brain 2004; 127(Pt 5):1172-1181.
90. Coutinho P, Cruz VT, Tuna A et al. Cerebellar ataxia with spasmodic cough: a new form of dominant ataxia. Arch Neurol 2006; 63 (4):553-555.
91. Knight MA, Hernandez D, Diede SJ et al. A duplication at chromosome 11q12.2-11q12.3 is associated with spinocerebellar ataxia type 20. Hum Mol Genet 2008; 17(24):3847-3853.
92. Devos D, Schraen-Maschke S, Vuillaume I et al. Clinical features and genetic analysis of a new form of spinocerebellar ataxia. Neurology 2001; 56(2):234-238.
93. Delplanque J, Devos D, Vuillaume I et al. Slowly progressive spinocerebellar ataxia with extrapyramidal signs and mild cognitive impairment (SCA21). Cerebellum 2008; 7(2):179-183.
94. Vuillaume I, Devos D, Schraen-Maschke S et al. A new locus for spinocerebellar ataxia (SCA21) maps to chromosome 7p21.3-p15.1. Ann Neurol 2002; 52(5):666-670.
95. Verbeek DS, van de Warrenburg BP, Wesseling P et al. Mapping of the SCA23 locus involved in autosomal dominant cerebellar ataxia to chromosome region 20p13-12.3. Brain 2004; 127(Pt 11):2551-2557.
96. Verbeek DS. Spinocerebellar ataxia type 23: a genetic update. Cerebellum 2009; 8(2):104-107.
97. Stevanin G, Broussolle E, Streichenberger N et al. Spinocerebellar ataxia with sensory neuropathy (SCA25). Cerebellum 2005; 4(1):58-61.
98. Yu GY, Howell MJ, Roller MJ et al. Spinocerebellar ataxia type 26 maps to chromosome 19p13.3 adjacent to SCA6. Ann Neurol 2005; 57(3):349-354.
99. van Swieten JC, Brusse E, de Graaf BM et al. A mutation in the fibroblast growth factor 14 gene is associated with autosomal dominant cerebral ataxia. Am J Hum Genet 2003; 72(1):191-199.
100. Brusse E, de Koning I, Maat-Kievit A et al. Spinocerebellar ataxia associated with a mutation in the fibroblast growth factor 14 gene (SCA27): A new phenotype. Mov Disord 2006; 21(3):396-401.
101. Dalski A, Atici J, Kreuz FR et al. Mutation analysis in the fibroblast growth factor 14 gene: frameshift mutation and polymorphisms in patients with inherited ataxias. Eur J Hum Genet 2005; 13(1):118-120.
102. Misceo D, Fannemel M, Baroy T et al. SCA27 caused by a chromosome translocation: further delineation of the phenotype. Neurogenetics 2009; 10(4):371-374.
103. Mariotti C, Brusco A, Di Bella D et al. Spinocerebellar ataxia type 28: a novel autosomal dominant cerebellar ataxia characterized by slow progression and ophthalmoparesis. Cerebellum 2008; 7(2):184-188.
104. Di Bella D, Lazzaro F, Brusco A et al. Mutations in the mitochondrial protease gene AFG3L2 cause dominant hereditary ataxia SCA28. Nat Genet 2010; 42(4):313-321.
105. Edener U, Wollner J, Hehr U et al. Early onset and slow progression of SCA28, a rare dominant ataxia in a large four-generation family with a novel AFG3L2 mutation. Eur J Hum Genet 2010; 18(8):965-968.
106. Furman JM, Baloh RW, Chugani H et al. Infantile cerebellar atrophy. Ann Neurol 1985; 17(4):399-402.
107. Tomiwa K, Baraitser M, Wilson J. Dominantly inherited congenital cerebellar ataxia with atrophy of the vermis. Pediatr Neurol 1987; 3(6):360-362.
108. Jen JC, Lee H, Cha YH et al. Genetic heterogeneity of autosomal dominant nonprogressive congenital ataxia. Neurology 2006; 67(9):1704-1706.
109. Dudding TE, Friend K, Schofield PW et al. Autosomal dominant congenital nonprogressive ataxia overlaps with the SCA15 locus. Neurology 2004; 63(12):2288-2292.
110. Storey E, Bahlo M, Fahey M et al. A new dominantly inherited pure cerebellar ataxia, SCA 30. J Neurol Neurosurg Psychiatry 2009; 80(4):408-411.
111. Nagaoka U, Takashima M, Ishikawa K et al. A gene on SCA4 locus causes dominantly inherited pure cerebellar ataxia. Neurology 2000; 54(10):1971-1975.
112. Owada K, Ishikawa K, Toru S et al. A clinical, genetic and neuropathologic study in a family with 16q-linked ADCA type III. Neurology 2005; 65(4):629-632.
113. Ishikawa K, Toru S, Tsunemi T et al. An autosomal dominant cerebellar ataxia linked to chromosome 16q22.1 is associated with a single-nucleotide substitution in the 5' untranslated region of the gene encoding a protein with spectrin repeat and Rho guanine-nucleotide exchange-factor domains. Am J Hum Genet 2005; 77(2):280-296.
114. Ouyang Y, Sakoe K, Shimazaki H et al 16q-linked autosomal dominant cerebellar ataxia: a clinical and genetic study. J Neurol Sci 2006; 247(2):180-186.
115. Hirano R, Takashima H, Okubo R et al. Clinical and genetic characterization of 16q-linked autosomal dominant spinocerebellar ataxia in South Kyushu, Japan. J Hum Genet 2009; 54(7):377-381.
116. Sato N, Amino T, Kobayashi K et al. Spinocerebellar ataxia type 31 is associated with "inserted" penta-nucleotide repeats containing (TGGAA)n. Am J Hum Genet 2009; 85(5):544-557.
117. Hirayama K, Takayanagi T, Nakamura R et al. Spinocerebellar degenerations in Japan: a nationwide epidemiological and clinical study. Acta Neurol Scand Suppl 1994; 153:1-22.
118. Goizet C, Lesca G, Durr A. Presymptomatic testing in Huntington's disease and autosomal dominant cerebellar ataxias. Neurology 2002; 59(9):1330-1336.

CHAPTER 28

TOURETTE SYNDROME

Andrea E. Cavanna*,[1,2] and Cristiano Termine[3]

[1]Department of Neuropsychiatry, University of Birmingham and BSMHFT, Birmingham, UK; [2]Sobell Department of Movement Disorders, Institute of Neurology, UCL, London, UK; [3]Child Neuropsychiatry Unit, Department of Experimental Medicine, University of Insubria, Varese, Italy
*Corresponding Author: Andrea E. Cavanna—Email: a.cavanna@ion.ucl.ac.uk

Abstract: Tourette syndrome (TS) is a neurodevelopmental disorder consisting of multiple motor and one or more vocal/phonic tics. TS is increasingly recognized as a common neuropsychiatric disorder usually diagnosed in early childhood and comorbid neuropsychiatric disorders occur in approximately 90% of patients, with attention deficit hyperactivity disorder (ADHD) and obsessive-compulsive disorder (OCD) being the most common ones. Moreover, a high prevalence of depression and personality disorders has been reported. Although the mainstream of tic management is represented by pharmacotherapy, different kinds of psychotherapy, along with neurosurgical interventions (especially deep brain stimulation, DBS) play a major role in the treatment of TS. The current diagnostic systems have dictated that TS is a unitary condition. However, recent studies have demonstrated that there may be more than one TS phenotype. In conclusion, it appears that TS probably should no longer be considered merely a motor disorder and, most importantly, that TS is no longer a unitary condition, as it was previously thought.

INTRODUCTION

Tourette syndrome (TS) is a neuropsychiatric disorder of childhood onset characterised by the presence of multiple motor tics and one or more vocal tics, not necessarily concurrently, but lasting for over one year.[1,2] Originally described over a century ago, TS is increasingly recognised as a relatively common disorder, with a privileged position at the borderlands of neurology and psychiatry. Both clinical and epidemiological studies suggest that associated behavioural problems are common in people with TS and it seems

---

Neurodegenerative Diseases, edited by Shamim I. Ahmad.
©2012 Landes Bioscience and Springer Science+Business Media.

likely that the investigation of the neurobiological bases of TS will shed light on the common brain mechanisms underlying movement and behaviour regulation.

EPIDEMIOLOGY

Tic disorders, usually referring to motor tics are much more common than TS. Although prevalence figures differ, depending on the population studied. They have been reported to occur with a point prevalence of between 1% to 29% depending on the study design and methods employed, the diagnostic criteria and whether or not the sample was a population sample or referral sample.[3]

The epidemiology of TS is more complex than once thought. Until fairly recently TS was considered to be a rare disorder.[4] The prevalence of TS depends, at least in part, on the definition of TS, the type of ascertainment and epidemiological study methods used.[3] The majority of studies agree that TS occurs three to four times more commonly in males than in females and that it is found in all social classes. Moreover, the main clinical features of individuals with TS appear to be uniform worldwide, irrespective of the country of origin, with possibly only some minor cultural differences.[1,5]

Prevalence estimates have largely been consistent, with a figure of between 0.6% and 1% likely in mainstream schoolchildren.[3] Of note, the majority of these individuals have mild, sub-clinical symptoms, which are unlikely to be disruptive to their lives. Studies conducted in special education environments have found a higher prevalence of TS in populations with learning difficulties and autistic spectrum disorders.[6,7]

AETIOLOGY

The aetiology of TS is much more complex than previously recognized, with strong genetic influences, some infections, possibly having effects in a subgroup of patients and pre and peri-natal difficulties also affecting the phenotype.[8] Neuroimaging studies have improved understanding of the pathophysiological processes involved in TS, with disturbances in the basal ganglia and frontal cortical circuits implicated in the development of tic symptoms.[9] One model suggests that decreased inhibitory output from the basal ganglia results in excessive activity in the frontal cortical areas, giving rise to the involuntary movements. A second model proposes that the basal ganglia facilitate desired behaviours by filtering out and preventing competing unwanted behaviours from interfering. A dysfunction in the basal ganglia, based on this model could result in the clinical spectra associated with TS: tics, obsessive compulsive behaviour (OCB) and attention deficit hyperactivity disorder (ADHD), all of which involve impaired inhibition of unwanted behaviours.

While the mode of inheritance is not simple, it is clear that TS has a significant genetic basis and that some individuals with TS, Chronic Motor Tics (CMT) and/or obsessive compulsive disorder (OCD) manifest variant expressions of the same genetic susceptibility factors. Genome scans have highlighted areas of interest in several chromosomes,[10] and recently rare sequence variants of the gene *SLITRK1* on chromosome 13q31.1 have been associated with TS in certain patients, though these findings have yet to be confirmed.[11] Thus, the genetics of TS is much more complicated than was previously thought and there is almost certainly genetic as well as clinical heterogeneity. It is therefore not surprising that other aetiological suggestions are becoming increasingly important.

Environmental factors thought also to affect the phenotype of TS include infections, perinatal problems and hormonal influences. Paediatric autoimmune neuropsychiatric disorders, associated with group A ß-haemolytic streptococcal infections (PANDAS) can be associated with the onset or exacerbation of tics and/or OCD symptoms,[12] and patients with Tourette syndrome are more likely to have had a group A ß-haemolytic streptococcal infection three months prior to onset than controls.[13] It has been suggested that an individual may inherit a susceptibility to TS and also to the way he reacts to infections.[14] Ultimately, this link requires further research and is currently too speculative to play a part in clinical practice.

CLINICAL ASPECTS AND DIAGNOSTIC CRITERIA

A tic is a sudden movement or vocalisation that is rapid, recurrent, nonrhythmic and stereotyped. Tics decrease with distraction and relaxation and increase with stress and anxiety. They can be suppressed through the patient's own voluntary control, though they will reappear when the patient's effort is relaxed. Tics may be preceded by premonitory feelings and patients often feel their tics are intentional responses executed in order to relieve this uncomfortable sensation.[15,16]

Simple motor tics typically involve isolated muscles, producing movements such as blinking or sniffing. Complex motor tics involve contractions in different muscle groups and coordinated movements resembling normal motor gestures. More complex motor phenomena include echopraxia (repetition of the movements of others) and copropraxia (making obscene gestures), both reported in up to 20% of the patients in specialist clinics.[1]

Vocal tics are typically nonverbal ('phonic tics'), involving sounds such as throat-clearing or barking. However, approximately 35% of vocal tics are verbal in nature.[17] Coprolalia (involuntary uttering of obscenities) is relatively uncommon, reported in less than a third of clinical cases and has a mean age of onset of 14 years.[1] Other vocal phenomena include palilalia (repetition of one's own sounds and words) and echolalia (imitation of sounds or words of others).

It is important to note that tics fluctuate in severity (wax and wane) and change characters within the same person; this variability of expression may contribute to diagnostic confusion and misdiagnosis. Jankovic elegantly described the intimate phenomenology of tics showing how most tics are semi-voluntary ('involuntary') or involuntary ('suppressible').[18] Moreover, tics and vocalizations may be suggestible and are characteristically aggravated by anxiety, stress, boredom, fatigue and excitement. On the other hand, sleep, alcohol, orgasm, fever, relaxation and concentration lead to temporary disappearance of symptoms.[1,18] The significance of these factors for a patho-physiological model of tics remains unclear.

TS is a neurodevelopmental disorder, the mean age at onset being 6-7 years. The onset of phonic tics is usually later, at around 11 years. Typically, tics occur many times a day nearly every day, or intermittently with a waxing and waning course. It has been reported that their anatomic location, number, frequency, complexity, type and severity usually changes over time.[18] The diagnosis of TS is made also by clinical and historical observations, as no specific laboratory or instrumental tests are currently available. Formal diagnostic criteria from the World Health Organisation and the American Psychiatric Association differ only slightly (Table 1).[19,20]

Table 1. Diagnostic criteria for Tourette syndrome

Criterion	DSM-IV-TR	ICD-10
Both multiple motor tics and one or more vocal tics present at some point during the disorder, though not necessarily concurrently.	Yes	Yes
Tics occur many times a day, almost daily, for more than a year.	Yes	Yes
Tic-free period must not last longer than:	3 months	2 months
Onset before age 18	Yes	Yes
The disorder is not due to a substance (e.g., stimulants) or general medical condition (e.g., Huntington's chorea)	Yes	Unmentioned

DSM-IV-TR diagnostic system can distinguishes TS from chronic tic disorders in which patients have either multiple motor or vocal tics (but not both).[20] Moreover, it has been suggested[1,21] to be useful to clinically subdivide TS into: (i) "pure TS", consisting primarily and almost solely of motor and phonic tics; (ii) "full blown TS" which includes coprophenomena, echophenomena and paliphenomena; (iii) "TS-plus" (originally coined by Packer),[22] in which an individual can also have ADHD, significant obsessive-compulsive symptoms or OCD, or Self Injourious Behaviors (SIB). Others presenting with severe comorbid neuropsychiatric conditions (e.g., depression, anxiety, personality disorders and other difficult and antisocial behaviours) may also be included in this group (see Table 2).

Approximately one in ten individuals with TS has no other behavioural problems, as evidenced by studies in both, epidemiological and clinical settings.[23,24] The most common comorbid psychiatric disorder in patients with TS is ADHD, followed by OCD and affective disorders.[25,26] Hence, it is important that a thorough assessment is conducted to ensure that the putative patient has in fact both, TS and ADHD. Some putative patients with TS

Table 2. Main clinical features of chronic tic disorder, pure GTS, full-blown GTS and GTS-plus

	Chronic Tic Disorder	Pure GTS	Full-Blown GTS	GTS-Plus
Motor tics	+/–	+	+	+
Vocal tic(s)	+/–	+	+	+
Echophenomena	-	-	+/–	+/–
Paliphenomena	-	-	+/–	+/–
Coprophenomena	-	-	+/–	+/–
NOSI	-	-	+/–	+/–
Forced touching	-	-	+/–	+/–
Stuttering	-	-	+/–	+/–
SIB	-	-	-	+/–
ADHD	-	-	-	+/–
OCD	-	-	-	+/–
Depression	-	-	-	+/–
Mood disorders	-	-	-	+/–
Personality disorders	-	-	-	+/–

Abbreviations: NOSI: non-obscene socially inappropriate behaviours; SIB: self-injurious behaviours.

are so fidgety with their tics, and when trying to suppress, they may appear to have poor concentrations. It has also been pointed out that in children with TS the symptoms of ADHD often contribute to the behavioural disturbances, poor school performance and impaired executive functioning test (IEFT).[27] Recently, some research groups have separated individuals with TS into subgroups, with and without ADHD, demonstrating significant differences.[28-31] These studies generally indicated that sufferers of TS-only did not differ from unaffected controls on several ratings, including aggression, delinquency and/or conduct difficulties. In contrast, children with TS+ADHD scored significantly higher on IEFT than unaffected controls and similar to those with ADHD only, on the indices of disruptive behaviours and this clearly has major management and prognostic implications.

Recent literatures indicate that OCD and TS are intimately related, although the percentage of patients with TS who also show OCD varies from 11% to 80%.[1] OCD bears strong similarities to TS, with the suggestion that certain obsessive-compulsive symptoms (OCS) or OCB form an alternative phenotypic expression of TS.[32,33] Several studies have documented that although the OCB encountered in TS are integral to GTS,[34] they are both clinically and statistically different to those encountered in "pure" or "primary" OCD.[35,36] Cluster analysis of the inventory responses, comparing OCD and TS groups, led to a group of questions that were preferentially endorsed by TS patients (blurting obscenities, counting compulsions, impulsions to hurt oneself) and another group of questions that elicited high scores from OCD patients (ordering, arranging, routines, rituals, touching one's body, obsessions about people hurting each other).[35] George et al showed that patients with TS+OCD had significantly more violent, sexual and symmetrical obsessions and more touching, blinking, counting and self-damaging compulsions, compared to patients with OCD-alone who had more obsessions concerning dirt or germs and more compulsions relating to cleaning; the subjects who had both disorders (i.e.,TS+OCD) reported that their compulsions arose spontaneously, whereas the subjects with OCD-alone reported that their compulsions were frequently preceded by cognitions.[36]

Depression has long been found in association with TS.[37] Good evidence supports the view that affective disorders are common in patients with TS, with a lifetime risk of 10% and prevalence of between 1.8% and 8.9%.[38] In specialist clinics, patients with TS, depression or depressive symptomatology was found to occur in between 13% and 76% of 5295 individuals. The clinical depression in patients with TS appear to correlates with tic's severity and duration, the presence of echo-phenomena and coprophenomena, premonitory sensations, sleep disturbances, OCD, SIB, aggression, childhood conduct disorder and possibly ADHD.[39] The depression in patients with TS has been shown to result in a reduced quality of life,[40] and may lead to hospitalisation and even suicide in few cases.[38] TS can be a distressing condition, particularly if tics are moderate to severe. Thus, depression in TS patients could be branded, at least in part as a socially disabling and stigmatising disease. Moreover, it has been clearly demonstrated that even normal children who have been bullied at school may become depressed[41,42] and children with TS if bullied, teased and given pejorative nicknames, inevitably develop depression.[1,38] The depression in patients with TS seen in specialist clinics may also be due to the side effects of antidopaminergic agents (both typical and atypical antipsychotics), as well as other medications commonly used in TS. Depression, for example, has been reported from the treatment by haloperidol, pimozide, fluphenazine, tiapride, sulpiride and risperidone, as well as tetrabenazine, the calcium antagonist flunarizine, mecamylamine and clonidine.[1]

Finally, it appears that in TS patients there is an increased number of personality disorders.[43-45] The range of personality disorders is across the board and not restricted to obsessive-compulsive personality type, but includes specific personality profiles such as schizotypal personality traits. The cause of this increase in personality disorders in TS patients may well be the result of the long-term outcome of childhood ADHD, referral bias or because of other kinds of childhood psychopathology.

ASSESSMENT AND MANAGEMENT

TS is diagnosed through detailed personal and family history, together with physical, neurological and mental state examination and there are no specific investigations that can be employed although in late-onset tic disorders it is important to rule out brain lesions with neuroimaging and to check serum levels of copper and caeruloplasmin in order to exclude Wilson disease.[46]

Reassurance or psychological intervention may be sufficient for patients with mild symptoms. Considerable evidence exists to support the use of habit reversal training, (a specific type of behaviour approach), as an alternative or adjunct treatment for some patients with TS.[47] For the majority of patients, however, pharmacological intervention will prove most useful in treating tic symptoms and associated behaviour disorders.[1] The choice of the medication should be based on the predominant symptoms (Table 3). For tic symptoms, neuroleptics drugs such as Haloperidol and Pimozide and atypical antipsychotics medications including Risperidone and Aripiprazole should be useful. In general, the atypicals drugs show a better tolerability profile. The cautious use of stimulants such as methylphenidate for the treatment of patients with both TS and ADHD has recently been advocated.[39] Clonidine has also been used successfully in patients with TS and ADHD.[48] Clonidine may also be useful as an adjunctive treatment along with stimulants, for example this drug has been suggested to extend stimulant effects while controlling their adverse side effects.[49] A single night-time dose of Clonidine can prove useful for the treatment of sleeping difficulties in children with neurodevelopmental disorders.[50] OCD responds to selective serotonin reuptake inhibitors (SSRIs), such as Sertraline and Fluoxetine, or the tricyclic antidepressant Clomipramine. Tic symptoms are usually unaffected by SSRIs. In patients with TS that does not respond to one SSRI, augmentation with a typical neuroleptic is indicated.[51] Patients undergoing pharmacological treatment may experience depression due to medications commonly used in Tourette syndrome, such as Haloperidol and Risperidone. Depression is also a common complication of OCD,[52] and is a major comorbid disorder in patients with ADHD.[53] Thus, the depressive symptoms may be secondary to comorbid OCD or ADHD disorders.

Table 3. Recommended medication based on predominant symptom

Predominant Symptom	Recommended Medication
Tics	Atypical antipsychotic (e.g., Risperidone, Aripiprazole) Neuroleptic (e.g., Haloperidol, Pimozide)
Tics and ADHD	Clonidine
ADHD	Stimulant (e.g., methylpehnidate in once-daily preparation)/Atomoxetine Add neuroleptic if tics worsen
OCD/OCB	SSRI (e.g., sertraline) + neuroleptics
Depression	SSRI (e.g., fluoxetine)

Although the mainstream of tic management is represented by pharmacotherapy, different kinds of psychotherapy, along with neurosurgical interventions (especially deep brain stimulation, DBS) can play a major role in the treatment of TS.

Behavioral therapy methods may be useful alone or in combination with medications for several aspects of TS. Recently, Habit Reversal Training (HRT) has been demonstrated to be significantly better than supportive psychotherapy in patients with TS.[54,55] HRT consists of awareness training, self-monitoring, relaxation training, competing response training and contingency management. Clearly for youngsters with TS these trainings with HRT are encouraging as they may obviate the need for medications with their adverse side effects. Less often used but successful treatments mainly in adults can be botulinum toxin injections to affected areas, e.g., vocal cords In the presence of loud distressing vocal tics and coprolalia.[56] Finally, DBS, which entails the implantation of stimulating electrodes in the deep structures of the brain, has recently been reported to be successful in a relatively small number of individuals with severe, treatment-refractory TS.[57] However, there is still debate as to which are the correct direction to follow.[58]

CONCLUSION

TS is now recognized as a relatively common neurodevelopmental disorder, with a privileged position at the borderlands of neurology and psychiatry. Both the World Health Organization and the American Psychiatric Association criteria have dictated that TS is a unitary condition.[19,20] However, recent studies, using hierarchical cluster and principal component factor analyses have demonstrated that there may be more than one TS phenotype.[59,60] The clinical and epidemiological studies suggest that associated behavioural problems are common in people with TS and it seems likely that the investigation of the neurobiological bases of TS will shed light on the common brain mechanisms underlying movement and behaviour regulation. TS is rarely considered in isolation, with its clinical picture often influenced by closely related conditions such as OCD and ADHD, impulse discontrol and affective disorders.

A major topic for future research is the precise nature of the relationship between these disorders. Those treating patients with TS must be prepared to deal with associated symptoms from comorbid disorders and should aim treatment at the predominant symptom. Disease-specific instruments, such as the recently developed Gilles de la Tourette syndrome—Quality of Life scale (GTS-QOL) could prove helpful in assisting physicians in the assessment of the symptoms which need to be prioritized.[61]

In conclusion, it appears that TS probably should no longer be considered merely a motor disorder and, most importantly, that TS is no longer a unitary condition, as it was previously thought. Future studies will demonstrate further etiological-phenotypic relationships.

REFERENCES

1. Robertson MM. Tourette syndrome, associated conditions and the complexities of treatment. Brain 2000; 123:425-462.
2. Robertson MM, Cavanna AE. Tourette Syndrome: The Facts. 2nd ed. Oxford: Oxford University Press, 2008.
3. Robertson MM. The prevalence and epidemiology of Gilles de la Tourette syndrome. Part 1: the epidemiological and prevalence studies. J Psychosom Res. 2008;65(5):461-72.
4. Robertson MM. Diagnosing Tourette Syndrome: Is it a common disorder? J Psychosom Res 2003; 55:3-6.

5. Staley D, Wand R, Shady G. Tourette disorder: a cross-cultural review. Compr Psychiatry 1997; 38:6-16.
6. Kurlan R, McDermott MP, Deeley C et al. Prevalence of tics in schoolchildren and association with placement in special education. Neurology 2001; 57:1383-1388.
7. Baron-Cohen S, Scahill VL, Izaguirre J et al. The prevalence of Gilles de la Tourette syndrome in children and adolescents with autism: a large scale study. Psychol Med 1999; 29:1151-1159.
8. Robertson MM. Tourette syndrome. In: Skuse D, ed. Child Psychiatry IV. Oxford: The Medicine Publishing Company Group, Psychiatry 2005; 4:92-97.
9. Mink JW. Neurobiology of basal ganglia and Tourette syndrome: basal ganglia circuits and thalamocortical outputs. Adv Neurol 2006; 99:89-98.
10. Keen-Kim D, Freimer NB. Genetics and epidemiology of Tourette's syndrome. J Child Neurol 2006; 21:665-671
11. Abelson JF, Kwan KY, O'Roak BJ et al. Sequence variants in SLITRK1 are associated with Tourette's syndrome. Science 2005; 310:317-320.
12. Swedo SE, Leonard HL, Mittleman BB et al. Identification of children with pediatric autoimmune neuropsychiatric disorders associated with streptococcal infections by a marker associated with rheumatic fever. Am J Psychiatry 1997; 154:110-112.
13. Mell LK, Davis RL, Owens D. Association between streptococcal infection and obsessive-compulsive disorder, Tourette's syndrome and tic disorder. Pediatrics 2005; 116:56-60.
14. Rizzo R, Gulisano M, Pavone P et al. Increased antistreptococcal antibody titers and anti-basal ganglia antibodies in patients with Tourette syndrome: controlled cross-sectional study. J Child Neurol 2006; 21:747-753.
15. Leckman JF, Bloch MH, Scahill L et al. Tourette syndrome: the self under siege. J Child Neurol 2006; 21:642-649.
16. Berardelli A, Currà A, Fabbrini G et al. Pathophysiology of tics and Tourette syndrome. J Neurol 2003; 250:781-787.
17. Ludlow CL, Polinsky RJ, Caine ED et al. Language and speech abnormalities in Tourette syndrome. Adv Neurol 1982; 35:351-361.
18. Jankovic J. Phenomenology and classification of tics. In Jankovic J, ed. Neurologic Clinics. Philadelphia: WH Saunders Company, 1997; 267-275.
19. World Health Organization. The ICD-10 Classification of Mental and Behavioural Disorders: Diagnostic Criteria for Research. Geneva: WHO, 1993.
20. American Psychiatric Association. Diagnostic and Statistical Manual of Mental Disorders. 4th ed, text revision (DSM-IV-TR). Washington, DC: APA, 2000.
21. Robertson MM. The heterogeneous psychopathology of Tourette Syndrome. In: Bedard MA, Agid Y, Chouinard S et al, eds. Mental and Behavioral Dysfunction in Movement Disorders. New Jersey: Humana Press, Totowa, 2003:443-466.
22. Packer LE. Social and educational resources for patients with Tourette syndrome. Neurol Clin 1997; 15:457-473.
23. Freeman RD, Fast DK, Burd L et al. An international perspective on Tourette syndrome: selected findings from 3,500 individuals in 22 countries. Dev Med Child Neurol 2000; 42:436-447.
24. Khalifa N, von Knorring AL. Tourette syndrome and other tic disorders in a total population of children: clinical assessment and background. Acta Paediatr 2005; 94:1608-1614.
25. Cavanna AE, Servo S, Monaco F et al. The behavioral spectrum of Gilles de la Tourette syndrome. J Neuropsy Clin Neurosci 2008; 21:13-23.
26. Termine C, Balottin U, Rossi G et al. Psychopathology in children and adolescents with Tourette's syndrome: a controlled study. Brain Dev 2006; 28:69-75.
27. Singer HS, Brown J, Quaskey S et al. The treatment of attention-deficit hyperactivity disorder in Tourette's syndrome: a double-blind placebo-controlled study with clonidine and desipramine. Pediatrics 1995; 95:74-81.
28. Spencer T, Biederman J, Harding M et al. Disentangling the overlap between Tourette's disorder and ADHD. J Child Psychol Psychiatr 1998; 39:1037-1044.
29. Stephens RJ, Sandor P. Aggressive behaviour in children with Tourette syndrome and comorbid attention-deficit hyperactivity disorder and obsessive-compulsive disorder. Can J Psychiatry 1999; 44:1036-1042.
30. Carter AS, O'Donnell DA, Schultz RT et al. Social and emotional adjustment in children affected with Gilles de la Tourette's syndrome: associations with attention deficit hyperactivity disorder and family functioning. J Child Psychol Psychiatry 2000; 41:215-223.
31. Sukhodolsky DG, Scahill L, Zhang H et al. Disruptive behavior in children with Tourette's syndrome: association with ADHD comorbidity, tic severity and functional impairment. J Am Acad Child Adolesc Psychiatry 2003; 42:98-105.

32. Pauls DL, Leckman J, Towbin KE et al. A possible genetic relationship exists between Tourette's syndrome and obsessive-compulsive disorder. Psychopharmacol Bull 1986; 22:730-733.

33. Eapen V, Pauls D, Robertson M. Evidence for autosomal dominant transmission in Tourette's syndrome. Br J Psychiatry 1993; 163:593-596.

34. Robertson MM, Trimble MR, Lees AJ. The psychopathology of the Gilles de la Tourette syndrome: a phenomenological analysis. Br J Psychiatry 1988; 152:383-390.

35. Frankel M, Cummings JL, Robertson MM et al. Obsessions and compulsions in Gilles de la Tourette's syndrome. Neurology 1986; 36:378-382.

36. George MS, Trimble MR, Ring HA et al. Obsessions in obsessive compulsive disorder with and without Gilles de la Tourette syndrome. Am J Psychiatry 1993; 150:93-97.

37. Montgomery MA, Clayton PJ, Friedhoff AJ. Psychiatric illness in Tourette syndrome patients and first-degree relatives. Adv Neurol 1982; 35:335-340.

38. Robertson MM. Mood disorders and Gilles de la Tourette's syndrome: an update on prevalence, etiology, comorbidity, clinical associations and implications. J Psychosom Res 2006; 61:349-358.

39. Robertson MM. Attention deficit hyperactivity disorder, tics and Tourette's syndrome: the relationship and treatment implications. Eur Child Adolesc Psychiatry 2006; 15:1-11.

40. Elstner K, Selai CE, Trimble MR et al. Quality of Life (QOL) of patients with Gilles de la Tourette's syndrome. Acta Psychiatr Scand 2001; 103:52-59.

41. Salmon G, James A, Smith DM. Bullying in schools: self reported anxiety, depression and self esteem in secondary school children. Br Med J 1998; 317:924-925.

42. Bond L, Carlin JB, Thomas L et al. Does bullying cause emotional problems? A prospective study of young teenagers. Br Med J 2001; 323:480-484.

43. Shapiro AK, Shapiro ES, Bruun RD et al. Gilles de la Tourette syndrome. New York: Raven Press;1978.

44. Robertson M, Banerjee S, Fox-Hiley PJ et al. Personality disorder and psychopathology in Tourette's syndrome: a controlled study. Br J Psychiatry 1997; 171:283-286.

45. Cavanna AE, Robertson MM, Critchley HD. Schizotypal personality traits in Gilles de la Tourette syndrome. Acta Neurol Scand 2007; 116:385-391.

46. Stern JS, Burza S, Robertson MM. Gilles de la Tourette's syndrome and its impact in the UK. Postgrad Med J 2005; 81:12-19.

47. Himle MB, Woods DW, Piacentini JC et al. Brief review of habit reversal training for Tourette syndrome. J Child Neurol 2006; 21:719-725.

48. Freeman RD. Attention deficit hyperactivity disorder in the presence of Tourette syndrome. Neurol Clin 1997; 15:411-420.

49. Tourette's Syndrome Study Group. Treatment of ADHD in children with tics: a randomized controlled trial. Neurology 2002; 58:527-536.

50. Ingrassia A, Turk J. The use of clonidine for severe and intractable sleep problems in children with neurodevelopmental disorders—a case series. Eur Child Adolesc Psychiatry 2005; 14:34-40.

51. Miguel EC, Shavitt RG, Ferrão YA et al. How to treat OCD in patients with Tourette syndrome. J Psychosom Res 2003; 55:49-57.

52. Nestadt G, Samuels J, Riddle MA et al. The relationship between obsessive-compulsive disorder and anxiety and affective disorders: results from the Johns Hopkins OCD Family Study. Psychol Med 2001; 31:481-487.

53. Milberger S, Biederman J, Faraone SV et al: Attention deficit hyperactivity disorder and comorbid disorders: issues of overlapping symptoms. Am J Psychiatry 1995; 152:1793-1799

54. Wilhelm S, Deckersbach T, Coffey BJ et al. Habit Reversal versus supportive therapy for Tourette's disorder: a randomised controlled trial. American Journal of Psychiatry 2003; 160:1175-1177.

55. Deckersbach T, Rauch S, Buhlmann U et al. Habit reversal versus supportive psychotherapy in Tourette's disorder: a randomized controlled trial and predictors of treatment response. Behavioral Research Therapy 2006; 160:1175-1177.

56. Porta M, Maggioni G, Ottaviani F et al. Treatment of phonic tics in patients with Tourette's syndrome using botulinum toxin type A. Neurological Sciences 2004; 24:420-423.

57. Servello D, Porta M, Sassi M et al. Deep brain stimulation in 18 patients with severe Gilles de la Tourette syndrome refractory to treatment : the surgery and stimulation. J Neurol Neurosurg Psychiatry 2008; 79:136-142.

58. Rickards H, Wood C, Cavanna AE. Hassler and Dieckmann's seminal paper on stereotactic thalamotomy for Gilles de la Tourette syndrome: translation and critical reappraisal. Mov Disord 2008; 23:1966-1972.

59. Mathews CA, Jang KL, Herrera LD et al. Tic symptom profiles in subjects with Tourette Syndrome from two genetically isolated populations. Biol Psychiatry 2007; 61:292-300.

60. Robertson MM, Cavanna AE. The Gilles de la Tourette syndrome: a principal component factor analytic study of a large pedigree. Psychiatr Genet 2007; 17:143-152.

61. Cavanna AE, Schrag A, Morley D et al. The Gilles de la Tourette syndrome-Quality of Life scale (GTS-QOL): development and validation. Neurology 2008; 71:1410-1416.

INDEX